MECHANICS OF MATERIALS

MECHANICS OF MATERIALS

K. PETE ARGES AND AUBREY E. PALMER

Department of Civil Engineering, Duke University

McGRAW-HILL BOOK COMPANY

New York San Francisco Toronto London

MECHANICS OF MATERIALS

 Library of Congress Catalog Card Number 62-15519

02199

3 4 5 6 7 8 9 – MAMM – 1

PREFACE

This text has been written for use with mechanics of materials courses at the sophomore-junior level, such as are given at most E.C.P.D. accredited colleges. The presentation assumes a thorough knowledge on the part of the student of engineering *statics* and mathematics through *calculus*. An attempt has been made to cover the subject in an orderly and concise manner, to derive and explain the basic principles of the subject, and to further the understanding of these principles with a generous number of examples (more than a hundred). The first example for each topic is explained in great detail, with successive examples in progressively less detail. It is hoped that this feature will free the instructor from some of the pressure normally associated with teaching mechanics and will enable him to spend more time discussing theory and answering questions. This should also make the text especially suitable for individual study.

The authors believe that, in order to obtain a true understanding of the principles of mechanics, the student must work a large number and variety of problems. Only in this way can he discover the full meaning and the many ramifications of the theory. For this reason, a large number of problems (more than eight hundred) have been provided. In order for the student to check himself, answers are given (at the end of the book) for approximately 35 per cent of the problems; these problems are identified by an asterisk preceding the problem number. Enough problems have been provided so that the instructor can make assignments for a minimum of three consecutive terms without repetition. In this connection, problems are listed in informal groups of two or three, the second and third being "revisions" of the first. Thus, essentially identical problems could be assigned for two or three successive terms without encouraging the use of "student files."

Since, in many cases, course time available will preclude covering the book in its entirety, it has been divided into a relatively large number of chapters. This arrangement provides greater flexibility in making assignments and should simplify the assimilation of the material by the student. In this connection, there is a current trend (unfortunately) to reduce, for certain curricula, the amount of time devoted to the study of mechanics of materials. It is believed that arranging the book into a large number of chapters (many of which could be omitted entirely) will make it particularly suitable for an abbreviated, but still meaningful, course designed to teach thoroughly only the most important principles. Chapters which could be omitted without serious loss of continuity are marked with an asterisk in the Table of Contents.

The authors decided against including chapters of "optional topics," since such chapters are of necessity fragmentary and because of their location are often ignored by both the instructor and the student. Instead, such topics are included as the latter articles of the chapters to which they most closely relate, where they can be more effectively covered (or conveniently omitted). These articles and others which can be omitted without serious loss of continuity are marked with an asterisk in the Table of Contents.

Mention should be made of the final chapter, which serves to introduce the subject of plastic analysis. This topic, not normally included in such detail in books of this sort, is presented because of its importance in providing for more economical utilization of materials. In the event that time is not available for covering this chapter in the formal mechanics of materials course, it is hoped that it will prove to be useful as a reference.

A book of this type reflects the influence and efforts of a great many people. For their knowledge of the subject, the authors are indebted to their teachers and to the authors of many books from which they have studied and taught. The authors are also indebted to their colleagues, both in teaching and in industry, and to their many former students; it is they who have exerted the greatest influence on the method of presenting the subject matter. Appreciation is also expressed to Mrs. Gerald P. Cantini and Miss Joanna Huffer who typed most of the final manuscript and to Mrs. Robert L. Fortune for clerical assistance. The authors are especially grateful to Professor John E. Brock of the U.S. Naval Postgraduate School, who has read the entire manuscript and made many valuable suggestions.

K. P. Arges
A. E. Palmer

CONTENTS

CHAPTERS AND ARTICLES MARKED WITH AN ASTERISK MAY BE OMITTED WITHOUT SERIOUS LOSS OF CONTINUITY.

LIST OF SYMBOLS

A	Area of cross section
A_M	Area of moment diagram
A_V	Area of shear diagram
a	Area
b	Breadth; subscript "bearing"
C	Constant (as "of integration"); compression
c	Maximum distance from neutral axis; subscript "compression"
cr	Subscript "critical"
D	Diameter
d	Depth; diameter
E	Modulus of elasticity (Young's modulus)
E_s	Modulus of rigidity (modulus of elasticity in shear)
e	Eccentricity
F	Force or concentrated load
FS	Factor of safety
g	Acceleration due to gravity
h	Height or depth
I, I_x, I_y	Rectangular moments of inertia of area
I/c	Section modulus of beam
J	Polar moment of inertia of area
J_L	Polar moment of inertia of line
k	Radius of gyration of area; stress-concentration factor
L	Length; subscript "left"
L/k	Slenderness ratio
M	Bending moment; moment of force or couple
M_p	Plastic moment
m	Subscript "maximum"

n	Number; load factor; ratio of moduli of elasticity of two materials; subscript "normal"
P	Internal force; force or concentrated load
p	Major principal stress; unit pressure; rivet pitch
Q	First moment of area with respect to neutral axis; force or concentrated load
q	Minor principal stress; shear flow (shearing force per linear inch)
R	Radius; reaction force; force per rivet; subscript "right"
r	Radius; subscript "restrained"
s	Normal unit stress; subscript "shearing"
s_s	Shearing unit stress
s_{xy}	Shearing stress on plane normal to x axis acting in y direction
s_{yx}	Shearing stress on plane normal to y axis acting in x direction
T	Torque (torsional moment); tension
t	Thickness; temperature; subscript "tension" or "temperature"
U	Energy
V	Shearing force; volume
v	Velocity
W	Weight; concentrated load; width
w	Load per unit length; width; weld size
$\bar{x},y$	Distances to centroid of area
y	Distance from neutral axis; deflection of beam; subscript "yield point" or "yield strength"
y'	dy/dx
y''	d^2y/dx^2
Z	Plastic modulus of beam
α	Angle; coefficient of thermal expansion
β	Angle
γ	Angle; angle of shearing strain; weight per unit volume
Δ	Displacement from tangent to elastic curve
δ	Deformation or total strain
δ_s	Shearing deformation
ϵ	Normal unit strain
ϵ_s	Shearing unit strain
θ	Angle; slope angle; angle of twist
ρ	Distance from polar axis
μ	Poisson's ratio

1

STRESSES AND STRAINS

1.1 INTRODUCTION TO SUBJECT

Mechanics of materials is the field of learning concerned with *the behavior of materials subjected to load.* The ultimate purpose of the study is to develop the ability to predict whether a member made of a particular material will be able to withstand a particular load condition without excessive deformation or fracture. Necessary to a proper understanding of the subject, which is also known as *strength of materials* and *resistance of materials,* is an understanding of statics and mathematics through calculus. Inasmuch as engineers of all fields must utilize load-carrying materials to accomplish their assigned tasks, the need for a fundamental knowledge of mechanics of materials by all engineering students should be readily apparent. Although some of the problems which will be studied may appear to be a bit involved, the student should strive to understand the subject as a series of simple principles to be thoroughly mastered. Such an understanding will enable a person to develop a feel for the subject so that he can proceed almost intuitively to the solution of a wide variety of engineering problems.

1.2 NORMAL AND SHEARING STRESS

Perhaps the term most fundamental to the study of mechanics of materials is that of unit stress. It will be recalled that the internal force at any section of a member can be determined by the use of an appropriate free-body diagram and the principles of statics. *Unit stress* may be defined as the *internal force per unit area* at any section of a member. The term *unit stress* is commonly shortened to the single word *stress.* Unfortunately, many people also refer to the internal force at a section as the stress (more properly, the total stress). In order to avoid confusion, the word stress will not be used in this text to denote a force; the two terms *stress* and *unit stress* will be used interchangeably.

As a matter of convenience, stresses are considered to be of two general types: *normal stresses*, which act perpendicular (or "normal") to a given section, and *shearing stresses*, which act parallel to the section. Actually, in the general case, the stress will be neither perpendicular nor parallel to the section; however, such an inclined stress can always be resolved into normal and shearing components.

Normal stress is usually designated as either tensile or compressive, depending on whether it tends to lengthen or shorten the material on which it acts. The meaning of normal stress will be demonstrated with respect to the section indicated on the two-force tension member shown in Fig. 1.1*a*. From a free-body diagram of the bottom part of the member, shown in Fig. 1.1*b*, it can be seen that an internal force P must act normal to the specified section. This force is, of course, distributed over the entire cross section. The internal force acting on a typical differential area is as shown in Fig. 1.1*c*. Thus the mathematical definition of normal stress is

$$s = \frac{dP}{dA} \tag{1.1a}$$

If normal stress is uniformly distributed over a complete cross section, as shown in Fig. 1.1*d*, it can be evaluated from the equation

$$s = \frac{P}{A} \tag{1.1b}$$

where P is the internal force and A is the area of the section on which P acts. A uniform stress distribution will occur, on a section only if (1) the section is not immediately adjacent to the point of application of a concentrated load, (2) the line of action of the

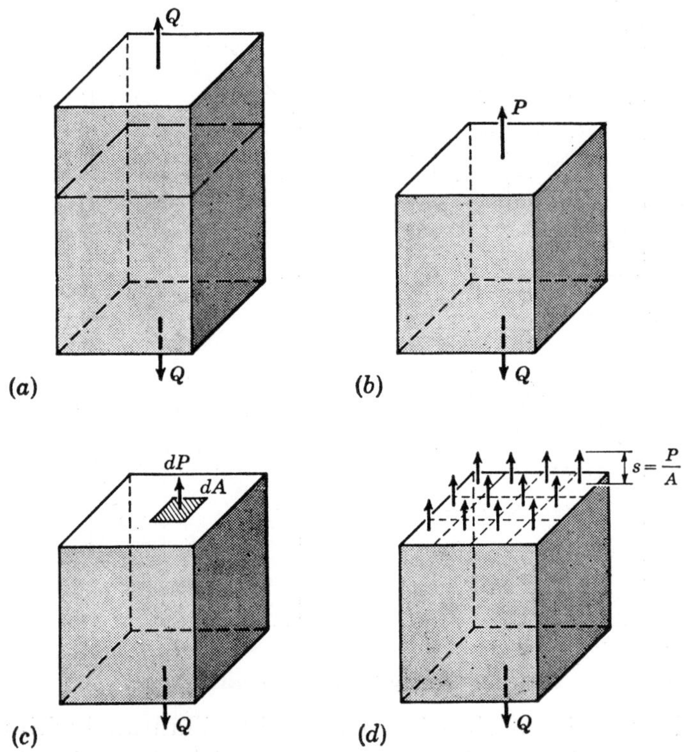

FIG. 1.1 Analysis of tensile stress.

internal force passes through the centroid of the area of the section, (3) there is no discontinuity, such as a notch or flaw, at the section, and (4) the member is made of a homogeneous material (as can be assumed in most cases). The distribution of normal stress when the conditions above are not met (or cannot safely be assumed) is discussed in other chapters throughout the book. It should be noted that, in the above explanation, the internal force was equal to the applied load (i.e. $P = Q$); these forces were designated by different letters merely to emphasize that unit stress is *internal* force per unit area. As indicated by Fig. 1.2, compressive stress differs from tensile stress only in the direction of the force vectors and can be evaluated by the same equations.

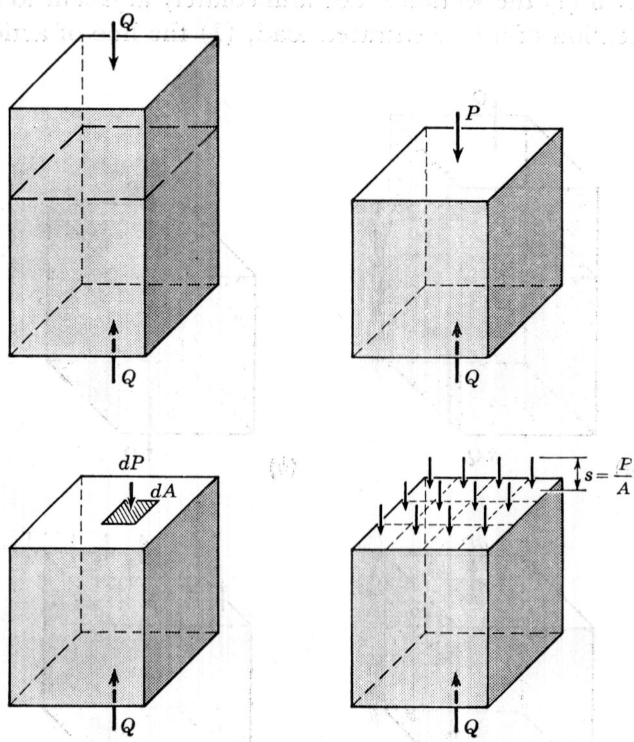

FIG. 1.2 Analysis of compressive stress.

The meaning of shearing stress will be illustrated with respect to the section indicated on the short member shown in Fig. 1.3*a*. The member is restrained at one end and supports a vertical load Q at the other. Figure 1.3*b* shows the portion of the member to the right of the section. In order for the summation of vertical forces to equal zero,† the internal force P must act parallel to the

† Figure 1.3*b* is not a complete free-body diagram; for equilibrium to exist an internal couple must also act on the section. This couple is not pertinent and is omitted for the sake of simplicity.

cross section and thus causes shearing stress. The mathematical definition of unit shearing stress, as indicated in Fig. 1.3*c*, is

$$s_s = \frac{dP}{dA} \tag{1.2a}$$

The *average* shearing stress can be expressed as

$$s_s = \frac{P}{A} \tag{1.2b}$$

where P is the internal shearing force and A is the area of the section on which P acts. Although shearing stress is seldom (if ever) uniformly distributed, it is common practice to assume a uniform shearing stress distribution *for very short members.* When such an

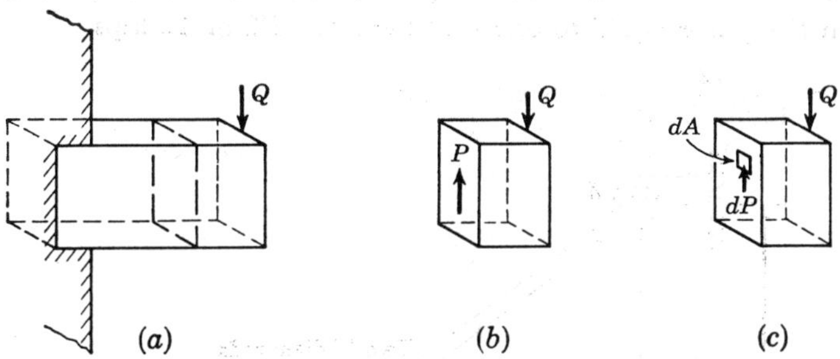

FIG. 1.3 Analysis of shearing stress.

assumption is justified, as in the case of rivets and bolts, for example, the shearing stress over the entire section can be determined by Eq. (1.2*b*).

Since force and area are commonly expressed as pounds and square inches, stress is usually expressed in units of pounds per square inch (psi). When the forces in a problem are measured in kips (1 kip = 1000 lb), it is usually convenient to express stress as kips per square inch (ksi).

Example 1.1. The pin-connected hanger shown in Fig. 1.4*a* is comprised of a 6- by 8-in. wooden member BC and a member AC consisting of two 1-in.-diam steel rods. The pin at B is 1.5 in. in diameter. Determine the normal stress in members AC and BC at typical cross sections (not adjacent to the pin connections) and the average shearing stress in pin B.

The internal forces in members AC and BC are found by using the "method of joints" at joint C; referring to Fig. 1.4b

$\Sigma F_y = 0 \qquad -18 + AC(0.6) = 0$
$AC = 30.0$ kips T (tension)
$\Sigma F_x = 0 \qquad +BC - 30(0.8) = 0$
$BC = 24.0$ kips C (compression)

For the normal stress in AC

$$s = \frac{P}{A} = \frac{30}{2(\pi/4)(1)^2} = 19.1 \text{ ksi } T \qquad \textit{Ans}$$

For the normal stress in BC

$$s = \frac{P}{A} = \frac{24}{6(8)} = 0.5 \text{ ksi } C \qquad \textit{Ans}$$

For the shearing stress in the pin at B, from Fig. 1.4c, the internal force on the pin is equal to one-half the force BC or 12 kips.

$$s_s = \frac{P}{A} = \frac{12}{(\pi/4)(1.5)^2} = 6.8 \text{ ksi} \qquad \textit{Ans}$$

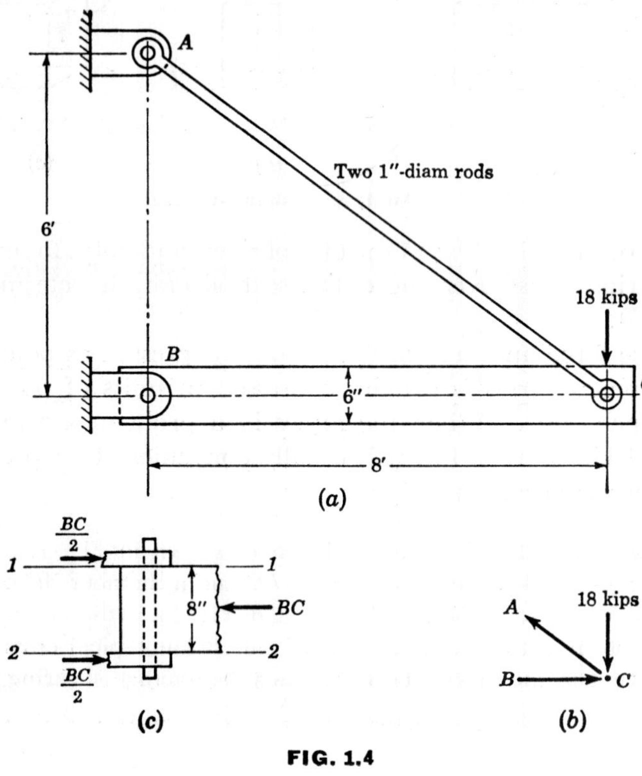

FIG. 1.4

Note that the 24 kips acting on the center portion of the pin is resisted by internal forces acting on two pin areas (planes 1-1 and 2-2). For this reason the pin is said to be in *double shear.*

PROBLEMS

***1.1.** For the truss shown in Fig. P1.1, the cross-sectional areas of members *DE, EJ,* and *DJ* are 9, 15, and 12 in.^{2}, respectively. Determine the unit stress in each of these members.

1.2. For the truss shown in Fig. P1.2, the cross-sectional areas of members *AB, AH, GH,* and *BH* are 20, 11, 9, and 5 in.^{2}, respectively. Determine the unit stress in each of these members.

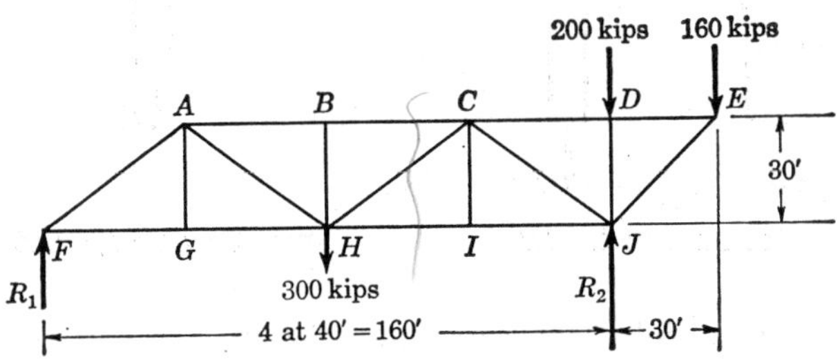

FIG. P1.1, P1.2, P1.3, P1.4

***1.3.** For the truss shown in Fig. P1.3, the cross-sectional areas of members *BC, CH, HI,* and *CI* are 20, 16, 5, and 5 in.^{2}, respectively. Determine the unit stress in each of these members.

1.4. For the truss shown Fig. P1.4, the cross-sectional areas of members *AF, AG, FG,* and *CJ* are 14, 5, 8, and 20 in.^{2}, respectively. Determine the unit stress in each of these members.

***1.5.** The tension member *A* is supported by the square bar *B*. If $h = \frac{1}{2}$ in., $t = \frac{1}{4}$ in., and $Q = 6000$ lb, determine (*a*) the average tensile stress at the net (i.e. the smallest) section of the 2-in. portion of *A*, (*b*) the tensile stress in the 1.5-in. portion of *A*, (*c*) the average shear stress in bar *B*, (*d*) the average shear stress in member *A*.

1.6. Same as Prob. 1.5 except let $h = \frac{3}{4}$ in., $t = \frac{3}{8}$ in., and $Q =$ 10,000 lb.

1.7. Same as Prob. 1.5 except let $h = 1$ in., $t = \frac{1}{2}$ in., and $Q =$ 13,000 lb.

***1.8.** A long bolt is supported by a 2-in.-thick floor as shown in Fig. P1.8. If $t = \frac{5}{16}$ in., $d = \frac{1}{2}$ in., $D = 1.5$ in., and $Q = 1500$ lb, find

* Answers to problems marked with an asterisk are listed at the end of the book.

(*a*) the average shear stress in the bolt head, (*b*) the average shear stress in the floor.

1.9. Same as Prob. 1.8 except let $t = \frac{1}{2}$ in., $d = \frac{3}{4}$ in., $D = 2$ in., and $Q = 2000$ lb.

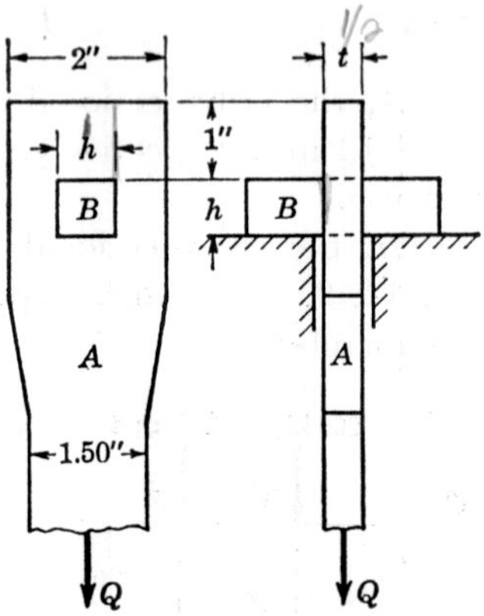

FIG. P1.5, P1.6, P1.7

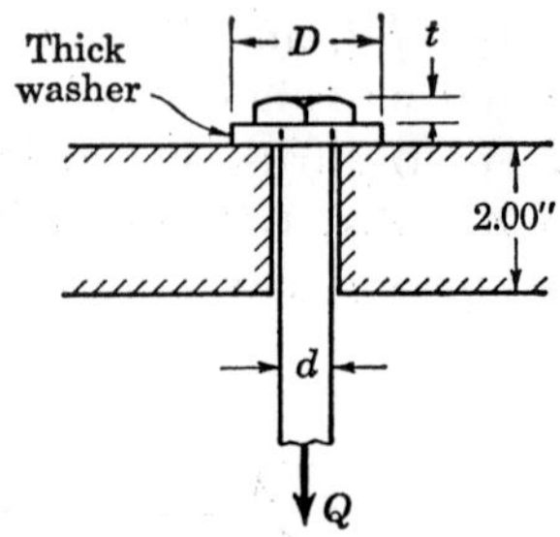

FIG. P1.8, P1.9, P1.10

1.10. Same as Prob. 1.8 except let $t = \frac{5}{8}$ in., $d = 1$ in., $D = 2.5$ in., and $Q = 2500$ lb.

1.3 NORMAL AND SHEARING STRAIN

An essential principle of mechanics of materials is that all materials deform when subjected to load. Whereas in statics loads are assumed to act on members which are perfectly rigid, in mechanics of materials this assumption can seldom be made. The importance to the student of this change in concept, from "rigid bodies" to deformable bodies, cannot be overemphasized. The deformation which a body undergoes is also known as *strain.* The amount of strain in a member depends on the magnitude and type of loading, the dimensions of the member, and the stiffness of the material. *Unit strain* is the amount of strain per unit length. As is the case with stress, strain is of two general types: normal and shearing. Of course, normal strain can be either tensile or compressive.

Normal unit strain will be illustrated by reference to the compression member shown in Fig. 1.5. The original length L of the member is shortened owing to the load Q by an amount δ (shown greatly exaggerated). Therefore, the deformation of any differential length dL is $d\delta$, and the normal unit strain ϵ can be expressed as

$$\epsilon = \frac{d\delta}{dL} \tag{1.3a}$$

The average unit strain for the entire member can be expressed as

$$\epsilon = \frac{\delta}{L} \tag{1.3b}$$

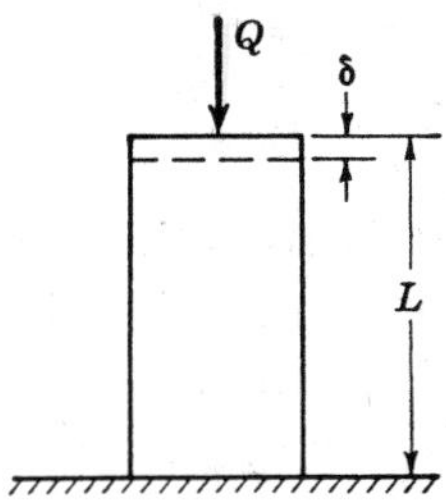

FIG. 1.5

where δ is the total strain and L is the unloaded length of the member. If a member is uniformly stressed throughout its length, the unit strain at any point will be equal to the average value as defined by Eq. (1.3*b*).

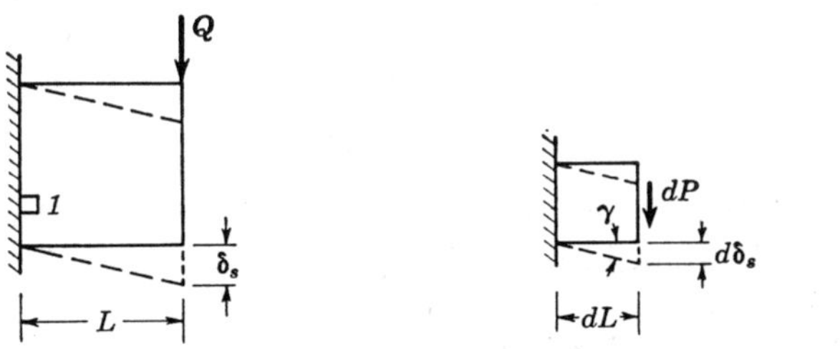

(*a*) Short member with shear load **(*b*) Element at *1* enlarged**

FIG. 1.6

Shearing unit strain will be illustrated by reference to the member shown in Fig. 1.6*a*. As indicated, the total shearing deformation δ_s is dependent on the length L perpendicular to the direction of the shearing force.† The shearing force and deformation of an element at point 1 are as shown in Fig. 1.6*b*. From this sketch, the unit shearing strain is defined in equation form as

$$\epsilon_s = \frac{d\delta_s}{dL} \tag{1.4a}$$

† The deformation shown is that due to shear only; in the interest of simplicity, the bending deformation of the member is not shown. Bending deformations are dealt with at length in later chapters.

If the member is short enough so that a uniform stress distribution can be assumed throughout, the average unit strain for the entire member is

$$\epsilon_s = \frac{\delta_s}{L} \tag{1.4b}$$

It is sometimes convenient to express unit strain as an angle in radians. For the element of Fig. 1.6*b* this angle is shown as γ; since the angle is very small, it is for practical purposes equal to its tangent. Thus it can be seen that $\gamma = \tan \gamma = d\delta_s/dL$ and $\gamma = \epsilon_s$.

Inasmuch as unit strain is merely the ratio of two lengths, δ and L, it is actually a dimensionless quantity. In practice, however, it is usually expressed as inches per inch (in./in.). For experimental work, as a matter of convenience, it is common practice to express (or measure) unit strain as microinches per inch (μin./in.). Thus, for example, a unit strain of 0.00053 in./in. could be expressed as 530 μin./in.

PROBLEMS

★1.11. A 100-ft steel member lengthens 0.960 in. under load. Determine the unit strain in (*a*) inches per inch, (*b*) feet per foot, (*c*) microinches per inch.

1.12. A 60-ft aluminum member shortens 1.75 in. under load. Express the unit strain in (*a*) inches per inch, (*b*) feet per foot, (*c*) microinches per inch.

★1.13. In Fig. 1.6, if L is 2 in. and δ_s is 0.0012 in., determine the average unit shear strain expressed (*a*) in inches per inch, (*b*) as an angle in radians.

1.14. In Fig. 1.6, if L is 3 in. and δ_s is 0.0060 in., determine the average unit shear strain expressed (*a*) in inches per inch, (*b*) as an angle in radians.

1.4 MODULUS OF ELASTICITY AND MODULUS OF RIGIDITY

A basic law of physics, known as Hooke's law,† states, in effect, that *within certain limits* the ratio of unit stress to unit strain *due to*

† Named after the English scientist Robert Hooke (1635–1703) who formulated the law from a series of experiments.

load is a constant. For the problems that follow, unless otherwise specified, all materials are assumed to obey Hooke's law. For normal stresses and strains the constant of proportionality is known as the *modulus of elasticity* or *Young's modulus.*† It is normally designated as E and can be expressed in equation form as

$$E = \frac{s}{\epsilon} \tag{1.5}$$

For shearing stresses and strains the constant of proportionality is known as the *modulus of rigidity* or the *modulus of elasticity in shear.* It is usually designated as either E_s or G and can be expressed in equation form as

$$E_s = \frac{s_s}{\epsilon_s} \tag{1.6}$$

The moduli of elasticity and rigidity are usually assigned the same units as used for unit stress, that is, either pounds per square inch or kips per square inch. Typical values of the two moduli for some engineering materials are listed in Table 1.1. Values equivalent to Young's modulus for concrete vary widely from about 1,500,000 to 6,000,000 psi, depending on composition of mix, age and other factors. Values of Young's modulus for timber vary

TABLE 1.1 Typical Values of Young's Modulus and Modulus of Rigidity for Some Engineering Materials

Material	Young's modulus, psi	Modulus of rigidity, psi
Steel	30,000,000	12,000,000
Aluminum	10,000,000	3,800,000
Magnesium	6,500,000	2,400,000
Copper	16,000,000	6,000,000
Brass, cast	9,000,000	3,400,000
Brass, wrought	15,000,000	5,700,000
Beryllium copper	18,000,000	7,000,000

† Named after the English scientist Thomas Young (1773–1829) who first defined the constant (though not in its presently used form). Young made many other valuable contributions to the field of mechanics of materials.

widely from about 1,000,000 to 2,000,000 psi, depending on the species of wood, moisture content, density, and other factors.

Example 1.2. The member shown in Fig. 1.7 is a 2-in.-diam aluminum rod with a 1-in.-diam concentric hole in one end. Determine the amount that the rod shortens as a result of an axial compressive load of 40,000 lb.

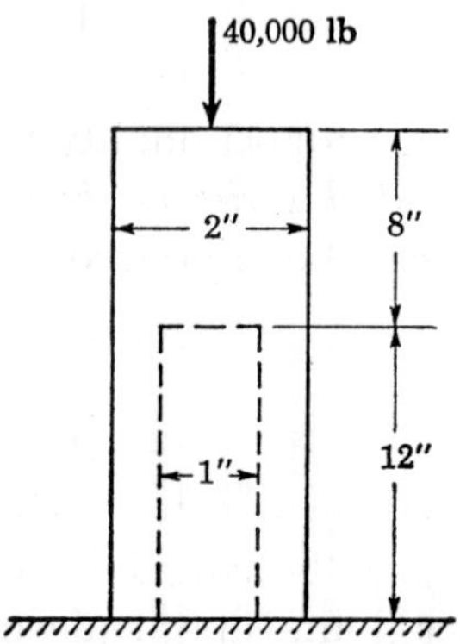

FIG. 1.7

For the top (solid) portion

$$s = \frac{P}{A} = \frac{40{,}000}{0.785(2)^2} = 12{,}740 \text{ psi}$$

$$E = \frac{s}{\epsilon} \qquad \epsilon = \frac{s}{E} = \frac{12{,}740}{10{,}000{,}000} = 0.001274 \text{ in./in.}$$

$$\epsilon = \frac{\delta}{L} \qquad \delta = \epsilon L = 0.001274(8) = 0.01019 \text{ in.}$$

For the bottom (hollow) portion

$$s = \frac{P}{A} = \frac{40{,}000}{0.785[(2)^2 - (1)^2]} = 17{,}000 \text{ psi}$$

$$\epsilon = \frac{s}{E} = \frac{17{,}000}{10{,}000{,}000} = 0.00170 \text{ in./in.}$$

$$\delta = \epsilon L = 0.00170(12) = 0.0204 \text{ in.}$$

$$\text{Total } \delta = 0.0102 + 0.0204 = 0.0306 \text{ in.} \qquad \textit{Ans}$$

Example 1.3. A weight W suspended from a steel wire 12 ft long and ⅛ in. in diameter causes it to lengthen 0.060 in. The same weight suspended from a titanium wire 1/16 in. in diameter and 8 ft long causes it to lengthen 0.320 in. Determine the modulus of elasticity of titanium.

To find the value of W (from the steel-wire data)

$$\epsilon = \frac{\delta}{L} = \frac{0.060}{12(12)} = 0.000417 \text{ in./in.}$$

$$E = \frac{s}{\epsilon} \qquad s = \epsilon E = 0.000417(30{,}000{,}000) = 12{,}500 \text{ psi}$$

$$s = \frac{P}{A} \qquad W = P = As = 0.785(0.125)^2(12{,}500) = 153.3 \text{ lb}$$

To find E for titanium

$$\epsilon = \frac{\delta}{L} = \frac{0.320}{8(12)} = 0.00333 \text{ in./in.}$$

$$s = \frac{P}{A} = \frac{153.3}{0.785(0.0625)^2} = 49{,}900 \text{ psi}$$

$$E = \frac{s}{\epsilon} = \frac{49{,}900}{0.00333} = 15{,}000{,}000 \text{ psi} \qquad \textit{Ans}$$

PROBLEMS

★1.15. What unit stress in an aluminum member would cause (*a*) a unit tensile strain of 0.002 in./in., (*b*) a unit shear strain of 0.002 in./in.?

1.16. How much would a ¼-in. wire 60 ft long lengthen because of a 1000-lb load if it were made of (*a*) steel, (*b*) aluminum, (*c*) magnesium?

1.17. A pipe 5 in. OD, 4 in. ID, and 4.5 ft long is subjected to an axial compressive load of 80 kips. How much will the pipe shorten if it is made of (*a*) steel, (*b*) aluminum, (*c*) magnesium?

★1.18. An eyebar 6 by 2 in. in cross section and 30 ft long stretches 0.240 in. under a tensile load. How great is the load if the bar is made of (*a*) steel, (*b*) aluminum, (*c*) magnesium?

1.19. A 4-ft length of pipe 4 in. OD and 3.5 in. ID shortens 0.030 in. under an axial compressive load. How great is the load if the pipe is made of (*a*) steel, (*b*) aluminum, (*c*) magnesium?

1.20. Determine the diameter required for a 25-ft steel rod which is to support a 50-kip tensile load without having the total deformation exceed 0.16 in. or a unit stress greater than 20,000 psi.

★1.21. Determine the required cross-sectional dimensions for a square timber post (E = 1,600,000 psi) 8 ft long which is to support an 80-kip compressive load without having a total deformation greater than 0.05 in. or a unit stress greater than 1000 psi.

1.5 LATERAL STRAIN; POISSON'S RATIO

A member subjected to an axial load changes dimensions not only in the axial direction but also in the lateral directions (that is, the directions perpendicular to the load). This behavior of materials is illustrated in Fig. 1.8, which shows a member under compressive load. The member shortens in the direction of the load and expands in the lateral directions. The ratio of the *lateral unit strain* to the *axial unit strain* is known as Poisson's ratio.† The theoretical value of the ratio for an ideal material, as first determined by Poisson, is 0.25. For practical purposes the value depends on the material and must be determined experimentally. Values for most metals fall in the range from 0.25 to 0.35. By contrast, the value for portland cement concrete ranges from 0.10 to 0.30.

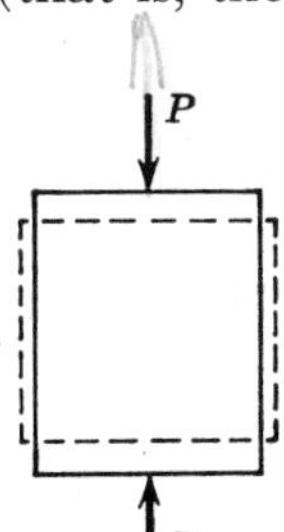

FIG. 1.8

† Named after the French mathematician S. D. Poisson (1781–1840).

In equation form Poisson's ratio can be expressed as

$$\mu = -\frac{\epsilon_l}{\epsilon_a} \tag{1.7}$$

where ϵ_a is the unit strain in the direction of the axial stress and ϵ_l is the unit strain in any direction perpendicular to the axial stress. The minus sign in the equation indicates that the lateral strain is of the opposite sign to the axial strain. In other words, compressive stress causes shortening in the axial direction and lengthening in the lateral direction; tensile stress causes lengthening in the axial direction and shortening in the lateral direction. It should be noted that this lateral strain is *not* accompanied by lateral stress.

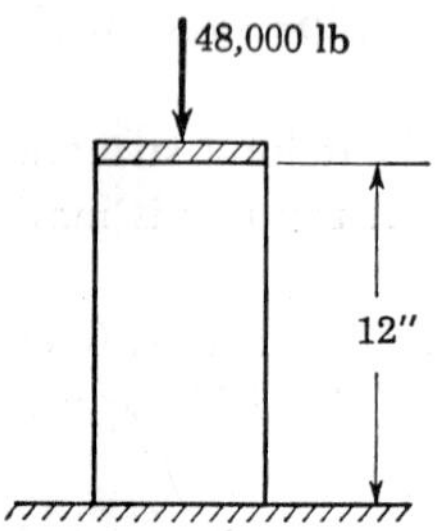

FIG. 1.9

Example 1.4. The concrete block shown in Fig. 1.9 is 12 in. long and 6 by 4 in. in cross section. $E = 4{,}000{,}000$ psi; $\mu = 0.20$. Determine the change in dimensions due to an axial load of 48,000 lb.

To find the axial unit strain

$$s = \frac{P}{A} = \frac{48{,}000}{6(4)} = 2000 \text{ psi (compression)}$$

$$E = \frac{s}{\epsilon} \qquad \epsilon = \frac{s}{E} = \frac{-2000}{4{,}000{,}000} = -0.0005 \text{ in./in.}$$

The change in length is

$$\epsilon = \frac{\delta}{L} \qquad \delta = \epsilon L = -0.0005(12) = -0.006 \text{ in.} \qquad \textit{Ans}$$

The lateral unit strain is

$$\mu = -\frac{\epsilon_l}{\epsilon_a} \qquad \epsilon_l = -\mu\epsilon_a = -0.20(-0.0005) = +0.0001 \text{ in./in.}$$

The change in the 6-in. dimension is

$$\delta_l = \epsilon_l(6) = +0.0001(6) = +0.0006 \text{ in.} \qquad \textit{Ans}$$

The change in the 4-in. dimension is

$$\delta_l = \epsilon_l(4) = +0.0001(4) = +0.0004 \text{ in.} \qquad \textit{Ans}$$

PROBLEMS

★**1.22.** A 2-in.-diam magnesium rod, 1.5 ft long, is subjected to a compressive load of 20,000 lb. Determine the change in length and diameter. Use $\mu = 0.30$.

1.23. A 10-ft length of plastic ($E = 600{,}000$ psi; $\mu = 0.25$) 0.12 by 1.52 in. in cross section is subjected to a tensile load of 1800 lb. Determine the change in length and cross-sectional dimensions.

★**1.24.** When a metal bar 2 by 4 in. in cross section and 8 in. long is subjected to a load of 160 kips, the length increases 0.0100 in. and the 4-in. side decreases in width 0.0015 in. Determine Poisson's ratio and Young's modulus.

1.25. Under an axial compressive load of 85,000 lb a standard concrete cylinder (12 in. long and 6 in. in diameter) shortens 0.0072 in. and increases in diameter 0.00036 in. Determine Poisson's ratio and Young's modulus.

1.6 BIAXIAL STRESSES AND STRAINS

Frequently members are loaded in such a way that a biaxial stress condition results. Such a condition exists at a point when normal stresses occur in two mutually perpendicular directions. Equations defining the normal strains due to biaxial stresses will be written with respect to the stressed element shown in Fig. 1.10. Using the principle of superposition,† the unit strain in the 1-1 direction is

$$\epsilon_1 = \frac{s_1}{E} - \mu \frac{s_2}{E} \tag{1.8a}$$

and the resulting strain in the 2-2 direction is

$$\epsilon_2 = \frac{s_2}{E} - \mu \frac{s_1}{E} \tag{1.8b}$$

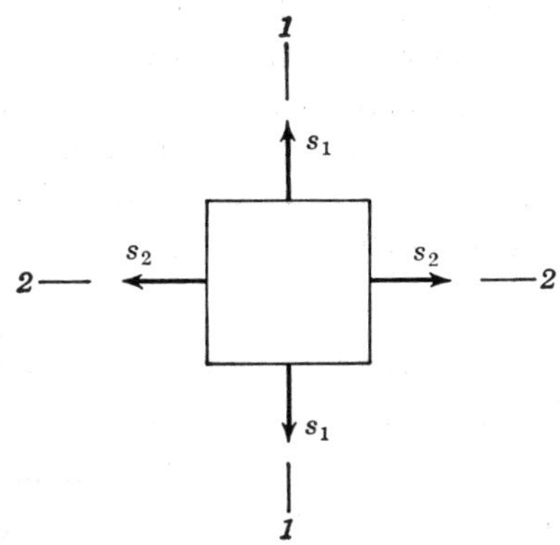

FIG. 1.10

When biaxial stresses at a point cannot be readily computed, they are sometimes found by measuring the strains in the direc-

† The principle of superposition states that effects can be added algebraically as long as a linear relationship exists between cause and effect.

tions of the stresses. When this is done, Eqs. (1.9*a*) and (1.9*b*) [obtained by solving Eqs. (1.8*a*) and (1.8*b*) simultaneously] are useful.

$$s_1 = \frac{E(\epsilon_1 + \mu\epsilon_2)}{1 - \mu^2} \tag{1.9a}$$

$$s_2 = \frac{E(\epsilon_2 + \mu\epsilon_1)}{1 - \mu^2} \tag{1.9b}$$

It should be noted that, since Eqs. (1.9*a*) and (1.9*b*) were written for tensile stresses, plus values (of stress and strain) in the equations indicate tension and minus values compression.

PROBLEMS

★1.26. A steel ($\mu = 0.27$) member which is 20 ft long (in the x direction) is subjected to uniform stresses of $s_x = 20{,}000$ psi compression and $s_y = 8000$ psi tension. Determine the change in length of the member.

1.27. An aluminum ($\mu = 0.33$) member which is 15 ft long (in the x direction) is subjected to uniform stresses of $s_x = 15{,}000$ psi tension and $s_y = 12{,}000$ psi compression. Determine the change in length of the member.

★1.28. A 1- by 2- by 4-in. aluminum ($\mu = 0.33$) block has a tensile force of 40,000 lb on each 1- by 2-in. face and a compressive force of 60,000 lb on each 1- by 4-in. face. Determine the change in dimensions.

1.29. A 1- by ½- by 2-in. steel ($\mu = 0.27$) block has a compressive force of 15,000 lb on each 1- by ½-in. face and a compressive force of 24,000 lb on each ½- by 2-in. face. Determine the change in dimensions.

1.7 THERMAL STRAINS AND STRESSES

It is known that all materials expand or contract with a change of temperature. Thus, heating a material causes it to expand, while lowering its temperature causes it to contract. A basic law of physics states that the change in length of a material is proportional to the change in its temperature. This law can also be expressed in equation form as

$$\epsilon = \alpha(\Delta t) \tag{1.10}$$

where ϵ is the change in length per unit length (that is, unit strain);

α is a constant, called the coefficient of thermal expansion; and Δt is the change in temperature. Thus, α is the unit strain per degree change of temperature, usually expressed in units of inches per inch per degree Fahrenheit. Since $\delta = \epsilon L$, the total change in length of a member can be expressed as

$$\delta = \alpha(\Delta t)L \tag{1.11}$$

The average coefficients of thermal expansion (for Fahrenheit temperature changes) for some commonly used engineering materials are listed in Table 1.2.

TABLE 1.2 Average Coefficients of Thermal Expansion for Some Engineering Materials

Material	*Thermal expansion coefficient, in./in./°F*
Steel	0.0000065
Aluminum	0.000013
Magnesium	0.000014
Copper	0.0000093
Concrete	0.0000055
Oak (parallel to grain)	0.0000027
Pine (parallel to grain)	0.0000030

It should be noted that, unless a material is restrained, the strain produced by temperature change is unaccompanied by stress. However, when materials are prevented from expanding or contracting, owing to temperature changes, stresses will be induced. These thermal stresses can be computed using the relationship $s = \epsilon E$, where ϵ represents the *restrained* unit strain. If provisions for thermal expansion (or contraction) were not made in the design of many structures and machines, large thermal stresses would result. In actual practice, thermal stresses have caused or contributed to the failure of many structures. Temperature stresses in composite structures are discussed in Chap. 3.

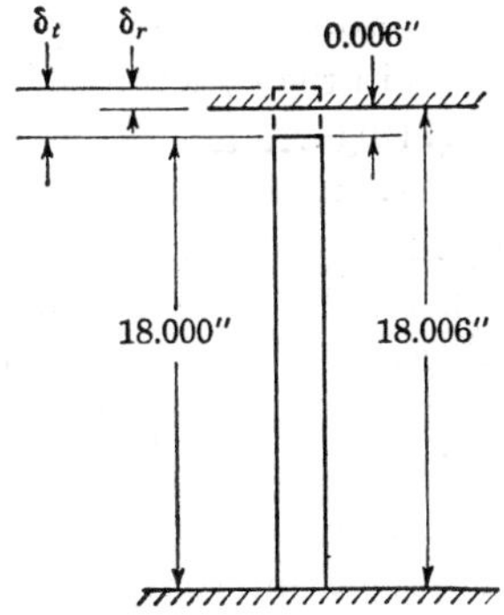

FIG. 1.11

Example 1.5. A 2- by 2-in. steel bar, 18 in. long at 20°F, is placed between two supports as shown in Fig. 1.11. The distance between supports is 18.006 in. Assuming that the supports are perfectly rigid (the 18.006 distance remains constant), determine the stress in the bar when its temperature is raised to 120°F.

The amount that the bar would expand if it were completely unrestrained is

$$\delta_t = \alpha(\Delta t)L = 0.0000065(120 - 20)(18) = 0.0117 \text{ in.}$$

Since the bar is prevented from expanding more than 0.006 in., the amount of restrained expansion is

$$\delta_r = 0.0117 - 0.006 = 0.0057 \text{ in.}$$

and the restrained unit strain is

$$\epsilon_r = \frac{\delta_r}{L} = \frac{0.0057}{18.0} = 0.000317 \text{ in./in.}$$

(The length in the above equation theoretically should be 18.0117 in.; however, using 18.0 in. does not affect the answer for practical purposes.)

The stress in the bar (due to its being prevented from expanding the full 0.0117 in.) is

$$s = \epsilon_r E = 0.000317(30{,}000{,}000) = 9510 \text{ psi} \qquad \textit{Ans}$$

PROBLEMS

★**1.30.** How much will a 1550-ft steel television tower increase in height when the temperature increases from 60°F at night to 97°F the next day?

1.31. A 5-ft concrete (E = 2,500,000 psi) sidewalk has 0.10-in. expansion joints every 30 ft at a temperature of 25°F. Determine the stress in the concrete (if any) when the temperature is (*a*) 60°F, (*b*) 80°F, (*c*) 110°F.

1.32. A concrete parking lot between two buildings 100 ft apart has ¼-in. expansion joints at each end when the temperature is 32°F. What will be the stress in the concrete when the temperature of the concrete is 120°F? Use E = 2,500,000 psi.

★**1.33.** The ends of a 4- by ½-in. aluminum bar are fixed to walls which are 20 ft apart at 192°F. Assuming that, when the temperature is 32°F, the bar springs the walls together a total of 0.20 in., what are the stress and internal force in the bar?

1.34. The ends of a 1.25-in.-diam steel rod are fixed to walls which are 30 ft apart at 158°F. Assuming that, when the temperature is 48°F, the bar springs the walls together a total of 0.10 in., what are the stress and internal force in the bar?

★1.35. A steel rim is to be mounted on a wooden wagon wheel so as to provide a durable wearing surface. At 68°F the inside diameter of the rim is 48.00 in. and the outside diameter of the wooden wheel is 48.04 in. (*a*) To what minimum temperature must the rim be heated so that it will just slip on the wheel? (*b*) When the temperature of both the wheel and the rim returns to 68°F, what will be the stress in the rim if the wooden wheel is assumed to shorten in diameter 0.01 in.? (*c*) What would be the stress in the rim if the wheel is assumed rigid? (HINT: The change in the circumference is proportional to the change in length of the diameter. Therefore, the circumferential unit strain is equal to the change per unit length of the diameter.)

1.36. Same as Prob. 1.35 except let the original inside diameter of the rim and outside diameter of the wheel be 36.00 and 36.03 in., respectively.

ADDITIONAL PROBLEMS

1.37. Show that a prismatic bar suspended vertically lengthens owing to its own weight an amount $WL/2AE$. (HINT: Write the equation of the elongation of a typical differential length and integrate.)

★1.38. A tapered aluminum bar 20 in. long is suspended vertically. The cross section at the top is 4 by 4 in.; the cross section at the bottom is 2 by 2 in. Determine the elongation of the bar due to an axial load of 20 kips applied at the lower end.

1.39. A tapered copper bar 30 in. long is suspended vertically. The diameter at the top is 3 in.; at the bottom the diameter is 2 in. Determine the elongation of the bar due to an axial load of 20 kips applied at the lower end.

★1.40. Determine the total elongation due to its own weight of the compound member shown in Fig. P1.40. Member A is of steel ($\gamma = 0.283$ lb/in.^{3}) and is 4 in. square; member B is of aluminum ($\gamma = 0.095$ lb/in.^{3}) and is 3 in. square. $L_1 = 4$ ft; $L_2 = 6$ ft. (The answer to Prob. 1.37 can be used in the solution.)

1.41. Determine the total elongation due to its own weight of the compound member shown in Fig. P1.41. Member A is of aluminum ($\gamma = 0.095$ lb/in.^{3}) and is 4 in. square; member B is of copper ($\gamma = 0.322$ lb/in.^{3}) and is 3 in. in diameter. $L_1 = 6$ ft; $L_2 = 4$ ft. (The answer to Prob. 1.37 can be used in the solution.)

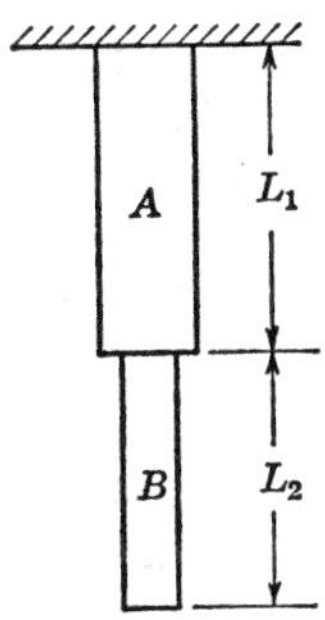

FIG. P1.40, P1.41

2

PROPERTIES OF MATERIALS

2.1 INTRODUCTION

In order to predict how a material will function in service, certain characteristics or *properties* of the material should be known. These characteristics may be classified into two general groups: mechanical properties and physical properties. *Mechanical properties* are those which measure or indicate the behavior of materials subjected to loads; other measurable characteristics are known as *physical properties*. Examples of physical properties are density, electrical conductivity, thermal conductivity, and coefficient of thermal expansion. Although a knowledge of both groups of properties is essential to the engineer, mechanical properties are of greater importance to the understanding of mechanics of materials.

2.2 STRESS-STRAIN DIAGRAM; TENSION TEST

The most used mechanical properties, but not the most important in all cases, are commonly determined from a Standard Tension Test.† This test consists essentially of gradually loading a standard-sized specimen of a material and noting corresponding values of load and elongation until the specimen fractures. The dimensions of a popular standard tension specimen are shown in Fig. 2.1. Load is applied and measured by a testing machine. *Values of unit stress* are computed by dividing the load values by the original cross-sectional area of the specimen. The elongation is measured by determining the amount

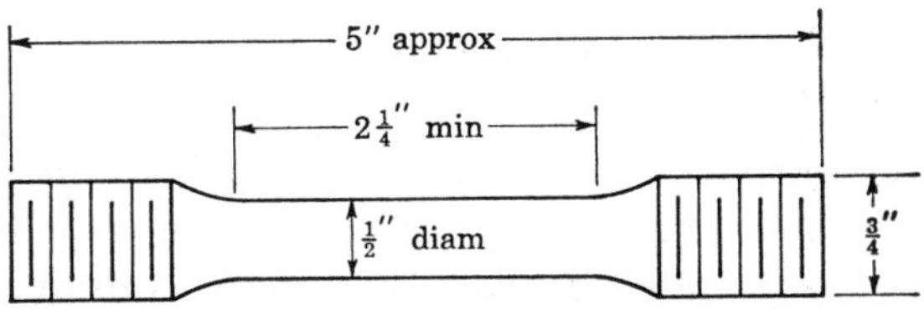

FIG. 2.1 A standard tension specimen.

that two reference points on the specimen are moved apart by the action of the machine; the original distance between the two reference points is known as the gage length. *Values of unit strain* are computed by dividing the elongation values by the gage length.

The plotted values of unit stress versus corresponding unit strain are used to draw the stress-strain diagram of the material tested. The curve that is drawn is a *faired curve;* that is, it does not necessarily pass through each point that is plotted. Thus, the curve has the advantage of indicating any erroneous readings or computations. Since the curve is an average curve, it more nearly represents the actual behavior of the material than would a curve passing through each plotted point. The stress-strain diagram has the further advantage of showing the results of the test at a glance; the significance of the results is much more readily apparent from the diagram than from the unplotted data.

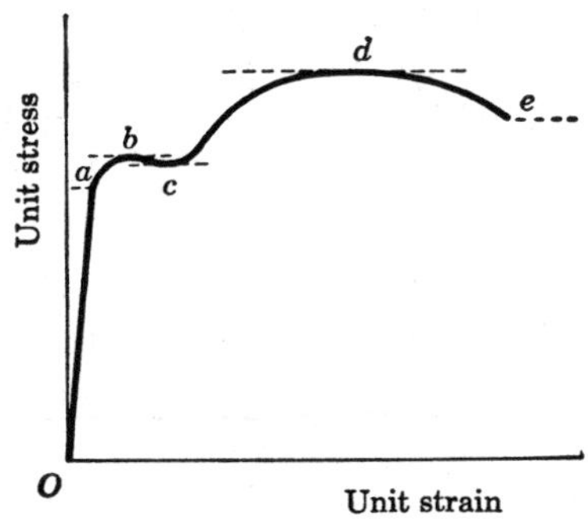

FIG. 2.2 Tension stress-strain diagram.

† ASTM Standards, Designation E8, American Society for Testing Materials, Philadelphia 3, Pa.

One type of stress-strain curve, not to scale, is shown in Fig. 2.2. It can be seen that stress and strain are proportional up to point *a*. As loading is continued beyond point *a*, the strain increases rapidly, relative to the increase of stress, to the extent that eventually, at point *b*, the curve becomes horizontal. Beyond point *b* a surprising relationship between stress and strain develops. The strain continues to increase rapidly, while the stress actually decreases until the curve again becomes horizontal at point *c*. Beyond this point stress and strain both increase until the curve becomes horizontal at point *d*. Thereafter the stress again decreases, with increasing strain, until fracture occurs at point *e*.

The stress-strain curve described above is of the type characteristic of structural steel. The stress-strain curves of other important structural materials may vary slightly or considerably in general shape.

2.3 MECHANICAL PROPERTIES

A stress-strain curve can be drawn for a material subjected to tension, compression, or shear loading. Such a curve can be used to determine many important mechanical properties. For ready reference these and other mechanical properties will be defined† and discussed in separate paragraphs.

Proportional Limit. The proportional limit is the largest stress for which Hooke's law applies. In other words, it is the stress at the end of the straight-line portion of the stress-strain curve. In Fig. 2.2 the proportional limit is the ordinate of point *a*.

Yield Point. The yield point is the first stress, short of the ultimate strength, at which a material undergoes a marked increase in strain without any increase in stress. On a stress-strain curve, it is the first stress, short of the ultimate strength, where a tangent to the stress-strain curve becomes horizontal. In Fig. 2.2 the yield point is the ordinate of point *b*. The ordinates of points *b* and *c* are sometimes referred to as the *upper yield point* and *lower yield point*, respectively. To the engineer, the yield point is a critically important stress in that it defines the stress at which considerable strain

† For standard definitions see ASTM Standards, Designation E6, American Society for Testing Materials, Philadelphia 3, Pa.

occurs. If stressed beyond the yield point to a significant degree, many structural members will no longer serve their intended purpose and therefore will have "failed."

Ultimate Strength. The ultimate strength is the largest stress, based on the original cross-sectional area, which a material can withstand. Thus it is the maximum ordinate of the stress-strain curve. In Fig. 2.2, the ultimate strength is the ordinate of point *d*. The ultimate strength as determined from a tension test is known as the *tensile strength*.

Breaking Strength. The breaking strength is the stress in a material, based on the original cross-sectional area, at the time it breaks. It is the final ordinate on a complete stress-strain curve. In Fig. 2.2 the breaking strength is the ordinate of point *e*. The breaking strength is also known as the *fracture strength* or *rupture strength*.

Modulus of Elasticity. The modulus of elasticity, also known as Young's modulus, is the constant of proportionality of stress to strain. It is equal to the slope of the straight-line portion of the stress-strain curve with respect to the strain axis. In Fig. 2.2 Young's modulus is equal to the slope of line Oa. The term modulus of elasticity is a misnomer; it is not a measure of the elasticity of a material. Rather, it is a measure of *stiffness*, or the ability of a material to withstand high stress while undergoing small elastic deformation. The stiffer a material, the higher will be its modulus of elasticity.

Elasticity. Elasticity is the ability of a material to deform under stress and to regain its original dimensions when the stress is removed.

Plasticity. Plasticity is the ability of a material to deform under stress and to retain this deformation after the stress has been removed.

Elastic Limit. The elastic limit is the largest stress which a material can withstand elastically. Thus, the elastic limit is the largest stress which a material can withstand without undergoing plastic deformation. The elastic limit is seldom actually measured, since to do so would require a large number of successive loadings and unloadings of the material. This loading and unloading procedure would have to be continued, each loading being slightly larger than the previous one, until a measurable strain remained with the material in an unstressed condition. This residual or plastic deformation is known as *permanent set.* Fortunately, for most engineering materials the elastic limit is equal to the proportional limit for all practical purposes. This fact accounts for the use of the term *proportional-elastic limit.*

Apparent Elastic Limit. As previously mentioned, the proportional limit can be taken as the stress at the end of the straight-line portion of the stress-strain curve. In practice, it is difficult to

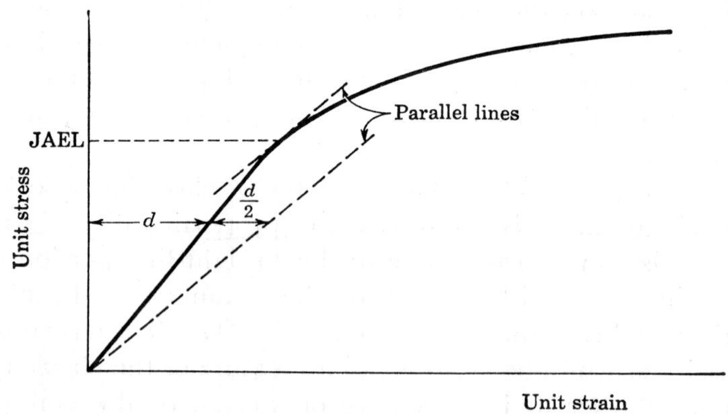

FIG. 2.3 Method of obtaining Johnson's apparent elastic limit.

judge where the straight-line portion ends with any degree of consistency. For this reason a value known as *Johnson's apparent elastic limit* is sometimes determined and used instead of the proportional limit or the elastic limit. Johnson's apparent elastic limit is the stress at the first point on the stress-strain curve which has a slope 50 per cent greater, *with respect to the stress axis,* than the slope of the straight-line portion of the curve. A rapid and convenient method of obtaining Johnson's apparent elastic limit, often designated as *JAEL* or the *apparent elastic limit,* is shown in Fig. 2.3.

Yield Strength. The importance of the yield point warrants defining a *comparable value* of stress for the many materials which have no yield point. These materials include many steels and most nonferrous and nonmetallic materials. The property which is determined in lieu of the yield point is called the yield strength. The yield strength is that stress which will cause the material to undergo a certain specified amount of *plastic strain.* Usually the yield strength is determined by the *offset method* as illustrated in Fig. 2.4 and described below. The plastic strain or offset is laid off from the origin, and a line parallel to the straight-line portion of the stress-strain curve is drawn through the offset point. The yield strength is the stress at the intersection of the parallel line and the curve. The specified offset for steel and aluminum is 0.2 per cent (the equivalent of 0.002 in./in. unit strain). Different offsets, usually ranging from 0.05 to 0.35 per cent, are specified for some other materials.

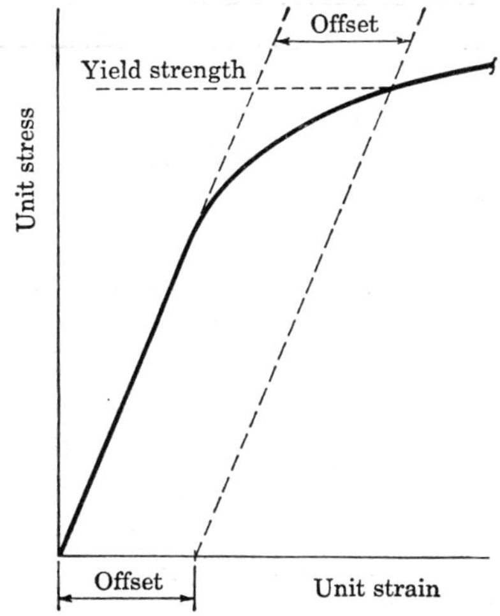

FIG. 2.4 Offset method of obtaining the yield strength.

Per Cent Elongation. The per cent elongation is obtained by comparing the total elongation at fracture to the gage length. The two pieces of the fractured specimen are fitted together, and the distance between the gage marks is measured; the total elongation is this distance minus the gage length. In equation form

$$\% \text{ elongation} = \frac{B - G}{G} 100 \tag{2.1}$$

where G is the gage length and B is the distance between gage points at fracture. These distances are illustrated in Fig. 2.5. The per cent elongation is frequently determined as a measure of ductility even when a stress-strain curve is not required.

Per Cent Reduction of Area. As a ductile material is stressed beyond the ultimate strength, its cross-sectional area is appreciably decreased until fracture occurs. This decrease of area is called *necking down* and is illustrated in Fig. 2.5. The per cent reduction of area is obtained by comparing the reduction of area at the smallest section of the fractured specimen with the original cross-sectional area. In equation form

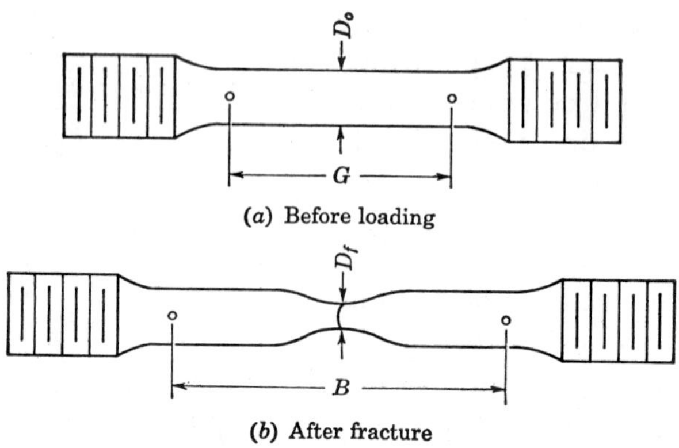

FIG. 2.5 Necking of a ductile steel tension specimen.

$$\% \text{ reduction of area} = \frac{A_o - A_f}{A_o} 100 \tag{2.2}$$

where A_o and A_f are the original and final minimum cross-sectional areas, respectively.

2.4 ADDITIONAL COMMENTS ON TENSION TESTS

When a material is tested in a testing machine of the mechanical or hydraulic type, the material is deformed by the action of the machine; the "load" is actually applied *by the material* by virtue of its resistance to being deformed. Therefore, the load can either increase or decrease with increasing deformation, depending on the inherent characteristics of the material. Conversely, if a material is loaded by dead weights, it must either support the entire load or fracture. Thus, the stress-strain curve obtained by loading a material in a testing machine does not necessarily describe the behavior of a similar material in actual use; it simply shows the results from a practical and conventional method of testing.

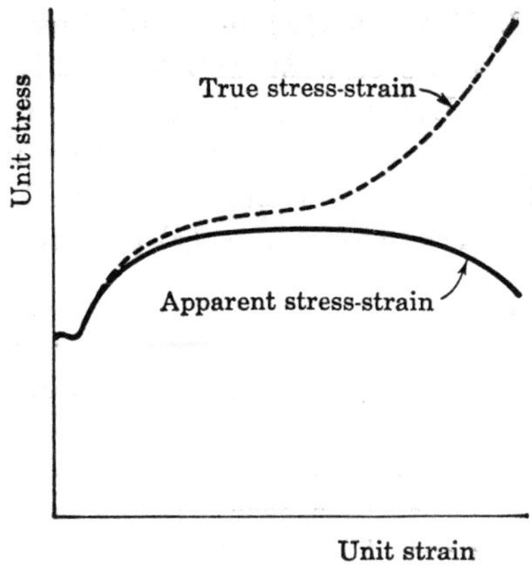

FIG. 2.6 Relation between true and apparent stress.

As previously stated, the stress from a tension test is defined as load divided by the original area. The stress thus computed is sometimes referred to as the *nominal stress;* the *true stress,* due to any particular load, is that load divided by the actual area. A material which does not fracture at its ultimate strength will exhibit a gradual and visible decrease in cross-sectional area before breaking. The true stress-strain curve for such a material at its narrowest section would be radically different from its nominal stress-strain curve, as shown in Fig. 2.6. The laborious process of computing true stresses is not generally justified, since members are designed on an original area basis.

2.5 DETERMINATION OF PROPERTIES FROM STRESS-STRAIN CURVES

The reason for drawing a curve, stress-strain or any other type, is to have it serve a purpose. This purpose may be to show the results readily, to emphasize a trend or rate of change, or to aid in the determination of specific values. Most stress-strain curves are drawn to facilitate the determination of mechanical properties. For this reason the curve from a single test is frequently drawn to two different scales. First, a small scale is used to depict the results of the entire test and to allow for the determination of certain properties such as the ultimate strength. Then, a portion of the curve is redrawn to a larger strain scale in order to facilitate a more accurate determination of other properties such as Young's modulus, apparent elastic limit, and yield strength.

Example 2.1. A test was conducted on a standard tension specimen of AISI 1015 HR steel. Original diameter = 0.505 in.; gage length = 2.00 in.; final diameter = 0.313 in.; distance between gage points at fracture =

2.81 in. Simultaneous readings of load and elongation are listed in the first two columns of Table E2.1.

Draw complete stress-strain diagram and determine the following: (*a*) Johnson's apparent elastic limit, (*b*) yield point or yield strength, (*c*) ultimate strength, (*d*) breaking strength, (*e*) Young's modulus, (*f*) per cent elongation, (*g*) per cent reduction of area.

TABLE E2.1

Load, lb	Elongation, in.	Unit stress, psi	Unit strain, in./in.
0	0.0000	0	0.0000
1,100	0.0004	5,500	0.0002
2,250	0.0008	11,250	0.0004
3,500	0.0012	17,500	0.0006
4,700	0.0016	23,500	0.0008
5,800	0.0020	29,000	0.0010
7,050	0.0025	35,250	0.0012
7,400	0.0030	37,000	0.0015
7,600	0.0046	38,000	0.0023
7,650	0.0066	38,250	0.0033
7,600	0.0080	38,000	0.0040
7,650	0.0100	38,250	0.0050
7,800	0.02	39,000	0.0100
7,700	0.04	38,500	0.02
8,400	0.06	42,000	0.03
9,700	0.10	48,500	0.05
11,000	0.15	55,000	0.08
11,600	0.20	58,000	0.10
12,100	0.29	60,500	0.14
12,250	0.40	61,250	0.20
12,200	0.52	61,000	0.26
12,100	0.61	60,500	0.30
11,400	0.66	57,000	0.33
11,000	0.72	55,000	0.36
10,100	0.77	50,500	0.38
9,250	0.80	46,250	0.40

The original cross-sectional area is computed to be 0.200 in.^{2}; values of unit stress and unit strain are computed and listed in the last two columns of the table. The complete curve is drawn to convenient stress and strain scales. The first part of the test results is replotted to the same stress scale but to a greatly enlarged strain scale. These plots are shown in Fig. 2.7.

The apparent elastic limit determined from the large-scale curve is 33,800 psi (*Ans*).

The yield point from the complete curve is 39,400 psi (*Ans*).

The ultimate strength from the curve is 61,300 psi (*Ans*).

The breaking strength from the curve is 46,000 psi (*Ans*).

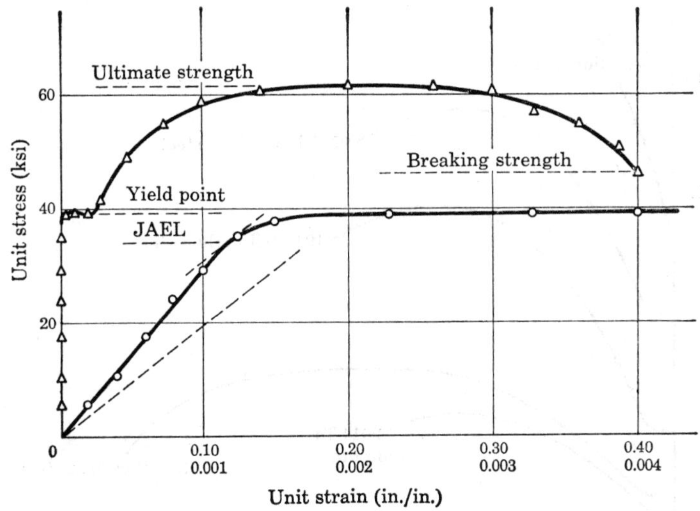

FIG. 2.7 Typical tension stress-strain diagram of a ductile steel.

Young's modulus is determined from the slope of the straight-line portion of the large-scale curve. Since this curve passes through the origin,

$$E = \frac{30.0 - 0.0}{0.00102 - 0.0} = 29{,}400 \text{ ksi or } 29{,}400{,}000 \text{ psi} \qquad Ans$$

The per cent elongation is

$$\frac{2.81 - 2.00}{2.00} 100 = 40.5\% \qquad Ans$$

The per cent reduction of area is

$$\frac{0.200 - 0.077}{0.200} 100 = 61.5\% \qquad Ans$$

2.6 DUCTILITY AND BRITTLENESS

Ductility is the ability of a material to deform plastically prior to fracture under *tensile* stress. *Malleability* is the ability of a material to deform *plastically* prior to fracture under *compressive* stress. Most engineering materials which are ductile are also malleable. *Brittleness* is the absence of ductility and malleability. Ductility and brittleness are actually relative terms. There is no clear demarcation between brittle and ductile materials; materials are usually *judged* to be ductile or brittle on the basis of their per cent elongation and per cent reduction of area.

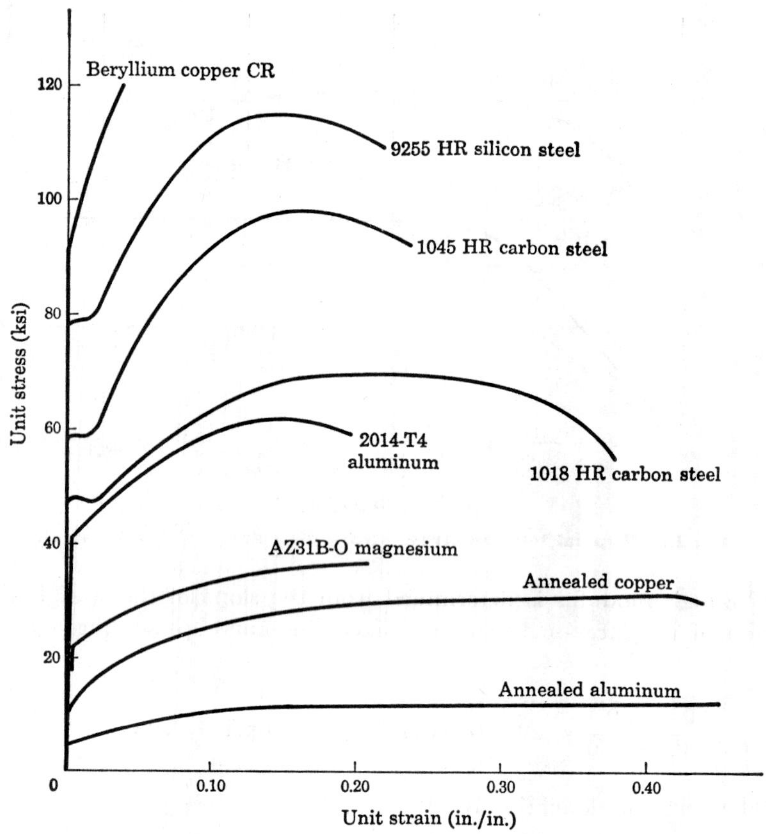

FIG. 2.8 Typical tension stress-strain curves of some metallic materials.

Ductility in a material is desirable for a number of reasons. In most members there exist, at certain points, unusually high stresses known as *stress concentrations.* Ductility enables the material to yield *locally,* thus preventing premature fracture at these points. As will be seen in Art. 2.7, ductility increases the ability of a material to withstand suddenly applied or impact loading. Ductile materials also have the advantage of usually giving visible evidence of impending fracture.

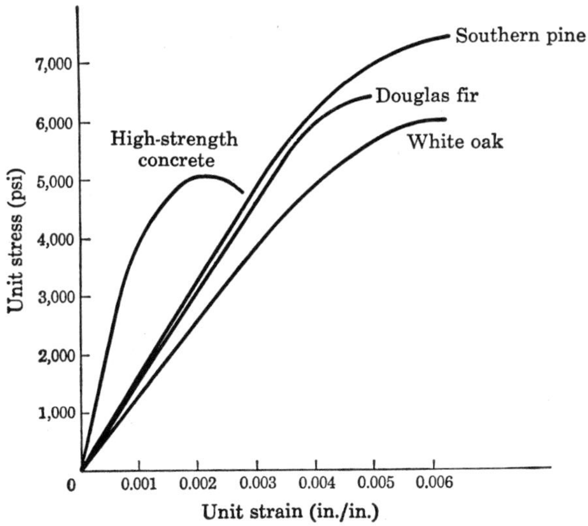

FIG. 2.9 Typical compression stress-strain curves of some nonmetallic materials.

High strength in materials is generally associated with low ductility, although many low-strength materials are extremely brittle. The association of high strength to low ductility is illustrated by the stress-strain curves of some representative metals in Fig. 2.8. Since both high strength and ductility are desirable qualities, the engineer is usually faced with a compromise between the two in his selection of a structural material.

Typical compression stress-strain curves of some brittle materials are shown in Fig. 2.9. Compressive stress-strain values are obtained from tests of relatively short specimens. The length of these specimens is limited to from two to ten times the least transverse dimension in order to minimize the effect of buckling on the test results. Brittle materials have a definite ultimate compressive strength as they fail by fracture. Ductile or malleable materials do not fracture when loaded in compression; they simply continue to deform until they lose their usefulness. The tension and compression stress-strain curves of most ductile materials are identical, for practical purposes, up to the yield strength.

2.7 RESISTANCE TO DYNAMIC LOADING

Dynamic loading occurs when load is applied to a member *rapidly* or with *impact*. An impact load is one which has velocity, in the direction of the load, at the instant of contact with the member. A rapidly applied load does not have velocity in the direction of the load at the instant of contact with the member. *Suddenly applied loading* is an extreme case of rapidly applied loading; this occurs when a vertical load is brought into contact with a member and then suddenly released.

To resist a dynamic loading a material must absorb the energy of the load. The ability of a material to absorb energy elastically is termed *resilience;* the ability to absorb energy without fracture is termed *toughness*. The mechanical properties which measure the ability of a material to resist dynamic loading are the modulus of resilience and the modulus of toughness.

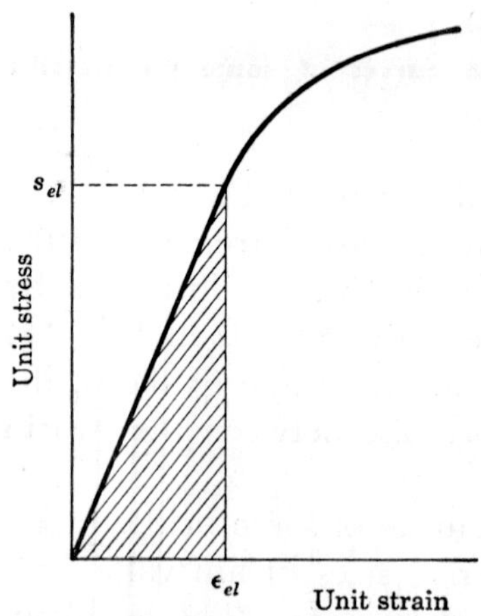

FIG. 2.10 Modulus of resilience from stress-strain curve.

The *modulus of resilience* is the amount of energy *per unit volume* which a material can absorb elastically. As shown below, it can be determined as the area under the stress-strain curve to the elastic limit. From the definition

$$\text{MR} = \frac{\text{energy}}{\text{volume}} = \frac{P_{\text{av}}\delta_{el}}{AL} = \frac{\frac{1}{2}P_{el}}{A}\frac{\delta_{el}}{L}$$

since the load varies linearly to the elastic limit. Therefore

$$\text{MR} = \tfrac{1}{2}\, s_{el}\epsilon_{el} \tag{2.3a}$$

Thus, the modulus of resilience is equal to the area under the stress-strain curve up to the elastic limit as illustrated in Fig. 2.10. When the stress-strain curve is not available, the modulus of resilience can be determined if expressed in another form:

$$MR = \frac{1}{2} s_{el}\epsilon_{el} = \frac{s_{el}}{2}\frac{s_{el}}{E} = \frac{s_{el}^2}{2E} \tag{2.3b}$$

For clarity of meaning, the modulus of resilience should have units consistent with its definition. The units usually assigned are inch-pounds per cubic inch.

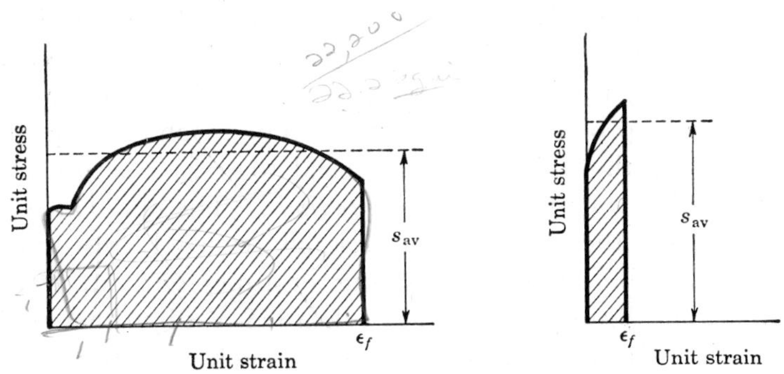

(*a*) Stress-strain curve for structural steel

(*b*) Stress-strain curve for malleable cast iron

FIG. 2.11 Modulus of toughness from stress-strain curve.

The *modulus of toughness* is the amount of energy *per unit volume* which a material can absorb prior to fracture. As shown below, it can be determined as the area under the complete stress-strain curve of the material. From the definition

$$MT = \frac{\text{energy}}{\text{volume}} = \frac{P_{av}}{A}\frac{\delta_f}{L} = s_{av}\epsilon_f \tag{2.4}$$

As can be seen from Fig. 2.11, the average stress s_{av} multiplied by the unit strain at fracture ϵ_f would give the area under the complete stress-strain curve. Therefore, the modulus of toughness must be equal to the area under the complete stress-strain curve. It should be emphasized that the stronger of two materials may have the lower modulus of toughness. This fact can easily be seen by comparing the shaded areas of the two stress-strain curves shown in

Fig. 2.11. The cast iron is stronger than the structural steel but has a much smaller modulus of toughness.

2.8 RESISTANCE TO REPEATED LOADING

Many machine and structural members are loaded and unloaded thousands or millions of times during their service life. Such members are said to be subjected to repeated loading. A repeating load will cause the unit stress to fluctuate at any point in the member. Depending on the loading cycle, the fluctuating normal stress may vary between two different tensile values, between two different compressive values, or between a value in tension and an equal or different value in compression. When the stress varies from a maximum tensile value to an equal maximum compressive value, it is said to be *completely reversed.* The mechanical property which measures the ability of a material to resist fracture due to repeated loading is called the *endurance limit.* The endurance limit can be defined as the maximum completely reversed stress to which a material can be subjected millions of times without fracture. The endurance limit is determined from *S-N* diagrams such as are shown in Fig. 2.12. In order to draw such a diagram a number of specimens of the material must be tested. Each specimen is subjected to a different maximum completely reversed stress, in cycles, until fracture. A cycle of stress is two consecutive complete stress reversals (from s_T to s_C and back to s_T). The testing of each specimen provides data for one point on the *S-N* curve, *S* being the maximum cyclic stress and *N* being the number of stress cycles "endured" before fracture. *S-N* diagrams are usually drawn on semilogarithmic paper in order to emphasize the stress-cycle relationship. Figure 2.12*a* illustrates the type of *S-N* curve which is typical of many ferrous products. With this type, the curve becomes horizontal after a few million cycles of stress. The endurance limit is therefore the ordinate of the horizontal portion of the *S-N* curve. Another type, illustrated by Fig. 2.12*b*, is typical of the *S-N* curve of many nonferrous materials. This type of curve does not become horizontal even after several hundred million cycles. For such a material the endurance limit is usually taken as the stress, from the curve, at a specified number of cycles.

In this country the specified number of cycles may be as high as 500,000,000.

Fatigue failure is the term generally used to denote fracture resulting from repeated stress. For this reason the endurance limit is also known as the fatigue limit. The process of failure due

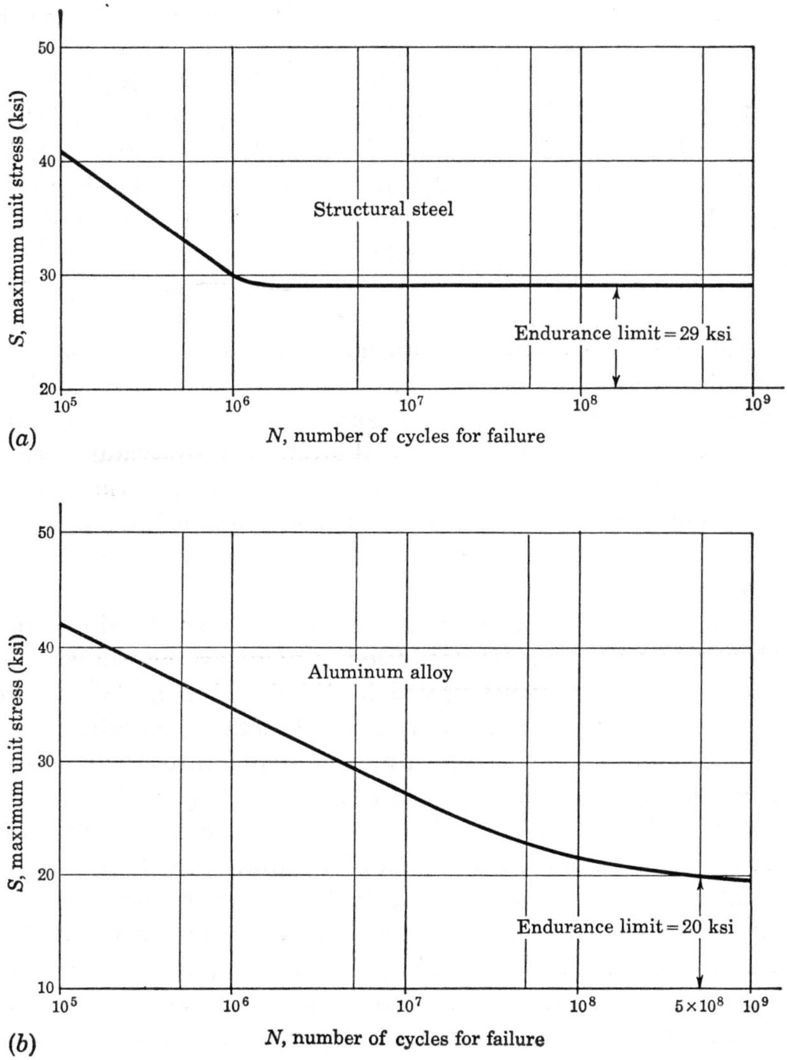

FIG. 2.12 Endurance limit from S-N curve.

to repeated stress is not fully understood. Several theories, such as "crystallization of the material," have been discarded. The current, and perhaps most logical, theory explains the failure as being one of progressive fracture. Even so-called "homogeneous materials" have small imperfections such as scratches, pits due to corrosion, and minute discontinuities due to slag and other impurities. The stress at such an imperfection is much greater than the "average" or "computed" stress for the point and causes a minute crack to develop. The stress at the base of a crack is also much greater than average, so that with repeated applications of stress the crack is progressively extended. Finally, when the effective area of the material has been sufficiently reduced, complete fracture suddenly occurs. Visible deformation prior to fracture is not apparent even for ductile materials.

2.9 EFFECT OF TEMPERATURE ON PROPERTIES

Mechanical properties of materials usually reported are obtained from tests conducted at normal atmospheric or room temperatures. Such properties may be materially altered by a significant increase or decrease in temperature. The usefulness of a material may be lost because of increased brittleness at exceptionally low temperatures or excessive deformation or loss of strength at excessively high temperatures.

At sufficiently high temperatures most materials will deform continuously, even though subjected to a steady constant stress. This continuing deformation, which is inelastic, is known as *creep* and must be limited if the material is to function properly.

Creep is a function of stress, time, and temperature; an increase in any one of these will increase the creep. Since a member is normally designed for a particular temperature and service life, the stress must be made small enough to limit creep to an acceptable value. This limiting stress is called the *creep strength* or *creep limit.* The creep strength can be defined as the maximum constant stress to which a material at a *given temperature* may be subjected for a *specified period of time* without having the *plastic deformation exceed a specified amount.* For many structural materials, the creep strength is that stress which will cause a plastic deformation of

1 per cent after 100,000 hr (approximately 11.4 years). The materials are not actually tested for such long periods of time. *Creep-time curves* are determined for the particular material and used to compute the deformations after the specified period of time.

Although creep is generally associated with elevated temperatures, some materials, such as lead and concrete, will creep at normal temperatures. Most metals, such as steel and aluminum, will not creep at normal temperatures.

2.10 PROPERTIES OF MATERIALS IN SHEAR

Shearing properties of materials differ appreciably from their tensile and compressive properties. Generally, the tension and compression stress-strain curves of a ductile material are quite similar;

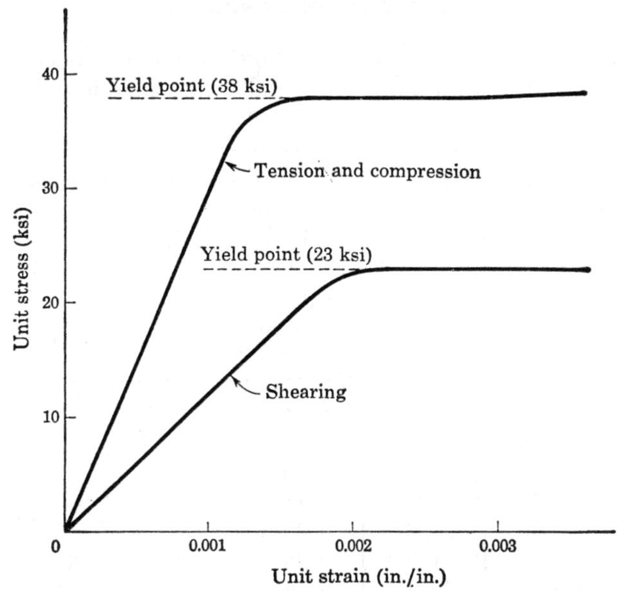

FIG. 2.13 Stress-strain diagrams to yield point; structural steel in tension, compression, and shear.

the shearing stress-strain curve, although having the same general shape, has considerably different critical values. Figure 2.13 illustrates how shearing-stress properties differ from comparable normal-stress properties. As can be seen, structural steel is appreciably weaker and less stiff in shear than in tension or compression. Values of Young's modulus and modulus of rigidity from the curves are 29,000,000 and 12,000,000 psi, respectively.

Shearing stress-strain curves are usually determined from torsion tests of cylindrical specimens. The method of computing stresses and strains for torsional loading is explained in Chap. 7. A useful relationship among Young's modulus, Poisson's ratio, and the modulus of rigidity† is

$$E_s = \frac{E}{2(1 + \mu)} \tag{2.5}$$

This expression is frequently used to compute the modulus of rigidity when torsion test data are not available.

QUESTIONS AND PROBLEMS

★2.1. A test was conducted on a standard tension specimen of AISI 1045 HR steel. Original diameter = 0.505 in.; gage length = 2.00 in.;

TABLE P2.1

Load, lb	Total elongation, in.	Load, lb	Total elongation, in.
0	0.0000	11,600	0.0130
1,400	0.0005	11,400	0.0150
3,000	0.0010	12,100	0.02
4,600	0.0015	13,300	0.03
6,200	0.0020	16,600	0.04
8,000	0.0026	17,800	0.05
10,100	0.0033	19,900	0.10
10,900	0.0036	20,900	0.15
11,600	0.0040	21,300	0.20
11,700	0.0050	21,300	0.26
11,800	0.0060	20,600	0.34
11,800	0.0080	19,500	0.40
11,800	0.0095	18,300	0.44
11,700	0.0110		

† For a proof of this relationship see S. P. Timoshenko and J. N. Goodier, "Theory of Elasticity," 2d ed., McGraw-Hill Book Company, Inc., New York, 1951.

final diameter = 0.367 in.; distance between gage points at fracture = 2.44 in. Simultaneous readings of load and elongation are listed in Table P2.1.

Draw the complete stress-strain curve using scales of approximately 1 in. = 20 ksi and 1 in. = 0.05 in./in. On the same sheet redraw the initial portion of the curve, using the same stress scale but to a strain scale of approximately 1 in. = 0.001 in.

Determine the following properties: (*a*) JAEL, (*b*) yield point or yield strength (0.2 per cent offset), (*c*) ultimate strength, (*d*) breaking strength, (*e*) Young's modulus, (*f*) per cent elongation, (*g*) per cent reduction of area.

2.2. A test was conducted on a standard tension specimen of AISI 1018 CD steel. Original diameter = 0.502 in.; gage length = 2.00 in.; final diameter = 0.310 in.; distance between gage points at fracture = 2.32 in. Simultaneous readings of load and elongation are listed in Table P2.2.

TABLE P2.2

Load, lb	Total elongation, in.	Load, lb	Total elongation, in.
0	0.0000	15,900	0.0105
1,700	0.0006	16,100	0.0115
3,300	0.0012	16,300	0.0125
5,000	0.0018	16,400	0.02
6,700	0.0024	16,500	0.03
8,400	0.0030	16,500	0.04
10,700	0.0040	16,600	0.05
12,500	0.0050	16,600	0.07
13,700	0.0060	16,400	0.10
14,500	0.0070	16,000	0.12
15,400	0.0085	13,900	0.24
15,700	0.0095	12,300	0.31

Draw the complete stress-strain curve using scales of approximately 1 in. = 20 ksi and 1 in. = 0.05 in./in. On the same sheet redraw the initial portion of the curve, using the same stress scale but to a strain scale of approximately 1 in. = 0.001 in./in.

Determine the following properties: (*a*) JAEL, (*b*) yield point or yield strength (0.2 per cent offset), (*c*) ultimate strength, (*d*) breaking strength, (*e*) Young's modulus, (*f*) per cent elongation, (*g*) per cent reduction of area.

2.3. A test was conducted on a standard tension brass specimen. Original diameter = 0.500 in.; gage length = 2.00 in.; final diameter = 0.357 in.; distance between gage points at fracture = 2.63 in. Simultaneous readings of load and elongation are listed in Table P2.3.

Draw the complete stress-strain diagram using scales of approximately 1 in. = 10 ksi and 1 in. = 0.05 in./in. On the same sheet redraw the initial portion of the curve, using the same stress scale but to a strain scale of approximately 1 in. = 0.002 in./in.

Determine the following properties: (*a*) JAEL, (*b*) yield point or yield strength (0.35 per cent offset), (*c*) ultimate strength, (*d*) breaking strength, (*e*) Young's modulus, (*f*) per cent elongation, (*g*) per cent reduction of area.

TABLE P2.3

Load, lb	Total elongation, in.	Load, lb	Total elongation, in.
0	0.0000	9,300	0.05
1,100	0.0009	9,600	0.07
2,000	0.0016	9,900	0.10
2,900	0.0024	10,100	0.13
4,000	0.0032	10,200	0.17
4,900	0.0040	10,500	0.22
5,800	0.0048	10,700	0.30
6,500	0.0056	10,700	0.35
7,100	0.0065	10,800	0.40
7,700	0.0075	10,800	0.46
8,200	0.0095	10,700	0.52
8,500	0.0115	10,400	0.57
8,700	0.0145	9,700	0.62
9,000	0.03		

★**2.4.** A standard concrete cylinder was tested in compression. Diameter = 6.00 in.; gage length = 8.00 in. Simultaneous readings of load and total strain are listed in Table P2.4.

Draw the complete stress-strain diagram using scales of approximately 1 in. = 1000 psi and 1 in. = 0.0002 in./in.

Determine the following properties: (*a*) proportional limit, (*b*) JAEL, (*c*) ultimate strength, (*d*) Young's modulus, (*e*) modulus of toughness.

TABLE P2.4

Load, kips	Total strain, in.	Load, kips	Total strain, in.
0	0.0000	80	0.0053
10	0.0005	90	0.0060
20	0.0014	100	0.0069
30	0.0021	110	0.0077
40	0.0027	120	0.0090
50	0.0033	130	0.0104
60	0.0038	136	Fracture
70	0.0044		

2.5. A compression test was conducted on a yellow pine specimen, load being applied parallel to the grain. Cross-sectional dimensions = 1.62 by 1.63 in.; gage length = 6.00 in. Simultaneous readings of load and total strain are listed in Table P2.5.

TABLE P2.5

Load, lb	Total strain, in.	Load, lb	Total strain, in.
0	0.0000	14,000	0.0173
1,000	0.0014	16,000	0.0206
2,000	0.0028	18,000	0.0244
4,000	0.0049	20,000	0.0293
6,000	0.0079	22,000	0.0386
8,000	0.0098	23,000	0.0482
10,000	0.0139	23,700	Fracture
12,000	0.0146		

Draw the complete stress-strain diagram using scales of approximately 1 in. = 2000 psi and 1 in. = 0.002 in./in.

Determine the following properties: (*a*) proportional limit, (*b*) JAEL, (*c*) ultimate strength, (*d*) Young's modulus, (*e*) modulus of resilience, (*f*) modulus of toughness.

2.6. What are the advantages of using a graph to show the results of a test?

2.7. What are the advantages of using a stress-strain curve in obtaining mechanical properties?

2.8. (*a*) What is the difference between the proportional limit and the elastic limit? (*b*) What is the difference between the yield point and the yield strength?

2.9. What is the difference between the elastic limit and the apparent elastic limit?

2.10. Explain how the breaking strength of a material can be lower than the ultimate strength.

2.11. What is meant by (*a*) elasticity, (*b*) plasticity, (*c*) ductility, and (*d*) brittleness?

2.12. List several reasons why ductility is an important consideration in the selection of a structural material.

2.13. What is the difference between resilience and toughness?

★**2.14.** (*a*) Determine the modulus of resilience and modulus of toughness of a steel having the stress-strain diagram shown in Fig. 2.7. (*b*) What must be the minimum length of a bar of this material having a cross-sectional area of 2 in.^{2} if it is to withstand elastically an impact load of 1800 in.-lb?

2.15. What characteristic of a material is measured by (*a*) modulus of elasticity, (*b*) per cent elongation, (*c*) endurance limit, (*d*) per cent reduction of area?

2.16. What property is determined from an *S-N* curve?

2.17. What is creep? On what three parameters is it dependent?

★**2.18.** Compute Poisson's ratio for a material having a Young's modulus and a modulus of rigidity of 15,000,000 and 5,800,000 psi, respectively.

2.19. A member made of steel similar to that of Example 2.1 is 2 by 3 in. in cross-sectional area and is 10 ft long. How much will the member lengthen under an axial load of (*a*) 150,000 lb, (*b*) 300,000 lb?

3

AXIALLY LOADED MEMBERS

3.1 ALLOWABLE STRESSES

The design engineer's work is generally governed by specifications and codes which set minimum standards with respect to quality of materials, structural theory, loading assumptions, and construction practices. These specifications and codes are usually compiled and prepared by committees of engineers representing professional, governmental, and industrial organizations. Examples of specifications in common use are those of the American Standards Association, the American Institute of Steel Construction, the American Society of Mechanical Engineers, and the American Association of State Highway Officials. Included in such specifications are allowable stresses for certain structural materials. *Allowable stress* is the maximum computed stress that is permitted in the material. The allowable stress is always less than the expected strength of the material in order to allow for the many uncertainties involved in its use for structural purposes. These uncertainties are many and varied but may be classified into three general groups: uncertainties regarding the material, uncertainties regarding the loading, and uncertainties regarding the theory used for computations. These three groups of uncertainties are discussed in separate paragraphs below.

The uncertainties with regard to materials are many. The accepted properties of a particular engineering material are "average properties" determined from a large number of test samples; they are not necessarily the properties of any one piece of that material. In addition, the properties of a piece of material may be altered adversely during fabrication or while it is in service. For example, the heat from welding may weaken or embrittle a metal, or an absorption of moisture may decrease the strength and stiffness of timber. Other uncertainties arise from the loss of effective material as a result of corrosion, decay, or spalling. Material may also be lost, sometimes at a most critical section, because of unplanned openings made by craftsmen such as electricians and plumbers.

The loads which will act on most structures must of necessity be *assumed*, since there is no way of predicting their exact magnitude or location. The actual loads may be static, dynamic, or repeating. In many cases the loading possibilities are so vague as to warrant the assumption of an extremely simplified loading condition. For a school or office building, for example, a uniformly distributed floor load is normally assumed in place of the actual but unpredictable loads due to people, furniture, and equipment. Loads resulting from wind, flooding, and earthquakes are even more uncertain, but they, too, must be assumed.

The stresses and deformations of a material are computed using theory based on certain assumptions. These assumptions are generally very nearly, but not exactly, true. Some of the assumptions *must* be made in order to solve the problem; other assumptions are made in order to simplify computations and save time without sacrifice of accuracy for practical purposes. Some of the assumptions made for the derivation of certain equations in this text *must* be made, while others would not be made in a more advanced treatment of the subject.

Allowable stresses are also referred to as *design stresses* and *work-*

ing stresses. More properly the term working stress should be used to denote the computed stress in a member. Thus the working stress in a member should be *equal to or less than* the allowable (or design) stress for the material.

3.2 FACTOR OF SAFETY

The ratio of the strength of a material to the allowable stress is known as the *factor of safety*. Generally the factor of safety is decided on as a matter of judgment, based on the prospective use of the material and the consequences of its failure. The allowable stress is then obtained by dividing the strength of the material by the factor of safety. The strength used to determine the allowable stress depends on what constitutes failure for the material when in use. Most machine and structural members fail when they are excessively deformed. Therefore the factor of safety should be applied to the yield point or yield strength to obtain the allowable stress. The engineer who usually works with allowable stresses from specifications or codes may lose sight of the basis for the factor of safety. Therefore, when occasions arise for him to determine or report a factor of safety, he may think in terms of ultimate strength when he should think in terms of yield strength. Example 3.1 is used to illustrate some of the implications of the term factor of safety.

Example 3.1. A structural steel has the following properties: ultimate strength = 66,000 psi; yield point = 33,000 psi; endurance limit = 28,000 psi. If the allowable stress were 20,000 psi, determine the factor of safety with respect to failure due to (*a*) fracture, (*b*) excessive deformation, (*c*) fatigue.

(*a*) With respect to fracture

$$\text{FS} = \frac{\text{ultimate strength}}{\text{allowable stress}} = \frac{66{,}000}{20{,}000} = 3.30 \qquad \textit{Ans}$$

(*b*) With respect to excessive deformation

$$\text{FS} = \frac{\text{yield point}}{\text{allowable stress}} = \frac{33{,}000}{20{,}000} = 1.65 \qquad \textit{Ans}$$

(*c*) With respect to fatigue

$$\text{FS} = \frac{\text{endurance limit}}{\text{allowable stress}} = \frac{28{,}000}{20{,}000} = 1.40 \qquad \textit{Ans}$$

It should be emphasized that, when allowable stresses have been used in the design of a structure or machine, the loading could not necessarily be increased in proportion to the factor of safety. The purpose of the factor of safety is to allow for the many uncertainties of materials, loading, and theory as described in the previous article. Neither should the factor of safety be used as an excuse for making crude or "rough" final computations or for making oversimplified assumptions.

As will be seen in later chapters (particularly Chaps. 15 and 17) factor of safety may be defined in other ways. For example, when columns are being analyzed, the factor of safety is defined as the ratio of the load which the column could theoretically support to its design load. Similarly, by plastic analysis the factor of safety is the ratio of the load which would theoretically cause a structure to collapse to its design load.

PROBLEMS

***3.1.** A punch which is 0.75 in. in diameter is made of an alloy steel having a yield strength of 90,000 psi. If the factor of safety for the punch is 3 (based on the yield strength), determine the maximum thickness of plate material having an ultimate shearing strength of 45,000 psi which can be punched.

3.2. An alloy-steel punch is to be used to punch holes in 0.25-in.-thick plate material having an ultimate shearing strength of 48,000 psi. The yield strength of the punch material is 75,000 psi. If the factor of safety for the punch is 2.5 (based on the yield strength), determine the minimum diameter of hole which can be punched.

3.3. A punch is to be used to make 0.5-in.-diam holes in 0.25-in.-thick plate material having an ultimate shearing strength of 50,000 psi. Determine the minimum yield strength for the punch material if a factor of safety of 2.5 (based on yield strength) is desired.

***3.4.** The frame shown in Fig. P3.4 consists of a wooden member *ABC* and a member *BD* made of two flat bars each 3 in. wide and 0.400 in. thick. (*a*) Compute the allowable load Q as governed by the stress in member BD. (*b*) Determine the required diameters of pins B and C. Allowable normal stress = 20,000 psi; allowable shearing stress = 15,000 psi.

3.5. The frame shown in Fig. P3.5 consists of a wooden member ABC and a member BD made of two flat bars each 2.75 in. wide. The diameter of pin B is 0.75 in. Determine (*a*) the allowable load Q as limited by the stress in pin B, (*b*) the required thickness of each of the two flat bars. Allowable normal stress = 20,000 psi; allowable shearing stress = 15,000 psi.

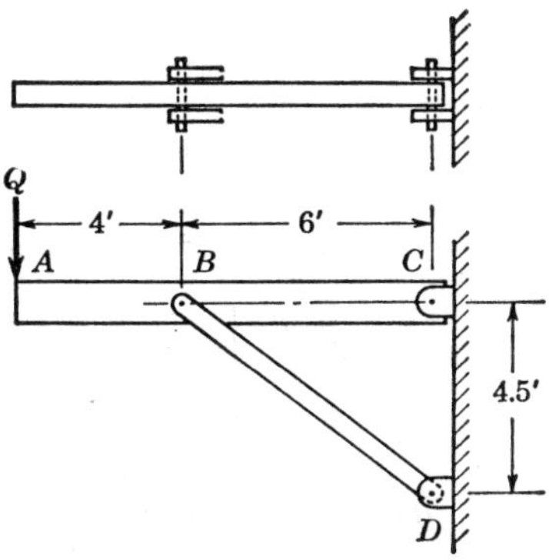

FIG. P3.4, P3.5, P3.6

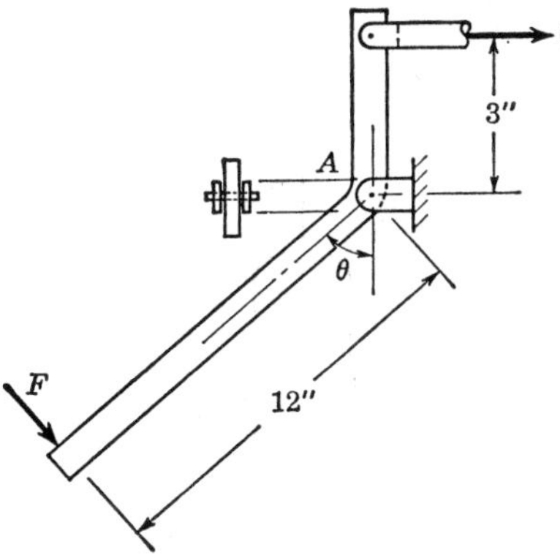

FIG. P3.7, P3.8, P3.9

3.6. The frame shown in Fig. P3.6 consists of a wooden member ABC and a member BD made of two flat bars each 3 in. wide. The diameter of pin C is 1 in. Determine (*a*) the allowable load Q as limited by the stress in pin C, (*b*) the required thickness of each of the two flat bars. Allowable normal stress = 20,000 psi; allowable shearing stress = 15,000 psi.

***3.7.** A force F equal to 250 lb is to be exerted on the pedal assembly shown in Fig. P3.7. If $\theta = 60°$, determine the required cross-sectional area of the horizontal bar and the diameter of pin A. Because of the possibility of dynamic loading and "fatigue" use a factor of safety of 6 applied to the following ultimate strengths: tension = 65,000 psi; shear = 45,000 psi.

3.8. On the pedal assembly shown in Fig. P3.8, pin A is 0.375 in. in diameter and the net area of the horizontal bar is 0.10 in.^{2} If $\theta = 45°$, determine the allowable force F based on a factor of safety of 6 applied to the following ultimate strengths: tension = 65,000 psi; shear = 45,000 psi.

3.9. A force $F = 50$ lb is to be exerted on the pedal assembly shown in Fig. P3.9. Pin A is 0.438 in. in diameter; the net area of the horizontal bar is 0.12 in.^{2} If $\theta = 50°$, determine the factor of safety based on the following yield strengths: tension = 35,000 psi; shear = 24,000 psi.

3.3 BEARING STRESSES

Bearing stress is *localized compressive stress* at the area of contact between two members. Bearing forces crush or tend to crush a member; they may be exerted through a body, fluid pressure, or another member. Bearing stresses can be illustrated using the

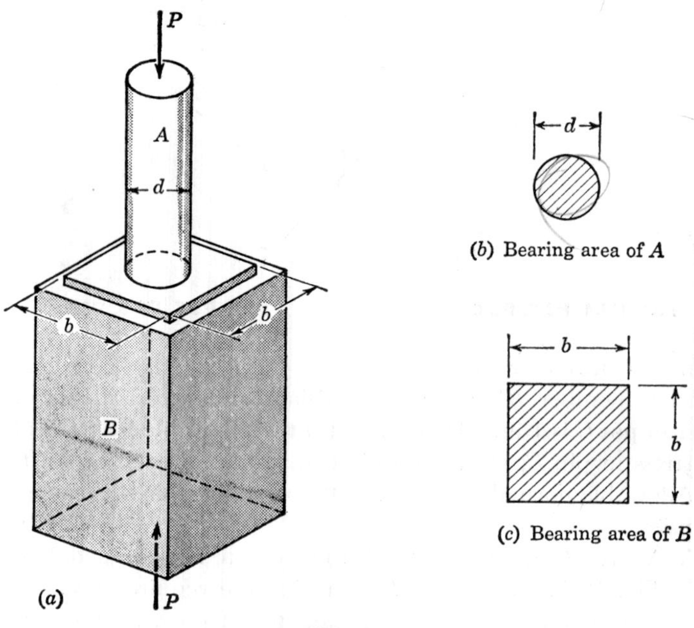

FIG. 3.1

structure shown in Fig. 3.1*a*. Member A transmits the load P through the flat "bearing plate" to the pedestal B. The bearing area of member A against the top of the plate is shown in Fig. 3.1*b*; the bearing area of member B against the bottom of the plate is shown in Fig. 3.1*c*. Generally speaking, allowable stresses for ductile materials are higher in bearing than in compression; for brittle materials the allowable stresses in compression and in bearing are practically the same.

PROBLEMS

★3.10. A timber post is to transmit an axial load Q through a steel bearing plate to a concrete footing as shown in Fig. P3.10. The post is 8 in. in diameter; the bearing plate is 9 in. square; the footing is 12 in. square at the top. Compute the allowable axial load based on the following allowable stresses: timber compression = 1000 psi; concrete compression = 450 psi; concrete bearing = 600 psi.

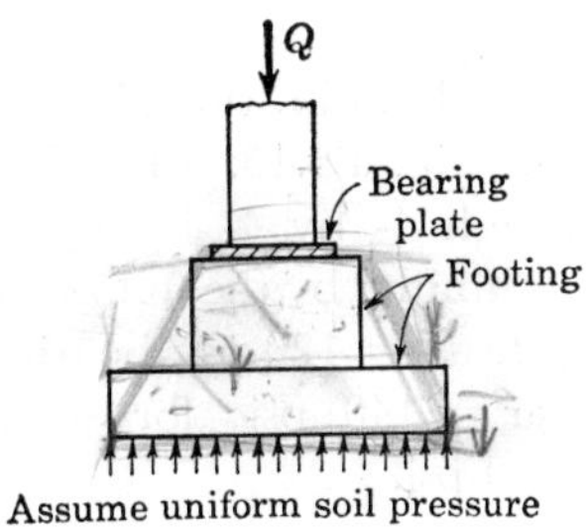

FIG. P3.10, P3.11, P3.12

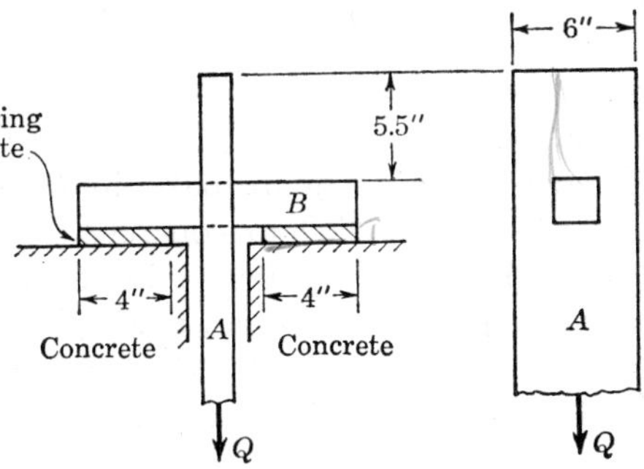

FIG. P3.13, P3.14, P3.15

3.11. A square timber post is to transmit an axial load Q of 100 kips through a square bearing plate to a concrete footing as shown in Fig. P3.11. The footing is to be square both at the top and at the bottom. The allowable soil pressure is 4000 psf (pounds per square foot). Allowable stresses are as follows: timber compression = 1000 psi; concrete compression = 450 psi; concrete bearing = 600 psi. Compute the minimum required cross-sectional dimensions of the post, of the bearing plate, and of the footing (top and bottom).

3.12. A square timber post is to transmit an axial load Q through a square bearing plate to a concrete footing as shown in Fig. P3.12. The footing is to be square at the top and bottom. If the allowable soil pressure is 3000 psf and the bottom of the footing is 40 in. square, determine the maximum allowable load and the minimum required cross-sectional dimensions of the post, the bearing plate, and the top of the footing. Allowable stresses are as follows: concrete compression = 675 psi; concrete bearing = 900 psi; timber compression = 1200 psi.

★**3.13.** An aluminum member A which is 6 in. wide and 0.75 in. thick is supported by a 2-in.-square steel pin B as shown in Fig. P3.13. The pin transmits the load Q to concrete supports through 4- by 3- by ½-in. bearing plates. Find the maximum value of load Q as limited by the following allowable stresses: for aluminum: bearing = 36,000 psi, tension = 20,000 psi, shear = 13,000 psi; for steel: bearing = 24,000 psi, shear = 15,000 psi; for concrete: bearing = 600 psi.

3.14. A wood member A which is 6 in. wide and 1 in. thick is supported by a 1.2-in.-square aluminum pin B as shown in Fig. P3.14. The pin transmits the load Q to concrete supports through 4- by 3.5- by ½-in. bearing plates. Find the maximum value of Q as limited by the following allowable stresses: for wood: tension = 2000 psi, shear = 180 psi, bearing = 2400 psi; for aluminum: bearing = 36,000 psi, shear = 13,000 psi; for steel: bearing = 24,000 psi; for concrete: bearing = 600 psi.

3.15. A pine member A which is 6 in. wide and 1.5 in. thick is supported by a 2-in.-square oak pin B as shown in Fig. P3.15. The pin transmits the load Q to concrete supports through 4- by 4- by ½-in. bearing plates. Find the maximum value of Q as limited by the following allowable stresses: for pine: tension = 1600 psi, shear = 120 psi, bearing = 1000 psi; for oak: bearing = 400 psi, shear = 140 psi; for steel: bearing = 24,000 psi; for concrete: bearing = 600 psi.

3.4 STATICALLY INDETERMINATE MEMBERS

A member is statically indeterminate if the internal force (and moment) on any section of the member cannot be computed from the principles of statics alone. The internal force in an axially loaded "indeterminate member" depends not only on the conditions of static equilibrium but also on the dimensions of the member and the stiffness of the material. There are many types of problems pertaining to statically indeterminate members; the solution of any of them is accomplished by using elastic properties to relate the geometry of deformation to the conditions of static equilibrium. The solution of these problems can usually be facilitated by the use of the following "step" procedure:

1. Draw the free-body diagrams needed, and write the appropriate equations of equilibrium.
2. Draw a sketch showing the deformation of the bodies involved, and write an equation relating these deformations.

3. Use the principles of elastic bodies to correlate the equations from the first two steps.

An expression of deformation which is convenient to use in the solution of indeterminate problems can be obtained from the definitions of unit stress, unit strain, and Young's modulus. Thus

$$\epsilon = \frac{\delta}{L} \qquad E = \frac{s}{\epsilon} \qquad s = \frac{P}{A}$$

$$\delta = \epsilon L = \frac{s}{E} L$$

therefore

$$\delta = \frac{PL}{AE} \tag{3.1}$$

There is no need to memorize this equation, since it can be obtained so easily from basic definitions.

Example 3.2. A uniform rigid block weighing 90 kips is to be supported on three bars as shown in Fig. 3.2*a*. Bars *A* and *C* are of aluminum alloy (E = 10,000 ksi); each is 12.000 in. long and 1.50 in. square.

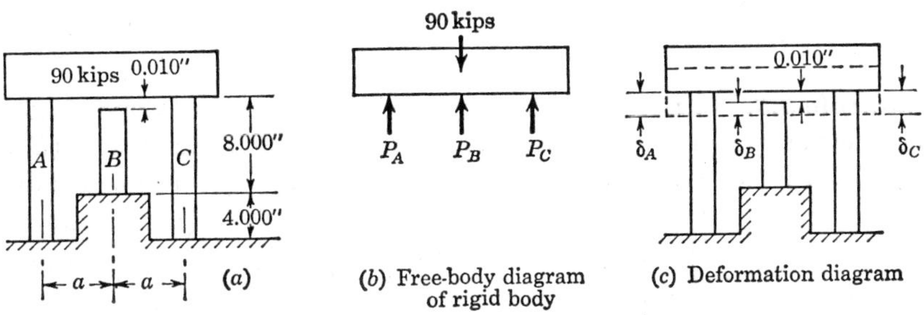

(*a*) (*b*) Free-body diagram of rigid body (*c*) Deformation diagram

FIG. 3.2

Bar *B* is of steel (E = 30,000 ksi), 7.990 in. long and 1.00 in. square. Compute the stress in each bar.

Unless the 90-kip load shortens the aluminum bars more than 0.010 in., there will be no stress in bar *B*. Checking,

$$\delta = \frac{PL}{AE} = \frac{90(12)}{2(2.25)(10{,}000)} = 0.024 \text{ in.}$$

Therefore part of the 90-kip load will be supported by the steel, and the problem is statically indeterminate.

Applying the laws of statics to the free-body diagram of the rigid block, as shown, and noting that $P_A = P_C$ (from symmetry),

$$2P_A + P_B = 90 \tag{a}$$

The deformation diagram is drawn as shown; from this

$$\delta_A = \delta_B + 0.010 \tag{b}$$

which can also be expressed as

$$\left(\frac{PL}{AE}\right)_A = \left(\frac{PL}{AE}\right)_B + 0.010 \qquad \text{or}$$

$$\frac{P_A(12)}{(2.25)(10{,}000)} = \frac{P_B(8)}{(1)(30{,}000)} + 0.010 \tag{b'}$$

Solving Eq. (b') with Eq. (a) gives

$$P_B = 26.2 \text{ kips}, \qquad \text{and} \qquad P_A = 31.9 \text{ kips}$$

Therefore

$$s_B = \frac{26.2}{1.00} = 26.2 \text{ ksi} \qquad Ans$$

$$s_A = s_C = \frac{31.9}{2.25} = 14.2 \text{ ksi} \qquad Ans$$

Example 3.3. A rigid beam is supported by a hinge and three vertical members as shown in Fig. 3.3*a*. The two 40-in. wires are of steel (E = 30,000 ksi), each wire having an area of 0.05 in.^{2} The 10-in. member

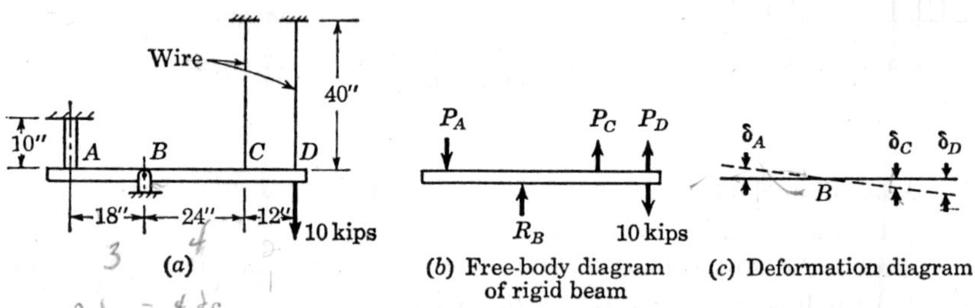

(*a*) (*b*) Free-body diagram of rigid beam (*c*) Deformation diagram

FIG. 3.3

is of aluminum alloy (E = 10,000 ksi) and has an area of 0.8 in.^{2} Determine the stress in each member and the reaction at B due to the 10-kip load.

From the free-body diagram shown, $\Sigma M_B = 0$; therefore

$$1.5P_A + 2P_C + 3P_D = 30 \tag{a}$$

From the deformation diagram shown

$$\delta_C = \tfrac{4}{3}\delta_A \quad \text{and} \tag{b}$$
$$\delta_D = 2\delta_A \tag{c}$$

From Eq. (*b*)

$$\frac{P_C(40)}{0.05(30{,}000)} = \frac{4P_A(10)}{3(0.8)(10{,}000)}$$

Therefore

$$P_C = 0.0625P_A$$

From Eq. (*c*)

$$\frac{P_D(40)}{0.05(30{,}000)} = \frac{2P_A(10)}{0.8(10{,}000)}$$

Therefore

$$P_D = 0.0938P_A$$

Use of the values of P_C and P_D in terms of P_A and Eq. (*a*) gives $P_A = 15.74$ kips; $P_C = 0.98$ kip; $P_D = 1.47$ kips. From $\Sigma F_y = 0$,

$$R_B = 23.29 \text{ kips} \qquad \textit{Ans}$$

$$s_A = \frac{15.74}{0.80} = 19.7 \text{ ksi (compression)} \qquad \textit{Ans}$$

$$s_C = \frac{0.98}{0.05} = 19.6 \text{ ksi (tension)} \qquad \textit{Ans}$$

$$s_D = \frac{1.47}{0.05} = 29.4 \text{ ksi (tension)} \qquad \textit{Ans}$$

Example 3.4. A 28-in. length of aluminum alloy bar is suspended from the ceiling so as to provide a clearance of 0.012 in. between it and a 10-in. length of steel, as shown in Fig. 3.4*a*. For the 28-in. bar: area = 2 in.^{2}; E = 10,000 ksi. For the 10-in. bar: area = 4 in.^{2}; E = 30,000 ksi. Determine the stress in the aluminum and in the steel due to a 60-kip load applied 20 in. from the ceiling.

Unless the load deforms the top 20-in. portion of the aluminum bar more than 0.012 in., there will be no stress in the steel or in the bottom 8 in. of the aluminum. Checking,

$$\delta = \frac{PL}{AE} = \frac{60(20)}{2(10{,}000)} = 0.06 \text{ in.}$$

Therefore the steel and all the aluminum will be stressed.

Since the aluminum bar will be stressed differently in the 20-in. portion from the way it is in the 8-in. portion, it will be considered for convenience of solution as two different members.

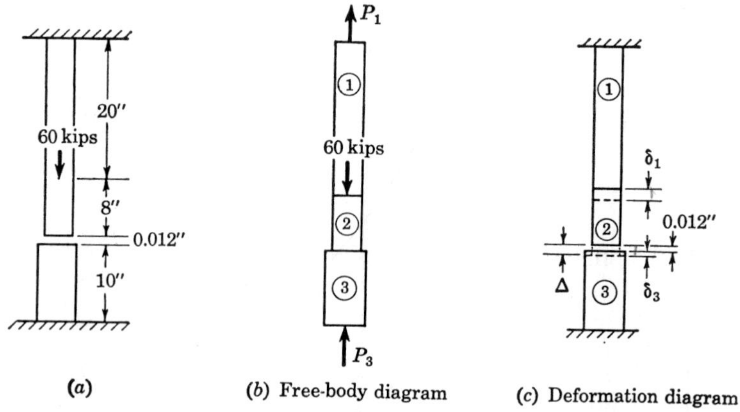

FIG. 3.4

From statics and the free-body diagram shown,

$$\Sigma F_y = 0 \qquad P_1 + P_3 = 60 \tag{a}$$

Also, the load in member 2 must be equal to the load in member 3, or

$$P_2 = P_3 \tag{b}$$

From the deformation diagram, it can be seen that the deformation of member 2 must be equal to the *difference in movement* of its two ends.

Assuming that δ_1 is larger than the distance Δ shown,

$$\delta_2 = \delta_1 - \Delta \qquad \text{or}$$
$$\delta_2 = \delta_1 - (\delta_3 + 0.012)$$

therefore

$$\delta_1 = \delta_2 + \delta_3 + 0.012 \tag{c}$$

or

$$\frac{P_1(20)}{2(10{,}000)} = \frac{P_2(8)}{2(10{,}000)} + \frac{P_3(10)}{4(30{,}000)} + 0.012 \qquad (c')$$

Solving Eqs. (*a*), (*b*), and (*c*′) simultaneously gives $P_1 = 27.6$ kips; $P_2 = P_3 = 32.4$ kips. Since all values of internal force are positive, the assumption as to the relative magnitudes of δ_1 and Δ was correct.

$$s_1 = \frac{27.6}{2} = 13.8 \text{ ksi (tension)} \qquad \textit{Ans}$$

$$s_2 = \frac{32.4}{2} = 16.2 \text{ ksi (compression)} \qquad \textit{Ans}$$

$$s_3 = \frac{32.4}{4} = 8.1 \text{ ksi (compression)} \qquad \textit{Ans}$$

PROBLEMS

***3.16.** An aluminum rod 15.006 in. long is placed inside a steel pipe 15.000 in. long and loaded with a weight $W = 50{,}000$ lb as shown in Fig. P3.16. Cross-sectional areas are as follows: aluminum rod $= 4$ in.^{2}; steel pipe $= 1.5$ in.^{2} For aluminum $E = 10{,}000$ ksi; for steel $E = 30{,}000$ ksi. Compute the stress in each material.

3.17. A brass rod 10.005 in. long is placed inside an aluminum pipe 10.000 in. long and loaded with a weight $W = 30$ kips as shown in Fig. P3.17. Cross-sectional areas are as follows: brass rod $= 2$ in.^{2}; aluminum pipe $= 1.2$ in.^{2} For brass $E = 15{,}000$ ksi; for aluminum $E = 10{,}000$ ksi. Compute the stress in each material.

3.18. An aluminum rod 15.006 in. long is placed inside a steel pipe 15.000 in. long and is to be loaded with a weight W as shown in Fig. P3.18. Cross-sectional areas are as follows: aluminum rod $= 4$ in.^{2}; steel pipe $= 1.5$ in.^{2} For aluminum $E = 10{,}000$ ksi; for steel $E = 30{,}000$ ksi. What value of W will cause the stress in the steel pipe to be twice the stress in the aluminum rod?

***3.19.** A rigid horizontal beam supported by two vertical bars as shown in Fig. P3.19 is to be loaded with a concentrated load $Q = 60$ kips. The left bar is of aluminum ($E = 10{,}000$ ksi), is 6 in. long, and has a cross-sectional area of 3 in.^{2}; the right bar is of steel ($E = 30{,}000$ ksi), is 10 in. long, and has a cross-sectional area of 2 in.^{2} Determine (*a*) the location of the load so that the rigid beam will remain horizontal, (*b*) the unit stress in each vertical bar when the load is so located.

3.20. A rigid horizontal beam supported by two vertical bars as shown in Fig. P3.20 is to be loaded with a concentrated load $Q = 20{,}000$ lb. The left bar is of brass ($E = 15{,}000$ ksi), is 18 in. long, and has a cross-sectional area of 1.5 in.^{2}; the right bar is of aluminum ($E = 10{,}000$ ksi), is 24 in. long, and has a cross-sectional area of 1.2 in.^{2} Determine (*a*) the location of the load so that the rigid beam will remain horizontal, (*b*) the unit stress in each vertical bar when the load is so located.

3.21. A rigid horizontal beam supported by two vertical bars as shown in Fig. P3.21 is to be loaded with a concentrated load $Q = 50$ kips. The left bar is of aluminum ($E = 10{,}000$ ksi), is 12 in. long, and has a cross-sectional area of 1.5 in.^{2}; the right bar is of steel ($E = 30{,}000$ ksi), is 30 in. long, and has a cross-sectional area of 2 in.^{2} Determine (*a*) the location of the load to make the deformation of the steel bar twice that of the aluminum bar, (*b*) the unit stress in each vertical bar when the load is so located.

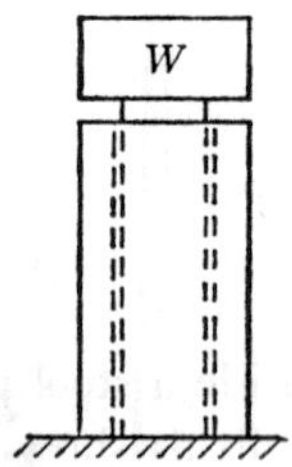

FIG. P3.16, P3.17, P3.18

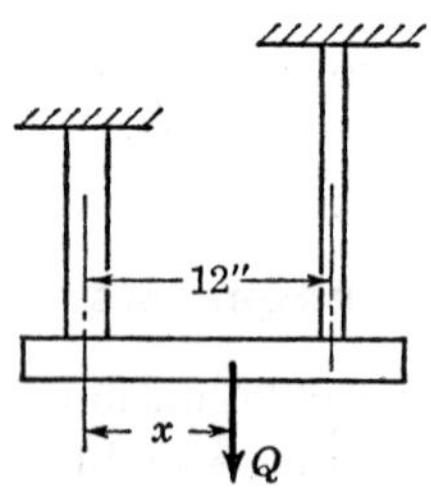

FIG. P3.19, P3.20, P3.21

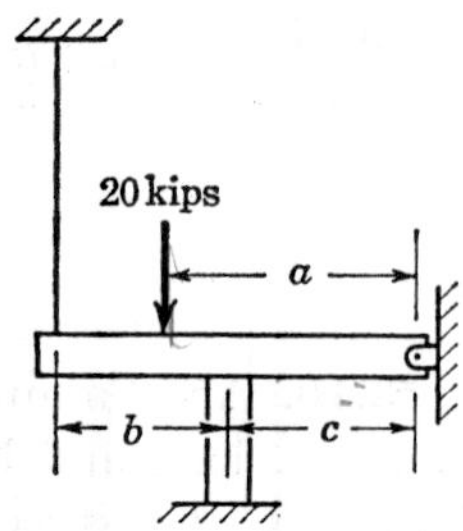

FIG. P3.22, P3.23, P3.24

★**3.22.** A rigid beam is supported by a hinge, a steel ($E = 30{,}000$ ksi) tension member, and a timber ($E = 2000$ ksi) compression member as shown in Fig. P3.22. The steel member is 5 ft long and has a cross-sectional area of 1 in.^{2}; the timber member is 2 ft long and has a cross-sectional area of 10 in.^{2} If $a = 6$ ft, $b = 4$ ft, and $c = 4$ ft, determine the stress in each of the two vertical members due to the 20-kip load.

3.23. A rigid beam is supported by a hinge, an aluminum ($E = 10{,}000$ ksi) tension member, and a cast-iron ($E = 15{,}000$ ksi) compression member as shown in Fig. P3.23. The aluminum member is 6 ft long and has a cross-sectional area of 0.6 in.^{2}; the cast-iron member is 2 ft long and has a cross-sectional area of 2 in.^{2} If $a = 4$ ft, $b = 2$ ft, and $c = 6$ ft, determine the stress in each of the two vertical members due to the 20-kip load.

3.24. A rigid beam is supported by a hinge, a steel ($E = 30{,}000$ ksi) tension member, and a concrete (E = 3500 ksi) compression member as shown in Fig. P3.24. The steel member is 8 ft long and has a cross-sectional area of 0.6 in.^{2}; the concrete member is 4 ft long and has a cross-sectional area of 20 in.^{2} If a = 4 ft, b = 6 ft, and c = 4 ft, determine the stress in each of the two vertical members.

★3.25. A 0.75-in.-diam steel (E = 30,000 ksi) bolt having 10 threads per inch is fitted snugly in a brass (E = 15,000 ksi) sleeve as shown in Fig. P3.25. The cross-sectional area of the sleeve is 0.05 in.^{2} Determine the stress in each material due to the nut being tightened one-tenth of a turn.

3.26. A 0.50-in.-diam brass (E = 15,000 ksi) bolt having 13 threads per inch is fitted snugly in an aluminum (E = 10,000 ksi) sleeve as shown in Fig. P3.26. The cross-sectional area of the sleeve is 0.30 in.^{2} Determine the stress in each material due to the nut being tightened one-quarter of a turn.

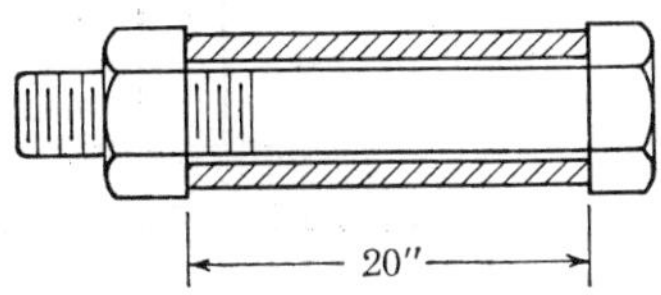

FIG. P3.25, P3.26, P3.27

3.27. A 0.375-in.-diam aluminum (E = 10,000 ksi) bolt having 16 threads per inch is fitted snugly in a brass (E = 15,000 ksi) sleeve as shown in Fig. P3.27. The cross-sectional area of the sleeve is 0.2 in.^{2} Determine the stress in each material due to the nut being tightened one-half of a turn.

★3.28. An aluminum (E = 10,000 ksi) rod is fastened to a steel (E = 30,000 ksi) pipe which is suspended as shown in Fig. P3.28. The rod is 15.000 in. long and has a cross-sectional area of 3 in.^{2}; the pipe is 25.000 in. long and has a cross-sectional area of 2 in.^{2} If the distance between the top support and the floor is 40.010 in., determine the stress in each material due to an axial load Q = 50 kips applied at the juncture of the rod and pipe.

3.29. A brass (E = 15,000 ksi) rod is fastened to an aluminum (E = 10,000 ksi) pipe which is suspended as shown in Fig. P3.29. The rod is 20.000 in. long and has a cross-sectional area of 4 in.^{2}; the pipe is 30.000 in. long and has a cross-sectional area of 3 in.^{2} If the distance between the top support and the floor is 50.050 in., determine the stress in each material due to an axial load Q = 80 kips applied at the juncture of the rod and pipe.

3.30. A brass (E = 15,000 ksi) rod is fastened to a steel (E = 30,000 ksi) pipe which is suspended as shown in Fig. P3.30. The rod is 18.000 in. long and has a cross-sectional area of 2.5 in.^{2}; the pipe is 12.000 in. long and has a cross-sectional area of 1.5 in.^{2} If the distance between the top support and the floor is 30.006 in., determine the stress in each material due to an axial load $Q = 40$ kips applied at the juncture of the pipe and rod.

***3.31.** A 38.000-in. length of steel ($E = 30{,}000$ ksi) rod having a cross-sectional area of 3 in.^{2} is fastened to rigid supports at each end as shown in Fig. P3.31. If $a = 16.000$ in., $b = 12.000$ in., $c = 10.000$ in., $Q_1 = 50$ kips, and $Q_2 = 30$ kips, determine the maximum tensile and compressive stresses.

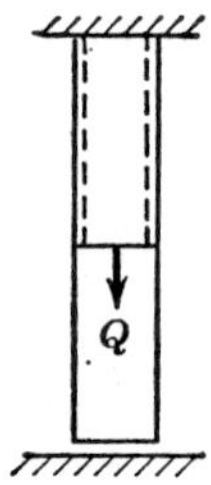

FIG. P3.28, P3.29, P3.30

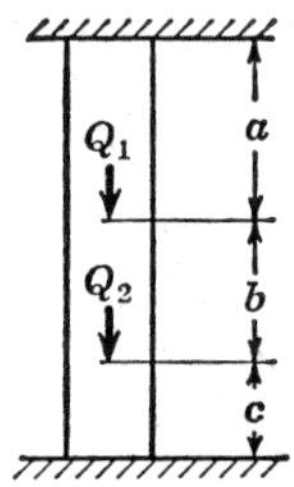

FIG. P3.31, P3.32, P3.33

3.32. A 47.000-in. length of aluminum ($E = 10{,}000$ ksi) rod having a cross-sectional area of 4 in.^{2} is fastened to rigid supports at each end as shown in Fig. P3.32. If $a = 20.000$ in., $b = 15.000$ in., $c = 12.000$ in., $Q_1 = 60$ kips, and $Q_2 = 40$ kips, determine the maximum tensile and compressive stresses.

3.33. A 35.000-in. length of magnesium-alloy ($E = 6500$ ksi) rod having a cross-sectional area of 2 in.^{2} is fastened to rigid supports at each end as shown in Fig. P3.33. If $a = 15.000$ in., $b = 12.000$ in., $c = 8.000$ in., $Q_1 = 30$ kips, and $Q_2 = 20$ kips, determine the maximum tensile and compressive stresses.

3.5 ADDITIONAL CASES OF THERMAL STRESS

All members in which thermal stresses are induced are statically indeterminate. In fact, the thermal-stress problems of Art. 1.7 are statically indeterminate. This article deals with slightly more challenging problems concerning the interaction of two or more members undergoing changes of temperature. The method of solution differs only slightly from that used for indeterminate problems not involving thermal stress. Generally, it is convenient to include two *fictitious steps* in the solution. First, the members are "separated" and allowed to *deform freely owing to the effect of tempera-*

ture change. Then the loading is computed which would cause the thermally deformed members to conform to the actual geometric conditions of the problem. This computed loading is equivalent to the loading induced in the members because of the temperature change.

Example 3.5. Two bars are unstressed and have lengths of 10.000 and 12.000 in. at 30°F as shown in Fig. 3.5*a*. Bar 1 is of aluminum alloy; bar 2 is of steel. Given the additional information in Table E3.5 and

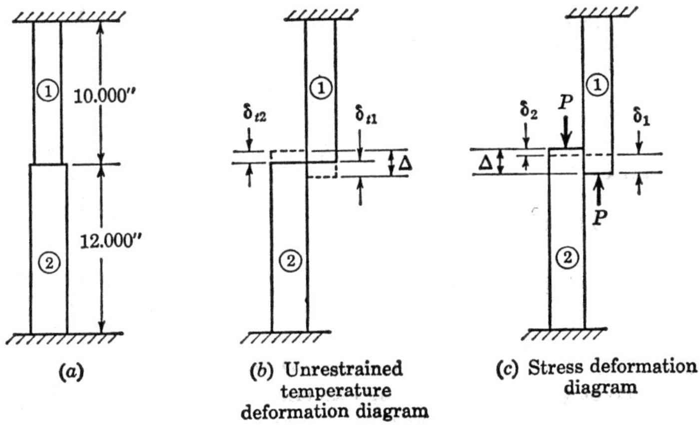

FIG. 3.5

TABLE E3.5

	Aluminum	Steel
α	0.000013	0.0000065
Area, in.^{2}	2	4
E, ksi	10,000	30,000

assuming that the top and bottom supports are rigid, compute the stress in each member when the temperature is 120°F.

First, the members are separated, as shown in Fig. 3.5*b*, and their unrestrained thermal deformations computed [using Eq. (1.11)]:

$$\delta_t = \alpha(\Delta t)L$$
$$\delta_{t1} = 0.000013(90)(10) = 0.0117 \text{ in.}$$
$$\delta_{t2} = 0.0000065(90)(12) = 0.00702 \text{ in.}$$

The distance shown as Δ represents the total amount that the expanded members must be compressed in order to fit between the rigid supports.

$$\Delta = \delta_{t1} + \delta_{t2} = 0.0117 + 0.00702 = 0.01872 \text{ in.}$$

Equal and opposite forces P are applied to the members to shorten them a total distance Δ, as shown in Fig. 3.5*c*. Thus

$$\delta_1 + \delta_2 = \Delta$$

$$\left(\frac{PL}{AE}\right)_1 + \left(\frac{PL}{AE}\right)_2 = \Delta$$

$$\frac{P(10)}{2(10{,}000)} + \frac{P(12)}{4(30{,}000)} = 0.01872$$

Therefore

$$P = 31.2 \text{ kips}$$

$$s_1 = \frac{31.2}{2} = 15.6 \text{ ksi} \qquad \textit{Ans}$$

$$s_2 = \frac{31.2}{4} = 7.8 \text{ ksi} \qquad \textit{Ans}$$

Example 3.6. Four 0.75-in. aluminum alloy plates are joined by a ¾-in. steel rivet (area = 0.442 in.²) as shown in Fig. 3.6*a*. At 30°F the materials fit tightly together but are unstressed. Compute the stress in

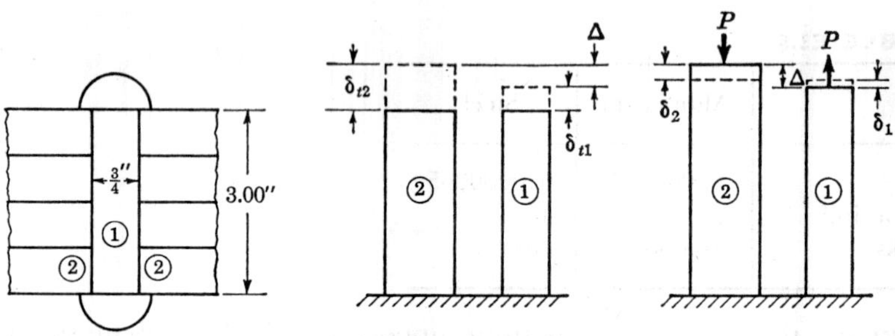

(*a*) (*b*) Unrestrained temperature deformation diagram (*c*) Stress deformation diagram

FIG. 3.6

each material when the temperature is 120°F, assuming the effective area of the aluminum to be 4 in.² For steel, $\alpha = 0.0000065$, $E = 30{,}000$ ksi. For the aluminum alloy, $\alpha = 0.0000133$, $E = 10{,}000$ ksi.

Considering the four plates as a single member 3 in. thick, the unrestrained thermal deformations are computed and sketched as shown.

$$\delta_{t1} = 0.0000065(120 - 30)(3) = 0.001755 \text{ in.}$$
$$\delta_{t2} = 0.0000133(120 - 30)(3) = 0.00359 \text{ in.}$$

The distance shown as Δ represents the amount which the "expanded" members must be deformed in order that the final length of the rivet will be equal to the final thickness of the plates.

$$\Delta = \delta_{t2} - \delta_{t1} = 0.00359 - 0.001755 = 0.001835 \text{ in.}$$

Equal and opposite forces P are applied, as shown, to compress the plates and stretch the rivet a total distance Δ.

$$\delta_1 + \delta_2 = \Delta$$
$$\frac{P(3)}{0.442(30{,}000)} + \frac{P(3)}{4(10{,}000)} = 0.001835$$

Therefore

$$P = 6.1 \text{ kips}$$
$$s_1 = \frac{6.1}{0.442} = 13.8 \text{ ksi (tension)} \qquad \textit{Ans}$$
$$s_2 = \frac{6.1}{4.00} = 1.52 \text{ ksi (compression)} \qquad \textit{Ans}$$

For a check, the final length of the rivet should be equal to the final thickness of the plates. From the geometry of the deformation diagrams

$$L_1 = 3 + \delta_{t1} + \delta_1 = 3 + 0.001755 + \frac{13.8(3)}{30{,}000}$$
$$L_1 = 3.00176 + 0.00138 = 3.00314 \text{ in.}$$
$$L_2 = 3 + \delta_{t2} - \delta_2 = 3 + 0.00359 - \frac{1.52(3)}{10{,}000}$$
$$L_2 = 3.00359 - 0.00046 = 3.00313 \text{ in.} \qquad \textit{Check}$$

PROBLEMS

***3.34.** An aluminum rod and a steel pipe are placed end to end between rigid supports 40.010 in. apart as shown in Fig. P3.34. At 30°F the pipe is 25.000 in. long and the rod is 15.000 in. long. Cross-sectional areas of the pipe and rod are 2 and 3 in.^{2}, respectively. For the steel,

$E = 30{,}000$ ksi and $\alpha = 0.0000065$; for the aluminum, $E = 10{,}000$ ksi and $\alpha = 0.000013$. Determine the stress in each material when the temperature is 120°F.

3.35. A brass rod and an aluminum pipe are placed end to end between rigid supports 50.025 in. apart as shown in Fig. P3.35. At 10°F the pipe is 30.000 in. long and the rod is 20.000 in. long. Cross-sectional areas of the pipe and rod are 3 and 4 in.^{2}, respectively. For the aluminum, $E = 10{,}000$ ksi and $\alpha = 0.000013$; for the brass, $E = 15{,}000$ ksi and $\alpha = 0.000010$. Determine the stress in each material when the temperature is 100°F.

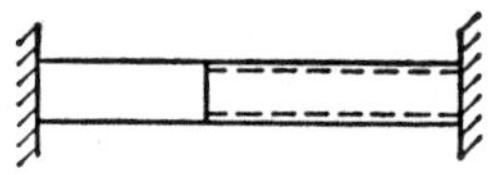

FIG. P3.34, P3.35, P3.36

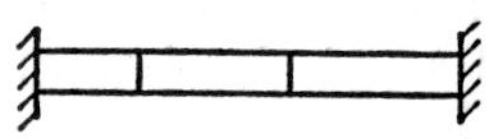

FIG. P3.37, P3.38, P3.39

3.36. A brass rod and a steel pipe are placed end to end between rigid supports 30.006 in. apart as shown in Fig. P3.36. At 20°F the pipe is 12.000 in. long and the rod is 18.000 in. long. The cross-sectional areas of the pipe and rod are 1.5 and 2.5 in.^{2}, respectively. For the steel, $E = 30{,}000$ ksi and $\alpha = 0.0000065$; for the brass, $E = 15{,}000$ ksi and $\alpha = 0.000010$. Determine the stress in each material when the temperature is 90°F.

★3.37. Bars of three different materials, each having a cross-sectional area of 2 in.^{2}, are fastened to one another and to rigid supports as shown in Fig. P3.37; the materials are unstressed at a temperature of 120°F. Given the information in Table P3.37, determine the stress in each material and the forces exerted on the rigid supports when the temperature drops to 40°F.

TABLE P3.37

	Steel	Brass	Aluminum
Length (at 120°F), ft	1.000	2.000	2.000
E, ksi	30,000	15,000	10,000
α	0.0000065	0.000010	0.000013

3.38. Bars of three different materials, each having a cross-sectional area of 1.5 in.^{2}, are fastened to one another and to rigid supports as shown in Fig. P3.38; the materials are unstressed at a temperature of

100°F. Given the information in Table P3.38, determine the stress in each material and the forces exerted on the rigid supports when the temperature drops to 10°F.

TABLE P3.38

	Brass	Aluminum	Invar steel
Length (at 100°F), in.	18.000	24.000	30.000
E, ksi	15,000	10,000	30,000
α	0.000010	0.000013	0.000000

3.39. Bars of three different materials, each having a cross-sectional area of 3 in.^{2}, are fastened to one another and to rigid supports as shown in Fig. P3.39; the materials are unstressed at a temperature of 10°F. Given the information in Table P3.39, determine the stress in each material and the forces exerted on the rigid supports when the temperature rises to 150°F.

TABLE P3.39

	Magnesium	Aluminum	Steel
Length (at 10°F), in.	10.000	15.000	20.000
E, ksi	6500	10,000	30,000
α	0.000014	0.000013	0.0000065

***3.40.** A 0.75-in.-diam steel bolt is placed in a brass sleeve having a cross-sectional area of 0.5 in.^{2} The bolt and sleeve fit snugly, as shown in Fig. P3.40, and are unstressed at a temperature of 10°F. Determine the stress and force in each material when the temperature is 150°F. For steel, $E = 30{,}000$ ksi and $\alpha = 0.0000065$; for brass, $E = 15{,}000$ ksi and $\alpha = 0.000010$.

3.41. A 0.5-in.-diam brass bolt is placed in an aluminum sleeve having a cross-sectional area of 0.3 in.^{2} The bolt and sleeve fit snugly, as shown in Fig. P3.41, and are unstressed at a temperature of 20°F. Determine the stress and force in each material when the temperature is 300°F. For brass, $E = 15{,}000$ ksi and $\alpha = 0.000010$; for aluminum, $E = 10{,}000$ ksi and $\alpha = 0.000013$.

3.42. A 0.375-in.-diam aluminum bolt is placed in a brass sleeve having a cross-sectional area of 0.2 in.^{2} The bolt and sleeve fit snugly, as shown in Fig. P3.42, and are unstressed at a temperature of 80°F. Determine the stress and force in each material when the temperature drops to −80°F. For aluminum, $E = 10{,}000$ ksi and $\alpha = 0.000013$; for brass, $E = 15{,}000$ ksi and $\alpha = 0.000010$.

★3.43. A 0.75-in.-diam steel bolt is placed in a brass sleeve (cross-sectional area = 0.5 in.^{2}), as shown in Fig. P3.43, and the nut tightened so as to stress the bolt 18,000 psi. Compute the minimum increase or decrease in temperature which will completely relieve the bolt (and sleeve) of stress. For steel, $E = 30{,}000$ ksi and $\alpha = 0.0000065$; for brass, $E = 15{,}000$ ksi, and $\alpha = 0.000010$.

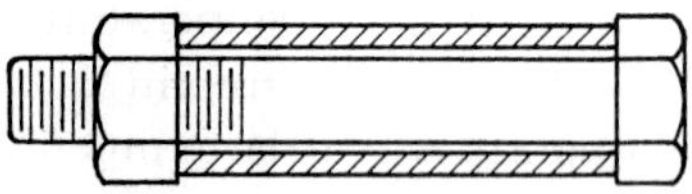

FIG. P3.40, P3.41, P3.42, P3.43, P3.44, P3.45

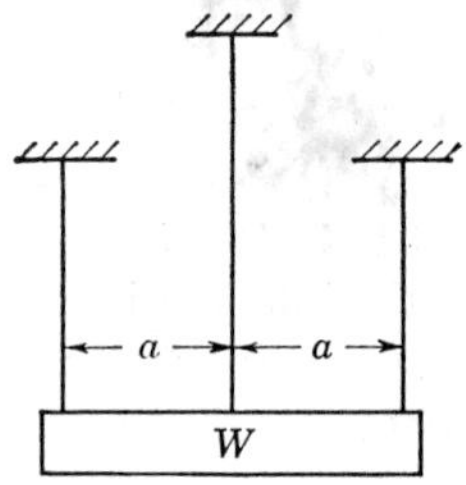

FIG. P3.46, P3.47, P3.48

3.44. A 0.5-in.-diam brass bolt is placed in an aluminum sleeve (cross-sectional area = 0.3 in.^{2}), as shown in Fig. P3.44, and the nut is tightened so as to stress the bolt 10,000 psi. Compute the minimum increase or decrease in temperature which will completely relieve the bolt (and sleeve) of stress. For brass, $E = 15{,}000$ ksi and $\alpha = 0.000010$; for aluminum, $E = 10{,}000$ ksi and $\alpha = 0.000013$.

3.45. A 0.375-in.-diam aluminum bolt is placed in a brass sleeve (cross-sectional area = 0.2 in.^{2}), as shown in Fig. P3.45, and the nut tightened so as to stress the bolt 14,000 psi. Compute the minimum increase or decrease in temperature which will completely relieve the bolt (and sleeve) of stress. For aluminum, $E = 10{,}000$ ksi and $\alpha = 0.000013$; for brass, $E = 15{,}000$ ksi and $\alpha = 0.000010$.

★3.46. A rigid beam of weight $W = 8000$ lb is supported by three vertical steel rods as shown in Fig. P3.46. Before being loaded, the middle rod was 10 ft long and 0.2 in.^{2} in cross-sectional area and the two outside rods were each 6 ft long and 0.15 in.^{2} in cross-sectional area. Compute the minimum increase or decrease in temperature to cause all the weight to be supported by the two outside rods. $E = 30{,}000$ ksi; $\alpha = 0.0000065$.

3.47. A rigid beam of weight $W = 15$ kips is supported by three vertical rods as shown in Fig. P3.47. Before being loaded, the middle rod (of aluminum) was 10 ft long and 0.8 in.^{2} in cross-sectional area and the two outside rods (of steel) were each 7 ft long and 0.2 in.^{2} in cross-sectional area. Compute the minimum increase or decrease in temperature to cause all the weight to be supported by the single aluminum rod. For aluminum, $E = 10{,}000$ ksi and $\alpha = 0.000013$; for steel, $E = 30{,}000$ ksi and $\alpha = 0.0000065$.

3.48. A rigid beam of weight $W = 24$ kips is supported by three vertical rods as shown in Fig. P3.48. Before being loaded, the middle rod (of steel) was 10 ft long and 0.8 in.^{2} in cross-sectional area and the two outside rods (of aluminum) were each 6 ft long and 0.4 in.^{2} in cross-sectional area. What increase or decrease of temperature would cause the three rods to be stressed equally? For steel, $E = 30{,}000$ ksi and $\alpha = 0.0000065$; for aluminum, $E = 10{,}000$ ksi and $\alpha = 0.000013$.

3.6 STRESS CONCENTRATION

Most structural members have, of necessity, certain discontinuities such as openings and abrupt changes of cross section. The stresses at these discontinuities may be much larger than would ordinarily be computed. At certain regular discontinuities (such as holes, fillets, and grooves) the maximum stress can be computed by using values known as *stress-concentration factors.* A stress-concentration factor can be defined as the ratio of the highest stress that exists at a discontinuity to the nominal stress computed on the basis of the net section. Usually designated by the letter k, the stress-concentration factor can be expressed in equation form as

$$k = \frac{s_{\max}}{s_{\text{nom}}} \tag{3.2}$$

where $s_{\max}$ is the maximum stress that exists at the discontinuity and s_{nom} is the "nominal stress" computed without considering the effects of stress concentration. The meaning of stress concentration is illustrated by Fig. 3.7.

For design purposes, stress-concentration factors for many types of regular discontinuities can be obtained from published diagrams† similar to Figs. 3.8 and 3.9. The factor depends not only on the relative dimensions of the member and the discontinuity but also

† R. E. Peterson, "Stress Concentration Design Factors," John Wiley & Sons, Inc., New York, 1953.

on the type of loading (such as axial, bending, or torsion). Some stress-concentration factors can be determined analytically by the science of *theory of elasticity*,† while many others must be determined experimentally.‡

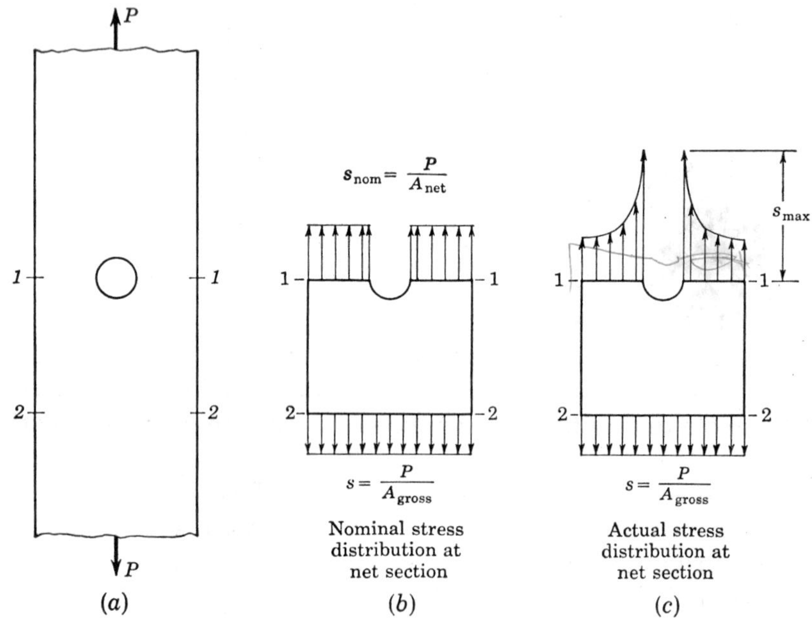

FIG. 3.7

The effect of stress concentrations should always be investigated when working with brittle materials or with dynamic or repeating load. With static (or "slowly applied") loading applied to a ductile material, a redistribution of stress occurs, so that the stresses at discontinuities do not normally exceed the yield strength. The manner in which this redistribution of stress occurs is illustrated by the sketches shown in Fig. 3.10. As load is slowly applied, the initial stress distribution is such that only elastic stresses exist throughout the section containing the discontinuity (Fig. 3.10*a*).

† S. P. Timoshenko and J. N. Goodier, "Theory of Elasticity," 2d ed., McGraw-Hill Book Company, Inc., New York, 1951.

‡ M. Hetenyi, "Handbook of Experimental Stress Analysis," John Wiley & Sons, Inc., New York, 1950.

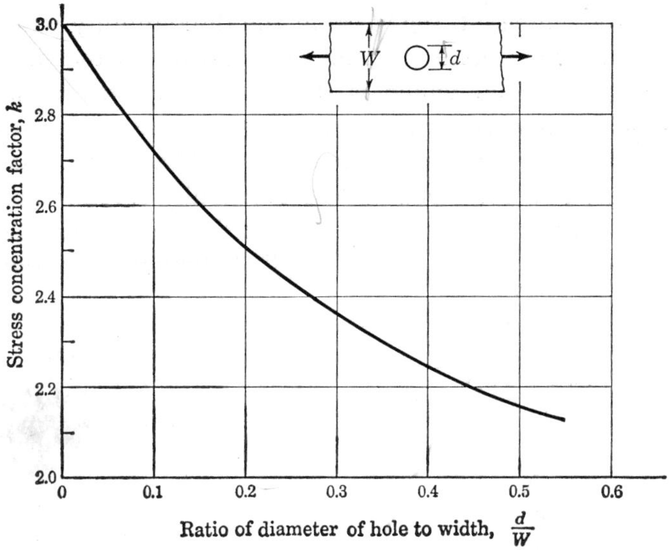

FIG. 3.8 Stress-concentration factors for axially loaded flat plate with centrally located round hole. (After Peterson, "Stress Concentration Design Factors," John Wiley & Sons, Inc., New York, 1953.)

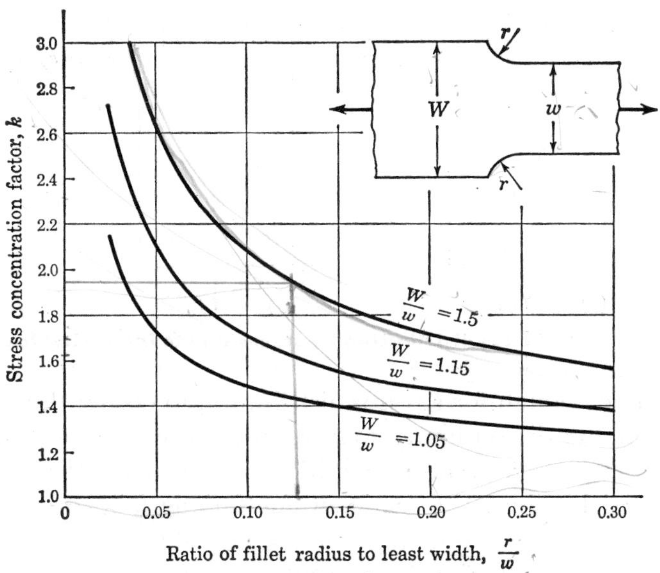

FIG. 3.9 Stress-concentration factors for axially loaded flat plates with symmetrical fillets. (After Peterson, "Stress Concentration Design Factors," John Wiley & Sons, Inc., New York, 1953.)

As loading is increased, the stress at the discontinuity will eventually reach the yield point (Fig. 3.10*b*). With a further increase of loading, the material which is stressed beyond the elastic limit deforms plastically without any increase of stress. The stress distribution is then similar to that shown as Fig. 3.10*c*. As load is increased even more, yielding will eventually spread throughout the section (Fig. 3.10*d*) and cause failure due to excessive deformation or fracture. When load is applied dynamically, there may

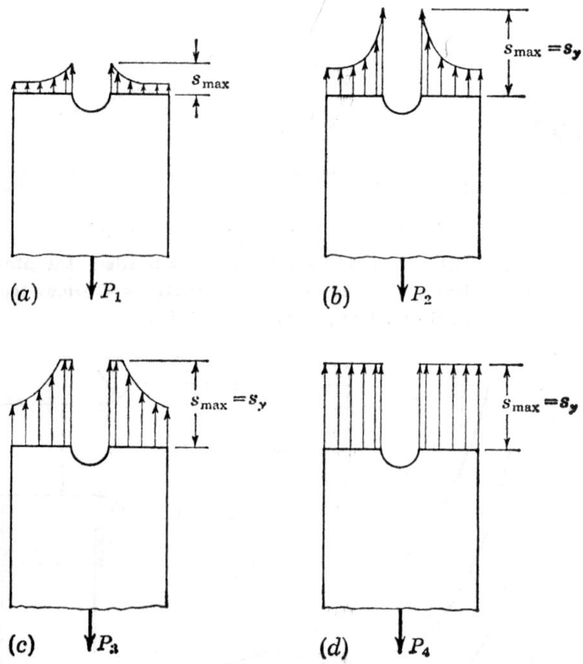

FIG. 3.10 Stress distribution by stages due to increasing load.

not be time enough for stress redistribution to occur; thus fracture starting at the discontinuity may result. Stress concentrations in ductile materials may also be critical if the material is subjected to repeated loading; the tendency for fatigue failure is increased owing to the fluctuating high stresses at the discontinuity. In summary, the effect of stress concentrations should be investigated except when designing with ductile materials subjected only to static loading.

PROBLEMS

★3.49. A 1-in.-diam hole is to be drilled in the middle of a bar 3 in. wide and 0.5 in. thick. The bar is to be subjected to an axial static load of 30 kips. Compute the maximum stress if the bar is made of (*a*) a ductile material having a yield point of 33,000 psi and an ultimate strength of 66,000 psi, (*b*) a brittle material having an ultimate strength of 55,000 psi.

3.50. A 1-in.-diam hole is to be drilled in the middle of a plate 6 in. wide and 0.75 in. thick. The plate is to be subjected to an axial static load of 125 kips. Compute the maximum stress if the plate is made of (*a*) a ductile material having a yield point of 33,000 psi and an ultimate strength of 66,000 psi, (*b*) a brittle material having an ultimate strength of 55,000 psi.

3.51. A 1-in.-diam hole is drilled in the middle of a bar 4 in. wide and 0.375 in. thick. The bar is made of a ductile steel having a yield point of 33,000 psi and an ultimate strength of 66,000 psi. Find the maximum stress in the bar due to an axial load of (*a*) 10 kips, (*b*) 20 kips.

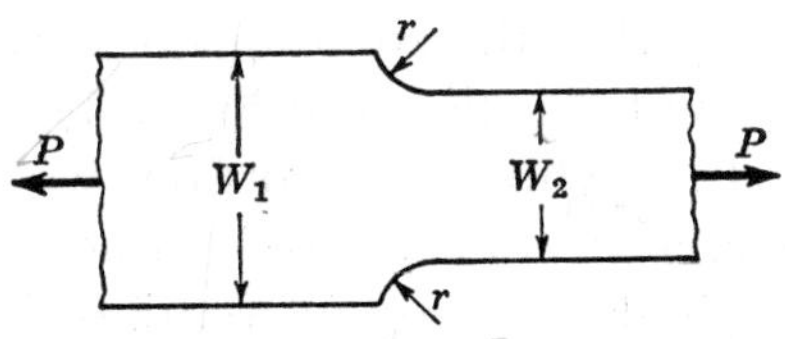

FIG. P3.52, P3.53, P3.54

★3.52. The member shown in Fig. P3.52 is to be 0.375 in. thick and have widths of $W_1 = 4$ in. and $W_2 = 3$ in. The radius of fillets r is to be 0.75 in. The member is to be subjected to an axial load of 30 kips. Compute the maximum stress if the member is made of (*a*) a ductile material having a yield point of 33,000 psi and an ultimate strength of 66,000 psi, (*b*) a brittle material having an ultimate strength of 55,000 psi.

3.53. The member shown in Fig. P3.53 is 0.75 in. thick and has widths of $W_1 = 6$ in. and $W_2 = 5$ in. The radius of fillets r is 0.5 in. The member is to be subjected to an axial load of 80 kips. Compute the maximum stress if the member is made of (*a*) a ductile material having a yield point of 33,000 psi and an ultimate strength of 66,000 psi, (*b*) a brittle material having an ultimate strength of 55,000 psi.

3.54. The member shown in Fig. P3.54 is 0.5 in. thick and has widths of $W_1 = 6$ in. and $W_2 = 4$ in. The radius of fillets r is 0.5 in. The member is made of a ductile steel having a yield point of 33,000 psi and an ultimate strength of 66,000 psi. Find the maximum stress due to an axial static load of (*a*) 50 kips, (*b*) 100 kips.

3.7 THIN-WALLED CYLINDRICAL PRESSURE VESSELS

A thin-walled cylindrical pressure vessel is one having a wall thickness which is relatively small in comparison with its diameter. In such vessels, the stress distribution throughout the plate thickness can be assumed to be uniform without significant error. For practical purposes, any pressure vessel having a diameter which is at least twenty times as large as its shell thickness can be classified as *thin-walled*. For *thick-walled* cylinders, the assumption of uniform stress distribution should not be made, since the maximum stress in the shell may be appreciably larger than the average.

When a thin-walled cylindrical vessel is subjected to internal pressure, the stress in the shell material can be readily computed by passing an appropriate section through the vessel and the pressure medium and applying the equations of equilibrium to a free-body diagram. If the tank shown in Fig. 3.11*a* were subjected to internal pressure, the cylindrical portion would lengthen and increase in circumference. Therefore, tensile stresses would exist in both the longitudinal and circumferential directions. To find the longitudinal stress, a free-body diagram to the left of section 1-1 is drawn and shown in Fig. 3.11*b*; the area of shell material which is stressed is shown in Fig. 3.11*c*; the area of the pressure medium cut by the section is shown in Fig. 3.11*d*. Since pressure acts normal to any area, the force acting to the left is the product of the pressure and area A'. Using $\Sigma F_x = 0$ gives

$$s_L A_C = pA' \tag{3.3a}$$

To find the circumferential stress, a free body of a *half-ring portion* of the tank is used. This free body can be obtained by first taking the ring between sections 1-1 and 2-2 and then passing plane 3-3 through a diameter. The free body of the half-ring portion to the left of plane 3-3 is shown in Fig. 3.11*g*; the area of shell material stressed circumferentially is shown in Fig. 3.11*h*; the area acted on by pressure is shown in Fig. 3.11*i*. Using $\Sigma F_x = 0$ gives

$$s_C A_L = pA'' \tag{3.3b}$$

It should be noted that Eqs. (3.3*a*) and (3.3*b*) are identical except for subscripts and primes; they simply state that, on a free body,

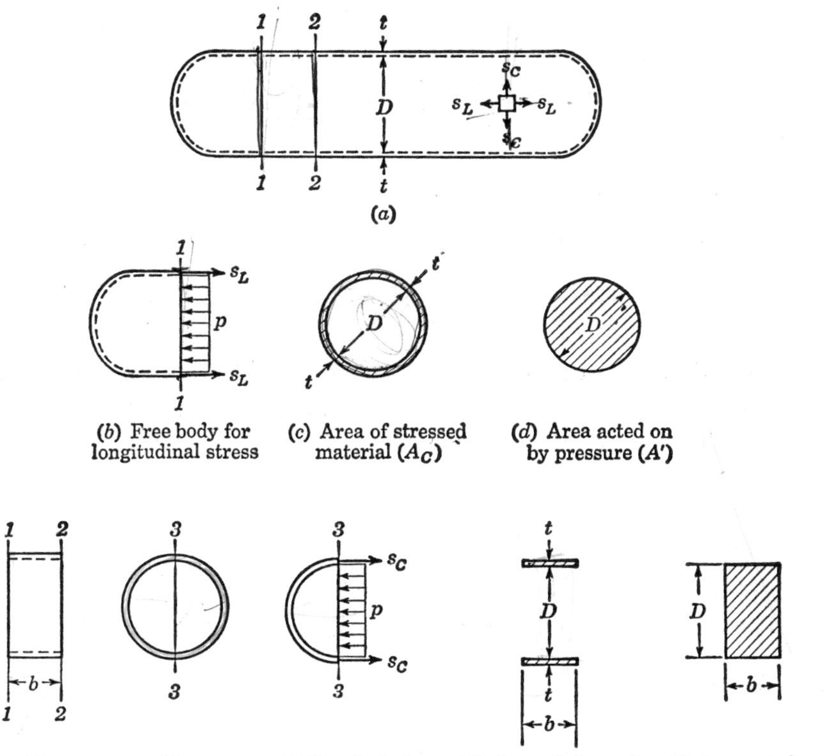

(*b*) Free body for longitudinal stress (*c*) Area of stressed material (A_C) (*d*) Area acted on by pressure (A')

(*e*) (*f*) (*g*) Free body for circumferential stress (*h*) Area of stressed material (A_L) (*i*) Area acted on by pressure (A'')

FIG. 3.11

the stress times the area of material cut is equal to the pressure times the area of pressure medium cut.

Example 3.7. A tank similar to that shown as Fig. 3.11*a* has an internal diameter of 36 in. and a plate thickness of 0.375 in. Using an allowable tensile stress of 18,000 psi, compute the allowable internal pressure.

The free-body diagram of Fig. 3.11*b* is used to determine the pressure as limited by longitudinal stress. The area of stressed material A_C can be assumed to be equal to the product of the inside circumference and the plate thickness.

$$\Sigma F_x = 0$$

$$18{,}000\pi(36)(0.375) = p\frac{\pi}{4}(36)^2$$

$$p = 750 \text{ psi (as limited by } s_L)$$

The free body shown in Fig. 3.11*g* is used to determine the pressure as limited by circumferential stress.

$$\Sigma F_x = 0$$
$$18{,}000(2)(0.375b) = p(36)(b)$$
$$p = 375 \text{ psi (as limited by } s_C)$$

Therefore, the circumferential stress controls and the allowable pressure is 375 psi (*Ans*).

Example 3.8. A wood-stave pipe having an inside diameter of 42 in. is held together by steel hoops, each having a net cross-sectional area of 0.89 in.^{2} The allowable stress in the hoops is 20,000 psi; the design pressure for the pipe is 135 psi. Compute the required spacing of the hoops.

Use *b* in. to designate the spacing of hoops, as shown in Fig. 3.12*a*. A free body of a half ring between sections 1-1 and 2-2 (0.5*b* each side of a typical hoop) is shown in Fig. 3.12*b*.

The allowable force *T* in each hoop is

$$T = 20{,}000(0.89) = 17{,}800 \text{ lb}$$

Therefore, from $\Sigma F_x = 0$,

$$2(17{,}800) = 135(42b)$$
$$b = 6.28 \text{ in.}$$

Use hoop spacing of 6 in. center to center (*Ans*).

The method of solution in this article applies only to thin-walled vessels subjected to *internal pressure.* External pressure applied to thin-walled vessels causes compression in the shell material and therefore tends to cause buckling. Unless these compressive stresses are quite low, the vessel must be reinforced with stiffeners.

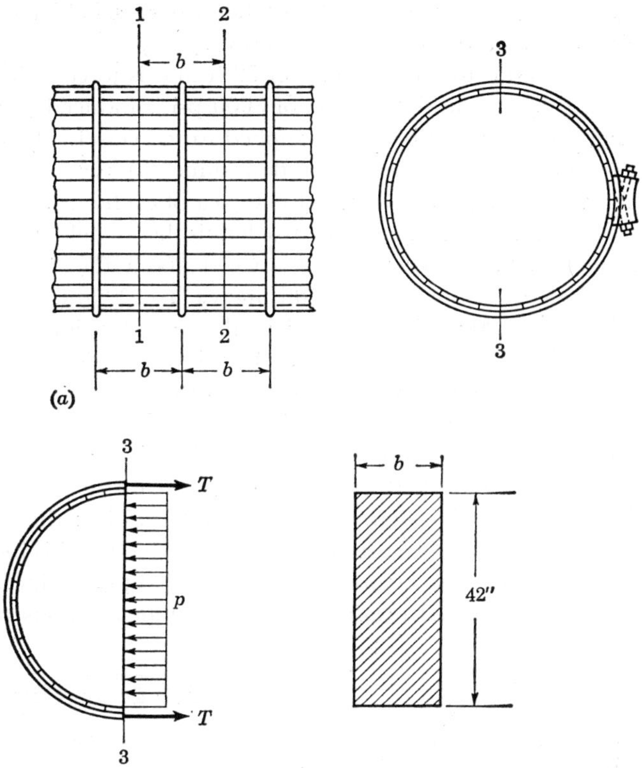

(*b*) Free body for hoop stress

(*c*) Area acted on by pressure

FIG. 3.12

PROBLEMS

***3.55.** A cylindrical tank is to have an inside diameter of 24 in. and be subjected to an internal pressure of 120 psi. Determine the required shell thickness if the allowable stress is 14,000 psi.

3.56. A cylindrical tank has inside and outside diameters of 30 and 31.5 in. Determine the allowable internal pressure based on an allowable tensile stress of 18,000 psi.

3.57. A cylindrical tank which is to contain an internal pressure of 60 psi is to be fabricated from 0.5-in. steel plate. Based on an allowable tensile stress of 16,000 psi, determine the maximum allowable inside diameter.

★3.58. A tank is to be made of two hemispheres held together, as shown in Fig. P3.58, by 120 bolts equally spaced about the flanges. The tank will have an inside diameter of 20 ft and is to contain gas at a pressure of 75 psi. An initial tension of 8000 psi is induced in the bolts by tightening them before pressure is applied to the vessel. Based on an allowable tensile stress of 15,000 psi, determine (*a*) the required wall thickness for the tank, (*b*) the required net area of each bolt.

3.59. A tank is to be made of two hemispheres held together, as shown in Fig. P3.59, by bolts equally spaced about the flanges. The inside and outside diameters of the hemispheres are 120 and 121 in. An initial tension of 10,000 psi is induced in the bolts by tightening them before pressure is applied to the vessel. Based on an allowable stress of 18,000 psi, determine the allowable pressure in the tank and the required number of 2-in. bolts (net area = 2.30 in.^{2}).

★3.60. A cover is to be bolted to one end of a pipe, as shown in Fig. P3.60, with equally spaced 0.75-in. bolts (net area = 0.30 in.^{2}). The pipe has inside and outside diameters of 20 and 20.5 in.; the allowable tensile stress is 18,000 psi. An initial tension of 10,000 psi is induced in the bolts by tightening them before pressure is applied to the pipe. Determine the allowable internal pressure and the number of bolts required.

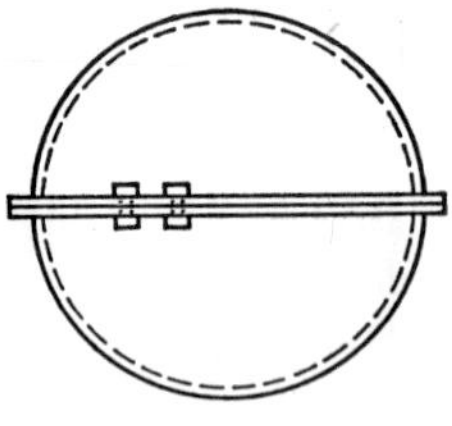

FIG. P3.58, P3.59

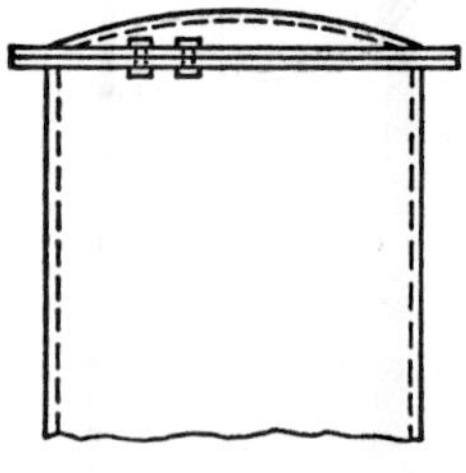

FIG. P3.60, P3.61

3.61. A cover is to be bolted to one end of a pipe, as shown in Fig. P3.61, with equally spaced 7/8-in. bolts (net area = 0.42 in.^{2}). The pipe is to have an inside diameter of 36 in. and is to contain pressure at 250 psi. An initial tension of 10,000 psi is induced in the bolts by tightening them before pressure is applied to the pipe. If the allowable tensile stress is 18,000 psi, determine the pipe wall thickness and the number of bolts required.

★3.62. A wood-stave pipe having an inside diameter of 42 in. is held together by threaded steel hoops spaced 6 in. center to center. The net cross-sectional area of each hoop is 0.42 in.^{2} Based on an allowable tensile stress in the hoops of 20,000 psi, determine the allowable water pressure.

3.63. A wood-stave pipe having an inside diameter of 5 ft is to be held together by equally spaced 1-in. threaded steel hoops (net area = 0.55 in.^{2}). The pipe is to contain water at 40 psi pressure. Determine the maximum center-to-center hoop spacing based on an allowable tensile stress of 20,000 psi.

3.64. A wood-stave pipe having an inside diameter of 6 ft is to be held together by threaded steel hoops spaced 6 in. center to center. The pipe is to contain water at a pressure of 60 psi. Determine the required net area of the hoops based on an allowable tensile stress of 18,000 psi.

4

SHEAR AND MOMENT IN BEAMS

4.1 BEAMS

A beam is a member which has the primary function of resisting transverse loading. Beams differ from axially loaded members in that they bend and thus have a nonuniform stress distribution on any cross section.

The deformations and stresses which result in a beam owing to a particular load or group of loads are dependent on the manner in which the beam is supported. The three basic types of supports for beams are *roller*, *pinned*, and *fixed-end;* the symbols commonly used for each type of support are shown in Table 4.1. A *roller support* is one which exerts a reactive force having a known line of action. As shown, this condition of support is commonly designated by a number of different symbols. A *pinned support* is one which allows the beam freedom to rotate but prevents it from any linear movement. The line of action of the reactive force at a pinned support is unknown; it is usually convenient to

TABLE 4.1 Symbols for Beam Supports

Types of support	Reaction symbols	Reaction components	Number of unknowns
Rollers	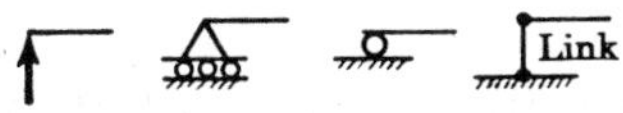	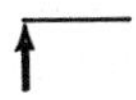	1
Pinned			2
Fixed	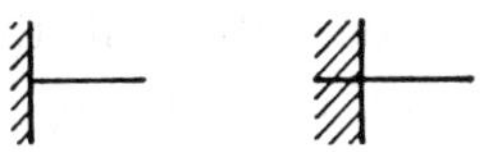	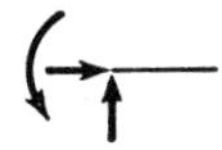	3

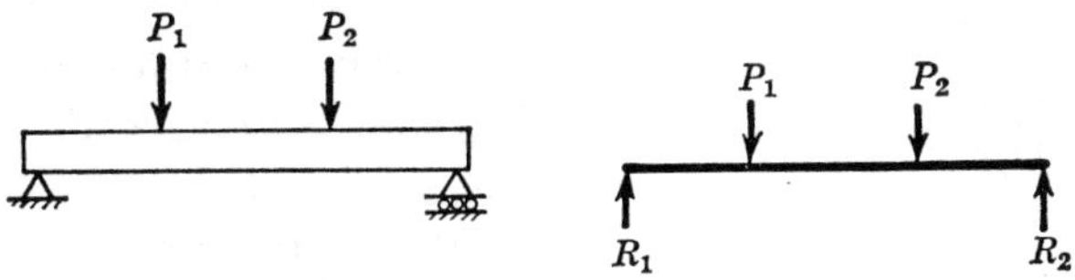

(*a*) Simply supported beam

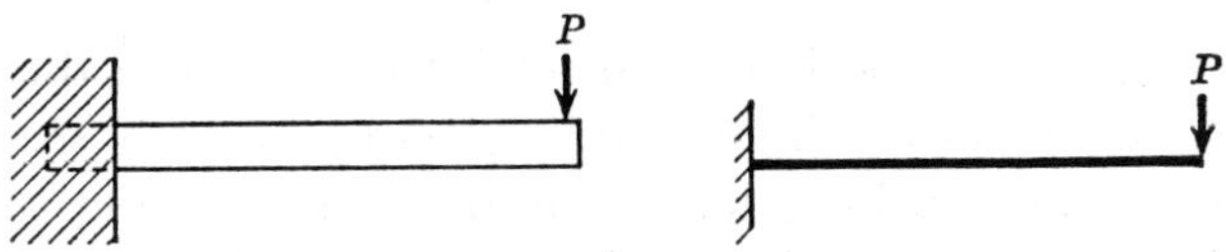

(*b*) Cantilever beam

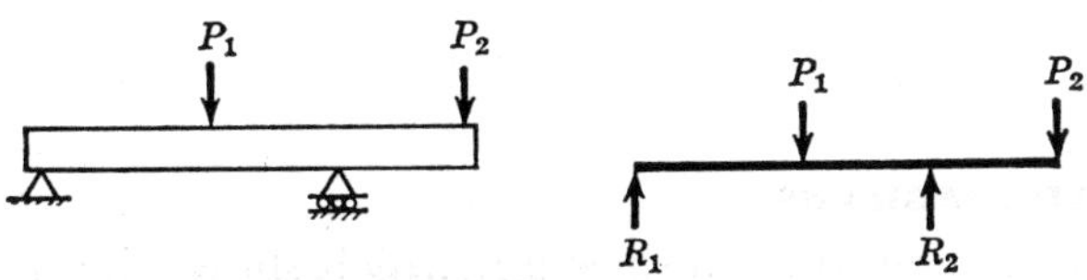

(*c*) Overhanging beam

FIG. 4.1

represent the reactive force by two mutually perpendicular components. A *fixed support* is one which prevents the beam from translating or rotating at the point of support. The reaction at such a support is unknown as to magnitude, direction, and line of action; it is usually represented by two mutually perpendicular force components and a couple.

Beams may be classified according to types and location of supports. Beams which are statically determinate are usually classified as *simply supported*, *cantilever*, or *overhanging*. Two methods of diagrammatically representing each type are shown in Fig. 4.1; for analytical purposes the single-line diagrams (shown to the right in each case) are usually preferred.

In order to determine the stresses that exist at any section of a beam, a free body of the portion to either side of the section can be used. For the beam shown in Fig. 4.2*a* the free-body diagram of the portion to the left of section 1-1 is shown in Fig. 4.2*b*. It can be seen that, if this part of the beam is to be in equilibrium, it must be acted on (by the right-hand part of the beam) by a vertical force V and a couple M. The internal force acting parallel to the section V is known as the *shear* at the section; the internal couple M is known as the *bending moment* at the section. The shear at a section of a beam must be known in order to compute the shearing stresses at the section. The bending moment at a section of a beam must be known in order to compute the bending (tension or compression) stresses at the section. The method of computing shearing and bending stresses in beams is the topic of the next chapter.

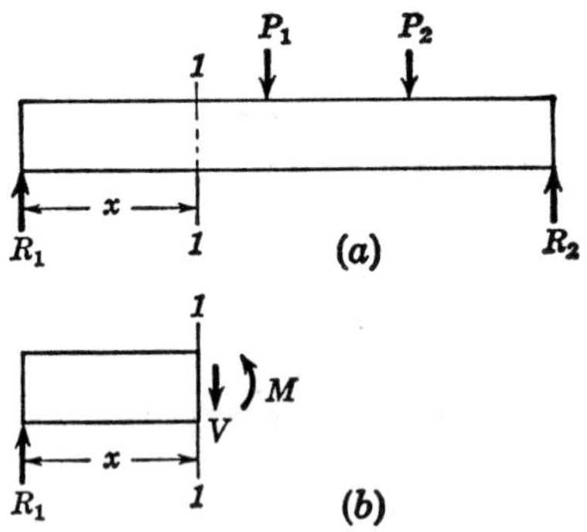

FIG. 4.2

4.2 SHEAR DIAGRAMS

A shear diagram is a diagram which shows the shear at every cross section of a beam. Such a diagram facilitates the determination of the critical shears in a beam. As will be seen in Art. 4.5, it also facilitates the determination of certain critical bending moments. The drawing of shear diagrams is a technique which can easily be mastered by learning the definition and sign convention for beam shear and realizing that a shear diagram is merely a plot of the shear for every cross section of the beam.

The shear at a section is equal to the algebraic sum of the external forces, acting normal to the beam, on either side of the section. As can be seen, the shear at section 1-1 of the beam in Fig. 4.3*a* can be determined either from the segment to the left of the section or from the segment to the right of the section. In either case, for equilibrium to exist, the shear must be *numerically equal* to the algebraic sum of the external forces on the particular segment. Thus, for the left segment

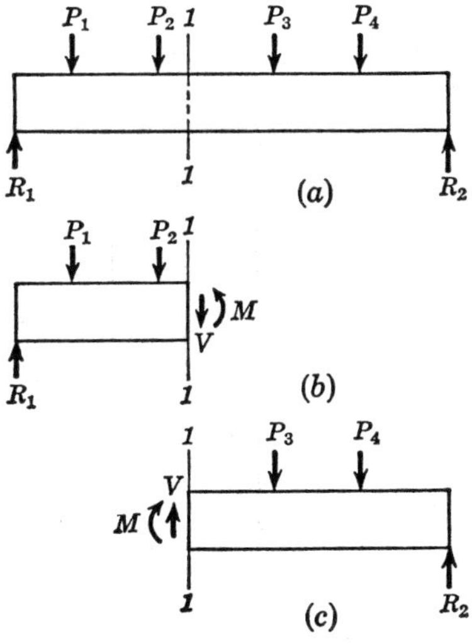

FIG. 4.3

$$V = R_1 - P_1 - P_2$$

or for the right segment

$$V = P_3 + P_4 - R_2$$

Of course, since the beam is in equilibrium, the two values of V must be numerically equal. In order to communicate with others in regard to shear and as a matter of convenience, a sign convention is necessary. The usual sign convention is as follows: *Upward external forces to the left of a section cause positive shear at the section.* It therefore follows that *downward external forces to the right of a section also cause positive shear at the section.*

The shear diagram for a beam is normally drawn immediately below the load diagram, with shear as ordinates and with abscissa distances projected down from the load diagram. Ordinates are usually drawn to approximate scale, and significant ordinate values shown on the diagram. The location of points of zero shear which are not readily apparent should be computed and shown on the diagrams.

Example 4.1. Draw the complete shear diagram for a simply supported beam having the load diagram shown in Fig. 4.4*a*.

From statics, the reactions R_A and R_D are determined to be as shown.

The equation for shear at any section between A and B is written, in terms of x measured from A, by summing external forces to the left of the various sections.

$$0 \leqq x \leqq 2 \qquad V = +1700 \text{ lb}$$

Thus the equation for shear between A and B represents a horizontal line.

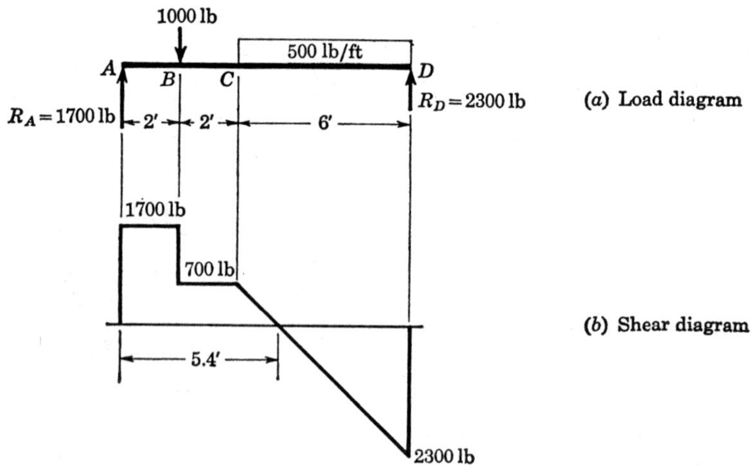

FIG. 4.4

The equation for shear between B and C is

$$2 \leqq x \leqq 4 \qquad V = +1700 - 1000 = +700 \text{ lb}$$

This equation also represents a horizontal line.

The equation for shear between C and D is

$$4 \leqq x \leqq 10 \qquad V = +1700 - 1000 - 500(x - 4)$$

or

$$V = +700 - 500(x - 4)$$

This equation represents a straight line which can be drawn if the coordinates of any two points on it are known.

At $x = 4$, $V = +700 - 500(4 - 4) = +700$ lb
At $x = 10$, $V = +700 - 500(10 - 4) = -2300$ lb

From the shear diagram, it is noted that there is a point of zero shear between points C and D. This point can be located by setting the appropriate shear equation equal to zero.

$$V = 0 = +700 - 500(x - 4)$$
$$x = 5.4 \text{ ft}$$

The complete shear diagram is shown in Fig. 4.4*b* (*Ans*).

Once the concept of shear diagrams is understood, there is little reason for writing shear equations. As can be seen from Example 4.1, the shear will always be constant between points of load application while the shear under a uniform load will be defined by a linear equation. It therefore follows that the shear diagram must be a horizontal line between points of load application and an inclined straight line under any uniform load. Therefore, the general shape of the shear diagram can be determined from observation of the load diagram. The ordinates for the shear diagram can be rapidly obtained by simply algebraically summing external loads from left to right and noting the subtotal at each point where an ordinate is desired.

Example 4.2. Draw the complete shear diagram for an overhanging beam having the load diagram shown in Fig. 4.5*a*.

To obtain the needed shear ordinates, an algebraic summation of external loads from left to right is made as follows:

		0
Left reaction		+1050
	Shear =	+1050
Uniform load	−200(6) =	−1200
	Shear =	− 150
Concentrated load		− 300
	Shear =	− 450
Concentrated load		− 700
	Shear =	−1150
Right reaction		+2350
	Shear =	+1200
Uniform load	−600(2) =	−1200
		0 *Check*

Noting that the shear curve will be a sloping straight line under the uniform loads and a horizontal line between loads, the shear diagram is drawn as shown in Fig. 4.5*b* (*Ans*).

It should be noted that in going from a section on one side of a concentrated load to a section on the other side, the shear changes an amount equal to the magnitude of the load.

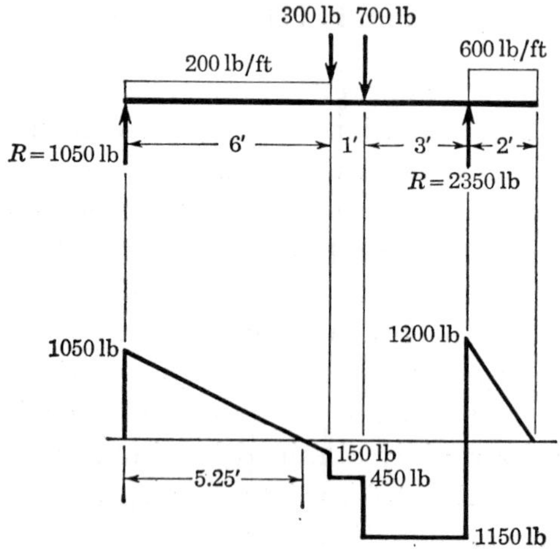

(*a*) Load diagram

(*b*) Shear diagram

FIG. 4.5

The location of the point of zero shear under the uniform load is computed by noting that the shear is +1050 lb at the left reaction and *decreases* at the rate of 200 lb/ft. Therefore the distance from the left reaction to the point of zero shear is

$$\frac{1050}{200} \text{ or } 5.25 \text{ ft}$$

PROBLEMS

4.1–4.6. For a beam having the load diagram shown in Figs. P4.1 to P4.6, write the shear equations for all sections and sketch the complete shear diagram. (*4.2 and *4.4)

4.7–4.12. For a beam having the load diagram shown in Figs. P4.7 to P4.12, compute shear ordinates by algebraically summing external loads from left to right and sketch the complete shear diagram. (*4.7 and *4.10)

4.13–4.15. Draw the load diagram for a beam having the shear diagram shown in Figs. P4.13 to P4.15. The bending moment at each end of the beam is zero.

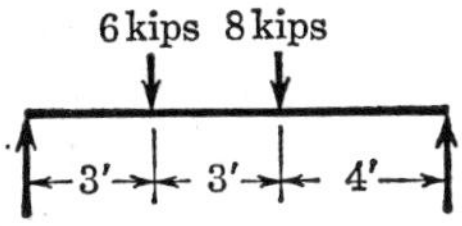

FIG. P4.1

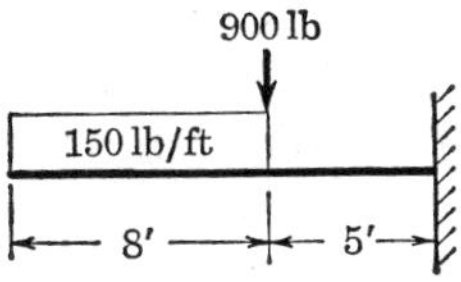

FIG. P4.2

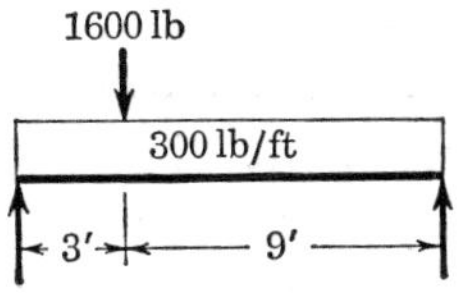

FIG. P4.3

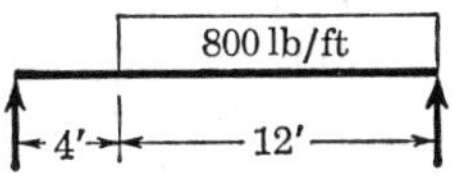

FIG. P4.4

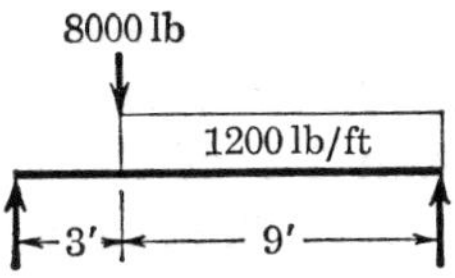

FIG. P4.5

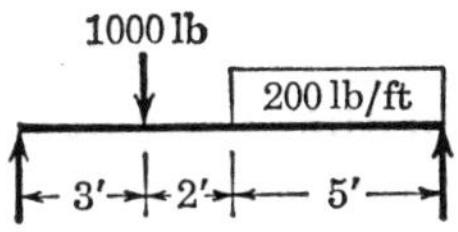

FIG. P4.6

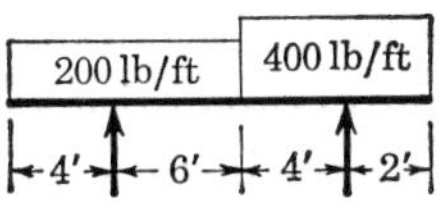

FIG. P4.7

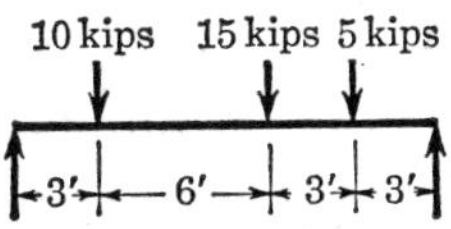

FIG. P4.8

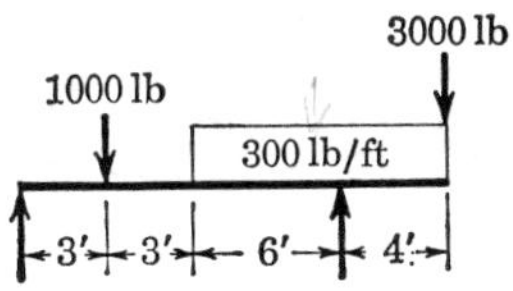

FIG. P4.9

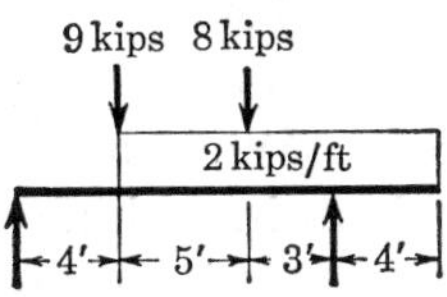

FIG. P4.10

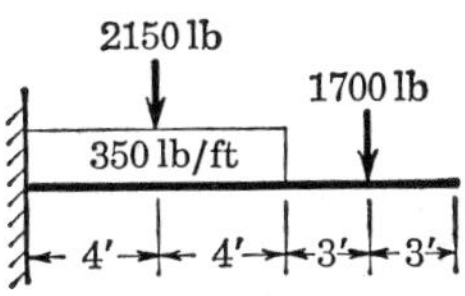

FIG. P4.11

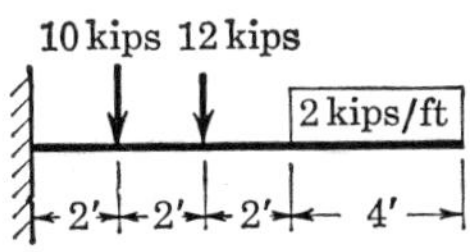

FIG. P4.12

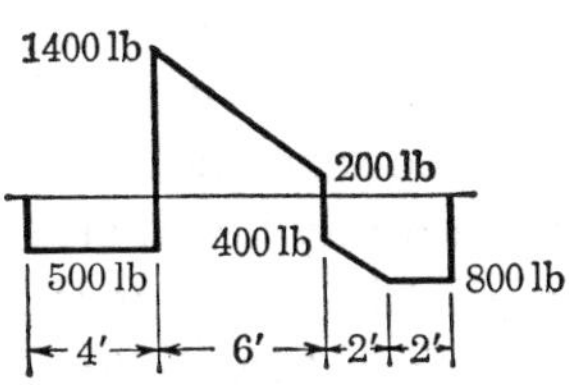

FIG. P4.13

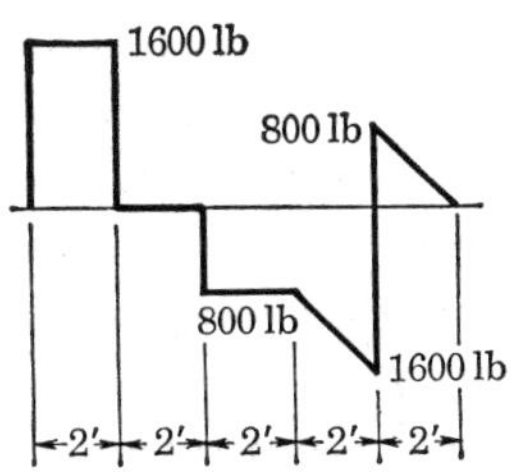

FIG. P4.14

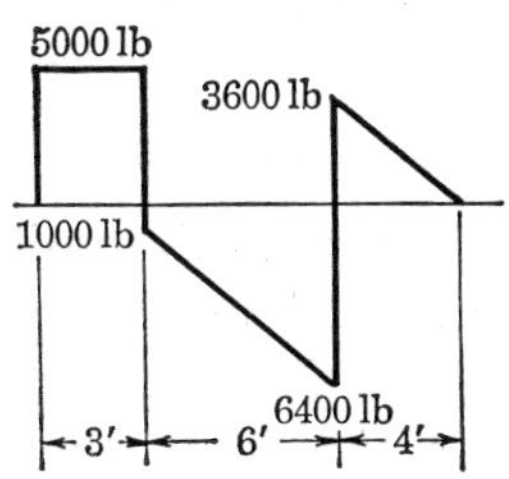

FIG. P4.15

4.3 MOMENT DIAGRAMS

A moment diagram is one which shows the bending moment at every cross section of a beam. Such a diagram facilitates the determination of the critical moments; a moment diagram is also useful for determining beam deflections by certain techniques.

The bending moment at a section is equal to the algebraic sum of the moments, about the centroidal axis of the section, of the external forces and couples acting on the segment to either side of the section. As can be seen, the moment at section 1-1 of the beam shown in Fig. 4.3*a* can be determined either from the segment to the left of the section or from the segment to the right of the section. In either case, for equilibrium to exist, the internal moment must be numerically equal to the algebraic sum of the moments of the external forces on the segment.

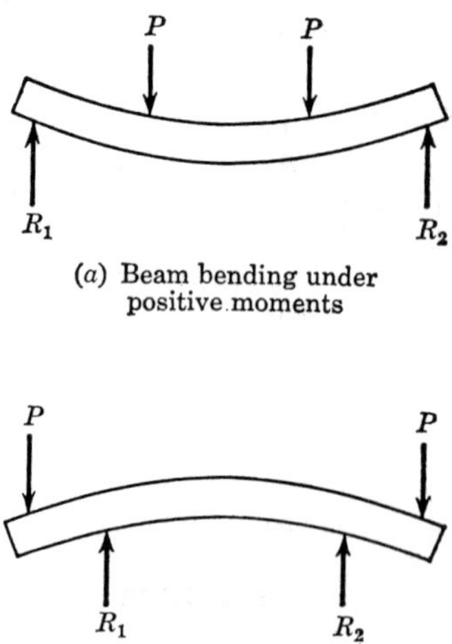

(*a*) Beam bending under positive moments

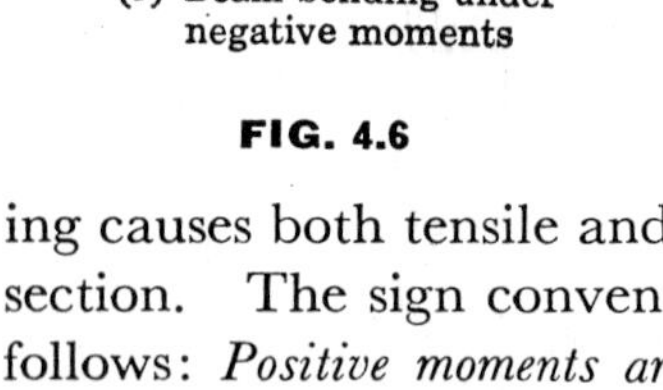

(*b*) Beam bending under negative moments

FIG. 4.6

A number of sign conventions are in common use for bending moment. The convention which is probably the most useful for learning purposes is dependent on whether tension bending stresses result at the top or at the bottom of the section. When a beam is loaded, it may be bent in such a way as to cause stretching at the bottom and shortening at the top, as shown in Fig. 4.6*a*; or it may be bent so as to cause stretching at the top and shortening at the bottom, as shown in Fig. 4.6*b*. Thus it can be seen that bending causes both tensile and compressive stresses on any beam cross-section. The sign convention which will be used in this text is as follows: *Positive moments are those which cause tension at the bottom of the section* (and, therefore, compression at the top). Conversely, negative moments are those which cause tension at the top of the section (and, therefore, compression at the bottom).

A moment diagram is simply a plot of the moment acting on every cross section of the beam. Moments are plotted as ordinates with abscissa distances projected down from the load diagram. Ordinates are usually drawn to approximate scale, and significant ordinate values shown on the diagram. The location of all points of maximum, minimum, and zero moment which are not readily apparent should be computed and shown on the diagram. The moment diagram is usually drawn immediately below the shear diagram (or immediately below the load diagram if the shear diagram is not drawn).

Example 4.3. Draw the complete moment diagram for a simply supported beam having the load diagram shown in Fig. 4.7*a*.

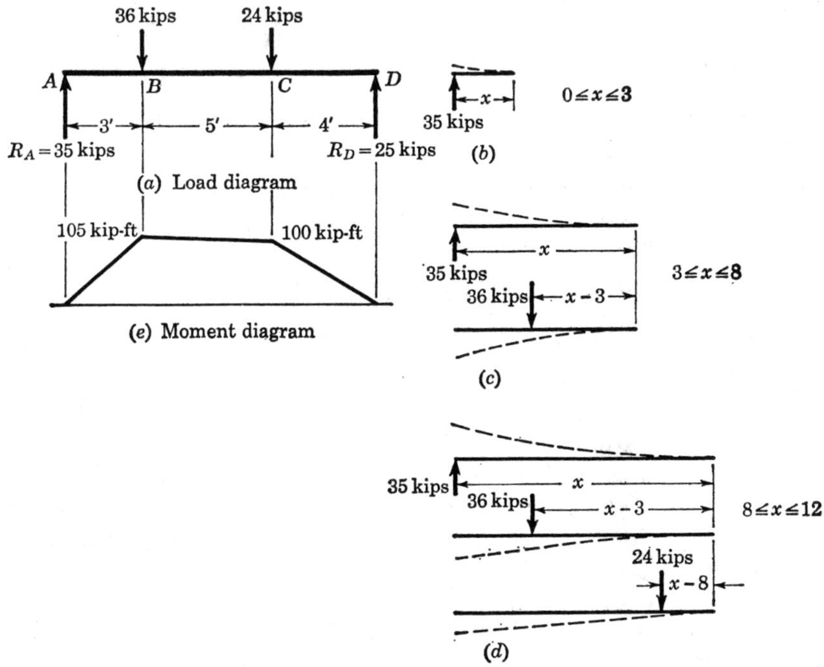

FIG. 4.7

As a matter of choice, moment equations will be written using only moments to the left of the various sections. Therefore, the equations are written in terms of x measured from A.

For any section between A and B the only force causing moment is the 35-kip reaction. As shown in Fig. 4.7*b*, this force tends to cause tension at the bottom of the section; therefore the bending moment is positive.

$$0 \leq x \leq 3 \qquad M = +35x$$

Thus the moment equation between points A and B represents a straight line which can be drawn if the coordinates of any two points are known.

At $x = 0$, $M = 0$
At $x = 3$, $M = +35(3) = +105$ kip-ft

For any section between B and C moment is caused by the 35-kip reaction *and* the 36-kip force. As shown in Fig. 4.7*c*, the 35-kip force tends to cause tension at the bottom, while the 36-kip force tends to cause tension at the top of the section.

$3 \leqq x \leqq 8 \qquad M = +35x - 36(x - 3)$

This equation also represents a straight line.

At $x = 3$, $M = +35(3) - 36(3 - 3) = +105$ kip-ft *Check*
At $x = 8$, $M = +35(8) - 36(8 - 3) = 280 - 180 = +100$ kip-ft

For any section between C and D moment is caused by the three forces shown in Fig. 4.7*d*.

$8 \leqq x \leqq 12 \qquad M = +35x - 36(x - 3) - 24(x - 8)$
At $x = 8$, $M = +35(8) - 36(8 - 3) - 24(8 - 8) = +100$ kip-ft *Check*
At $x = 12$, $M = +35(12) - 36(12 - 3) - 24(12 - 8) = 0$ *Check*

The complete moment diagram is shown as Fig. 4.7*e* (*Ans*).

Example 4.4. Draw the complete moment diagram for an overhanging beam having the load diagram shown as Fig. 4.8*a*.

In order to reduce computations, moment equations will be written from the left and from the right to a common section at C. For sections between A and C, x will be taken as positive to the right from A; for sections between E and C, x will be taken as positive to the left from E.

Between A and B, using forces to the left of the section

$0 \leqq x \leqq 2 \qquad M = +800x$
At $x = 0$, $M = 0$
At $x = 2$, $M = +800(2) = +1600$ lb-ft

Between B and C, using forces to the left of the section

$2 \leqq x \leqq 5 \qquad M = +800x - 600(x - 2)$
At $x = 2$, $M = +800(2) - 0 = +1600$ lb-ft *Check*
At $x = 5$, $M = +800(5) - 600(5 - 2) = +2200$ lb-ft

Between E and D, using forces *to the right* of the section, as shown in Fig. 4.8*b*,

$0 \leqq x \leqq 3 \qquad M = -400x\dfrac{x}{2} = -200x^2$

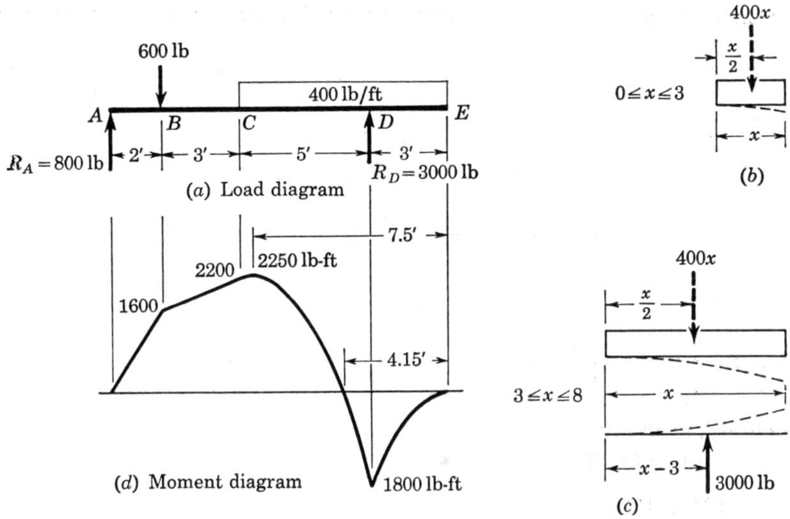

FIG. 4.8

Since the moment equation is nonlinear, more than two points are needed for sketching the curve.

At $x = 0$, $M = 0$
At $x = 1.5$, $M = -200(1.5)^2 = -450$ lb-ft
At $x = 3$, $M = -200(3)^2 = -1800$ lb-ft

Between D and C, using forces to the right of the section, as shown in Fig. 4.8*c*,

$3 \leqq x \leqq 8 \qquad M = -200x^2 + 3000(x - 3)$
At $x = 3$, $M = -200(3)^2 + 3000(3 - 3) = -1800$ lb-ft *Check*
At $x = 8$, $M = -200(8)^2 + 3000(8 - 3) = +2200$ lb-ft *Check*

Since the moment equation between D and C is nonlinear, it is necessary to locate the points of maximum and minimum moments. This is done by setting the derivative of the moment equation equal to zero.

$$\frac{dM}{dx} = -400x + 3000 = 0$$

$x = 7.5$ ft
At $x = 7.5$, $M = -200(7.5)^2 + 3000(7.5 - 3) = +2250$ lb-ft

The 7.5-ft distance locates the point of maximum moment; the minimum moment is -1800 lb-ft (at point D).

Since the moment changes from minus to plus between points D and C, a point of zero moment must exist in this portion of the beam.

This point is located by setting the appropriate moment equation equal to zero.

$$M = -200x^2 + 3000(x - 3) = 0$$
$$x = +4.15 \text{ ft}, +10.85 \text{ ft}$$

Note that the 10.85-ft value is invalid, since it is outside the limits of the equation.

The complete moment diagram is shown as Fig. 4.8*d* (*Ans*).

PROBLEMS

4.16–4.26. For a beam having the load diagram shown in Figs. P4.16 to P4.26 write bending moment equations for all sections and sketch the complete moment diagram. (*4.17 and *4.20)

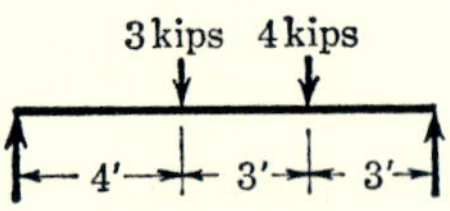

FIG. P4.16

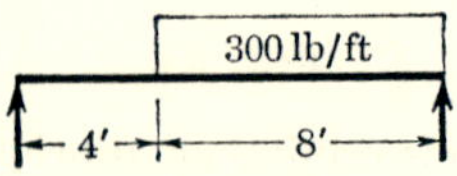

FIG. P4.17

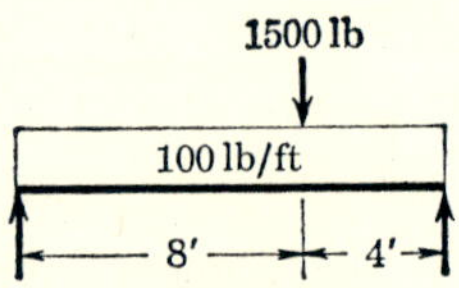

FIG. P4.18

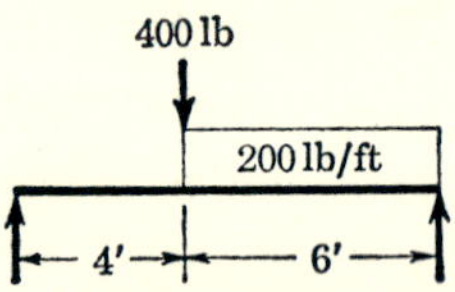

FIG. P4.19

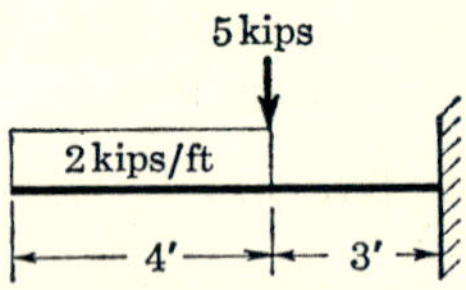

FIG. P4.20

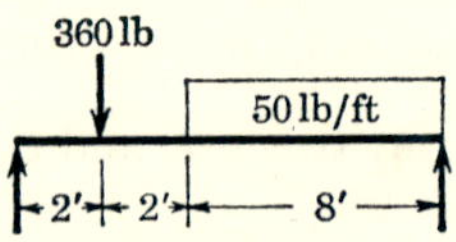

FIG. P4.21

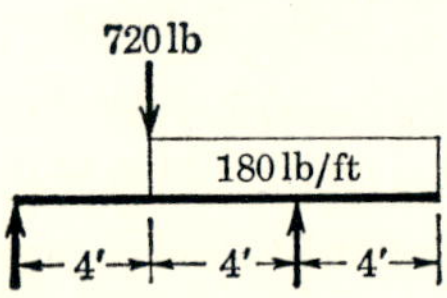

FIG. P4.22

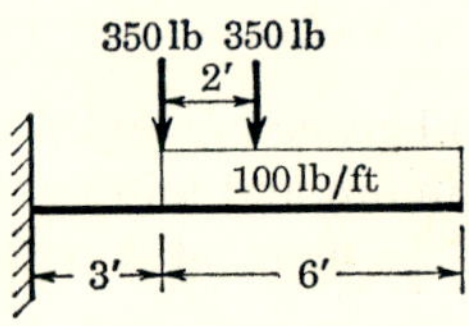

FIG. P4.23

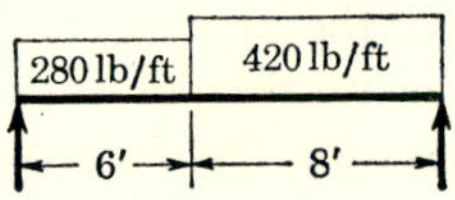

FIG. P4.24

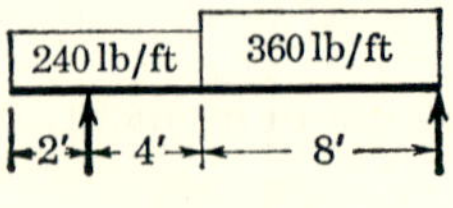

FIG. P4.25

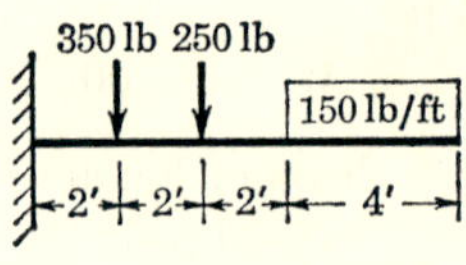

FIG. P4.26

4.27. Show that the maximum moment for a simply supported beam of length L subjected to a midspan load P is equal to $PL/4$.

4.28. Show that the maximum moment for a simply supported beam supporting a uniform load w lb/ft over its entire length L is equal to $wL^2/8$.

4.29. Show that the largest moment for a cantilever beam supporting a uniform load w lb/ft over its entire length L is equal to $-wL^2/2$.

4.30. A simply supported beam of length L supports a single load P positioned a distance from the left support and b distance from the right support. Show that the maximum moment is equal to Pab/L.

4.4 GENERAL MOMENT EQUATION

The *general moment equation* relates the moment at one section of a beam to the moment and shear at a second section. It will be

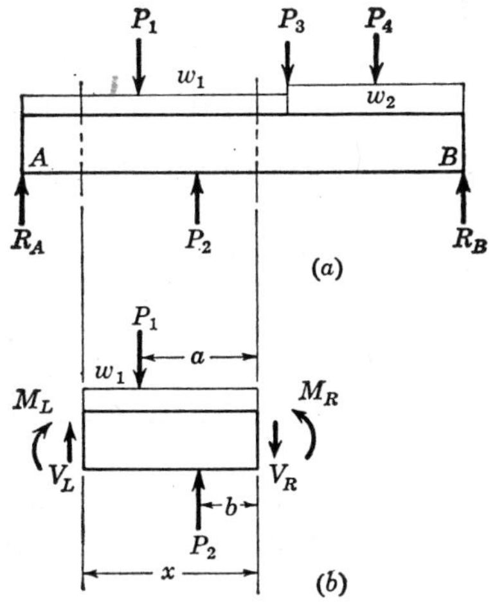

FIG. 4.9

derived using the beam with a somewhat general loading shown as Fig. 4.9*a*. Applying statics to a free-body diagram of the portion of the beam between the sections indicated in Fig. 4.9*b*,

ΣM about the right section $= 0$

$$M_R - M_L - V_L x + P_1 a - P_2 b + \frac{w_1 x^2}{2} = 0$$

$$M_R = M_L + V_L x - P_1 a + P_2 b - \frac{w_1 x^2}{2}$$

Noting that the last three terms of the above equation denote the *bending moment* of the external forces about the right-hand section, the general moment equation can be written as

$$M_R = M_L + V_L x + \Sigma M_{ef} \tag{4.1}$$

In the above equation, M_R is the bending moment at the right section, M_L and V_L are the bending moment and the shear at the left section, x is the distance between sections, and ΣM_{ef} is the algebraic sum of the bending moments of the external forces *with respect to the right section.* M_R, M_L, and V_L were positive as used in the derivation; therefore, the proper sign must be used when the values of any of these are substituted in the equation. The signs of the moments of the external forces are those previously adopted for bending moment.

For beams subjected to a large number of loads, use of the general moment equation substantially reduces the amount of computations required for drawing the moment diagrams.

Example 4.5. Figure 4.10 shows the loading of a *segment* of a beam. It is known that the moment at E is equal to +18.0 kip-ft and that the shear at C is +43 kips. Determine the moment at G and at C.

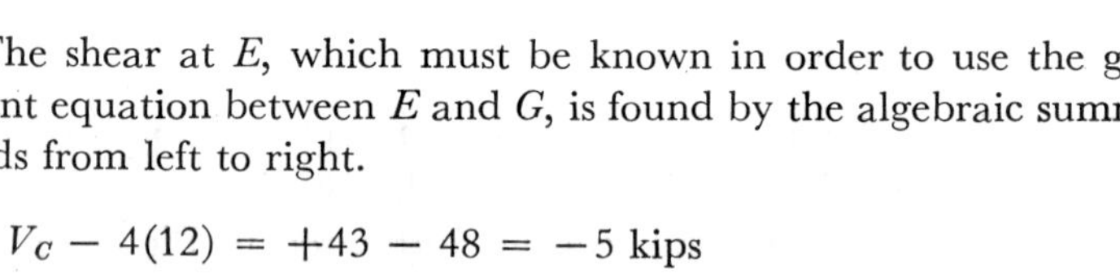

FIG. 4.10

The shear at E, which must be known in order to use the general moment equation between E and G, is found by the algebraic summation of loads from left to right.

$$V_E = V_C - 4(12) = +43 - 48 = -5 \text{ kips}$$

Using the general moment equation between sections at E and G,

$$M_R = M_L + V_L x + \Sigma M_{ef}$$
$$M_G = +18 + (-5)(5) + (-36)(2)$$
$$M_G = +18 - 25 - 72 = -79 \text{ kip-ft} \qquad \textit{Ans}$$

Using the general moment equation between sections at C and E,

$$M_E = M_C + V_C x + \Sigma M_{ef}$$
$$+18 = M_C + (+43)(14) + (-4)(12)(6)$$
$$+18 = M_C + 602 - 288$$
$$M_C = +18 - 602 + 288 = -296 \text{ kip-ft} \qquad \textit{Ans}$$

PROBLEMS

★4.31. On the beam *segment* shown in Fig. P4.31 let $P = 6$ kips and $w = 1$ kip/ft. If $M_C = +20$ kip-ft and $V_C = +4$ kips, compute the shear and moment at section D.

4.32. On the beam *segment* shown in Fig. P4.32 let $P = 2000$ lb and $w = 400$ lb/ft. If $M_C = -4000$ lb-ft and $V_C = +1500$ lb, compute the shear and moment at section D.

4.33. On the beam *segment* shown in Fig. P4.33 let $P = 500$ lb and $w = 200$ lb/ft. If $M_C = +1200$ lb-ft and $V_C = -900$ lb, compute the shear and moment at section D.

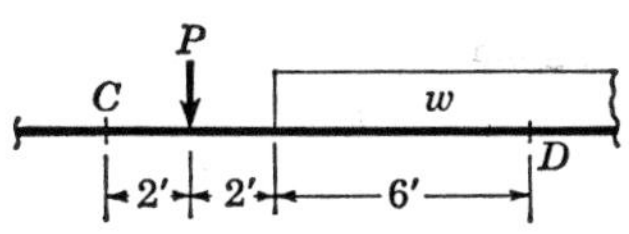

FIG. P4.31, P4.32, P4.33

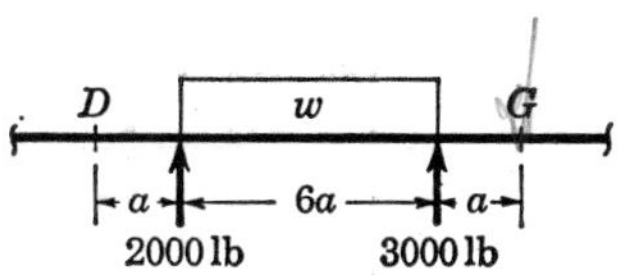

FIG. P4.34, P4.35, P4.36

★4.34. On the beam *segment* shown in Fig. P4.34 let $w = 300$ lb/ft and $a = 2$ ft. If $M_D = +4000$ lb-ft and $V_G = +3000$ lb, compute V_D and M_G.

4.35. On the beam *segment* shown in Fig. P4.35 let $w = 600$ lb/ft and $a = 3$ ft. If $M_G = -6000$ lb-ft and $V_G = -4000$ lb, compute the shear and moment at section D.

4.36. On the beam *segment* shown in Fig. P4.36 let $w = 1000$ lb/ft and $a = 1$ ft. If $M_G = -9000$ lb-ft and $V_D = +5000$ lb, compute V_G and M_D.

★4.37. Span CD of a continuous beam is 12 ft long and supports a uniform load of 200 lb/ft. If $M_C = -4000$ lb-ft and $M_D = -6400$ lb-ft, compute the shear at each support.

4.38. Span DE of a continuous beam is 10 ft long and supports a load of 12 kips at 4 ft from support D. If $M_D = -10$ kip-ft and $M_E = -8$ kip-ft, compute the shear at each support.

4.5 RELATION BETWEEN SHEAR AND MOMENT

An alternate method of determining bending moments and the shape of moment diagrams is based on certain relationships between shears and moments in beams. When fully understood, this method is extremely rapid and useful, especially for beams supporting a large number of loads. The relationships needed will

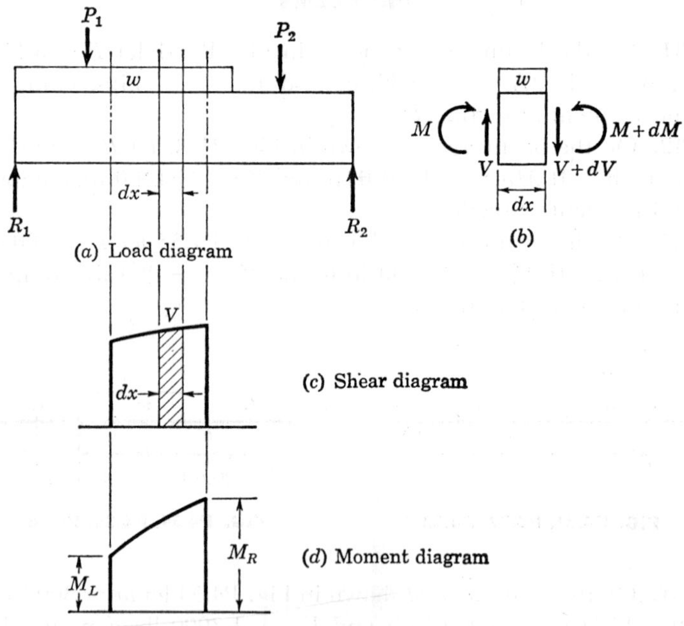

FIG. 4.11

be derived using a small segment of the beam shown as Fig. 4.11*a*. A free-body diagram of a dx length of the beam is drawn as shown in Fig. 4.11*b*. Setting the sum of moments about the right face of the segment equal to zero,

$$M + dM - M - V\,dx + w\,dx\,\frac{dx}{2} = 0 \qquad \text{or}$$

$$dM = V\,dx - \frac{w(dx)^2}{2}$$

The term containing the square of a differential is extremely small in comparison with the other two terms and is discarded. Therefore

$$dM = V\,dx \qquad \text{or}$$

$$V = \frac{dM}{dx} \tag{4.2}$$

Verbally, Eq. (4.2) states that *the slope of the moment diagram at any point is equal to the shear at that section.* This relationship is of much value in determining the *shape* of the moment diagram.

An examination of the equation $dM = V\,dx$ reveals that $V\,dx$ represents a differential area of shear diagram, as shown in Fig. 4.11*c*. Therefore, the *change in moment* between any two sections (such as those shown in Fig. 4.11) would be equal to $\int V\,dx$, which represents the area under the shear diagram between the two sections. In equation form this relationship is expressed as

$$M_R - M_L = A_V \tag{4.3}$$

where M_R and M_L are the moments at the right and left sections, respectively (as shown in Fig. 4.11*d*), and A_V is the area under the shear diagram between the two sections.

With an understanding of the relationships expressed by Eqs. (4.2) and (4.3), moment diagrams can be more easily drawn, generally without writing moment equations. Areas under shear diagrams are used to compute moment-diagram ordinates; the shapes of shear diagrams are used to determine the shapes of moment diagrams. Since $dM/dx = V$, the moment diagram will be a straight line where the shear diagram is horizontal; where the shear diagram is a sloping straight line, the moment diagram will be a segment of a parabola having its vertical axis through its point of zero slope. An abrupt change of shear (at a concentrated load) indicates a discontinuity of moment diagram slope. Since $M_R - M_L = A_V$, moment-diagram ordinates can be rapidly determined by algebraically summing shear-diagram areas from a section of known moment. The equation $dM/dx = V$ is of special value in determining the points of maximum and minimum moment. Since these points occur where dM/dx is equal to zero, they must occur where the shear is equal to zero.

As will be seen in Example 4.7, a couple acting on a beam will cause an abrupt ordinate change in the moment diagram *not indicated* by the shear diagram.

Example 4.6. Draw complete shear and moment diagrams for an overhanging beam having the load diagram shown as Fig. 4.12*a*.

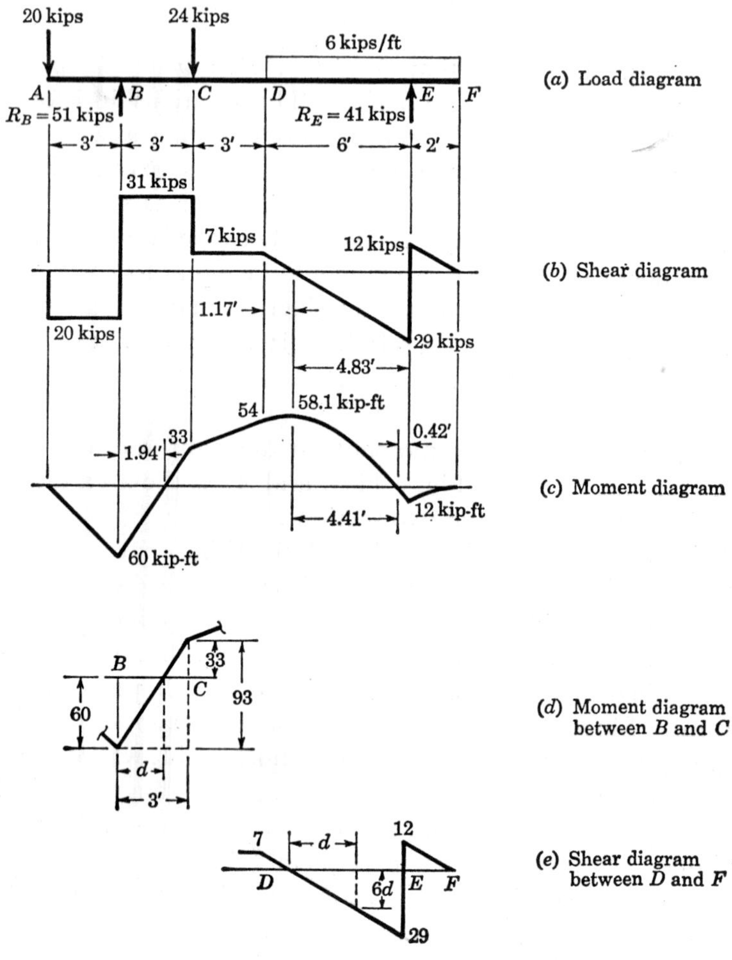

FIG. 4.12

Using the algebraic summation of external loads from left to right, the shear diagram is drawn as shown in Fig. 4.12*b*.

Using Eq. (4.3) in the form $M_R = M_L + A_V$, the required ordinates are computed as follows:

$$M_A = 0.0$$

$$-20(3) = \underline{-60.0}$$

	M_B	$= -60.0$
$+31(3)$		$= +93.0$
	M_C	$= +33.0$
$+7(3)$		$= +21.0$
	M_D	$= +54.0$
$+7(1.17)(\frac{1}{2})$		$= +\ 4.1$
	$M_{\max}$	$= +58.1$
$-29(4.83)(\frac{1}{2})$		$= -70.1$
	M_E	$= -12.0$
$+12(2)(\frac{1}{2})$		$= +12.0$
	M_F	$= \quad 0.0$ *Check*

The moment diagram is drawn as shown in Fig. 4.12*c*, using the shear diagram and the relationship $dM/dx = V$ as a guide. The moment diagram consists of three straight-line segments between A and D. The parabolic segment between D and E must be tangent to the straight line at D and horizontal at the point of zero shear and have a relatively steep negative slope at E. Since the parabolic segment has a vertical axis (through the point of zero slope), the moment at points which are equidistant from the vertical axis must be equal. Therefore, the moment at 1.17 ft to the right of the maximum ordinate must also be +54 kip-ft. The parabolic segment between E and F must have a relatively steep slope at E and a zero slope at F.

The point of zero moment between B and C can be located using similar triangles. As shown in Fig. 4.12*d*,

$$\frac{d}{3} = \frac{60}{93}$$

therefore

$$d = \frac{3(60)}{93} = 1.94 \text{ ft}$$

The point of zero moment between D and E can be found using the equation $M_R - M_L = A_V$ between the point of maximum moment and the point of zero moment. As shown in Fig. 4.12*e*, A_V is equal to $-6d(d)/2$. Therefore

$$M_R - M_L = A_V$$
$$0 - 58.1 = -3d^2$$
$$d = \pm 4.41 \text{ ft}$$

The -4.41-ft value is invalid, since it is outside the range of the parabolic segment.

Example 4.7. Draw the complete shear and moment diagrams for a simply supported beam having the load diagram shown as Fig. 4.13*a*.

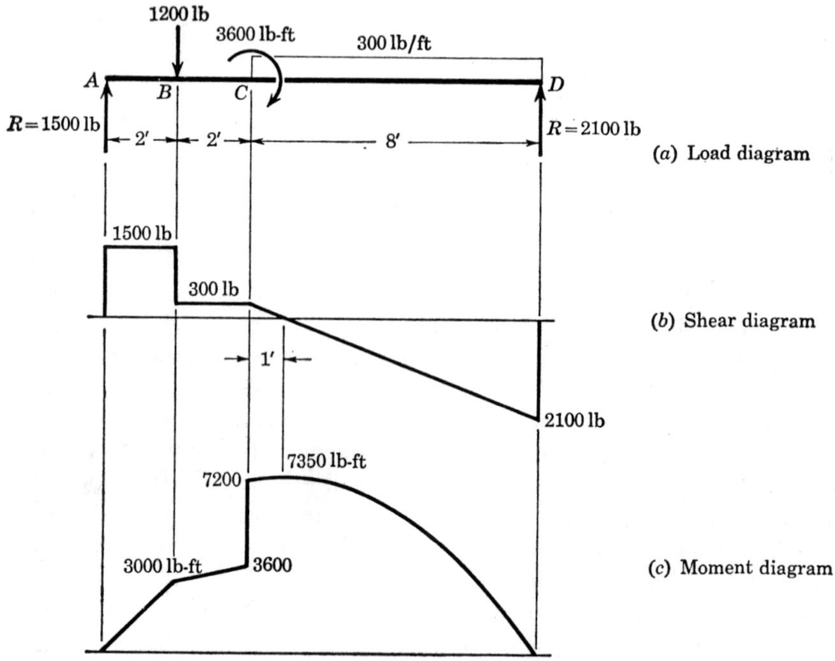

FIG. 4.13

Using the algebraic summation of external forces from left to right, the shear diagram is drawn as shown in Fig. 4.13*b*. The required ordinates of the moment diagram are computed using the area under the shear diagram and, *in addition*, the effect of the couple. The couple acts in such a way as to cause tension in the bottom of any section to the right of its point of application (point *C*); therefore the couple causes an abrupt positive change in the moment diagram at point *C*.

	M_A	$= 0$
$+1500(2)$		$= +3000$
	M_B	$= +3000$
$+300(2)$		$= +\ 600$
M_C (left of couple)		$= +3600$
Effect of couple		$= +3600$

$$M_C \text{ (right of couple)} \qquad = +7200$$
$$+300(1)(\tfrac{1}{2}) \qquad = +\ 150$$
$$M_{\max} = +7350$$
$$-2100(7)(\tfrac{1}{2}) \qquad = -7350$$
$$M_D \quad = \quad 0 \qquad \textit{Check}$$

The moment diagram is drawn as shown in Fig. 4.13*c* (*Ans*).

PROBLEMS

4.39–4.57. For a beam having the load diagram shown in Figs. P4.39 to P4.57, sketch the complete shear and moment diagrams. (*4.39, *4.42, *4.44, *4.47, *4.49, *4.54, and *4.57)

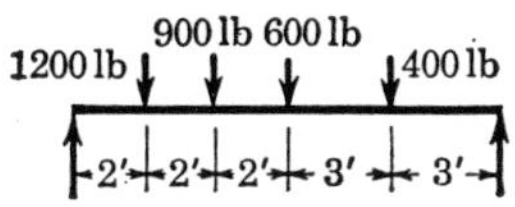

FIG. P4.39

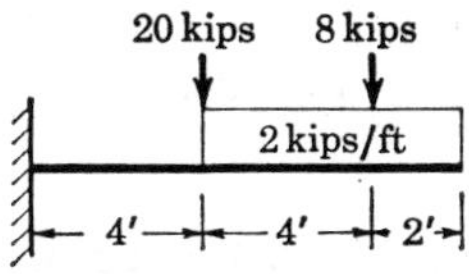

FIG. P4.40

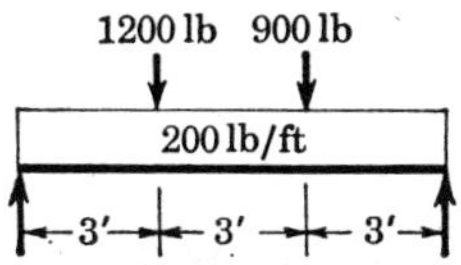

FIG. P4.41

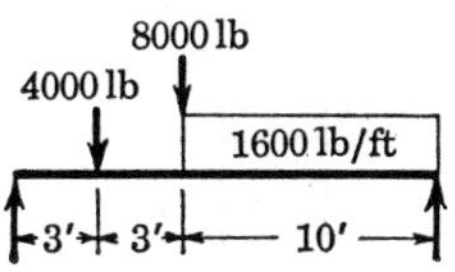

FIG. P4.42

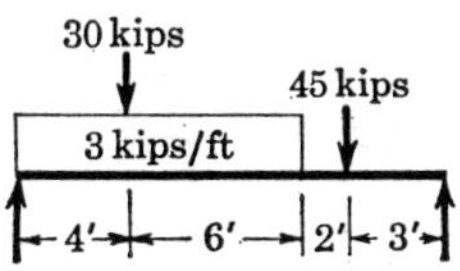

FIG. P4.43

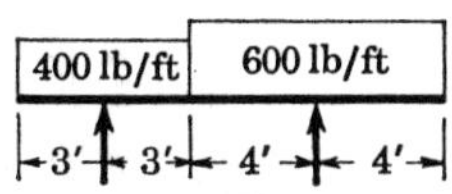

FIG. P4.44

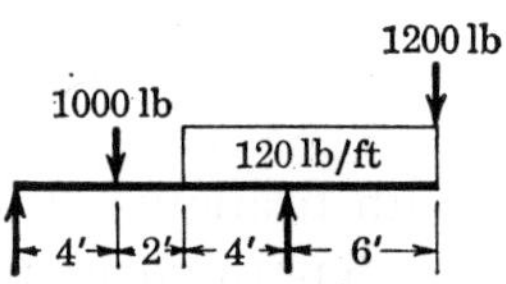

FIG. P4.45

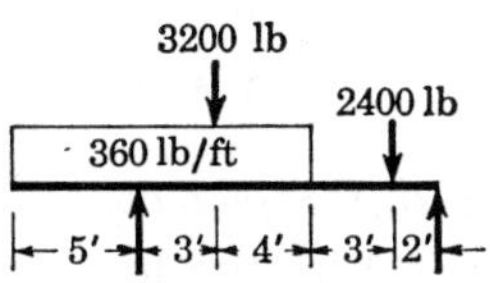

FIG. P4.46

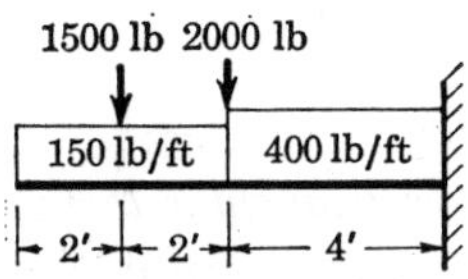

FIG. P4.47

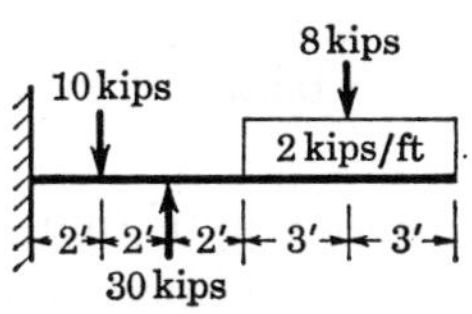

FIG. P4.48

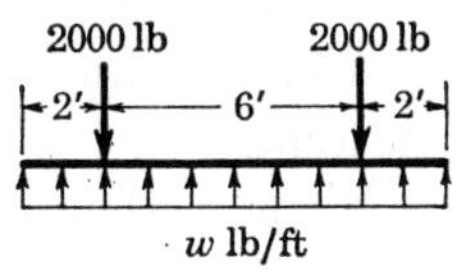

FIG. P4.49

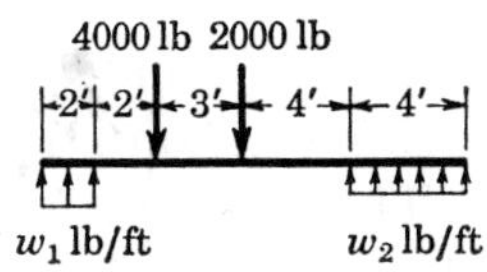

FIG. P4.50

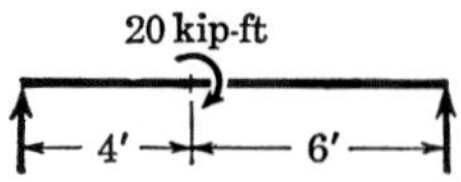

FIG. P4.51

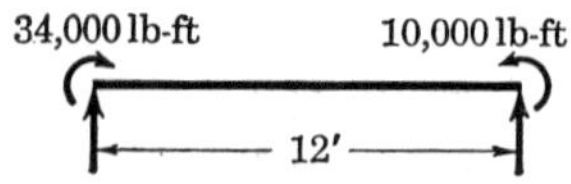

FIG. P4.52

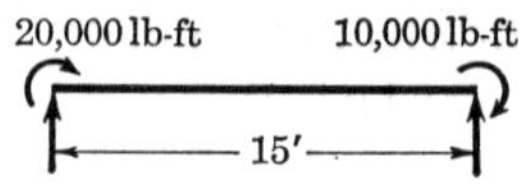

FIG. P4.53

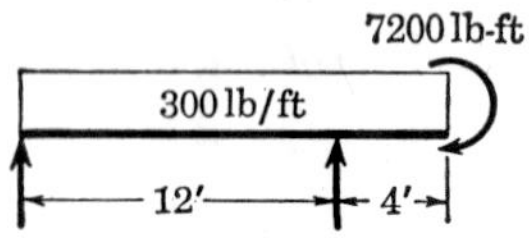

FIG. P4.54

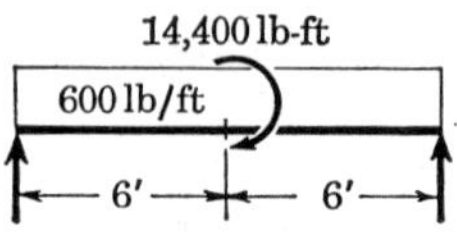

FIG. P4.55

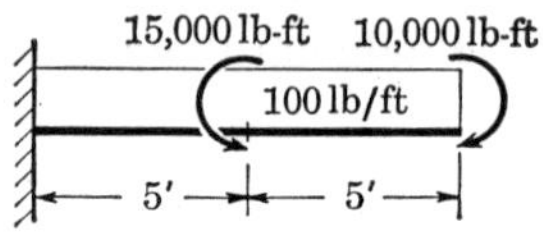

FIG. P4.56

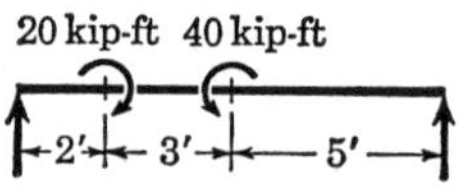

FIG. P4.57

4.6 BEAMS WITH UNIFORMLY VARYING LOADS

Uniformly increasing loads act or are assumed to act on many engineering structures. This type of loading is commonly referred

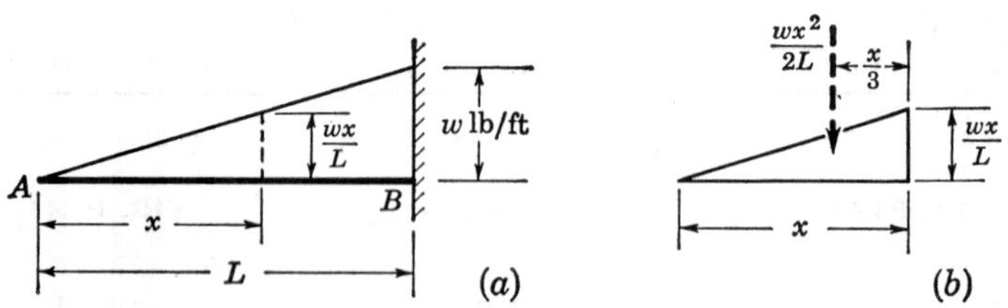

FIG. 4.14

to as *triangular loading* or *trapezoidal loading.* With triangular loading, as illustrated in Fig. 4.14*a*, the load per unit length varies uniformly from zero to a maximum value; with trapezoidal loading, as illustrated in Fig. 4.15*a*, the load per unit length varies uniformly from a minimum value to a maximum value. As shown in Fig. 4.15*b*, a trapezoidal load can be considered to be a triangular load superimposed on a uniform load.

The computations for shear and moment under a triangular load are only slightly different from those made for a uniform load. This difference can be illustrated using the cantilever beam shown in Fig. 4.14*a*. The shear and moment equations can be easily determined using the segment of the beam shown in Fig. 4.14*b*.

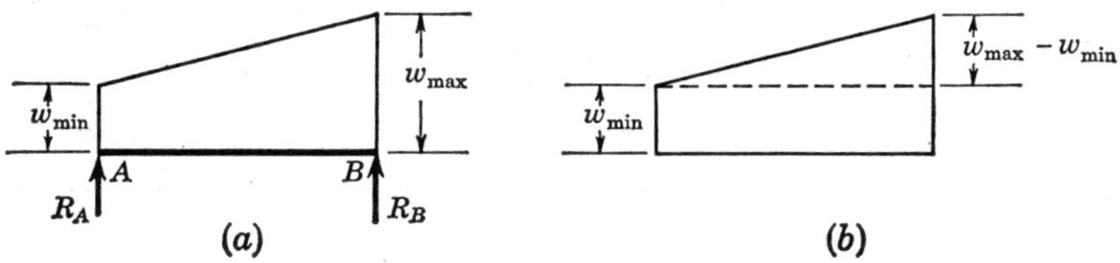

FIG. 4.15

With the use of similar triangles, the load per foot at a section x from the free end is determined as wx/L. The shear at x must be

$$V = -\frac{1}{2}\frac{wx}{L}x = -\frac{wx^2}{2L}$$

The moment at x, as shown, must be

$$M = -\frac{wx^2}{2L}\frac{x}{3} = -\frac{wx^3}{6L}$$

Thus it is seen that the shear and moment diagrams under a triangular load must be second- and third-degree curves, respectively. Although the computations for shear and moment under uniformly increasing loading are slightly more complex, it should be kept in mind that the principles used are the same regardless of the type of loads.

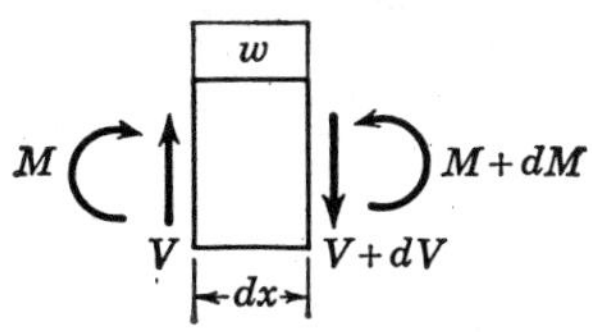

FIG. 4.16

The relationship between load per unit length and shear at a section is quite useful for determining the shape of the shear diagram under a uniformly increasing load. As will be seen, this relationship parallels that which exists between shear and moment at a section (Art. 4.5). Figure 4.16 (a repetition of Fig. 4.11*b*) shows a free body of a segment (*dx* in length) from a beam under load. Since the sum of the forces acting on the segment must be equal to zero

$$+V - (V + dV) - w\,dx = 0$$

$$-dV = w\,dx$$

$$\frac{dV}{dx} = -w \qquad (4.4)$$

Thus it is seen that the slope of the shear diagram at any section is numerically equal to the load per unit length at that section. The minus sign in the equation merely confirms that the slope of the shear diagram is negative under a downward load.

Example 4.8. Draw the complete shear and moment diagrams for a simply supported beam having the load diagram shown as Fig. 4.17*a*.

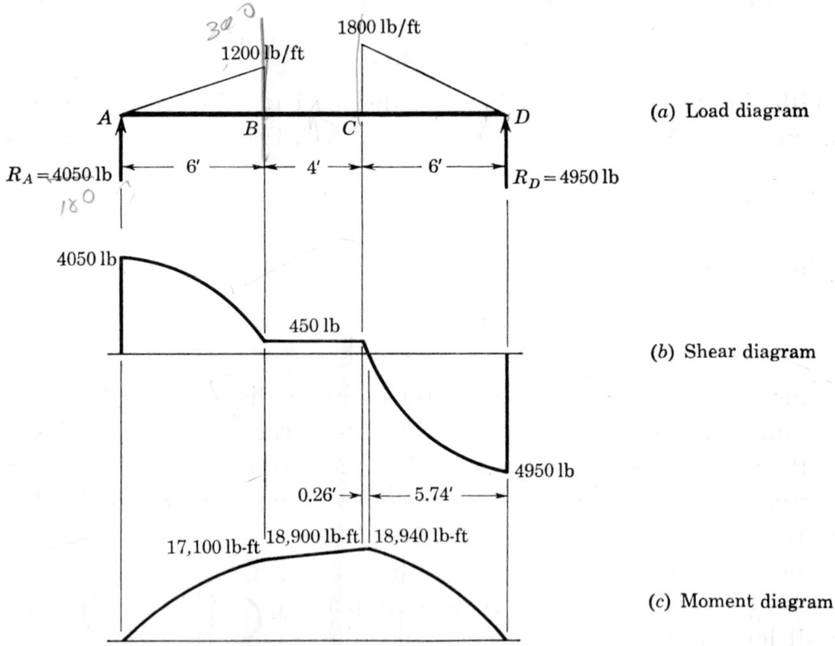

FIG. 4.17

The reactions are computed using the resultants of the two triangular loads (3600 lb at 4 ft from A and 5400 lb at 4 ft from D).

Measuring x from the left end, the equation for shear between A and B is

$$0 \leqq x \leqq 6 \qquad V = +4050 - \frac{1200x}{6} x \frac{1}{2} \qquad \text{or}$$

$$V = +4050 - 100x^2$$
At $x = 0$, $V = +4050$ lb
At $x = 6$, $V = +4050 - 100(36) = +450$ lb

Measuring x from the right end, the equation for shear between D and C is

$$0 \leqq x \leqq 6 \qquad V = -4950 + \frac{1800x}{6} x \frac{1}{2} \qquad \text{or}$$

$V = -4950 + 150x^2$
At $x = 0$, $V = -4950$ lb
At $x = 6$, $V = -4950 + 150(36) = +450$ lb

With the use of Eq. (4.4) and the load diagram as a guide, the shear diagram is drawn as shown in Fig. 4.17*b*. Between *A* and *B* the shear diagram is a parabolic segment having zero slope at *A* and a relatively steep slope at *B*. There being no load between *B* and *C*, the slope of the shear diagram between the two points is zero. Between *C* and *D* the shear diagram is a parabolic segment having a steep negative slope at *C* and zero slope at *D*.

The point of zero shear is found by setting the appropriate shear equation equal to zero.

$0 = -4950 + 150x^2$
$x^2 = 33.0$
$x = 5.74$ ft

Measuring x from the left end, the moment equation between *A* and *B* is

$$0 \leqq x \leqq 6 \qquad M = +4050x - 100x^2 \frac{x}{3}$$

$M = +4050x - 33.3x^3$
At $x = 0$, $M = 0$
At $x = 6$, $M = +4050(6) - 33.3(216) = +17{,}100$ lb-ft

Measuring x from the right end, the moment equation between *D* and *C* is

$$0 \leqq x \leqq 6 \qquad M = +4950x - 150x^2 \frac{x}{3}$$

$M = +4950x - 50x^3$
At $x = 0$, $M = 0$
At $x = 5.74$, $M = +4950(5.74) - 50(189.4) = +18{,}940$ lb-ft
At $x = 6$, $M = +4950(6) - 50(216) = +18{,}900$ lb-ft

The area under the shear diagram between *B* and *C* is used as a check.

$M_C - M_B = A_V$
$+18{,}900 - 17{,}100 = +450(4)$
$1800 = 1800$ *Check*

Using the shear diagram as a guide for slope, the moment diagram is drawn as shown in Fig. 4.17*c* (*Ans*).

PROBLEMS

4.58–4.63. For a beam having the load diagram shown in Figs. P4.58 to P4.63, sketch the complete shear and moment diagrams. (*4.60 and *4.62)

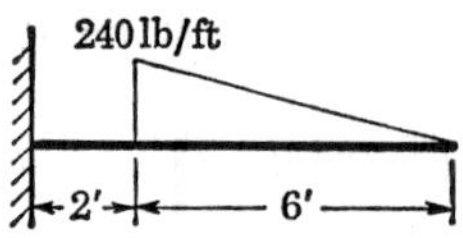

FIG. P4.58

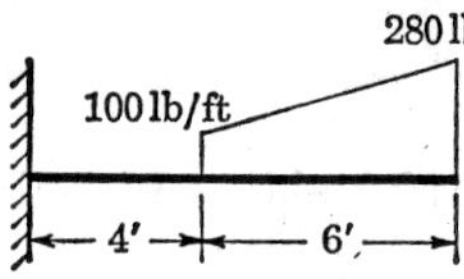

FIG. P4.59

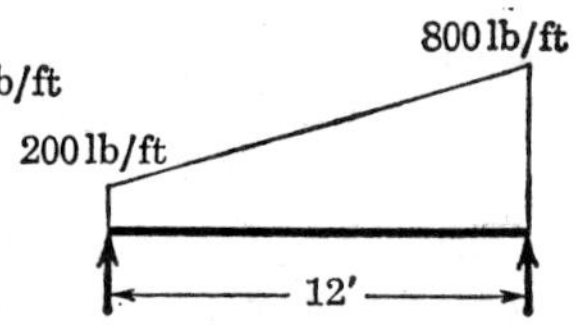

FIG. P4.60

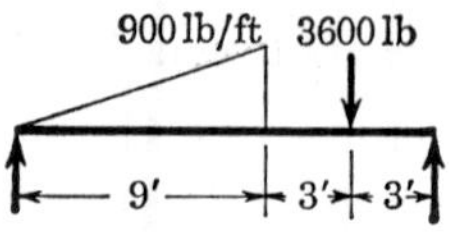

FIG. P4.61

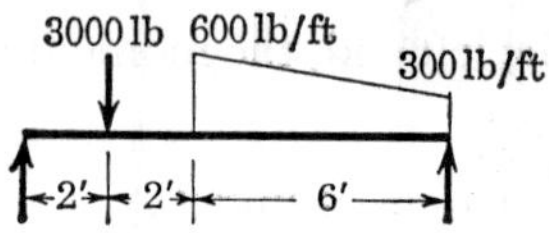

FIG. P4.62

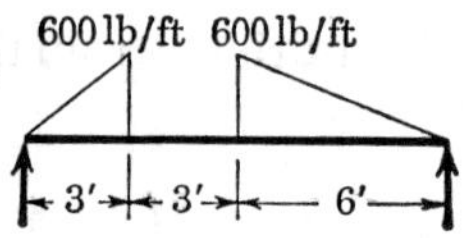

FIG. P4.63

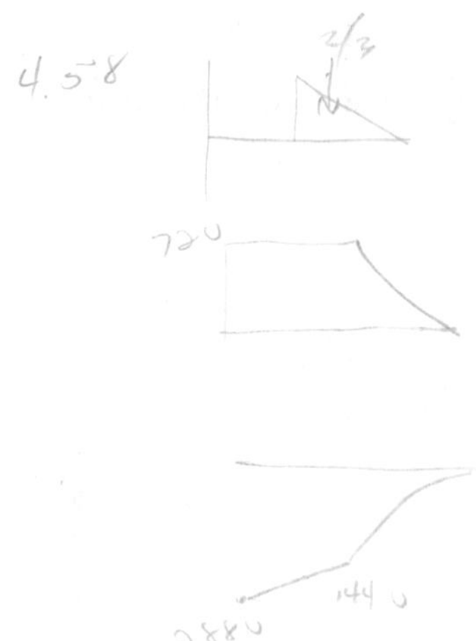

5

STRESSES IN BEAMS

5.1 THE FLEXURE FORMULA

As was mentioned in the previous chapter, when a beam is loaded, a nonuniform stress distribution will exist at any cross section. As a matter of fact, the sign convention for bending moment was based on the presence of both tensile and compressive stresses at any section of a beam. The *flexure formula* defines the bending stress, at a point, in terms of the bending moment at the section and the geometry of the cross section.

The derivation of the flexure formula will be based on the following assumptions:†

1. Stress is proportional to strain (that is, the beam material obeys Hooke's law).

2. The beam material is homogeneous and has the same Young's modulus in tension and compression.

3. Any cross-sectional plane of the beam (before bending) remains a plane after bending.

4. The beam has a uniform cross section

† Actually *all* the assumptions here listed need *not* be made in order for the flexure formula to be applicable to a given beam. For example, a beam need not be symmetrical about the plane of loading so long as the loading does not cause the beam to twist (see Art. 7.9).

throughout its length and is symmetrical about a vertical longitudinal plane.

5. The beam is straight before bending, and the loads acting on it pass through the vertical longitudinal plane of symmetry.

Essential to the derivation of the flexure formula is a knowledge of the stress distribution at any beam cross section which is

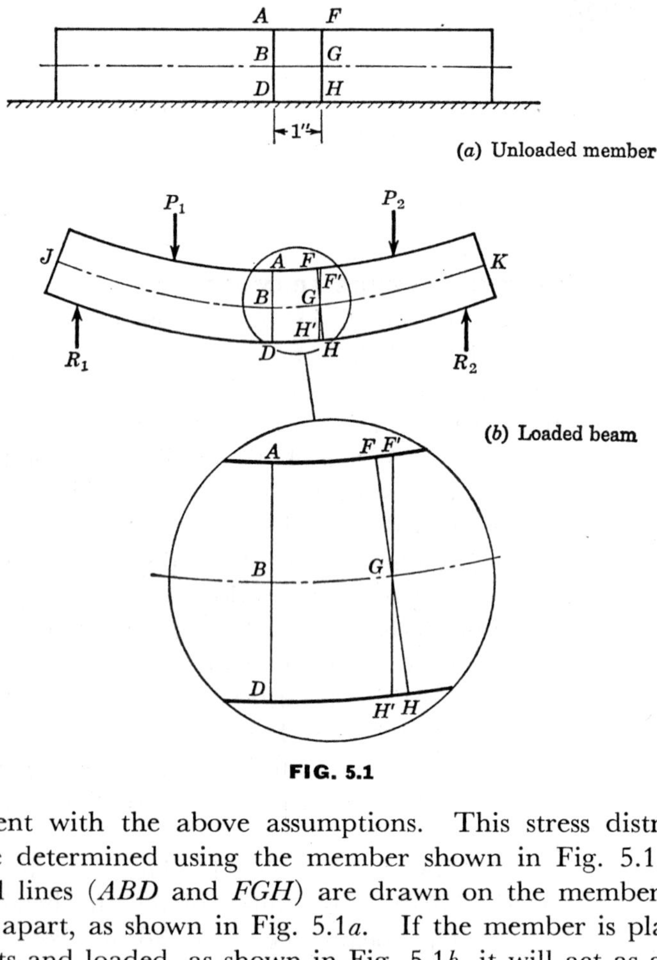

FIG. 5.1

consistent with the above assumptions. This stress distribution will be determined using the member shown in Fig. 5.1. Two vertical lines (*ABD* and *FGH*) are drawn on the member a unit length apart, as shown in Fig. 5.1*a*. If the member is placed on supports and loaded, as shown in Fig. 5.1*b*, it will act as a beam. As a result of the bending which occurs, the distance between *A* and *F* decreases, the distance between *D* and *H* increases, and the distance between *B* and *G* is unchanged. If line *F′GH′* is drawn parallel to line *ABD*, it is seen that the unit strain varies linearly along the depth of the beam from a maximum compressive value to a maximum tensile value. Since it is assumed that the beam material obeys Hooke's law and has the same modulus of elasticity in tension and compression, it follows that *the bending stress varies linearly along the depth of the beam*, from a maximum compressive value to a maximum tensile value. It should be noted that the stress at points *B* and *G* must be zero, since there is no strain at these points. The locus of points in a loaded beam having zero bending stress is termed the *neutral surface* of the beam. The edge view of the neutral surface of the beam in Fig. 5.1 is shown as line

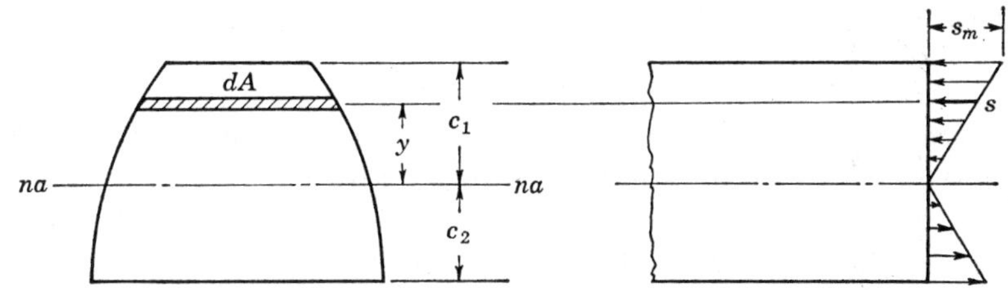

(*a*) Cross section

(*b*) Bending stress distribution at a typical section

FIG. 5.2

JBGK. The line defining the intersection of the neutral surface and any cross section of a beam is termed *the neutral axis* at that section.

The derivation of the flexure formula will be made using Fig. 5.2, which shows the cross-sectional view and stress distribution at a section of a beam where positive bending moment exists. The stress s (at any point y distance from the neutral axis) acting on a differential area dA is equivalent to a differential force $dF = s\,dA$. Such a differential force has a moment about the neutral axis of $dM = dF\,y = s\,dA\,y$. The internal moment at the section must be equal to the summation of the differential moments over the entire section; thus, $M = \int dM = \int s\,dA\,y$. As can be seen, the maximum stress s_m acts at a distance c_1 from the neutral axis; therefore, the stress at any point can be expressed in terms of this maximum stress $s = s_m(y/c_1)$. Substituting for s in terms of the maximum stress gives

$$M = \int \frac{s_m y}{c_1}\, dA\, y$$

Since s_m and c_1 are constants

$$M = \frac{s_m}{c_1} \int_{c_2}^{c_1} y^2\, dA$$

The integral term should be recognized as the moment of inertia of the cross-sectional area with respect to the reference (neutral) axis; therefore, $M = s_m I_{na}/c_1$ or $s_m = Mc_1/I_{na}$. The stress at *any point* on the section can be determined as

$$s = \frac{s_m y}{c_1} = \frac{Mc_1}{I_{na}} \frac{y}{c_1} = \frac{My}{I_{na}}$$

The stress distribution used in the above derivation was that due to bending alone (that is, without any normal stress due to axial or longitudinal loading); therefore, the summation of the differential forces $\int dF$ on the section must be equal to zero. Therefore

$$\int s\, dA = \int \frac{s_m y}{c_1}\, dA = 0 \qquad \text{or} \qquad \frac{s_m}{c_1} \int_{c_2}^{c_1} y\, dA = 0$$

Since the integral term is the *first moment* of the cross-sectional area about the neutral axis, the equation can be written as $(s_m/c_1)\bar{y}A = 0$. Since s_m/c_1 and A cannot be equal to zero, $\bar{y}$ must be equal to zero. Thus it is seen that, for the assumptions made, *the neutral axis must coincide with the centroidal axis of the cross-sectional area.*

The flexure formula is usually written, without subscripts, as

$$s = \frac{My}{I} \tag{5.1a}$$

or, in the case of the maximum tension or compression bending stress, as

$$s = \frac{Mc}{I} \tag{5.1b}$$

where s is the bending stress at a distance y or c from the centroidal axis, M is the bending moment at the section, and I is the moment of inertia of the cross-sectional area about its centroidal axis.

5.2 BENDING STRESSES; BEAMS OF SYMMETRICAL CROSS SECTION

In using the flexure formula care must be taken to avoid errors due to inconsistency of units. Consistent units usually assigned are

Bending moment M	lb-in.
Distance from neutral axis y or c	in.
Moment of inertia I	in.^{4}
Unit stress s	psi

The bending moment, as obtained from the moment diagram (or computed directly from the load diagram), frequently has units of pound-feet; in such cases it must be converted to units of pound-inches for use in the flexure formula.

A majority of the beams in common use are symmetrical about the neutral axis and therefore have a balanced tensile and compressive stress distribution at any cross section. Such beams are most economical when the beam material has tensile and compressive strengths which are practically equal.

Example 5.1. For a beam having the cross section and load diagram shown in Fig. 5.3, compute (*a*) the maximum tensile and compressive bending stresses, (*b*) the bending stress at a point on the beam 2.5 in. from the top and 2 ft from the left end.

The shear and moment diagrams are drawn as shown.

The moment of inertia can be most easily computed by considering the cross section as a 6- by 12-in. rectangle *less* two 2- by 8-in. rectangles. In doing this, the transfer formula for moment of inertia is not required, since the centroidal axis of each of the rectangles coincides with the neutral axis.

$$I = \frac{6(12)^3}{12} - 2\,\frac{2(8)^3}{12} = 864 - 171 = 693 \text{ in.}^4$$

(*a*) Both maximum bending stresses occur at midspan, where the moment is a maximum.

$$s = \frac{Mc}{I} = \frac{[14{,}400(12)]6}{693} = 1496 \text{ psi}$$

The 12 in the numerator serves to convert the bending moment from pound-feet to pound-inches.

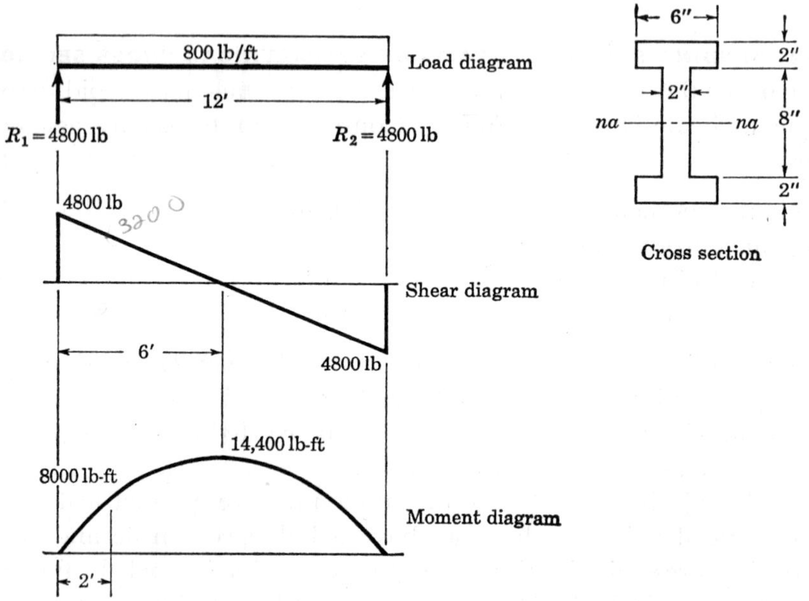

FIG. 5.3

Therefore

s = 1496 psi tension (at bottom) *Ans*
s = 1496 psi compression (at top) *Ans*

(*b*) At 2.5 in. from the top and 2 ft from the left end

M = 8000 lb-ft and
$y = 6 - 2.5 = 3.5$ in.

therefore

$$s = \frac{My}{I} = \frac{[8000(12)]3.5}{693} = 485 \text{ psi compression} \qquad Ans$$

Example 5.2. A cantilever beam has the cross section and load diagram shown in Fig. 5.4. If the allowable bending stress is 1800 psi, compute the allowable uniform and concentrated loads.

$$I = \frac{8(10)^3}{12} - \frac{4(6)^3}{12} = 667 - 72 = 595 \text{ in.}^4$$

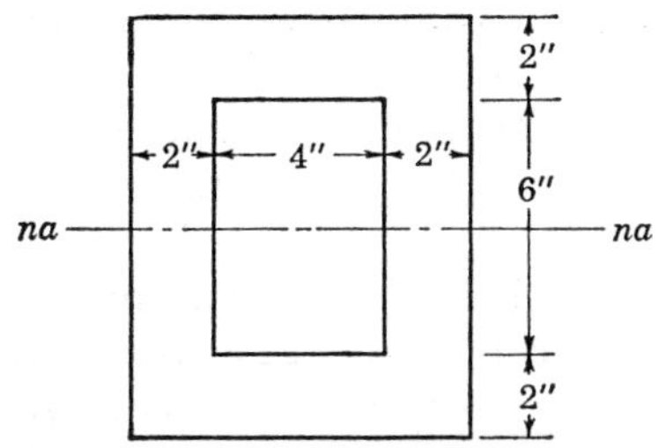

(*a*) Cross section

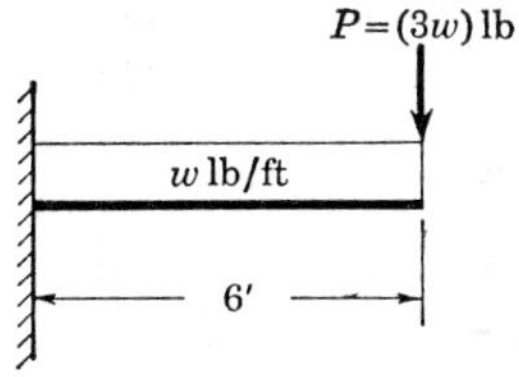

(*b*) Load diagram

FIG. 5.4

By inspection of the load diagram, it is seen that the maximum bending moment occurs at the point of support.

$$M = -(6w)(3) - (3w)(6) = -18w - 18w = -36w \text{ lb-ft}$$

$$s = \frac{Mc}{I} \qquad 1800 = \frac{[36w(12)]5}{595}$$

$$w = 496 \text{ lb/ft} \qquad \textit{Ans}$$

$$P = 3w = 3(496) = 1488 \text{ lb} \qquad \textit{Ans}$$

Thus the beam can support simultaneously a uniform load of 496 lb/ft (over its entire length) and a concentrated load of 1488 lb (at the free end).

PROBLEMS

***5.1.** A beam of rectangular cross section is 6 in. deep and 3 in. thick. At a section where the moment is +2400 lb-ft, compute (*a*) the maximum tensile and compressive bending stresses, (*b*) the bending stress 1 in. from the top of the section.

5.2. A 2.50-in.-diam rod is subjected to bending. At a section where the moment is −1200 lb-ft, compute (*a*) the maximum tensile and compressive bending stresses, (*b*) the bending stress ½ in. from the top of the section.

5.3. A pipe of 4.00-in. OD and 3.50-in. ID is subjected to bending. If the bending stress is not to exceed 16,000 psi, compute the allowable bending moment.

***5.4.** Four 6- by 2-in. planks (actual size) are securely fastened together to make a (hollow) box beam 10 in. deep and 6 in. wide. If the bending stress is not to exceed 1750 psi, compute the allowable bending moment.

5.5. A beam of rectangular cross section is to withstand a maximum bending moment of 3600 lb-ft. If the depth is to be twice the thickness and the allowable bending stress is 2000 psi, compute the minimum cross-sectional dimensions.

5.6. A round bar is to withstand a maximum bending moment of 2400 lb-ft. If the allowable bending stress is 18,000 psi, compute the required diameter of the bar.

★5.7. A 10-ft simply supported beam is loaded with a uniform load of 200 lb/ft over its entire length and a concentrated load of 1000 lb at 3 ft from the left reaction. The beam is rectangular in cross section, 4 in. deep, and 2 in. thick. Compute (*a*) the maximum tensile and compressive bending stresses, (*b*) the bending stress at a point 2 ft from the left reaction and 1.5 in. from the top.

5.8. A 12-ft simply supported beam supports two concentrated loads: 2000 lb at 3 ft from the left end and 1000 lb at 3 ft from the right end. The beam is rectangular in cross section, 8 in. deep, and 6 in. thick. Compute (*a*) the maximum tensile and compressive bending stresses, (*b*) the bending stress at a point 4 ft from the left end and 2 in. from the top.

5.9. A 3.00-in.-diam rod, 8 ft long, acts as a simply supported beam subjected to a uniform load of 400 lb/ft over its entire length and a concentrated load of 1600 lb at 2 ft from the left end. Compute (*a*) the maximum tensile and compressive bending stresses, (*b*) the bending stress at a point 2 ft from the left end and 1 in. from the bottom.

★5.10. A 12-ft simply supported beam having the cross section shown in Fig. P5.10 is to support a uniform load (*w* lb/ft) *over the left half of its length.* If the allowable bending stress is 2000 psi, compute the allowable uniform load.

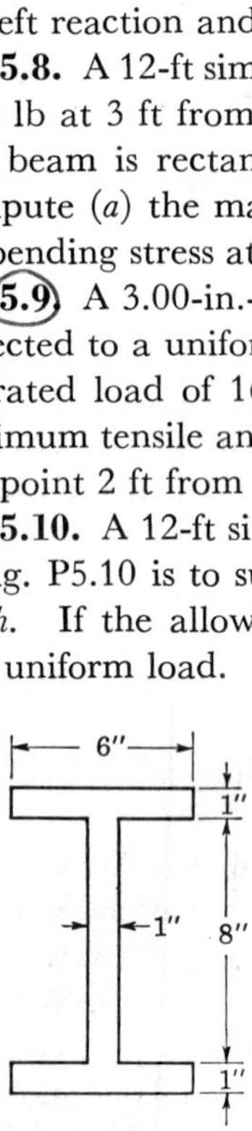

FIG. P5.10

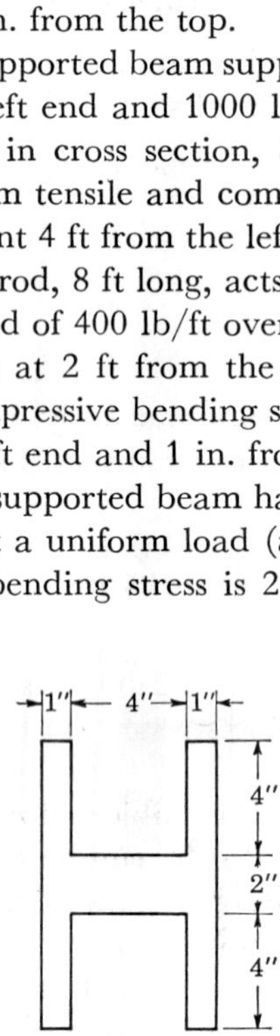

FIG. P5.11

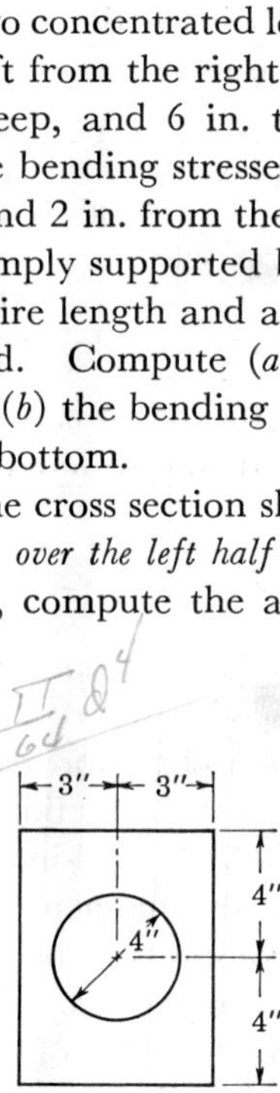

FIG. P5.12

5.11. A 12-ft simply supported beam having the cross section shown in Fig. P5.11 is to be loaded with three concentrated loads: *P* at mid-span, 2*P* at 3 ft from the left end, and 4*P* at 3 ft from the right end. If the allowable bending stress is 4000 psi, compute the allowable value of *P*,

5.12 A simply supported beam having the cross section shown in Fig. P5.12 is to support a uniform load of 600 lb/ft over its entire length. If the allowable bending stress is 1800 psi, compute the maximum allowable length for the beam.

5.3 STANDARD SECTIONS; SECTION MODULUS

Many beams in common use consist in whole or in part of standard sections manufactured and sold on a mass-production basis. Perhaps the most common sections are those of "hot-rolled" steel; standard rolled steel sections are available in a variety of shapes, such as those having the cross sections shown in Fig. 5.5. Certain

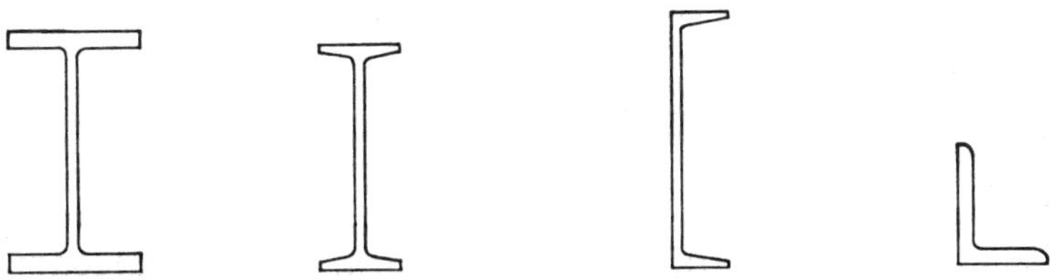

(*a*) Wide flange beam (*b*) American Standard beam (*c*) American Standard channel (*d*) Unequal leg angle

FIG. 5.5

properties of the cross-sectional areas of structural shapes are readily available from tables, such as in the Appendix. In order to save time and avoid confusion, designations for the various shapes are standardized. Thus a designation of 18WF60 defines a *wide-flange* beam having a *nominal depth* of 18 in. and weighing 60 lb/lin ft; a designation of 12I31.8 defines an American Standard beam (or "I beam") having a nominal depth of 12 in. and weighing 31.8 lb/lin ft. The actual depth and other pertinent properties of the section, such as flange and web dimensions, must be obtained from tables. Standard sections are also available in nonferrous metals such as aluminum and magnesium.

Included in tables of properties for standard sections are values of moment of inertia I about appropriate axes. Also tabulated are values of I/c, which are useful when the flexure formula is expressed as $s = M/(I/c)$. The expression I/c is known as the *section modulus* and is sometimes designated by the letters S or Z. In this text, in order to emphasize the meaning of the term, the section modulus will be designated as I/c.

Example 5.3. A 21WF82 simply supported beam is 16 ft long and supports a uniform load of 1000 lb/ft (including the weight of the beam) over its entire length. Find the maximum bending stress in the beam if it is positioned so that its web is (*a*) vertical, (*b*) horizontal.

The maximum bending moment is computed to be 32,000 lb-ft.

(*a*) From Table A.4, when the web is vertical, I/c is 168.0 in.^{3}

$$s = \frac{M}{I/c} = \frac{32{,}000(12)}{168} = 2290 \text{ psi} \qquad \textit{Ans}$$

(*b*) When the web is horizontal, I/c is 20.0 in.^{3}

$$s = \frac{M}{I/c} = \frac{32{,}000(12)}{20.0} = 19{,}200 \text{ psi} \qquad \textit{Ans}$$

The above example emphasizes that wide-flange and I beams are meant to be used with their webs vertical (for vertical loading). In the problems which follow, unless otherwise noted, it should be assumed that they are so oriented.

PROBLEMS

***5.13.** A 12-ft simply supported beam is to be loaded with a uniform load (w lb/ft) over its entire span. The allowable bending stress is 20,000 psi. Compute the allowable uniform load, if the beam is a (*a*) 12WF50 section, (*b*) 12I50 section.

5.14. A 15-ft simply supported beam is to support equal concentrated loads at the third points. The allowable bending stress is 20,000 psi. Neglecting the weight of the beam, compute the maximum allowable loads if the beam is a (*a*) 16WF50 section, (*b*) 12WF50 section.

5.15. A 12-ft simply supported beam is to be loaded with equal concentrated loads at midspan and at each quarter point. The allowable bending stress is 20,000 psi. Neglecting the weight of the beam, compute the maximum allowable loads if the beam is (*a*) made of two 12-in. 20.7-lb channel sections placed back to back with webs vertical, (*b*) a 12WF40 section.

***5.16.** A beam is made up of two 12-in. 20.7-lb channel sections and an 18WF50 section arranged as shown in Fig. P5.16. If the allowable bending stress is 20,000 psi, compute the allowable bending moment.

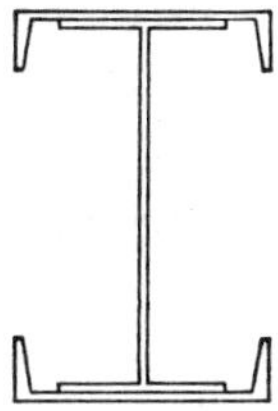

FIG. P5.16, P5.17

5.17. A beam is made up of two 10-in. 15.3-lb channel sections and a 16WF36 section arranged as shown in Fig. P5.17. If the allowable bending stress is 20,000 psi, compute the allowable bending moment.

5.18. If the channel sections in Prob. 5.17 are replaced by plates each 10 by 0.25 in. in cross section, what will be the allowable bending moment?

5.4 BENDING STRESSES; BEAMS WITH CROSS SECTIONS UNSYMMETRICAL ABOUT NEUTRAL AXIS

As a result of practical considerations encountered in the design of machines and structures, beams having cross sections which are unsymmetrical about the neutral axis are frequently used. These beams may be *wrought*, *cast*, or *cut* into final shape, or they may be *built up* of standard sections. When a beam is to be made of a material having substantially different tensile and compressive strengths, an unsymmetrical (about the neutral axis) cross section is generally used; the beam cross section is so proportioned as to utilize efficiently both strengths of the material.

Example 5.4. For a beam having the cross section and load diagram shown in Fig. 5.6, compute (*a*) the maximum tensile bending stress, (*b*) the maximum compressive bending stress.

The shear and moment diagrams are drawn as shown.

The section is considered to be made of a 1- by 4-in. rectangle and a 1- by 8-in. rectangle. The neutral axis, which coincides with the

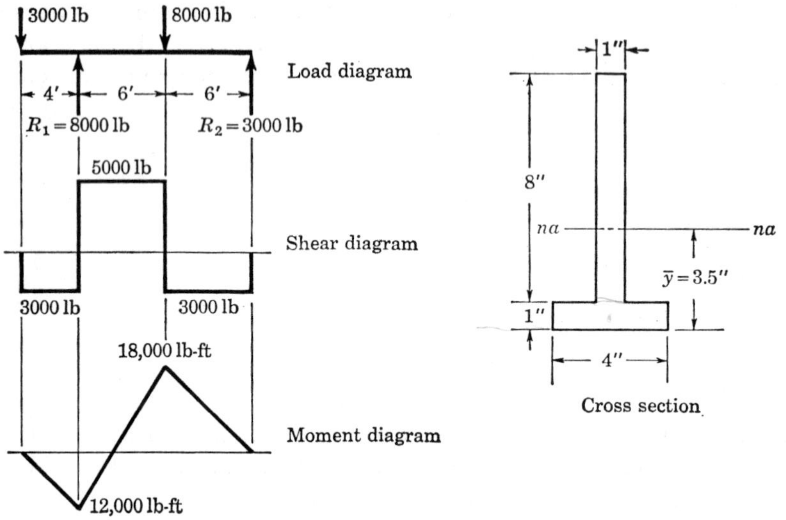

FIG. 5.6

centroidal axis of the composite section, is determined, using $\bar{y}A = \Sigma ay$ in tabular form. Table E5.4 is extended and used to compute the composite moment of inertia about the neutral axis.

TABLE E5.4

	a	y	ay	I_o	d	ad^2
1 by 4	4	0.5	2	0.3	3.0	36.0
1 by 8	8	5.0	40	42.7	1.5	18.0
Total	12		42	43.0		54.0

$$\bar{y} = \frac{\Sigma ay}{A} = \frac{42}{12} = 3.5 \text{ in.}$$

$$I = \Sigma(I_o + ad^2) = 43.0 + 54.0 = 97 \text{ in.}^4$$

(*a*) Since the maximum bending stress is a function of both M and c, the maximum tensile stress may be either at the bottom of the section where the moment is +18,000 lb-ft or at the top of the section where the moment is −12,000 lb-ft. Investigating at both places,

$$s_t = \frac{[18{,}000(12)]3.5}{97} = 7790 \text{ psi}$$

$$s_t = \frac{[12{,}000(12)]5.5}{97} = 8160 \text{ psi}$$

The maximum tensile bending stress is 8160 psi, occurring at the top of the section where the moment is −12,000 lb-ft (*Ans*).

(*b*) The maximum compressive stress at the left reaction need not be computed, since both M and c (for maximum compressive stress) are greater at the section where M is +18,000 lb-ft.

$$s_c = \frac{[18{,}000(12)]5.5}{97} = 12{,}250 \text{ psi} \qquad \textit{Ans}$$

PROBLEMS

★5.19. A cantilever beam 5 ft long supports a concentrated load of 200 lb at the free end. The beam is triangular in cross section, 3 in. deep, and 2 in. wide at the bottom. Compute (*a*) the maximum tensile and compressive bending stresses, (*b*) the bending stress at a point 3 ft from the free end and 1 in. from the top.

5.20. For a beam having the cross section and load diagram shown in Fig. P5.20, compute (*a*) the maximum tensile and compressive bending stresses, (*b*) the bending stress at a point 3 ft from the right end and 2 in. from the bottom.

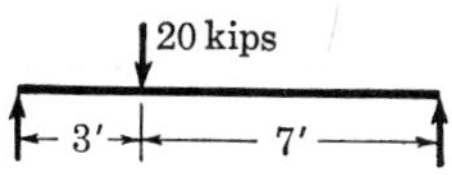

FIG. P5.20*a*

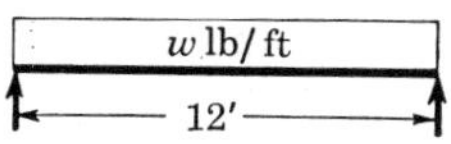

FIG. P5.23*a*

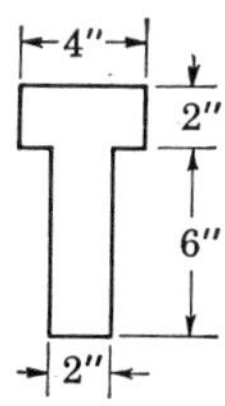

FIG. P5.20*b*, P5.23*b*

★5.21. For a beam having the cross section and load diagram shown in Fig. P5.21, compute the maximum tensile and compressive bending stresses.

5.22. For a beam having the cross section and load diagram shown in Fig. P5.22, compute (*a*) the maximum tensile and compressive bending stresses, (*b*) the bending stress at a point 2 ft from the left end and 2 in. from the bottom.

5.23. For a beam having the cross section and load diagram shown in Fig. P5.23, compute the allowable uniform load as limited by each of the following allowable stresses: $s_t = 20{,}000$ psi; $s_c = 16{,}000$ psi.

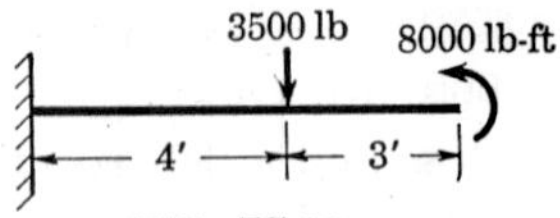

FIG. P5.21*a*

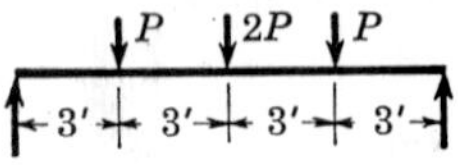

FIG. P5.24*a*

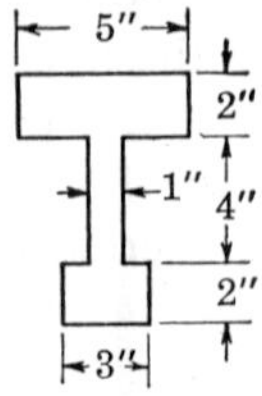

FIG. P5.21*b*, P5.24*b*

5.24. For a beam having the cross section and load diagram shown in Fig. P5.24, compute the maximum value of P as limited by each of the following allowable stresses: $s_t = 4000$ psi; $s_c = 12{,}000$ psi.

★5.25. For a beam having the cross section and load diagram shown in Fig. P5.25, compute the maximum value of P as limited by each of the following allowable stresses: $s_t = 20{,}000$ psi; $s_c = 17{,}000$ psi.

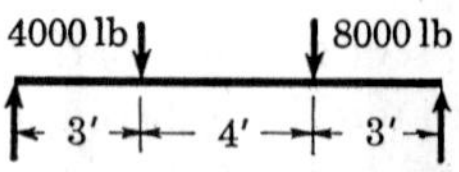

FIG. P5.22*a*

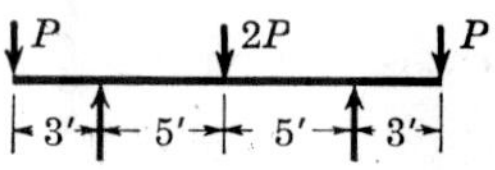

FIG. P5.25*a*

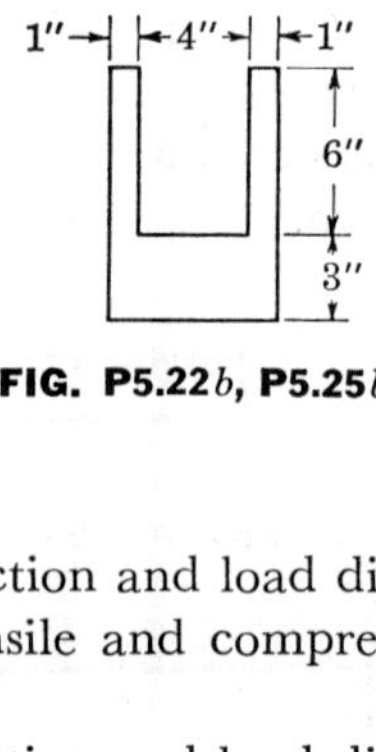

FIG. P5.22*b*, P5.25*b*

5.26. For a beam having the cross section and load diagram shown in Fig. P5.26, compute the maximum tensile and compressive bending stresses.

★5.27. For a beam having the cross section and load diagram shown in Fig. P5.27, compute the allowable uniform load as limited by each of the following allowable stresses: $s_t = 4000$ psi; $s_c = 10{,}000$ psi.

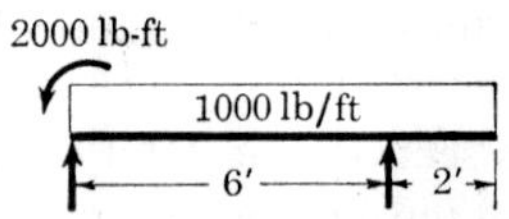

FIG. P5.26*a*

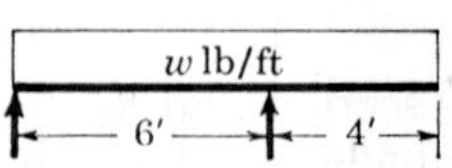

FIG. P5.27*a*

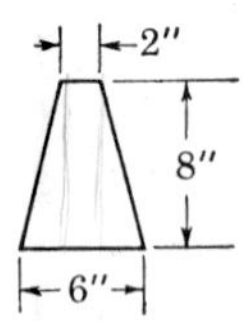

FIG. P5.26*b*, P5.27*b*

5.28. A beam is made of a 12-in. 25-lb channel and a 16WF40 section arranged as shown in Fig. P5.28. If L, on the load diagram shown, is 16 ft, compute the allowable uniform load as limited by each of the following allowable stresses: $s_t = 20{,}000$ psi; $s_c = 18{,}000$ psi.

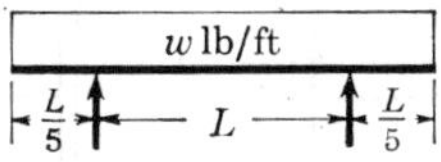

FIG. P5.28a, P5.29a

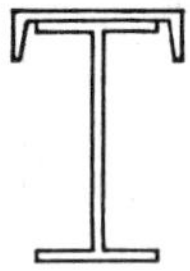

FIG. P5.28b, P5.29b

★5.29. A beam is made of an 8-in. 11.5-lb channel and a 12I50 section arranged as shown in Fig. P5.29. If the uniform load shown is 8000 lb/ft, compute the maximum allowable length between supports L as limited by each of the following allowable stresses: $s_t = 20{,}000$ psi; $s_c = 16{,}000$ psi.

5.5 SHEARING STRESSES IN TWO MUTUALLY PERPENDICULAR DIRECTIONS

In order to determine shearing stresses in beams, it is necessary to know the relationship between the shearing stresses which exist in any two mutually perpendicular directions. This relationship will be determined using a differential parallelepiped in a generalized biaxial stress condition. The dimensions of the parallelepiped are shown in Fig. 5.7*a*; the general biaxial stress condition is shown in Fig. 5.7*b*. The *forces* acting on the differential body, expressed as the product of stress and area, are shown in Fig. 5.7*c*. If the terms containing the squares of differentials are discarded (as being extremely small in comparison with the other terms), the free-body diagram will be as shown in Fig. 5.7*d*. Setting the sum of the moments of the forces about the lower left-hand corner of the body equal to zero,

$$+(s_{xy}\,dy\,t)\,dx - (s_{yx}\,dx\,t)\,dy = 0 \qquad \text{or}$$

$$s_{xy} = s_{yx} \tag{5.2}$$

Thus it has been shown that at any point the *shearing stresses in the two mutually perpendicular directions must be equal.* It should also be noted from Fig. 5.7*d* that the forces due to shearing stresses form equal and opposing couples. Thus if the direction of the shearing

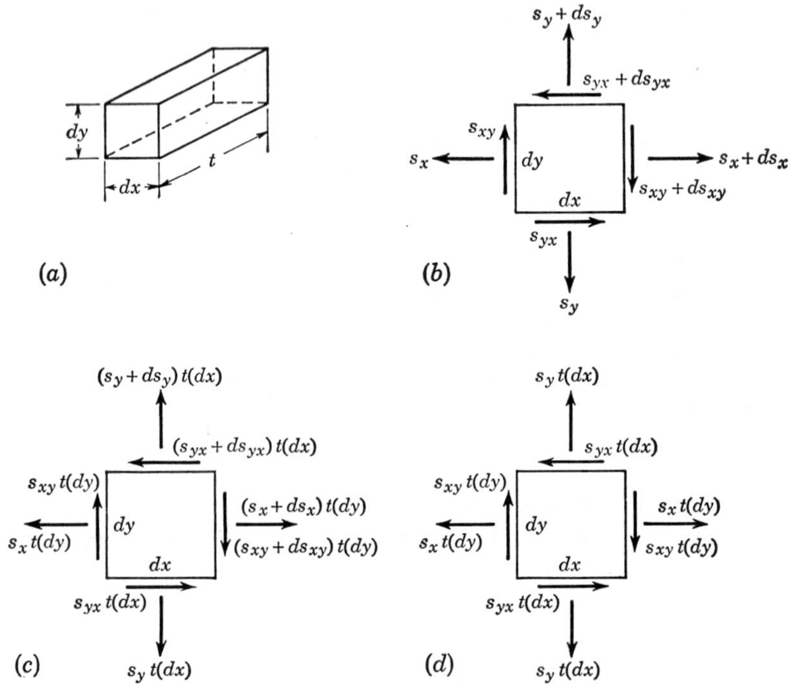

FIG. 5.7

stress on one face of an element is known, the direction of the shearing stresses on the other three faces can be readily determined.

5.6 SHEARING STRESSES IN BEAMS

As mentioned in Chap. 1, the equation $s_s = P/A$ gives the *average* shearing stress on a cross section due to transverse loading. However, as will be seen, the shearing-stress distribution over any beam cross section is far from uniform, so that the maximum and minimum shearing stresses vary considerably from the average. The shearing stress at any point on a beam cross section can be determined by the *formula for horizontal shearing stress* which is derived in this article. Of course, since vertical and horizontal shearing stresses

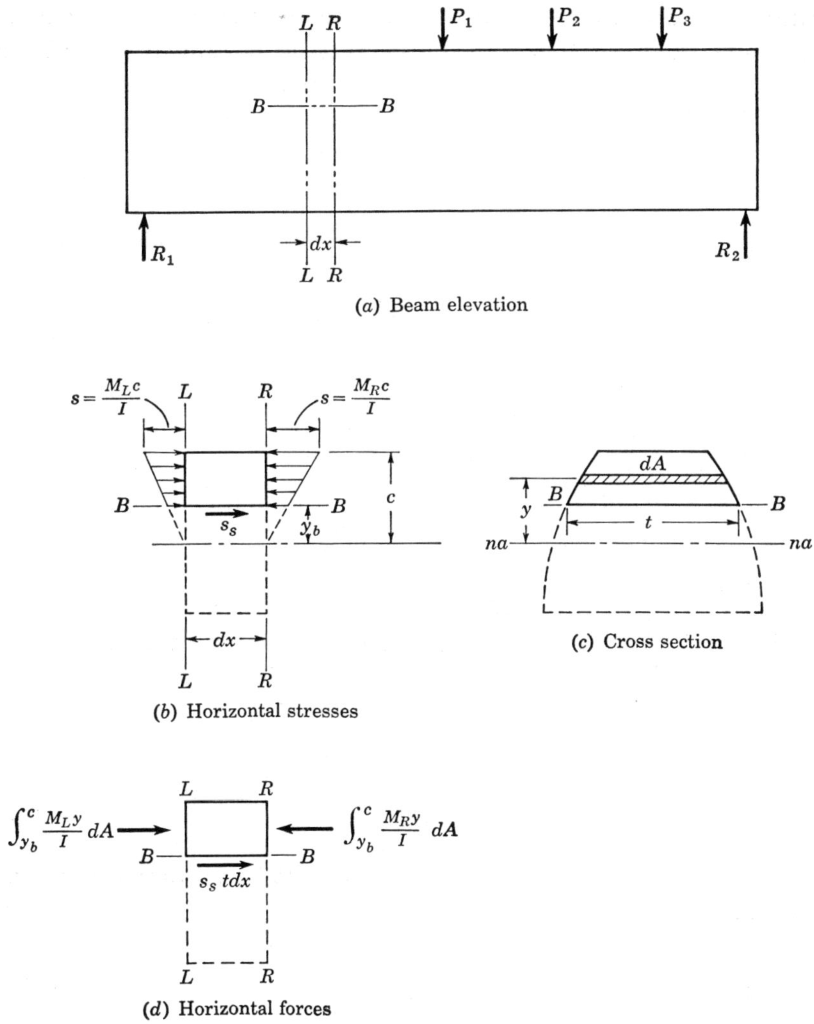

FIG. 5.8

act in mutually perpendicular directions, the equation also gives the vertical shearing stress at the point.

The equation for horizontal shearing stress will be derived from the stressed condition of a portion of the beam shown in Fig. 5.8*a*. Figure 5.8*b* shows the *horizontal* stresses acting on the differential piece of the beam defined by sections *LL*, *RR*, and *BB*.

Unless the end forces (resulting from bending stresses) are equal, horizontal shearing stress must exist at the bottom of the piece. Since bending stress acting on any differential area, such as dA shown in Fig. 5.8c, causes a differential force $s\,dA$, the resultant force on the left face can be expressed as $\int_{y_b}^{c} \frac{M_L y}{I}\,dA$. Similarly, the resultant force on the right face is $\int_{y_b}^{c} \frac{M_R y}{I}\,dA$. The force due to the horizontal shearing stress is $s_s\,dx\,t$. The three horizontal forces acting on the differential piece are shown in Fig. 5.8d; setting their sum equal to zero,

$$s_s t\,dx + \int_{y_b}^{c} \frac{M_L y}{I}\,dA - \int_{y_b}^{c} \frac{M_R y}{I}\,dA = 0$$

Since M_R, M_L, and I are constants, the equation can be written as

$$s_s t\,dx = \frac{M_R - M_L}{I} \int_{y_b}^{c} y\,dA$$

The integral term should be recognized as the first moment of the area between the limits shown and will be designated as Q. Therefore

$$s_s = \frac{M_R - M_L}{dx}\,\frac{Q}{It}$$

The term $(M_R - M_L)/dx$ represents the change in moment in a differential length, or dM/dx. Since $dM/dx = V$, the equation is written as

$$s_s = \frac{VQ}{It} \tag{5.3}$$

where s_s is the horizontal (*or vertical*) shearing stress at a point; V is the shear at the section; Q is the first moment, *with respect to the neutral axis*, of the area between the top (or bottom) of the section and the point where the stress is being determined; I is the centroidal moment of inertia of the *entire* cross-sectional area; and t is the thickness of the beam *at the particular point where the stress is being determined.* The usual units of t, Q, and I are inches, inches^3, and inches^4; the usual units of s_s and V are pounds per square inch and pounds; the equation is dimensionally consistent with these units.

An examination of Eq. (5.3) will reveal that the shearing stress will be a maximum at the section of a beam where the shear is maximum. Since I is constant, the shearing stress at any particular section will vary directly along the depth with the ratio Q/t. Therefore, since Q is a maximum at the neutral axis, a beam with constant thickness will have maximum shearing stress at the neutral axis.

Example 5.5. Given a beam having the cross section shown in Fig. 5.9*a*. At a section where the shear is 1000 lb, (*a*) compute the shearing stresses at 1-in. intervals along the depth; (*b*) plot the shearing stress distribution.

$$I = \frac{6(12)^3}{12} - \frac{4(8)^3}{12} = 864 - 171 = 693 \text{ in.}^4$$

For the stress at 1 in. from the top, Q will be of the area shown in Fig. 5.9*b*.

$$s_s = \frac{VQ}{It} = \frac{1000[6(1)(5.5)]}{693(6)} = 8 \text{ psi} \qquad \textit{Ans}$$

At 2 in. from the top *with a thickness of* 6 *in.*

$$s_s = \frac{VQ}{It} = \frac{1000[6(2)(5)]}{693(6)} = 14 \text{ psi} \qquad \textit{Ans}$$

At 2 in. from the top *with a thickness of* 2 *in.*

$$s_s = \frac{VQ}{It} = \frac{1000[6(2)(5)]}{693(2)} = 43 \text{ psi} \qquad \textit{Ans}$$

At 3 in. from the top, Q will be of the area shown in Fig. 5.9*c*.

$$s_s = \frac{VQ}{It} = \frac{1000[6(2)(5) + 2(1)(3.5)]}{693(2)} = 48 \text{ psi} \qquad \textit{Ans}$$

At 4 in. from the top

$$s_s = \frac{VQ}{It} = \frac{1000[6(2)(5) + 2(2)(3)]}{693(2)} = 52 \text{ psi} \qquad \textit{Ans}$$

At 5 in. from the top

$$s_s = \frac{VQ}{It} = \frac{1000[6(2)(5) + 2(3)(2.5)]}{693(2)} = 54 \text{ psi} \qquad \textit{Ans}$$

At the neutral axis

$$s_s = \frac{VQ}{It} = \frac{1000[6(2)(5) + 2(4)(2)]}{693(2)} = 55 \text{ psi} \qquad \textit{Ans}$$

Since the cross section is symmetrical about the neutral axis, the shear-stress distribution must be symmetrical about the neutral axis.

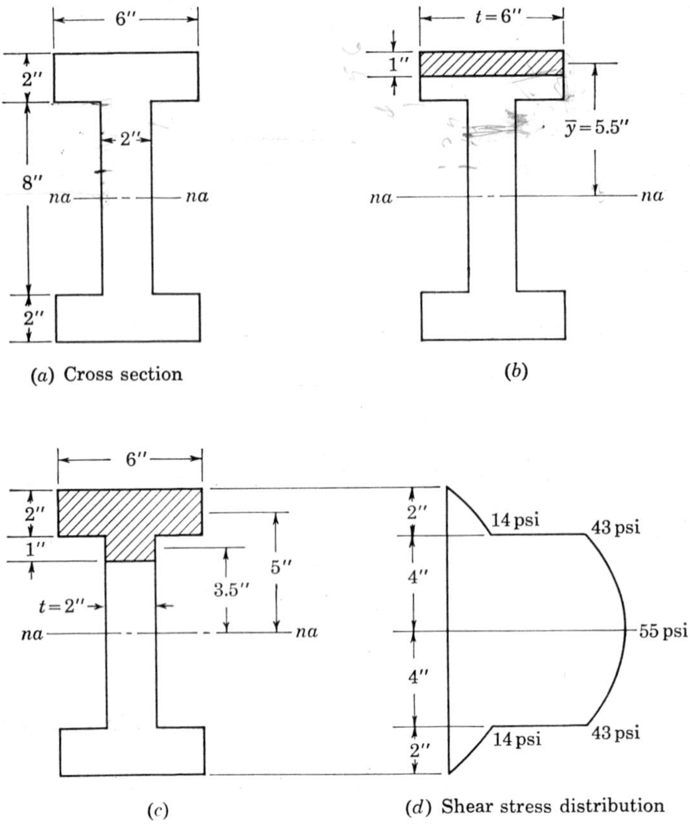

FIG. 5.9

It should be noted that, if stresses were computed at 1-in. intervals from the bottom (using areas on the cross section from the bottom), the corresponding computations would be identical with those above.

The shear-stress distribution is plotted as shown in Fig. 5.9*d* (*Ans*). It should be noted that the shearing stress is zero at the top and bottom of the section (since Q is zero at these points). At 2 in. from the top (and from the bottom), the stress changes abruptly owing to the abrupt change of thickness. It may also be noted that the shearing stress on the inside faces of the flanges must be zero, since there is no way for horizontal shearing stress to be resisted on these surfaces. Thus the stress distribution shown is somewhat idealized.

Example 5.6. Determine the maximum shearing stress in a beam having the load diagram and cross section shown in Fig. 5.10.

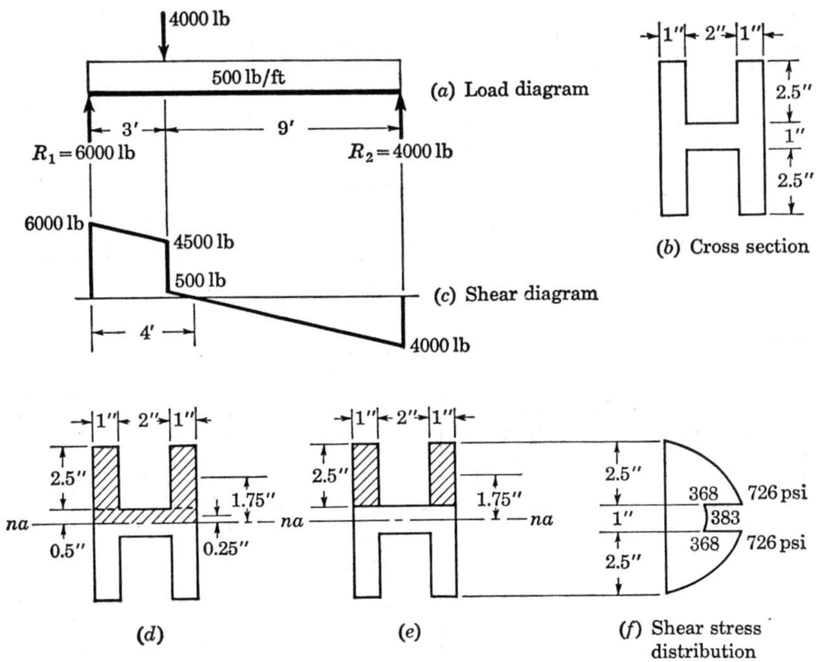

FIG. 5.10

From the shear diagram shown in Fig. 5.10*c*, it can be seen that the maximum shear is 6000 lb.

The moment of inertia is computed considering the cross section to be composed of two vertical and one horizontal rectangles.

$$I = 2\,\frac{(1)(6)^3}{12} + \frac{2(1)^3}{12} = 36.0 + 0.2 = 36.2 \text{ in.}^4$$

The maximum shearing stress will occur on the section of maximum shear at the point where Q/t is a maximum. Q/t must be a maximum *either* at the neutral axis (where Q is a maximum) *or* at the point nearest to the neutral axis where t is a minimum.

To compute the stress at the neutral axis, Q is of the area shown in Fig. 5.10*d*.

$$s_s = \frac{VQ}{It} = \frac{6000[2(1)(2.5)(1.75) + (4)(0.5)(0.25)]}{36.2(4)} = 383 \text{ psi}$$

To compute the stress at the point where the vertical and horizontal rectangles intersect, Q is of the area shown in Fig. 5.10*e*.

$$s_s = \frac{VQ}{It} = \frac{6000[2(1)(2.5)(1.75)]}{36.2(1 + 1)} = 726 \text{ psi} \qquad \textit{Ans}$$

The shearing-stress distribution on the section of maximum shear is sketched as shown in Fig. 5.10*f*.

Example 5.7. An overhanging beam has the cross section and load diagram shown in Fig. 5.11. Determine the maximum uniform load which will not cause any of the following stresses to be exceeded: tension = 6000 psi; compression = 10,000 psi; shear = 1000 psi.

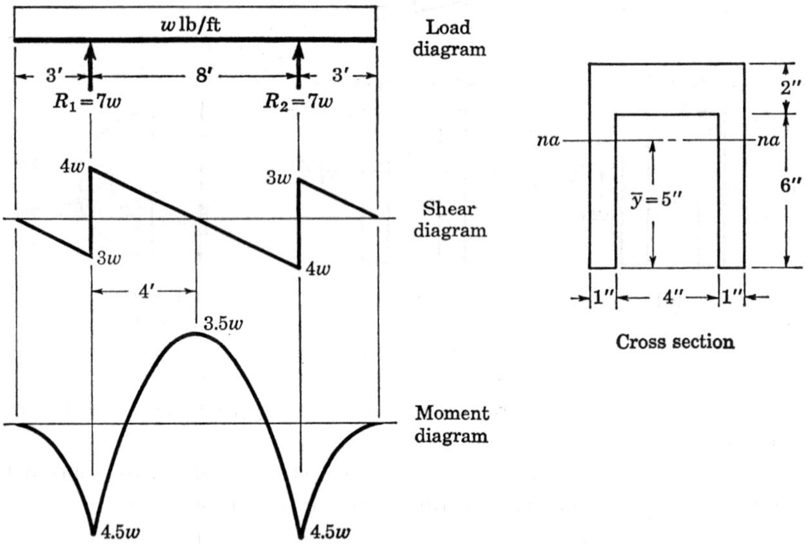

FIG. 5.11

The shear and moment diagrams are sketched as shown. The moment of iner ia is computed as shown in Table E5.7.

TABLE E5.7

	a	y	ay	I_o	d	ad^2
2(1 by 6)	12	3	36	36	2	48
2 by 6	12	7	84	4	2	48
Total	24		120	40		96

$$\bar{y} = \frac{120}{24} = 5.0 \text{ in.} \qquad I = 40 + 96 = 136 \text{ in.}^4$$

The maximum tensile stress will occur *either* at the bottom of the beam at midspan (where $c = 5$ in. and $M = 3.5w$ lb-ft) *or* at the top of the beam at each support (where $c = 3$ in. and $M = 4.5w$ lb-ft). Determining the allowable load as limited by tension at each place,

$$s_t = \frac{Mc}{I} \qquad 6000 = \frac{[3.5w(12)]5}{136}$$

$$w = 3890 \text{ lb/ft}$$

$$s_t = \frac{Mc}{I} \qquad 6000 = \frac{[4.5w(12)]3}{136}$$

$$w = 5040 \text{ lb/ft}$$

The maximum compressive stress must occur at the bottom of the section at either reaction, where both M and c are maximum.

$$s_c = \frac{Mc}{I} \qquad 10{,}000 = \frac{[4.5w(12)]5}{136}$$

$$w = 5040 \text{ lb/ft}$$

The maximum shearing stress will occur at the neutral axis at the sections where the shear is $4w$ lb.

$$s_s = \frac{VQ}{It} \qquad 1000 = \frac{4w[2(1)(5)(2.5)]}{136(1 + 1)}$$

$$w = 2720 \text{ lb/ft} \qquad \textit{Ans}$$

For simplicity, the area between the bottom of the section and the neutral axis was used for Q; the area between the top of the section and the neutral axis would yield the same value.

PROBLEMS

***5.30.** A beam of rectangular cross section is 8 in. deep and 4 in. thick. At a section where the shear is 5000 lb, (*a*) compute the shearing stress at 1-in. intervals along the depth, (*b*) plot the shearing-stress distribution.

5.31. A beam of rectangular cross section is 10 in. deep and 5 in. thick. At a section where the shear is 4000 lb, (*a*) compute the shearing stress at 1-in. intervals along the depth, (*b*) plot the shearing-stress distribution.

5.32. Four 6- by 2-in. planks (actual size) are fastened together to make a (hollow) box beam 10 in. deep and 6 in. wide. At a section where the shear is 5000 lb, (*a*) compute the shearing stress at 1-in. intervals along the depth, (*b*) plot the shearing-stress distribution.

***5.33.** A 10-ft beam of rectangular cross section is simply supported and is to support a uniform load of 200 lb/ft over its entire length. If the depth is to be twice the thickness and the allowable shearing stress is 180 psi, compute the minimum cross-sectional dimensions.

5.34. A 10-ft simply supported beam is loaded with a uniform load of 200 lb/ft over its entire length and a concentrated load of 1000 lb at 3 ft from the left reaction. The beam is rectangular in cross section, 4 in. deep, and 2 in. thick. Compute (*a*) the maximum shearing stress, (*b*) the shearing stress at a point 2 ft from the left reaction and 1 in. from the top.

5.35. A 12-ft simply supported beam supports two concentrated loads: 2000 lb at 3 ft from the left end and 1000 lb at 3 ft from the right end. The beam is rectangular in cross section, 8 in. deep, and 6 in. thick. Compute (*a*) the maximum shearing stress, (*b*) the shearing stress at a point 4 ft from the left end and 2 in. from the bottom.

***5.36.** A 12-ft simply supported beam having the cross section shown in Fig. P5.36 is to be loaded with three concentrated loads: P at midspan, $2P$ at 3 ft from the left end, and $4P$ at 3 ft from the right end. If the allowable shearing stress is 3000 psi, compute the allowable concentrated loads.

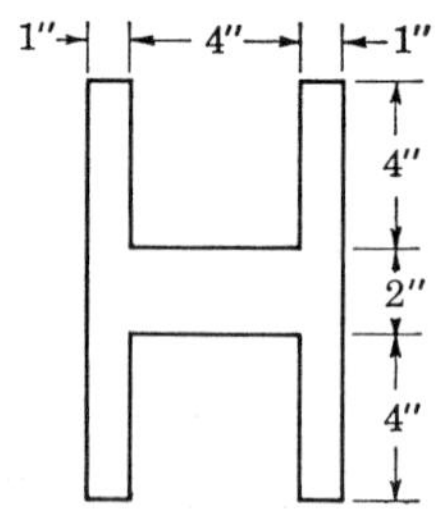

FIG. P5.36

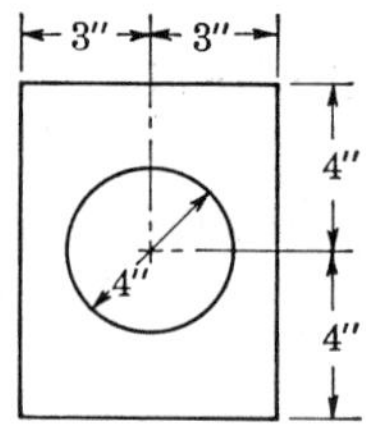

FIG. P5.37

5.37. A simply supported beam having the cross section shown in Fig. P5.37 is to support a uniform load of 600 lb/ft over its entire length. If the allowable shearing stress is 140 psi, compute the maximum allowable length of beam.

5.38. For a beam having the cross section and load diagram shown in Fig. P5.38, compute (*a*) the maximum shearing stress, (*b*) the shearing stress at a point 2 ft from the left end and 2 in. from the bottom.

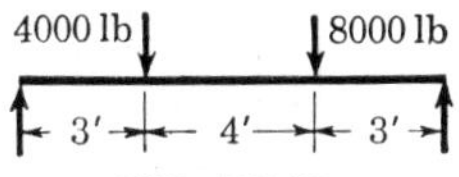

FIG. P5.38*a*

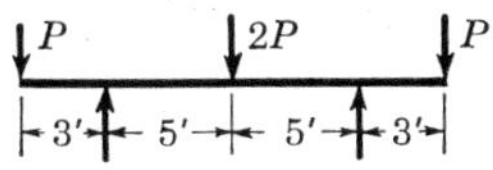

FIG. P5.39*a*

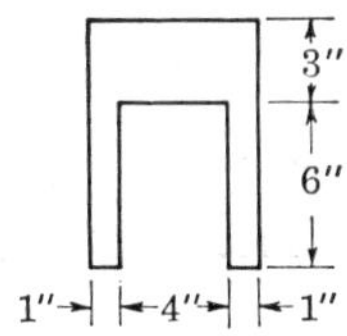

FIG. P5.38*b*, P5.39*b*

5.39. For a beam having the cross section and load diagram shown in Fig. P5.39, compute the maximum value of P as limited by an allowable shearing stress of 15,000 psi.

★**5.40.** A beam of rectangular cross section, 12 in. deep and 4 in. thick, is to be loaded with a uniform load of 500 lb/ft over its entire length. Compute the maximum allowable length of beam as limited by each of the following allowable stresses: shear = 140 psi; bending = 1600 psi.

5.41. An 8-ft cantilever beam is to support a uniform load of 500 lb/ft over its entire length. The cross section of the beam is to be rectangular with the depth three times the thickness. Compute the minimum cross-sectional dimensions as limited by each of the following allowable stresses: shear = 140 psi; bending = 1600 psi.

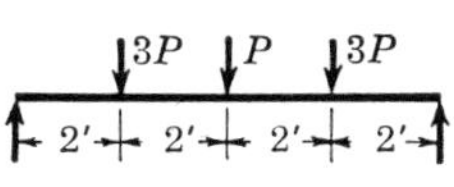

FIG. P5.42*a*

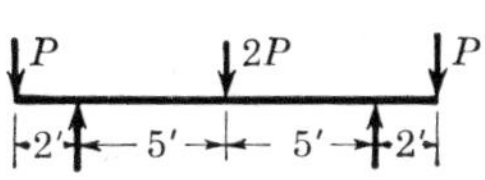

FIG. P5.43*a*

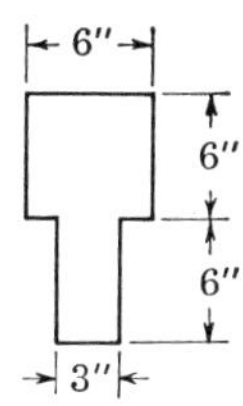

FIG. P5.42*b*, P5.43*b*

★**5.42–5.43.** For the beam having the cross section and load diagram shown in Figs. P5.42 and P5.43, compute the maximum value of P as limited by each of the following allowable stresses: tension = 20,000 psi; compression = 18,000 psi; shear = 15,000 psi.

5.44–5.45. For the beam having the cross section and load diagram shown in Figs. P5.44 and P5.45, compute the maximum value of P as limited by each of the following allowable stresses: tension = 4000 psi; compression = 10,000 psi; shear = 3000 psi.

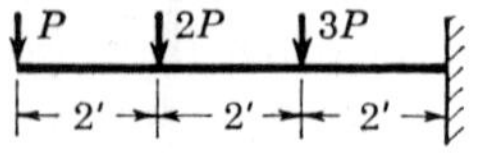

FIG. P5.44*a*

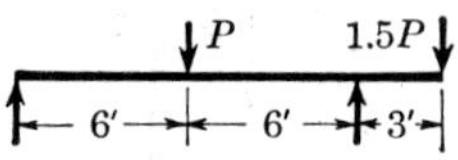

FIG. P5.45*a*

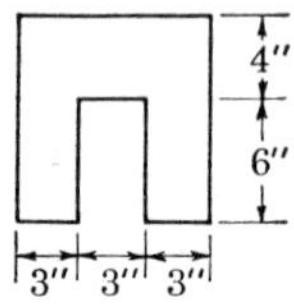

FIG. P5.44*b*, **P5.45***b*

★5.46–5.47. For the beam having the cross section and load diagram shown in Figs. P5.46 and P5.47, compute the maximum uniform load as limited by each of the following allowable stresses: bending = 2000 psi; shear = 200 psi.

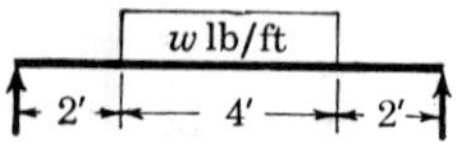

FIG. P5.46*a*

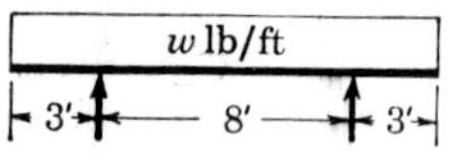

FIG. P5.47*a*

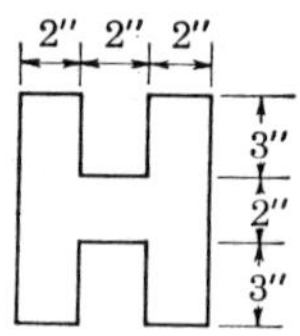

FIG. P5.46*b*, **P5.47***b*

5.48. For the beam having the cross section and load diagram shown in Fig. P5.48, compute (*a*) the value of P as limited by an allowable shearing stress of 200 psi, (*b*) the maximum length L as limited by an allowable bending stress of 2400 psi.

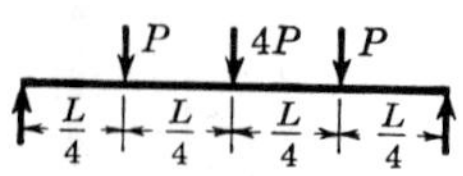

FIG. P5.48*a*

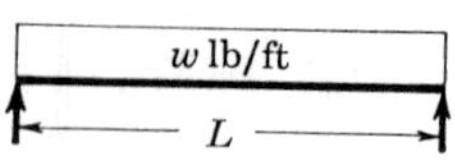

FIG. P5.49*a*

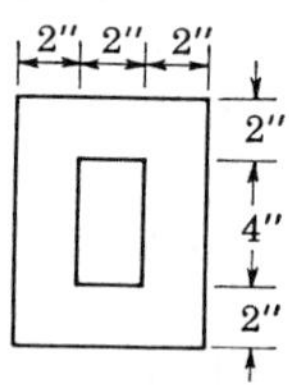

FIG. P5.48*b*, **P5.49***b*

5.49. For the beam having the cross section and load diagram shown in Fig. P5.49, compute the combination of length L and uniform load *w* which will cause the *maximum* bending stress to be 2400 psi when the *maximum* shearing stress is 200 psi.

★5.50. For the beam having the cross section and load diagram shown in Fig. P5.50, compute the maximum tensile, compressive, and shearing stresses.

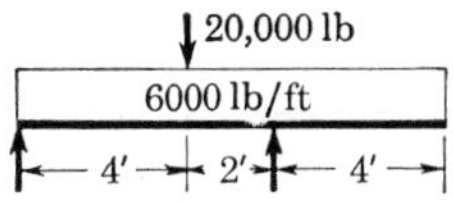

FIG. P5.50*a*

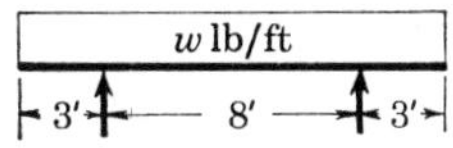

FIG. P5.51*a*

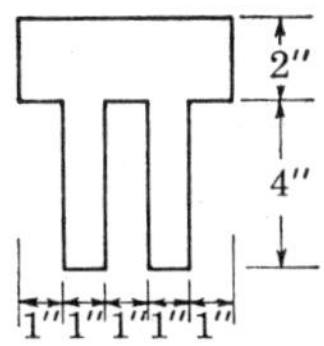

FIG. P5.50*b*, **P5.51***b*

5.51. For the beam having the cross section and load diagram shown in Fig. P5.51, compute the allowable uniform load as limited by each of the following allowable stresses: tension = 20,000 psi; compression = 16,000 psi; shear = 15,000 psi.

5.7 BEAMS OF TWO MATERIALS

In the design of structures, it occasionally becomes a practical necessity to use beams made up of two (or more) different materials securely fastened to each other. Inasmuch as the flexure formula is based on an assumption of homogeneity of materials, it follows that it cannot be applied directly to such beams. However, the bending stresses in each of the materials can be easily determined by the expedient of applying the flexure formula to an *equivalent section* of a single material. The method of using an equivalent section (also frequently referred to as a *transformed section*) is based on the ratio of the Young's moduli of the different materials; it will be explained with respect to the section shown in Fig. 5.12*a*. It will be recalled that, in the derivation of the flexure formula, the assumption was made that a cross-sectional plane before bending remains a plane after bending. This assumption can also be made for a beam of two materials provided, of course, that the two materials are securely fastened to each other Therefore, as is the case with homogeneous beams, the unit strain at any section of a beam of two materials must vary linearly along the depth. Thus, for a beam having the cross section shown in Fig. 5.12*a*, the unit-strain distribution would be as shown in Fig. 5.12*b*. The corresponding stress distribution, however, would be linear only within each material; as indicated in Fig. 5.12*c*, it would be dependent not only on the strain distribution but also on the relative Young's moduli of the two materials.

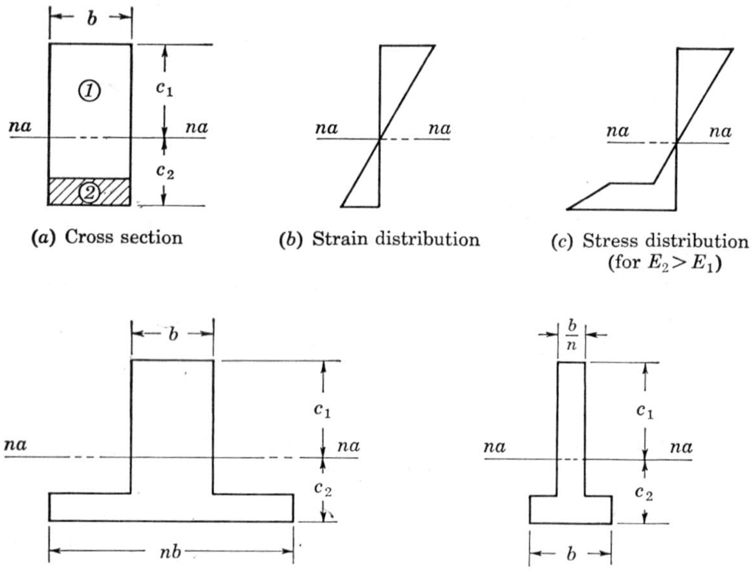

(*a*) Cross section (*b*) Strain distribution (*c*) Stress distribution (for $E_2 > E_1$)

(*d*) Equivalent section of material *1* (for $E_2 > E_1$) (*e*) Equivalent section of material *2* (for $E_2 > E_1$)

FIG. 5.12

To determine the equivalent section (of a single material) for the section shown in Fig. 5.12*a*, the bending forces on differential areas of each material will be compared. Thus for material 1

$$dF_1 = s_1\,dA = \epsilon E_1\,dA \tag{a}$$

and for material 2

$$dF_2 = s_2\,dA = \epsilon E_2\,dA \tag{b}$$

Designating the ratio of Young's moduli as n and letting $n = E_2/E_1$, the force dF_2 can be written in terms of E_1 as

$$dF_2 = \epsilon E_1 n\,dA \tag{b'}$$

A comparison of Eqs. (*a*) and (*b'*) reveals that dF_1 and dF_2 are identical except for the factor n. Thus if the area of material 2 is weighted n times, the section can be treated as one of a single material having a Young's modulus of E_1 (that is, an equivalent section of material 1). Similarly, if the area of material 1 is weighted $1/n$ times, the section can be treated as one of a single material having a Young's modulus of E_2 (that is, an equivalent section of material 2). The equivalent sections of materials 1 and 2 are shown in Fig. 5.12*d* and *e*, respectively; in each case it has been assumed that E_2 is greater than E_1 and that, therefore, $n > 1$.

Example 5.8. A standard 6- by 12-in. timber ($E = 1{,}500{,}000$ psi) section is reinforced with a ¼-in. steel ($E = 30{,}000{,}000$ psi) plate as shown in Fig. 5.13. Based on allowable bending stresses of 1800 psi for the timber and 20,000 psi for the steel, determine the allowable bending moment.

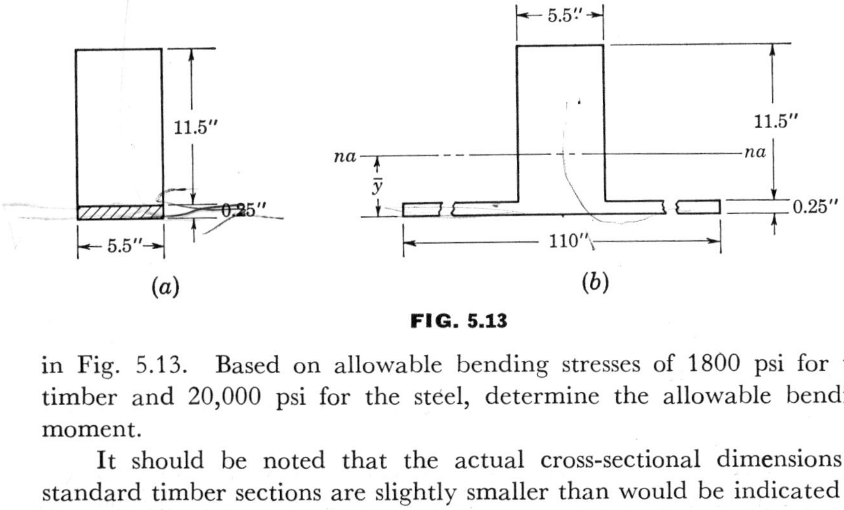

FIG. 5.13

It should be noted that the actual cross-sectional dimensions of standard timber sections are slightly smaller than would be indicated by their nominal size; the actual dimensions, as well as other useful information, can be obtained from tables such as Table A.1.

Using $n = 30{,}000{,}000/1{,}500{,}000 = 20$, the equivalent timber section is sketched as shown in Fig. 5.13*b*. The neutral axis and the moment of inertia of the equivalent timber section are determined in Table E5.8.

The allowable moment as limited by bending at the top of the section is

$$M = \frac{sI}{c} = \frac{1800(1360)}{7.53} = 325{,}000 \text{ lb-in.}$$

TABLE E5.8

	a	y	ay	I_o	d	ad^2
5.5 by 11.5	63.3	6	380	697	1.78	201
0.25 by 110	27.5	0.125	3	—	4.10	462
Total	90.8		383	697		663

$$\bar{y} = \frac{383}{90.8} = 4.22 \text{ in.} \qquad I = 697 + 663 = 1360 \text{ in.}^4$$

The allowable stress for the steel, when converted to equivalent timber, is 20,000/n or 1000 psi; thus the allowable moment as limited by the bending at the bottom of the section is

$$M = \frac{sI}{c} = \frac{1000(1360)}{4.22} = 322{,}000 \text{ lb-in.}$$

Therefore, the allowable bending moment is limited by the stress in the steel and is

$$\frac{322{,}000}{12} = 26{,}800 \text{ lb-ft} \qquad \textit{Ans}$$

PROBLEMS

NOTE: For all problems in this group, use the following values of Young's modulus: steel = 30,000,000 psi; timber = 1,500,000 psi; aluminum = 10,000,000 psi. Assume that all materials are securely fastened together.

5.52. Determine the allowable bending moment for the section of Example 5.8 if the steel plate is replaced by an aluminum plate of the same size.

5.53. A simply supported beam is comprised of two 2- by 10-in. standard timber sections and a 9.5- by ¼-in. steel plate arranged as shown in Fig. P5.53. Determine the maximum bending stress in the timber and in the steel due to a uniform load of 800 lb/ft over the entire 10-ft length of the beam.

5.54. A simply supported beam is comprised of two 2- by 12-in. standard timber sections and an 11.5- by ⅜-in. steel plate arranged as shown in Fig. P5.54. Determine the maximum bending stress in the steel and in the timber due to a uniform load of 1000 lb/ft over the entire 12-ft length of the beam.

***5.55.** A 12-ft simply supported beam is comprised of a standard 6- by 12-in. timber section and two 5.5- by ⅜-in. aluminum plates arranged as shown in Fig. P5.55. Based on allowable bending stresses of 2000 and 18,000 psi for the timber and the aluminum, respectively, determine the allowable uniform load which can be applied over the entire length of the beam.

5.56. A 16-ft simply supported beam is comprised of a standard 8- by 16-in. timber section and two 7.5- by ½-in. steel plates arranged as shown in Fig. P5.56. Based on allowable bending stresses of 1800 and 20,000 psi for the timber and the steel, respectively, determine the allowable uniform load which can be applied over the entire length of the beam.

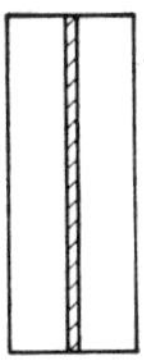

FIG. P5.53, P5.54

FIG. P5.55, P5.56

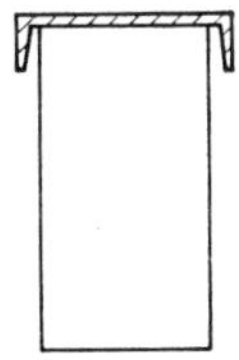

FIG. P5.58, P5.59

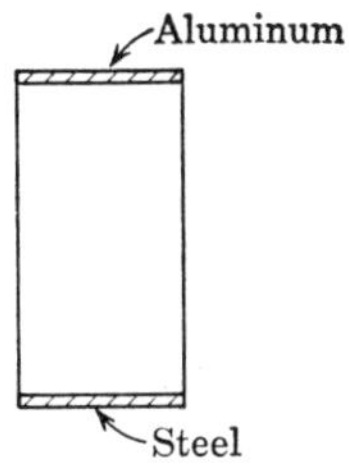

FIG. P5.60, P5.61

5.57. Solve Example 5.8 using an equivalent steel section (instead of an equivalent timber section).

★5.58. A 12-ft simply supported beam is comprised of a standard 6- by 12-in. timber section capped with a steel 8-in. 18.75-lb channel section as shown in Fig. P5.58. Based on allowable bending stresses of 20,000 and 1800 psi for the steel and timber, respectively, determine the allowable midspan concentrated load for the beam.

5.59. A 14-ft simply supported beam is comprised of a standard 8- by 14-in. timber section capped with a steel 9-in. 20.0-lb channel section as shown in Fig. P5.59. Based on allowable bending stresses of 20,000 and 1800 psi for the steel and timber, respectively, determine the allowable midspan concentrated load for the beam.

★5.60. A 12-ft simply supported beam is comprised of a standard 6- by 12-in. timber section reinforced as shown in Fig. P5.60 with a 5.5- by ⅜-in. aluminum plate at the top and a 5.5- by ⅜-in. steel plate at the bottom. Determine the maximum bending stress in each material due to a uniform load of 2 kips/ft over the entire length of the beam.

5.61. A 10-ft simply supported beam is comprised of a standard 6- by 10-in. timber section reinforced as shown in Fig. P5.61 with a 5.5- by ½-in. steel plate at the bottom and a 5.5- by ½-in. aluminum plate at the top. Determine the maximum bending stress in each material due to a uniform load of 3 kips/ft over the entire length of the beam.

6

DESIGN OF BEAMS

6.1 INTRODUCTION

Beam design consists of determining the minimum practical cross section required to prevent a beam from being excessively stressed. Fundamentally the procedure is one of using allowable stresses to determine suitable cross-sectional dimensions for a beam under a given load condition. Rather than basing the design on two or more allowable stresses simultaneously, it is usually much simpler to select a cross section on the basis of one allowable stress and to check against the others. Since the load-carrying capacity of most beams is limited by bending, the cross section is usually selected on a bending-stress basis followed by a check of other stresses. However, in the case of very short beams or beams with heavy loads near the supports, it is frequently advantageous to select the cross section on the basis of shearing stress. The cross sections of beams made of materials which are extremely weak in shear, such as timber, may also be selected on a shearing-stress basis.

Since a member must support its own weight in addition to the loads applied to it, nearly all design is a trial-and-error procedure. A reasonably close initial estimate of the weight of a member will, of course, reduce the number of trials which must be made. The ability to make close initial weight estimates comes with experience. An experienced designer will generally need only a single trial, while most beginners will not require more than two.

6.2 ROLLED SECTIONS

Because of the availability of pertinent properties in tables, the selection of rolled sections to act as beams is a relatively simple

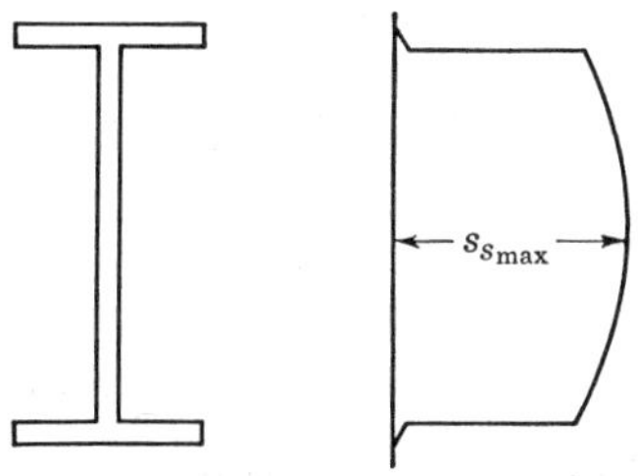

FIG. 6.1 Shear-stress distribution on thin web section.

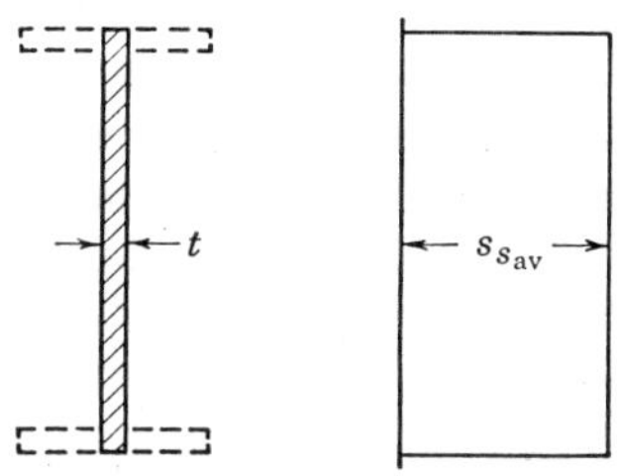

FIG. 6.2 Average shear stress on web area.

process. These tables list not only the significant dimensions but also such items as cross-sectional area, moment of inertia, and section modulus.

In the design of rolled sections, the standard practice is to use an allowable *average shearing stress on the web area* rather than an allowable (maximum) shearing stress for the total section. Figure 6.1 shows the shearing-stress distribution on a typical wide-flange or I section; as can be seen, the distribution is quite uniform except at the flanges. This uniformity, together with satisfactory results from tests, justifies the use of the allowable average shearing stress for design purposes. In equation form, the average shearing stress is expressed as

$$s_{s,av} = \frac{V}{A_{web}} \tag{6.1}$$

where V is the shear at the section being investigated, and A_{web} is the product of the web thickness and the overall depth of the section (as shown in Fig. 6.2). It should be emphasized that the allowable average shearing stress for a particular material will be somewhat less than the allowable (maximum) shearing stress; for example, for structural steel the values are 13,000 and 15,000 psi, respectively.

Example 6.1. Select the lightest wide-flange section to act as a 20.0-ft-long simply supported beam with a 20-kip midspan load. Allowable stresses are as follows: bending = 20,000 psi; average shearing = 13,000 psi.

Computing the maximum moment due to the applied load,

$$M = \frac{PL}{4} = \frac{20{,}000(20)}{4} = 100{,}000 \text{ lb-ft}$$

Assuming that the beam will weigh 60 lb/ft, the estimated maximum moment due to beam weight will be

$$M = \frac{wL^2}{8} = \frac{60(400)}{8} = 3000 \text{ lb-ft}$$

Therefore, for the first trial, the required section modulus will be based on a maximum moment of 100,000 plus 3000 or 103,000 lb-ft.

$$\frac{I}{c} = \frac{103{,}000(12)}{20{,}000} = 61.8 \text{ in.}^3$$

Using Table A.4, the lightest wide-flange section having enough I/c is a 16WF40 with an I/c of 64.4 in.^{3} Since this beam weighs less per foot than was estimated, it is satisfactory for bending. The difference in moment due to the actual and assumed beam weights is approximately 1 per cent and would not lower the required section modulus enough to allow the use of a lighter section.

Checking for shearing stress,

$$V_{max} = \frac{20{,}000}{2} + \frac{40(20)}{2} = 10{,}400 \text{ lb}$$

$$s_{s,av} = \frac{V}{A_{web}} = \frac{10{,}400}{(16.00)(0.307)} = 2120 \text{ psi} \qquad \textit{Check}$$

Use 16WF40 (*Ans*).

PROBLEMS

NOTE: For all problems in this group, use the following allowable stresses: bending = 20,000 psi; average shearing = 13,000 psi.

★6.1. Select the lightest wide-flange section to act as a 24-ft simply supported beam with a uniform load of 4000 lb/ft over its entire length.

6.2. Select the lightest wide-flange section to act as a 16-ft simply supported beam with a uniform load of 3500 lb/ft over its entire length.

6.3. Select the lightest wide-flange section to act as a 15-ft simply supported beam with a uniform load of 2000 lb/ft over its entire length.

★6.4. An 18-ft simply supported beam is to be loaded with 40-kip loads at midspan and at each quarter point. Select the lightest wide-flange section for this beam.

6.5. A 14-ft simply supported beam is to be loaded with 30-kip loads at midspan and at each quarter point. Select the lightest wide-flange section for this beam.

6.6. A 12-ft simply supported beam is to be loaded with 10-kip loads at midspan and at each quarter point. Select the lightest wide-flange section for this beam.

6.3 RECTANGULAR BEAMS

The use of beams of rectangular cross section is frequently economically justified even though they do not provide for the most efficient use of material (to withstand bending). Most beams of timber, for example, are made rectangular in cross section because of the resulting savings in manufacturing, handling, and synthesis. Many small metallic bending members, such as are used for machine parts, are made with rectangular cross sections for the sake of simplicity of fabrication and connection.

For rectangular beams the flexure formula can be modified into an expression which is sometimes more convenient to use. Since, in the equation $s_m = M/(I/c)$, I for a rectangle is $bh^3/12$ and c is $h/2$, $I/c = bh^2/6$. Therefore

$$s_m = \frac{M}{bh^2/6} \tag{6.2}$$

where s_m is the maximum bending stress at a section where the bending moment is M, b is the thickness and h is the depth of the section.

The equation for *maximum* shearing stress at a section of a rectangular beam can also be modified into a more convenient form. Using the equation

$$s_s = \frac{VQ}{It} \qquad I = \frac{bh^3}{12} \qquad \text{and} \qquad Q = \frac{bh}{2}\frac{h}{4} = \frac{bh^2}{8}$$

Therefore

$$s_{s,\max} = \frac{V(bh^2/8)}{(bh^3/12)b} = \frac{12V}{8(bh)} \qquad \text{or} \qquad s_{s,\max} = \frac{3}{2}\frac{V}{A} \tag{6.3}$$

where $s_{s,\max}$ is the maximum shearing stress at a section where the shear is V and A is the area of the section.

In designing with timber, it should be noted that standard nominal sizes of structural members begin with the so-called "2 by 4" and increase in either dimension in 2-in. increments. Also, all timber members are slightly smaller in cross section than their nominal size. As a general rule for dressed timber, all cross-sectional dimensions up to and including 4 in. are actually ⅜ in. smaller than nominal; all cross-sectional dimensions 8 in. or larger are actually ½ in. smaller than nominal. For example, a nominal dressed "2 by 4" is actually 1⅝ by 3⅝ in. in cross section; a nominal dressed "4 by 8" is actually 3⅝ by 7½ in. in cross section. Design of timber beams is facilitated by tables, such as Table A.1, which furnish such useful information as actual dimensions, weight per foot, and section modulus.

Example 6.2. Select the lightest standard southern pine member to act as a 16-ft simply supported beam having a uniform load of 1200 lb/ft over its entire span. Allowable stresses are as follows: bending = 1750 psi; shear = 120 psi.

Since shear so often governs the required dimensions of timber beams, the section will be selected on a shearing-stress basis and checked for bending stress.

The maximum shear due to the applied load will be 1200(16)/2 or 9600 lb. Assuming that the beam will weigh 40 lb/ft, the maximum shear due to the beam weight will be 40(16)/2 = 320 lb. Therefore, for the first trial, the cross section will be selected for a shear of 9600 plus 320 or 9920 lb.

$$s_{s,\max} = \frac{3V}{2A} \qquad 120 = \frac{3(9920)}{2A} \qquad \text{or} \qquad A = 124 \text{ in.}^2$$

The lightest standard section having the required area is a "10 by 14" having an area of 128 in.^{2} and weighing 35.6 lb/ft.

Checking for bending stress,

$$M = \frac{wL^2}{8} = \frac{1236(256)}{8} = 39{,}600 \text{ lb-ft}$$

$$s_m = \frac{M}{I/c} = \frac{39{,}600(12)}{289} = 1642 \text{ psi} \qquad \textit{Check}$$

Use standard "10 by 14" section (*Ans*).

PROBLEMS

NOTE: For all problems in this group, use the following allowable stresses: bending = 1750 psi; shearing = 120 psi.

★**6.7.** A 20-ft simply supported beam is to be loaded with a uniform load of 800 lb/ft over its entire length. Select the lightest standard timber section *which has the width approximately equal to one-half the depth.*

6.8. An 8-ft simply supported beam is to be loaded with a uniform load of 600 lb/ft over its entire length. Select the lightest standard timber section *which has the width approximately equal to one-half the depth.*

6.9. A 4-ft simply supported beam is to be loaded with a uniform load of 500 lb/ft over its entire length. Select the lightest standard timber section *which has the width approximately equal to one-half the depth.*

★**6.10.** A 16-ft simply supported beam is to be loaded with 1600-lb concentrated loads applied 7 ft from each reaction. If the depth of the beam cannot exceed 12 in., select the lightest standard timber section for this beam.

6.11. A 16-ft simply supported beam is to be loaded with 6500-lb concentrated loads applied 5 ft from each reaction. If the depth of the beam cannot exceed 12 in., select the lightest standard timber section for this beam.

6.12. A 16-ft simply supported beam is to be loaded with 10,000-lb concentrated loads applied 3 ft from each reaction. If the depth of the beam cannot exceed 12 in., select the lightest standard timber section for this beam.

6.4 BUILT-UP SECTIONS

In spite of the mass-production economy and convenience of using standard sections, it is sometimes necessary, economical, or expedient to use a beam section which is especially made for a particular purpose. Such beams are "built-up" from standard sections such as wide flanges, angles, and flat plates.

Most beams are loaded in such a manner that the moment is a maximum at only one section and is much smaller over a large portion of the span. With beams of constant cross section, the section modulus required by the maximum moment determines the section of the entire span; however, with built-up beams the section modulus can frequently be reduced over a considerable part of the length, and significant savings of material and weight realized. Built-up sections are most practical for beams which are very long or very heavily loaded.

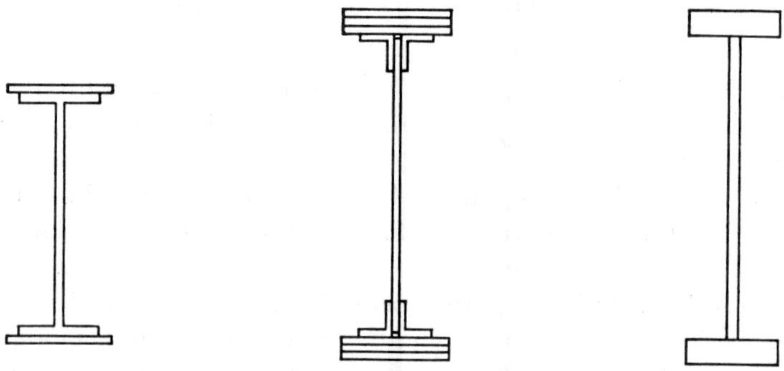

(*a*) (*b*) (*c*)

FIG. 6.3 Typical built-up sections.

Shown schematically in Fig. 6.3 are the cross sections of several types of built-up sections. Figure 6.3*a* is a wide-flange beam reinforced with a single flat plate at top and bottom. These plates, known as *cover plates*, can be "cut off" and used on only that part of the span where a relatively large section modulus is needed. Of course, in order for the member to function as a unit, the cover plates must be adequately fastened (either by riveting or welding) to the wide flange. Shown as Fig. 6.3*b* is a riveted-type plate girder such as is used for railroad trestles. For this section the web is a single plate, with each flange being composed of two angles and an appropriate number of cover plates. The cover plates are normally of varying lengths, thus providing approximately only the section modulus needed at different parts of the span. A welded plate girder, such as is shown in Fig. 6.3*c*, consists simply of a plate for the web and a single plate for each flange. Within practical limits, the *thickness* of the flange plates can be varied to provide only the section modulus needed for different portions of the span.

Example 6.3. Design a wide-flange or built-up section to act as a 30.0-ft simply supported beam with a uniform load of 10 kips/ft over its entire span. The overall depth of beam must not exceed 27 in.; cover plates are available, in ⅛-in. increments, in all thicknesses from ⅜ to 1 in. Allowable stresses are as follows: bending = 20,000 psi; average shear = 13,000 psi.

Assuming that the beam will weigh 300 lb/ft, the estimated maximum moment will be

$$M = \frac{wL^2}{8} = \frac{10{,}300(900)}{8} = 1{,}160{,}000 \text{ lb-ft}$$

The required section modulus will be

$$\frac{I}{c} = \frac{M}{s} = \frac{1{,}160{,}000(12)}{20{,}000} = 696 \text{ in.}^3$$

Table A.4 shows that there is no wide-flange section 27 in. deep or smaller which has the required section modulus. Therefore a 24-in. wide flange with cover plates will be used.

Using a 24WF160 (having $I = 5110$ and $I/c = 413.5$) and noting that the maximum I required is dependent on the overall depth of section, a thickness for the cover plates is first assumed. Thus if the cover plates were each 1 in. thick, the required moment of inertia for the composite section would be

$$I = \frac{I}{c}c = 696(12.36 + 1) = 9300 \text{ in.}^4$$

Therefore, the approximate I that the cover plates must furnish is 9300 less 5110 or 4190 in.^{4} Using a cover-plate width of 16 in. (slightly wider than the flanges of the WF section) and neglecting the moment of inertia of the plates about their own centroidal axes, the required plate thickness is computed. Referring to Fig. 6.4*a*,

$$I = 2(0 + Ad^2)$$

$$4190 = 2[0 + (16t)(12.86)^2]$$

$$t = \frac{4190}{32(165.5)} = 0.79 \text{ in. (say ⅞ in.)}$$

Checking the weight per foot at maximum section

$$\text{Wide flange} = 160$$

$$\text{Plate} = 2\,\frac{(16)(0.875)}{144}\,(490) = \underline{95}$$

$$255 \text{ lb/ft} \qquad \textit{Check}$$

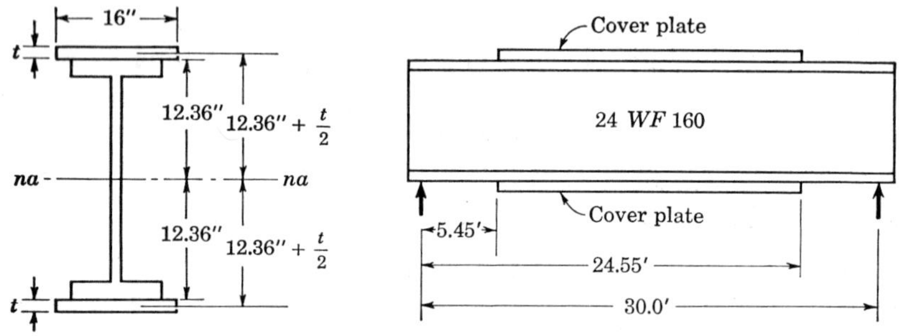

(*a*) Cross section (*b*) Elevation

FIG. 6.4

Computing the *actual* moment of inertia of the composite section as shown in Fig. 6.4*a*,

$$\text{WF section (from table)} = 5110$$

$$\text{Cover plates} = 2[(16)(0.875)(12.80)^2] = \underline{4590}$$

$$\text{Total } I \qquad 9700 \text{ in.}^4$$

Checking the maximum bending stress,

$$s = \frac{Mc}{I} = \frac{[(1{,}160{,}000)(12)](12.36 + 0.88)}{9700} = 19{,}000 \text{ psi} \qquad \textit{Check}$$

Checking the average shearing stress on the web at the critical section,

$$V_{\text{max}} = \frac{10{,}300(30)}{2} = 154{,}500 \text{ lb}$$

$$s_{s,\text{av}} = \frac{V}{A_{\text{web}}} = \frac{154{,}500}{(24.72)(0.656)} = 9540 \text{ psi} \qquad \textit{Check}$$

The cover plates are not needed over the entire span and can be eliminated or "cut off" where the section modulus of the WF is sufficient. The moment at the points of cutoff will be

$$M = s\,\frac{I}{c} = \frac{20{,}000(413.5)}{12} = 689{,}000 \text{ lb-ft}$$

Letting x be the distance from the left support to the point of cutoff,

$$689{,}000 = +154{,}500x - \frac{10{,}300x^2}{2}$$

Therefore

$$x = 5.45 \text{ or } 24.55 \text{ ft}$$

Use 24WF160 with 16- by ⅞-in. cover plates as shown in Fig. 6.4*b* (*Ans*).

PROBLEMS

NOTE: For all problems in this group, use the following allowable stresses: bending = 20,000 psi; average shearing = 13,000 psi.

★**6.13.** A 35-ft simply supported beam is to be loaded with 9000 lb/ft over its entire length. The beam is to consist of a 27WF177 section reinforced with a single 16-in.-wide cover plate on each flange. Plates are available in ⅛-in. increments in all thicknesses from ⅜ in. up. Determine the required thickness and minimum length of each cover plate.

6.14. A 30-ft simply supported beam which is to support a midspan load of 80 kips is to consist of an 18WF114 section reinforced with a single 14-in.-wide cover plate on each flange. Plates are available in ⅛-in. increments in all thicknesses from ⅜ in. up. Determine the required thickness and minimum length for each cover plate.

★**6.15.** Determine the required section for a welded plate girder to act as a 36-ft simply supported beam loaded with 160-kip concentrated loads applied 12 ft from each reaction. The web plate is to be 36 in. deep and is to have a minimum thickness of ⅜ in.; flange plates are to be 16 in. wide and have a constant thickness throughout the length of the beam. Plate material is available in ⅛-in. increments in all thicknesses from ⅜ in. up.

6.16. Determine the required section for a 60-ft welded plate girder to act as a simply supported beam with 70-kip loads applied at midspan and at each quarter point. The web plate is to be 48 in. deep, and is to have a minimum thickness of ⅜ in.; flange plates are to be 16 in. wide and have a constant thickness throughout the length of the beam. Plate material is available in ⅛-in. increments in all thicknesses from ⅜ in. up.

6.17. Design the necessary sections for a 35-ft welded plate girder to act as a simply supported beam with a uniform load of 6000 lb/ft over its entire span. The web plate is to be 30 in. deep and have a minimum thickness of ⅜ in. Flange plates are to be 12 in. wide; plate material is available in ⅛-in. increments in all thicknesses from ⅜ in. up. Where permissible, flange plates are to be reduced in thickness to even ½-in. sizes (specify the points at which these reductions are to be made).

6.18. Design the necessary sections for a 40-ft welded plate girder to act as a simply supported beam with a 50-kip midspan load. The web plate is to be 24 in. deep and have a minimum thickness of ⅜ in. Flange plates are to be 12 in. wide; plate material is available in ⅛-in. increments in all thicknesses from ⅜ in. up. Where permissible, flange plates are to be reduced in thickness to even ½-in. sizes (specify the points at which these reductions are to be made).

6.5 ADDITIONAL DESIGN FACTORS TO BE CONSIDERED

The previous articles of this chapter have dealt with only the most useful principles of beam design; a detailed coverage of all items which could be considered is, of course, beyond the scope and purpose of this text. However, it is desirable that the student be apprised of some of the factors which should be considered in a complete beam design. These factors arise from the fact that a beam may fail in ways other than in bending or shear. The design criteria against these additional modes of failure are generally governed by specifications for the particular beam material.† Such specifications are empirical to a large extent, being based on test results and past performances of actual structures as well as on theory.

As will be seen in Chap. 15, compressive stress in a member is accompanied by a tendency of the member to bend out of line, or to *buckle*. The greater the length of the member, the lower will be the stress at which it will buckle. Of importance in beam design is the fact that the compressive flange of a beam is susceptible to *lateral buckling*. This tendency of the compressive flange to buckle laterally (sideways) is resisted not only by the stiffness of the flange itself but also by the resistance to twisting of the entire

† American Institute of Steel Construction, "Steel Construction Manual," 5th ed., American Institute of Steel Construction, New York, 1961.

W. F. Scofield and W. H. O'Brien, "Modern Timber Engineering," 4th ed., Southern Pine Association, New Orleans, 1954.

The Aluminum Association, "Aluminum Construction Manual," The Aluminum Association, New York, 1959.

section. For the prevention of lateral buckling the AISC specifications, under certain conditions, require a reduction of the allowable stress (from 20,000 psi). This reduction must be made if the term Ld/bt exceeds 600, where L is the laterally unsupported length (in inches), d is the overall depth of the beam, and b is the width and t is the thickness of the compression flange. For values of Ld/bt greater than 600 the allowable compressive stress is obtained from the equation

$$s = \frac{12{,}000{,}000}{Ld/bt} \tag{6.4}$$

Since most beams are sufficiently laterally supported (either throughout their entire length as by a floor, or by a number of intersecting members), reduction of the allowable compressive stress is not generally required.

Example 6.4. Determine the allowable bending compressive stress for a simply supported 16WF36 beam which is 18.0 ft long if it is laterally supported (*a*) at the ends only, (*b*) at the ends and at midspan.

$$(a) \quad \frac{Ld}{bt} = \frac{[(18)(12)](15.85)}{(6.992)(0.428)} = 1145$$

$$\text{Allowable } s_c = \frac{12{,}000{,}000}{1145} = 10{,}480 \text{ psi} \qquad \textit{Ans}$$

$$(b) \quad \frac{Ld}{bt} = \frac{[(9)(12)](15.85)}{(6.992)(0.428)} = 572$$

Since Ld/bt is less than 600, use allowable $s_c = 20{,}000$ psi (*Ans*).

Web crippling is a localized failure of the web resulting from compressive stress due to a concentrated load applied at the flange. Failure occurs by yielding of the web material near the flange and under the load. Web crippling can be avoided by (1) distributing the load over a greater length of the beam (with a longer bearing plate), (2) selecting a beam section with a thicker web, or (3) reinforcing the beam web with vertical "bearing" stiffeners. Beams are most frequently susceptible to web crippling at the reactions.

Under certain conditions *buckling of the web* of a plate girder may result. This buckling can occur in either a vertical or inclined direction depending on the stress condition in the web. The names given to these types of failure are, respectively, *vertical web buckling* and *diagonal web buckling*. They can be prevented by the proper use of web stiffeners as dictated by appropriate specifications.

The design factors discussed in this article are not the only ones which may be encountered; they are merely the ones most frequently considered in steel design. Other factors, as required by the appropriate specifications for the particular beam material used, may require consideration.

PROBLEMS

★**6.19.** Determine the allowable bending compressive stress for a simply supported 24WF76 beam which is 30 ft long if it is laterally supported (*a*) at the ends only; (*b*) at the ends and at midspan.

6.20. An 18WF50 section is to be used as a simply supported beam which is 36 ft long. (*a*) Compute the allowable bending compressive stress if the beam is laterally supported at the ends only. (*b*) At what points must the beam be laterally supported in order to justify the use of 20,000 psi as the allowable bending compressive stress?

★**6.21.** Two sections, an 18I70 and an 18WF70, are being considered for use as a 30-ft-long beam to be laterally supported at the ends only. Compare the allowable bending compressive stresses for the two sections.

6.22. Two sections, a 12I50 and a 12WF50, are being considered for use as a beam to be laterally supported at the ends only. Compare the allowable bending compressive stresses for the two sections if the beam is (*a*) 24 ft long, (*b*) 18 ft long.

7

TORSIONAL STRESSES AND DEFORMATIONS

7.1 INTRODUCTION

Moment may be applied to a member in such a manner that *twisting* rather than bending will result. Such moments are known as *torsional moments*, or *torques*. A torque may be defined as a couple which acts in a plane normal to the longitudinal axis of a member. A convenient method of representing a couple is with a vector perpendicular to the plane in which the couple acts. Couple vectors will be designated by *double-headed arrows* to distinguish them from force vectors. In this text the direction of a couple will be determined by the "right-hand rule"; by this rule, if the thumb of the closed right hand is pointed in the direction of the (double-headed) vector, the fingers will indicate the direction of the couple. An example of this sign convention is shown in Fig. 7.1.

This chapter is primarily concerned with the torsional stresses and deformations of cylindrical members (i.e., members of circular cross section).

Although the theory for members of noncircular cross section is too complex for inclusion in this text, the stresses in such members are briefly discussed in Art. 7.7. The concept of torsional shear flow applicable to closed thin-walled members (of any configuration) is presented in Art. 7.8.

7.2 TORSIONAL STRAIN

An expression for shearing strain in a cylindrical shaft subjected to torsion can be derived based on the assumptions that upon application of torque (1) any plane (circular) cross section will not

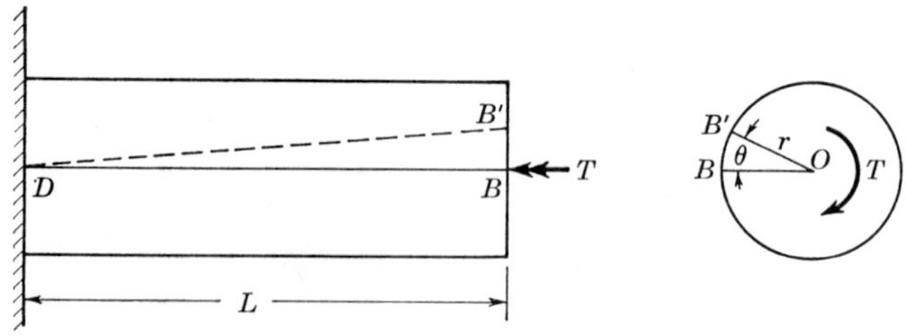

(*a*) Front view (*b*) End view

FIG. 7.1

warp or change configuration and (2) any radius of any cross section will remain a straight line. Tests have shown that these assumptions are valid, for practical purposes, provided that the proportional limit has not been exceeded. Figure 7.1 shows a cylindrical member fixed at one end and loaded with a torque T at the other. The torque causes the free end to rotate through the angle θ shown, which means that radius OB is rotated to the position OB'. Since the internal torque is constant throughout the length, element DB on the surface of the cylinder is twisted into the helix DB'. Since BB' is the total shearing strain occurring in length L, the unit shearing strain must be BB'/L. Noting that BB' is equal to $r\theta$,

$$\epsilon_s = \frac{r\theta}{L} \tag{7.1}$$

where ϵ_s = unit shearing strain at outer surface
r = radius of cylinder
θ = angle of twist which occurs over length L, radians

It should be noted that the units for L, in the above equation, must be the same as for r. Since the total strain at any cross section varies directly with the distance from the center of the section, the unit strain at a distance ρ from the center will be ρ/r times the unit strain at the surface (or $\rho\theta/L$).

Example 7.1. A 1.20-in.-diam aluminum (E_s = 4,000,000 psi) bar is twisted through an angle of 4° in a length of 2.00 ft. Determine the unit shearing strain and the unit shearing stress at the outer surface.

$$\theta = 4\frac{2\pi}{360} = \frac{\pi}{45} \qquad \text{radians}$$

$$\epsilon_s = \frac{r\theta}{L} = \frac{0.60}{2(12)}\frac{\pi}{45} = 0.001745 \text{ in./in.} \qquad \textit{Ans}$$

$$s_s = \epsilon_s E_s = 0.001745(4{,}000{,}000) = 6980 \text{ psi} \qquad \textit{Ans}$$

PROBLEMS

★7.1. Because of torsion, a steel (E_s = 12,000,000 psi) bar which is 1.00 in. in diameter and 6 ft long is stressed to 12,000 psi at its outer surface. How much will one end of the bar twist with respect to the other?

7.2. Because of torsion, a magnesium (E_s = 2,400,000 psi) bar which is 1.20 in. in diameter and 8 ft long is stressed to 10,000 psi at its outer surface. How much will one end of the bar twist with respect to the other?

★7.3. Determine the maximum allowable length for a 1.20-in.-diam steel (E_s = 12,000,000 psi) bar which is to be subjected to torsion without exceeding a shearing stress of 14,000 psi or an angle of twist of 6°.

7.4. Determine the maximum allowable length for a 1.40-in.-diam magnesium (E_s = 2,400,000 psi) bar which is to be subjected to torsion without exceeding a shearing stress of 8000 psi or an angle of twist of 5°.

★7.5. Determine the required diameter for a steel (E_s = 12,000,000 psi) bar which is to be subjected to torsion without exceeding a shearing stress of 12,000 psi or a twist per foot of 1°.

7.6. Determine the required diameter for a titanium (E_s = 5,800,000 psi) bar which is to be subjected to torsion without exceeding a shearing stress of 15,000 psi or a twist per foot of 2°.

7.3 TORSIONAL STRESS

The derivation of the equation for torsional stress is based on the assumptions made in the previous article and on the unit strain distribution determined therein. Figure 7.2*a* shows a cylindrical member fixed at one end and having a torque T applied at the other. Assuming that Hooke's law applies, the unit stress will be

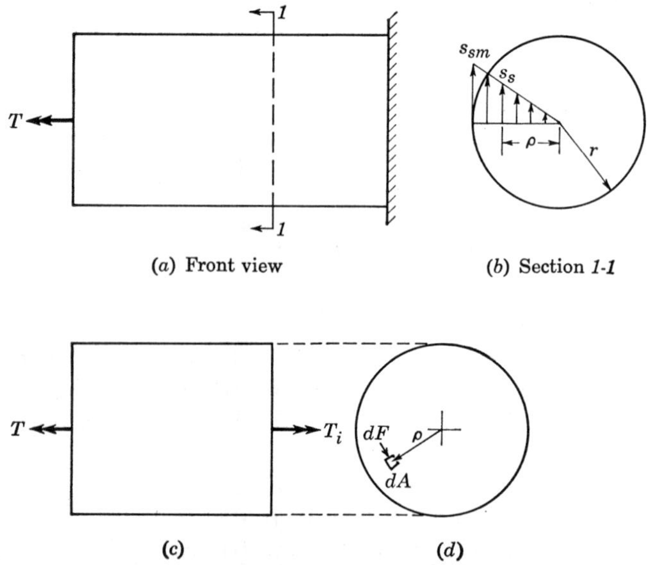

FIG. 7.2

proportional to the unit strain and the stress distribution will be as shown in Fig. 7.2*b*. Shown in Fig. 7.2*c* is a free body of the portion of the cylinder to the left of section 1-1. The internal torque, which is equal to the external torque, must be equal to the sum of the differential torques exerted on the differential areas of the cross section. As shown in Fig. 7.2*d*, the torque on a typical differential area is equal to $dF\,\rho$. Since $dF = s_s\,dA$,

$$dT = (s_s\,dA)\rho$$

Therefore

$$T = \int dT = \int s_s\,dA\,\rho$$

From Fig. 7.2*b*, it is seen that the stress s_s can be expressed in terms of the maximum stress s_{sm}. Thus

$$\frac{s_s}{\rho} = \frac{s_{sm}}{r} \qquad \text{or} \qquad s_s = \frac{s_{sm}\rho}{r}$$

and

$$T = \int \frac{s_{sm}\rho}{r}\, dA\, \rho = \frac{s_{sm}}{r} \int \rho^2\, dA$$

Since the expression $\int \rho^2\, dA$ represents the polar moment of inertia J of the cross section

$$T = \frac{s_{sm}J}{r} \qquad \text{or} \qquad \frac{T}{J} = \frac{s_{sm}}{r} = \frac{s_s}{\rho}$$

Therefore the shearing stress at any point on the section is

$$s_s = \frac{T\rho}{J} \tag{7.2a}$$

The maximum shearing stress is usually written as

$$s_s = \frac{Tr}{J} \tag{7.2b}$$

In the equations shown above, s_s is the shearing stress at a distance ρ or r from the longitudinal axis of the cylinder, T is the internal torque at the section, and J is the centroidal polar moment of inertia of the cross-sectional area. If ρ and r are measured in inches, J will have units of inches^4; therefore T should be expressed in units of pound-inches in order for the stress to be in psi.

Example 7.2. A hollow shaft, having inside and outside diameters of 4.00 and 5.00 in., is subjected to torsion as indicated by the load diagram shown as Fig. 7.3*a*. Determine (*a*) the maximum shearing stress, (*b*) the shearing stress at the inner surface of the section 3 ft from the left end of the shaft.

A torque diagram, Fig. 7.3*b*, is drawn showing the internal torque at every section of the shaft. Since the internal torque at any section must equal the algebraic sum of the external torques to either side of the section, a torque diagram can be easily drawn by summing external torques from either end and plotting the subtotals as ordinates. In Fig. 7.3*b* external torques were summed from left to right with the clockwise direction as positive.

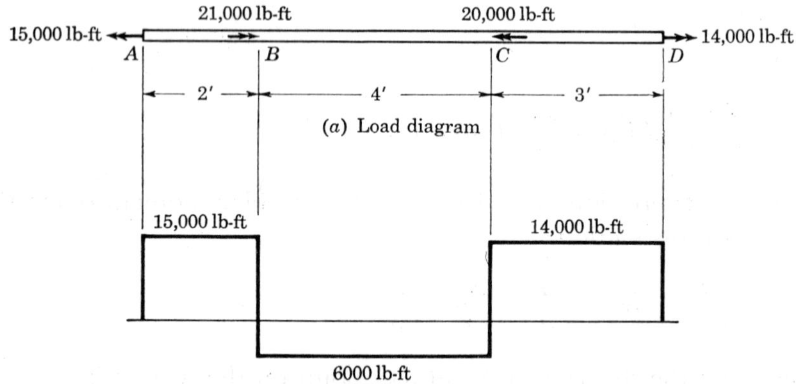

(*b*) Torque diagram

FIG. 7.3

The polar moment of inertia is computed by considering the cross section as consisting of a 5-in.-diam circle minus a 4-in.-diam circle.

$$J = \frac{\pi r_o^4}{2} - \frac{\pi r_i^4}{2} = \frac{\pi}{2}(2.5^4 - 2^4) = 36.3 \text{ in.}^4$$

(*a*) From the torque diagram, it is seen that the maximum shearing stress will occur in segment *AB*.

$$s_s = \frac{Tr}{J} = \frac{[15{,}000(12)]2.5}{36.3} = 12{,}400 \text{ psi} \qquad \textit{Ans}$$

(*b*) From the torque diagram, the torque at the section 3 ft from the left end is 6000 lb-ft.

$$s_s = \frac{T\rho}{J} = \frac{[6000(12)]2}{36.3} = 3970 \text{ psi} \qquad \textit{Ans}$$

PROBLEMS

★**7.7.** A hollow shaft has inside and outside diameters of 6 and 8 in. Determine the maximum shearing stress at a section where the torque is 80,000 lb-ft.

7.8. A hollow shaft has inside and outside diameters of 5 and 6 in. Determine the maximum shearing stress at a section where the torque is 20,000 lb-ft.

7.9. A hollow shaft has inside and outside diameters of 3 and 4 in. Determine the maximum shearing stress at a section where the torque is 9000 lb-ft.

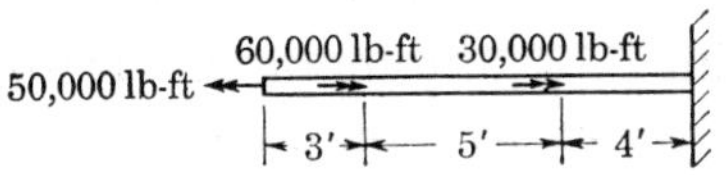

FIG. P7.10

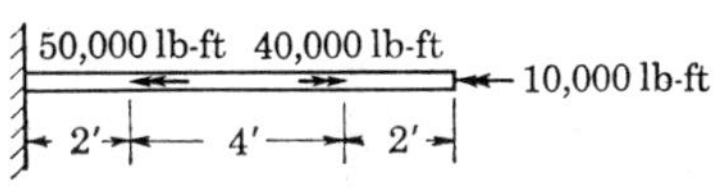

FIG. P7.11

7.10–7.12. Based on an allowable shearing stress of 15,000 psi, determine the required diameter for a solid shaft having the load diagram shown in Figs. P7.10 to P7.12. (*7.10)

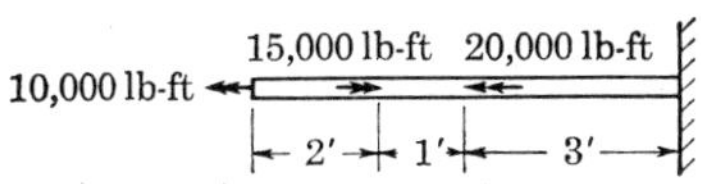

FIG. P7.12

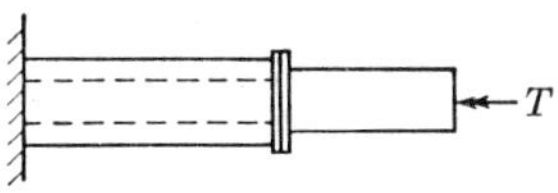

FIG. P7.13, P7.14

★**7.13.** A hollow aluminum cylinder, having outside and inside diameters of 4 and 2 in., is fixed at one end and securely fastened to a solid steel cylinder at the other. Allowable shearing stresses are as follows: aluminum = 8000 psi; steel = 12,000 psi. Compute the allowable torque T applied as shown in Fig. P7.13 and the required diameter of the steel cylinder.

7.14. A hollow brass cylinder, having outside and inside diameters of 3 and 2 in., is fixed at one end and securely fastened to a solid steel cylinder at the other. Allowable shearing stresses are as follows: brass = 10,000 psi; steel = 15,000 psi. Compute the allowable torque T applied as shown in Fig. P7.14 and the required diameter of the steel cylinder.

7.15. Two steel shafts, one hollow and the other solid, have equal cross-sectional areas and therefore weigh the same amount per foot. The hollow shaft has inside and outside diameters of 4 and 5 in.; the solid shaft is 3 in. in diameter. Based on an allowable shearing stress of 15,000 psi, compare the allowable torque for the two shafts.

7.16. Two aluminum shafts, one hollow and the other solid, have equal cross-sectional areas and therefore weigh the same amount per foot. The hollow shaft has inside and outside diameters of 12 and 13 in.; the solid shaft is 5.00 in. in diameter. Based on an allowable shearing stress of 10,000 psi, compare the allowable torque for the two shafts.

7.4 ANGLE OF TWIST

In order for a torsion member to serve its intended purpose, it must, in general, be relatively rigid; that is, the amount of twist that it undergoes, as well as its maximum shearing stress, must be limited. On the other hand, however, there are some cases when a certain amount of torsional flexibility may be desirable; an example of this is the use of torsion springs for automobiles. Design formulas for the angle of twist can be easily obtained from the equations previously derived in this chapter. Using Eq. (7.1) and Hooke's law, the angle of twist in terms of maximum shearing stress can be expressed as

$$\theta = \frac{\epsilon_s L}{r} = \frac{s_s}{E_s}\frac{L}{r} \qquad \text{or}$$

$$\theta = \frac{s_s L}{E_s r} \tag{7.3}$$

To express the angle of twist in terms of torque, the value of s_s in Eq. (7.2) is substituted in Eq. (7.3):

$$\theta = \frac{Tr}{J}\frac{L}{rE_s} \qquad \text{or}$$

$$\theta = \frac{TL}{E_s J} \tag{7.4}$$

Example 7.3. A hollow steel ($E_s = 12{,}000{,}000$ psi) shaft has inside and outside diameters of 3.5 and 4 in. and is 6 ft long. Determine the maximum horsepower which the shaft can transmit at 400 rpm without exceeding a shearing stress of 12,000 psi or an angle of twist of 1.5°.

To find the allowable torque as limited by shearing stress,

$$J = \frac{\pi}{2}(2.0^4 - 1.75^4) = 10.4 \text{ in.}^4$$

$$s_s = \frac{Tr}{J} \qquad T = \frac{s_s J}{r} = \frac{12{,}000(10.4)}{2} = 62{,}400 \text{ lb-in.}$$

To find the allowable torque as limited by twisting,

$$\theta = 1.5\,\frac{2\pi}{360} = 0.0262 \text{ radian}$$

$$\theta = \frac{TL}{E_s J} \qquad T = \frac{\theta E_s J}{L} = \frac{0.0262(12{,}000{,}000)(10.4)}{6(12)} = 45{,}400 \text{ lb-in.}$$

Therefore the allowable torque is 45,400 lb-in. as limited by the angle of twist.

A horsepower is defined as 33,000 ft-lb/min of work; the horsepower which can be transmitted by a torque T (in pound-inches) at n rpm is

$$hp = \frac{[T(1/12)](2\pi)n}{33{,}000} \qquad \text{or}$$

$$hp = \frac{Tn}{63{,}000}$$

Therefore

$$hp = \frac{(45{,}400)(400)}{63{,}000} = 288 \qquad \textit{Ans}$$

PROBLEMS

★7.17. Compute the maximum torque which can be applied to a hollow steel (E_s = 12,000,000 psi) shaft having outside and inside diameters of 6 and 5 in. without exceeding (*a*) a shearing stress of 15,000 psi, (*b*) a twist of 0.2° per foot.

7.18. Compute the maximum torque which can be applied to a hollow aluminum (E_s = 4,000,000 psi) shaft having outside and inside diameters of 5 and 4 in. without exceeding (*a*) a shearing stress of 12,000 psi, (*b*) a twist of 0.3° per foot.

7.19. A hollow steel (E_s = 12,000,000 psi) shaft having outside and inside diameters of 4 and 2 in. is twisted through an angle of 4° in a length of 12 ft. Compute the shearing stress at the outside and inside surfaces.

7.20. A hollow aluminum (E_s = 4,000,000 psi) shaft having outside and inside diameters of 5 and 3 in. is twisted through an angle of 6° in a length of 9 ft. Compute the shearing stress at the outside and inside surfaces.

7.21. A hollow steel (E_s = 12,000,000 psi) shaft is to deliver 120 hp at 875 rpm. (*a*) Based on an allowable shearing stress of 12,000 psi, determine the required diameters if the inside diameter is to be 0.7 of the outside. (*b*) Compute the maximum length for the shaft if the angle of twist is not to exceed 2°.

★7.22. A hollow steel (E_s = 12,000,000 psi) shaft is to deliver 100 hp at 1750 rpm. (*a*) Based on an allowable shearing stress of 15,000 psi, determine the required diameters if the inside diameter is to be 0.8 of the outside. (*b*) Compute the maximum length for the shaft if the angle of twist is not to exceed 4°.

7.23. A 2-in.-diam steel (E_s = 12,000,000 psi) shaft is loaded as shown in Fig. P7.23. Determine the net twist, in degrees, of the free end with respect to the fixed end.

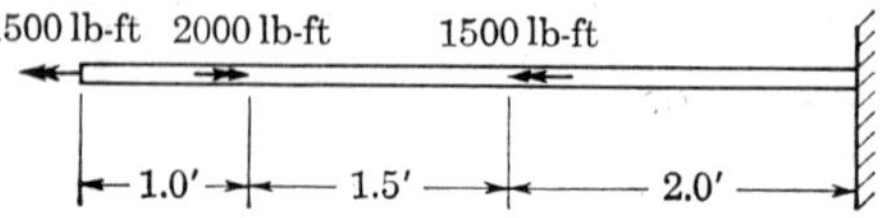

FIG. P7.23

7.24. A 4-in.-diam aluminum (E_s = 4,000,000 psi) shaft is loaded as shown in Fig. P7.24. Determine the net twist, in degrees, of the free end with respect to the fixed end.

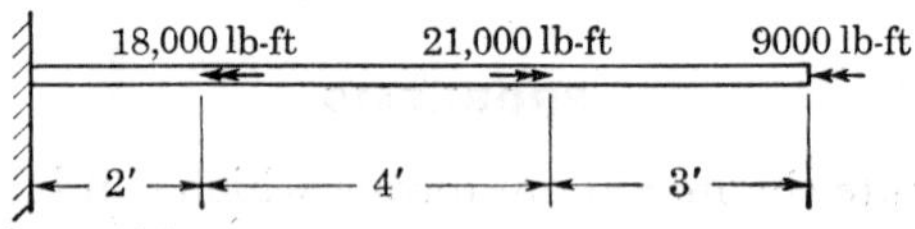

FIG. P7.24

7.25. A 6-in.-diam steel (E_s = 12,000,000 psi) shaft is loaded as shown in Fig. P7.25. Determine the net twist, in degrees, of the free end with respect to the fixed end.

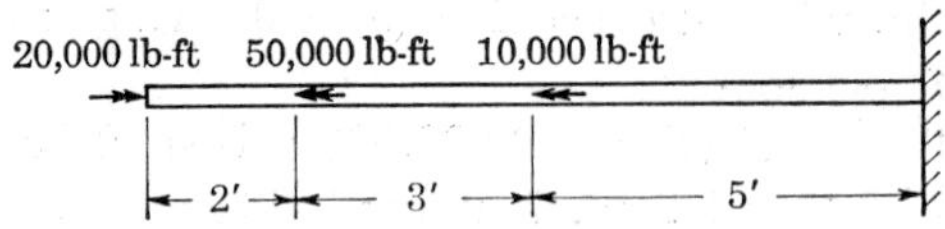

FIG. P7.25

7.5 INDETERMINATE SHAFTS

The reactions of a torsionally loaded member which is restrained at both ends are dependent not only on the conditions of equilibrium but also on the relative stiffness of the various shaft segments. The reactions of such statically indeterminate shafts can be determined by relating the angles of twist of the various shaft segments to the equations of equilibrium. Indeterminate shaft problems are included here primarily as a learning device, since they serve to underscore the interrelationship of torsional stresses and deformations.

Example 7.4. A 3.5-ft-long hollow aluminum (E_s = 4,000,000 psi) shaft having outside and inside diameters of 4 and 2 in. is connected end to end with a solid steel (E_s = 12,000,000 psi) shaft which is 4.0 ft long and 3.00 in. in diameter. The extreme ends of the combined shaft are fixed, and a torque of 12,000 lb-ft is applied to the connection as shown in Fig. 7.4*a*. Compute the maximum shearing stress in each material.

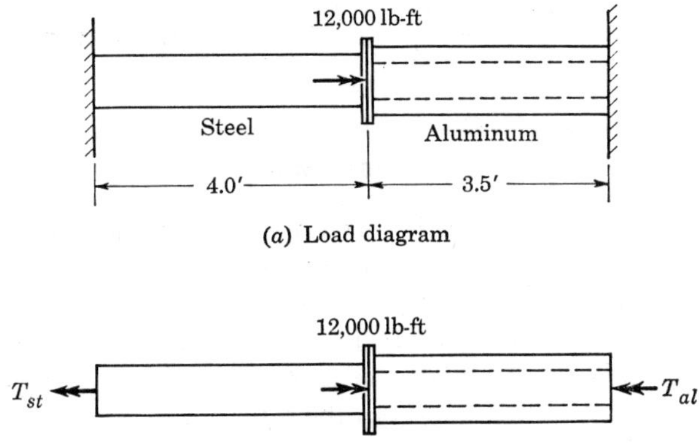

(*a*) Load diagram

(*b*) Free-body diagram

FIG. 7.4

From the free-body diagram, shown as Fig. 7.4*b*, it is found that the reaction torques T_{st} and T_{al} are statically indeterminate and that

$$T_{st} + T_{al} = 12{,}000$$

Therefore a relation of the stiffnesses of the two segments must also be used in order to determine the reactions. The polar moments of inertia are

$$J_{st} = \frac{\pi}{2}(1.5)^4 = 7.95 \text{ in.}^4$$

$$J_{al} = \frac{\pi}{2}(2^4 - 1^4) = 23.6 \text{ in.}^4$$

Since neither end of the combined shaft can rotate, the angle of twist must be the same for each of the two segments or $\theta_{st} = \theta_{al}$.

$$\left(\frac{TL}{E_sJ}\right)_{st} = \left(\frac{TL}{E_sJ}\right)_{al}$$

Substituting in the above equations and using

$$T_{st} = 12{,}000 - T_{al}$$

$$\frac{[(12{,}000 - T_{al})(12)][(4)(12)]}{(12{,}000{,}000)(7.95)} = \frac{[T_{al}(12)][(3.5)(12)]}{(4{,}000{,}000)(23.6)}$$

$$T_{al} = 6360 \text{ lb-ft}$$

$$T_{st} = 12{,}000 - 6360 = 5640 \text{ lb-ft}$$

Computing the maximum shearing stresses

For steel, $s_s = \frac{Tr}{J} = \frac{5640(12)(1.5)}{7.95} = 12{,}780 \text{ psi}$ *Ans*

For aluminum, $s_s = \frac{Tr}{J} = \frac{6360(12)(2)}{23.6} = 6470 \text{ psi}$ *Ans*

PROBLEMS

★7.26. A steel shaft which is 3 in. in diameter and 5 ft long is fixed at both ends. Determine the maximum shearing stress due to a 10,000 lb-ft torque applied 2 ft from the left end.

7.27. An aluminum shaft which is 4 in. in diameter and 7 ft long is fixed at both ends. Determine the maximum shearing stress due to a 20,000 lb-ft torque applied 3 ft from the left end.

7.28. A solid shaft of length L is fixed at both ends and loaded with an off-center torque T as shown. Show that the left and right reactions are equal to Tb/L and Ta/L, respectively.

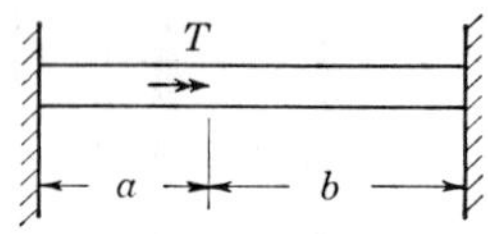

FIG. P7.28

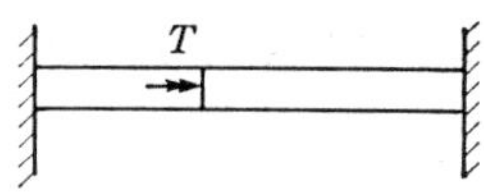

FIG. P7.29, P7.30

★7.29. A steel (E_s = 12,000,000 psi) shaft and an aluminum (E_s = 4,000,000 psi) shaft, each 4 in. in diameter, are connected end to end. The extreme ends of the combined shaft are fixed, and a 15,000 lb-ft torque is applied to the connection as shown. The steel member is 3 ft long; the aluminum member is 5 ft long. Compute the maximum shearing stress in each material.

7.30. An aluminum (E_s = 4,000,000 psi) shaft and a brass (E_s = 6,000,000 psi) shaft, each 6 in. in diameter, are connected end to end. The extreme ends of the combined shaft are fixed, and a 75,000 lb-ft torque is applied to the connection as shown. The aluminum member is 4 ft long; the brass member is 5 ft long. Compute the maximum shearing stress in each material.

7.31. A 4-ft-long hollow steel (E_s = 12,000,000 psi) shaft having outside and inside diameters of 5 and 3 in. is connected end to end with a solid aluminum (E_s = 4,000,000 psi) shaft which is 6 ft long and 4 in. in diameter. The extreme ends of the combined shaft are fixed, and a torque of 25,000 lb-ft is applied to the connection as shown. Compute the maximum shearing stress in each material.

★**7.32.** A 3-ft-long hollow aluminum (E_s = 4,000,000 psi) shaft having outside and inside diameters of 4 and 2 in. is connected end to end with a solid brass (E_s = 6,000,000 psi) shaft which is 6 ft long and 3 in. in diameter. The extreme ends of the combined shaft are fixed, and a torque of 16,000 lb-ft is applied to the connection as shown. Compute the maximum shearing stress in each material.

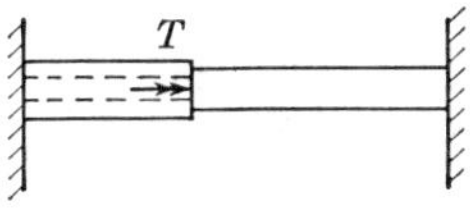

FIG. P7.31, P7.32

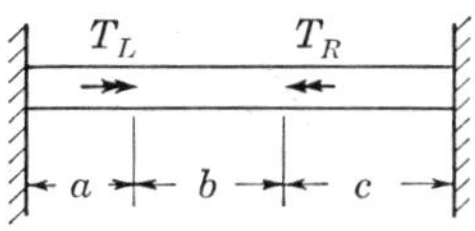

FIG. P7.33, P7.34

7.33. A steel shaft which is 4 in. in diameter and 9 ft long is fixed at both ends and loaded with the two torques shown. If a = 2 ft, b = 3 ft, T_L = 18,000 lb-ft, and T_R = 6000 lb-ft, determine the maximum shearing stress.

7.34. An aluminum shaft which is 5 in. in diameter and 12 ft long is fixed at both ends and loaded with the two torques shown. If a = 3 ft, b = 4 ft, T_L = 30,000 lb-ft, and T_R = 45,000 lb-ft, determine the maximum shearing stress.

7.6 HELICAL SPRINGS

Helical springs are often used for purposes of exerting steady loads or for absorbing energy loading. They are formed, usually from lengths of high-strength alloy wire or rod, into the shape of a helix. If the wire or rod is circular in cross section, the shearing stress and the spring deflection can be readily computed using principles previously discussed in this chapter.

To derive a formula for the maximum shearing stress, reference is made to Fig. 7.5*a* showing a helical spring having a mean coil radius R and a cross-sectional radius r and loaded with a central load P. A vertical section is passed through point A in Fig. 7.5*a*; the free-body diagram of the portion of the spring below point A is drawn as shown in Fig. 7.5*b*. In order for equilibrium to exist, the section must be acted on by an internal shear force P and an internal torque T which is equal to PR. The maximum shearing stress *due to torsion* is

$$s_s = \frac{Tr}{J} = \frac{(PR)r}{\pi r^4/2} \qquad \text{or}$$

$$s_s = \frac{2PR}{\pi r^3} \tag{7.5}$$

For a spring having a cross-sectional radius which is relatively small in comparison with the coil radius, the added shearing stress

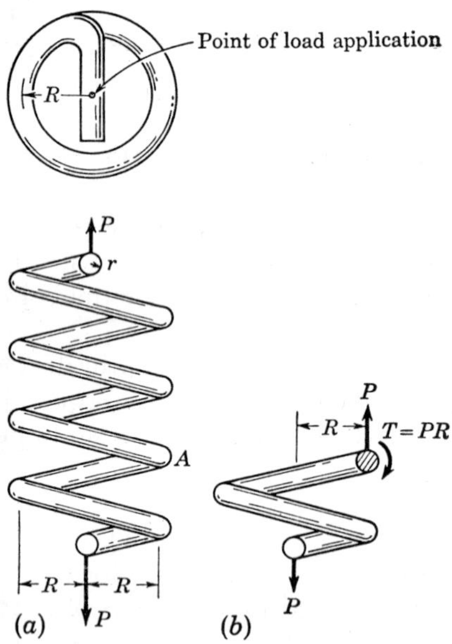

FIG. 7.5

due to the internal force can be neglected. However, if the cross-sectional radius is relatively large, the additional stress due to the internal force should be considered. Thus, adding the average shearing stress P/A due to the internal force,

$$s_s = \frac{P}{A} + \frac{Tr}{J} = \frac{P}{\pi r^2} + \frac{2PR}{\pi r^3} \qquad \text{or}$$

$$s_s = \frac{P}{A}\left(1 + 2\frac{R}{r}\right) \tag{7.6}$$

where A is the cross-sectional area and r is the radius of the wire or rod and R is the coil radius.

An examination of Eq. (7.6) reveals that, if the ratio R/r is as much as 10 to 1, the error involved in neglecting the stress due to the internal force [that is, in using Eq. (7.5)] will be less than 5 per cent. Actually Eq. (7.6) is so simple to use that there is little point in not using it exclusively.

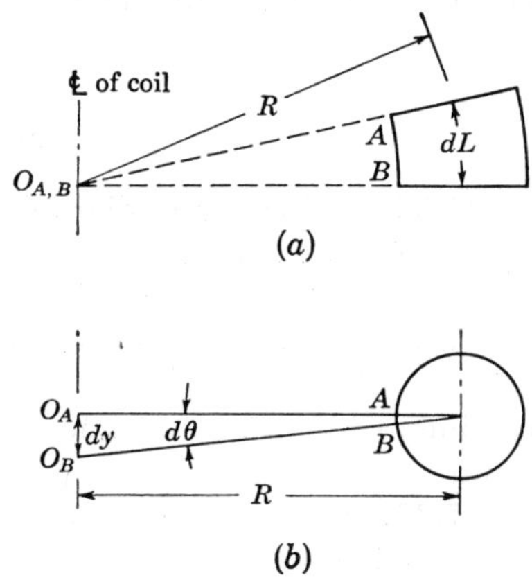

FIG. 7.6

The deflection of a helical spring, the amount it will lengthen or shorten, can also be readily computed using the principles previously discussed. Figure 7.6 shows two views of an elemental segment of a helical spring subjected to a centric load. The amount that one end of the element tends to twist with respect to the other is shown as $d\theta$; the vertical displacement *of the spring* dy due to this twisting tendency will be equal to $R\,d\theta$, as shown. Expressing $d\theta$ with its equivalent [from Eq. (7.4)],

$$dy = R\,d\theta = R\frac{T\,dL}{E_sJ} = \frac{R(PR)\,dL}{E_sJ}$$

The total deflection of the spring is equal to the summation of the vertical displacements, due to rotation, of all the elemental lengths:

$$y = \int dy = \int_0^L \frac{PR^2\,dL}{E_sJ} = \frac{PR^2L}{E_sJ}$$

Expressing the total length of coil as $N(2\pi R)$, where N is the number of coils,

$$y = \frac{PR^2(N2\pi R)}{E_sJ} = \frac{2\pi PR^3N}{E_sJ} \qquad \text{or}$$

$$y = \frac{4PR^3N}{E_sr^4} \tag{7.7}$$

It should be remembered that the equations of this article apply only to helical springs made of wire or rod material having a circular cross section and for which the helix angle is small. It should also be mentioned that the equations are not "exact," since their derivations were based on theory for straight rather than curved cylindrical members.

PROBLEMS

★7.35. A 5-in.-mean-diam helical spring having 10 coils is made of 0.50-in.-diam alloy steel ($E_s = 12{,}000{,}000$ psi) rod. Compute the maximum shearing stress and spring deflection due to a centric load of 500 lb.

7.36. A 3-in.-mean-diam helical spring having 8 coils is made of 0.375-in.-diam alloy steel ($E_s = 12{,}000{,}000$ psi) rod. Compute the maximum shearing stress and spring deflection due to a centric load of 350 lb.

7.37. A 2-in.-mean-diam helical spring having 6 coils is made of 0.25-in.-diam alloy steel ($E_s = 12{,}000{,}000$ psi) rod. Compute the maximum shearing stress and spring deflection due to a centric load of 150 lb.

★7.38. An alloy steel ($E_s = 12{,}000{,}000$ psi) helical spring is to have a 6-in. mean coil diameter and support a 600-lb centric load. (*a*) Determine the required rod diameter based on an allowable shearing stress of 60,000 psi; rod material is available in all diameter sizes from 3/8 to 1 in. in 1/8-in. increments. (*b*) Determine the required number of coils if the spring deflection is not to exceed 3.0 in.

7.39. An alloy steel ($E_s = 12{,}000{,}000$ psi) helical spring is to have a 4-in. mean coil diameter and support a 500-lb centric load. (*a*) Determine the required rod diameter based on an allowable shearing stress of 60,000 psi; rod material is available in all diameter sizes from 3/8 to 1 in. in 1/8-in. increments. (*b*) Determine the required number of coils if the spring deflection is not to exceed 2.0 in.

7.40. An alloy steel ($E_s = 12{,}000{,}000$ psi) helical spring is to have a 2.50-in. mean coil diameter and support a 200-lb centric load. (*a*) Determine the required rod diameter based on an allowable shearing stress of 60,000 psi; rod material is available in all diameter sizes up to 1 in. in 1/8-in. increments. (*b*) Determine the required number of coils if the spring deflection is not to exceed 1.5 in.

7.7 MEMBERS OF NONCIRCULAR CROSS SECTION

As has been seen, the theory for torsional stress and deformation of cylindrical members is based on the absence of warping of any cross section. When torque is applied to noncircular members, however, warping of the cross section will occur; for this reason the stress-deformation theory for such members is quite complex. The stresses for members of a few relatively simple cross sections

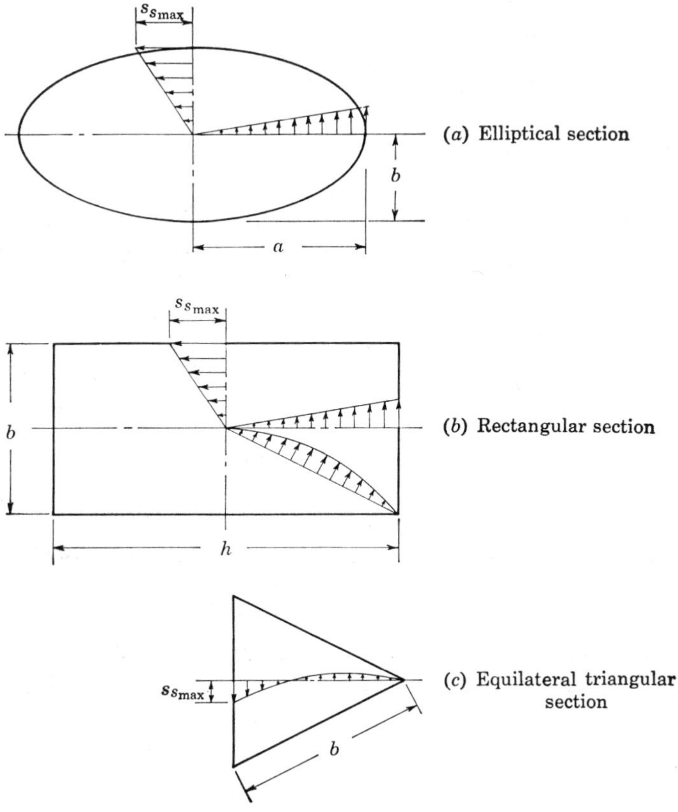

FIG. 7.7 Torsional stress distribution in solid sections.

can be determined using advanced theory.† For the cases of more complicated cross sections, approximate or empirical methods must be used.‡

The torsional shearing-stress distribution for members of several simple cross sections is shown in Fig. 7.7. It is interesting to

† S. Timoshenko and J. N. Goodier, "Theory of Elasticity," 2d ed., McGraw-Hill Book Company, Inc., New York, 1951.

‡ Glenn Murphy, "Advanced Mechanics of Materials," McGraw-Hill Book Company, Inc., New York, 1946.

note that, in each case, the maximum stress occurs at the outer surface which is nearest the axis of the member. It is also of interest, in the case of the rectangular and triangular sections, that the stress is zero at the corners (where the distance from the axis of the member is a maximum!).

The maximum torsional shearing stress in a member of elliptical cross section is defined by the theoretical equation†

$$s_s = \frac{2T}{\pi a b^2} \tag{7.8}$$

where a and b are the major and minor semiaxes, respectively. As can be seen from Fig. 7.7a, this maximum stress will occur at the ends of the minor axis.

The maximum torsional shearing stress in a member of rectangular cross section is given by the approximate equation†

$$s_s = \frac{(15h + 9b)T}{5h^2b^2} \tag{7.9}$$

where h and b represent the long and short dimensions, respectively. As can be seen from Fig. 7.7b, this stress occurs at the middle of the long side.

For a member having a cross section of the form of an equilateral triangle, the maximum torsional shearing stress is given by the theoretical equation†

$$s_s = \frac{20T}{b^3} \tag{7.10}$$

where b is the length of one leg of the triangle. As can be seen from Fig. 7.7c, this maximum shearing stress occurs at the middle of each leg.

PROBLEMS

★7.41. A steel shaft of square cross section is to be loaded with a torque of 60,000 lb-ft. Based on an allowable shearing stress of 12,000 psi, determine the required size.

† Raymond J. Roark, "Formulas for Stress and Strain," 3d ed., McGraw-Hill Book Company, Inc., New York, 1954.

7.42. An aluminum member having a cross section in the form of an equilateral triangle is to be loaded with a torque of 20,000 lb-ft. Based on an allowable shearing stress of 10,000 psi, determine the required size.

7.43. (*a*) Based on an allowable shearing stress of 12,000 psi, determine the allowable torque for a member 4 in. square in cross section. (*b*) What would be the allowable torque for a solid cylindrical member of the same material having the same weight per foot?

***7.44.** (*a*) Based on an allowable shearing stress of 12,000 psi, determine the allowable torque for a solid elliptical shaft having major and minor axes of 3 and 2 in., respectively. (*b*) What would be the allowable torque for a solid cylindrical member of the same material having the same weight per foot?

7.45. (*a*) Based on an allowable shearing stress of 12,000 psi, compare the allowable torques for two steel members having the cross sections shown in Fig. P7.45. (*b*) Which cross section utilizes the material most efficiently? Prove.

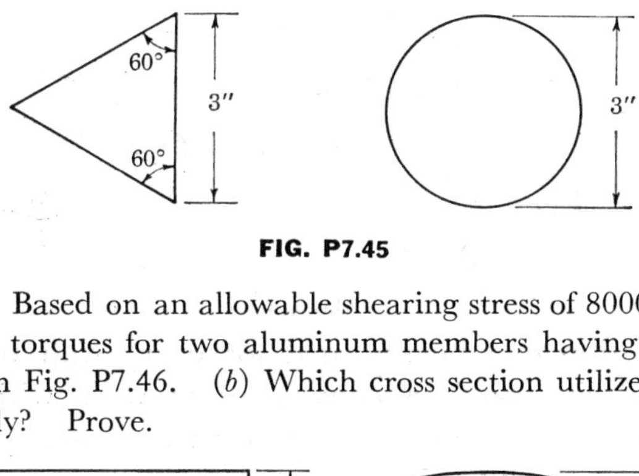

FIG. P7.45

7.46. (*a*) Based on an allowable shearing stress of 8000 psi, compare the allowable torques for two aluminum members having the cross sections shown in Fig. P7.46. (*b*) Which cross section utilizes the material most efficiently? Prove.

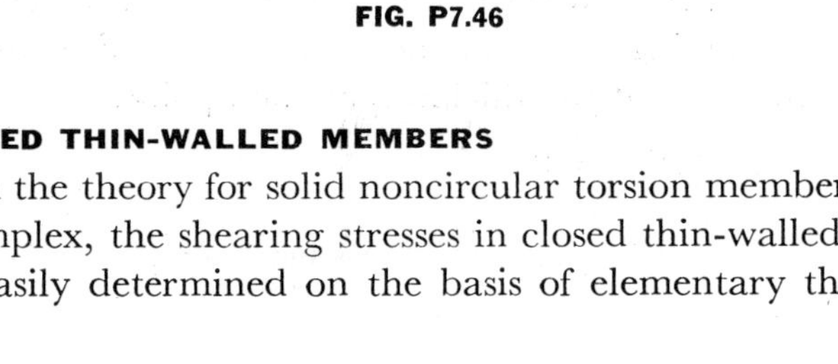

FIG. P7.46

7.8 CLOSED THIN-WALLED MEMBERS

Although the theory for solid noncircular torsion members is somewhat complex, the shearing stresses in closed thin-walled members can be easily determined on the basis of elementary theory. (A *closed* member is one which has a continuous cross-sectional periphery such as a tube.) An equation will be derived which defines the torsional shearing stress in any closed thin-walled member, regardless of the configuration of the section or the nonuniformity of the wall thickness. The equation is limited, however, to members made of material which is sufficiently thin so that a uniform

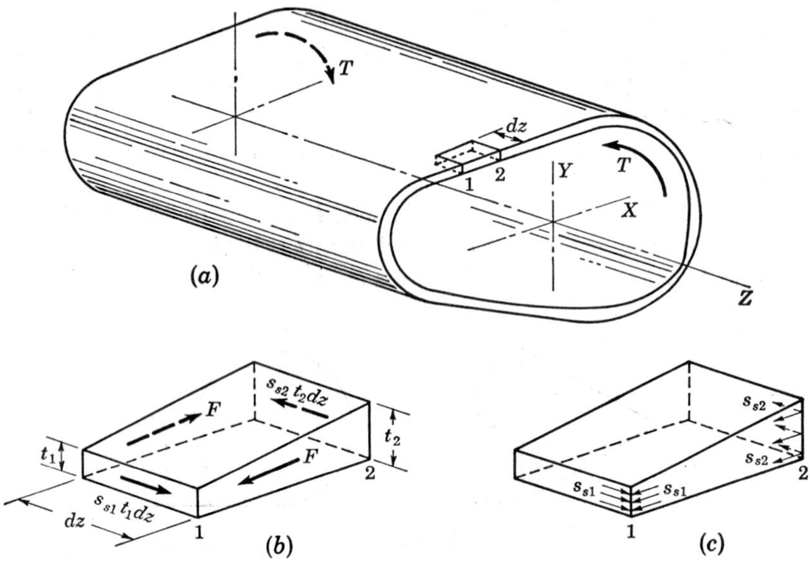

FIG. 7.8

stress distribution can be assumed for every wall thickness. In making the derivation it is convenient to work with the *internal shear force per linear inch* of cross section; this quantity is known as the *shear flow* and will be designated as q.

The derivation will be made with respect to the closed thin-walled member of general configuration and variable wall thickness shown in Fig. 7.8*a*. In order to determine the shearing-stress distribution about the periphery, a free-body diagram of an element (dz in width) between points 1 and 2 is drawn as shown in Fig. 7.8*b*. Summing forces in the z direction and assuming uniform stress throughout the thickness at each point,

$$+s_{s_1}(t_1\,dz) - s_{s_2}(t_2\,dz) = 0$$

whence

$$s_{s_1}t_1 = s_{s_2}t_2$$

Since the shearing stress on any two mutually perpendicular planes must be equal (Art. 5.5), it follows that the peripheral shearing stresses, as shown in Fig. 7.8*c*, must also be related by the above equation. Expressing the product of s_s and t at any point as the shear flow q gives

$$q_1 = q_2$$

Thus it has been shown that the shear flow about any closed thin-walled member is constant.

As shown in Fig. 7.9, the internal torque per differential length of wall area with respect to any point O is

$$dT = q\,dL\,\rho$$

and since q is constant, the total internal torque is

$$T = q\int \rho\,dL$$

Since the term $\rho\,dL$ represents twice the area of the shaded differential triangle shown, the term $\int \rho\,dL$ must represent twice the

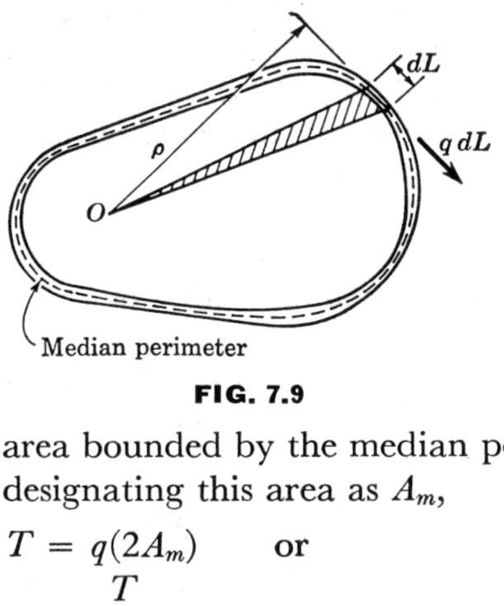

FIG. 7.9

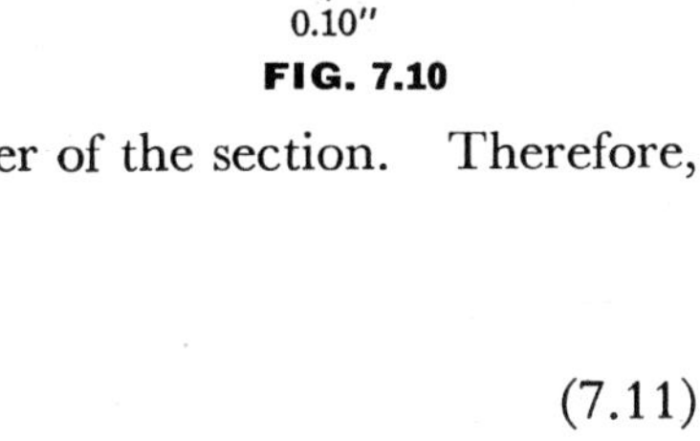

FIG. 7.10

area bounded by the median perimeter of the section. Therefore, designating this area as A_m,

$$T = q(2A_m) \qquad \text{or}$$

$$q = \frac{T}{2A_m} \tag{7.11}$$

Since at any point $q = s_s t$, the shearing stress at any point is

$$s_s = \frac{T}{2A_m t} \tag{7.12}$$

Example 7.5. A member having the cross section shown in Fig. 7.10 is subjected to a torque of 20,000 lb-ft. Determine the maximum shearing stress.

The area bounded by the median perimeter

$$A_m = 2\frac{\pi(6.03)^2}{2} + 10(12.1) = 235 \text{ in.}^2$$

The maximum stress occurs in the thinnest material (i.e. in the semicircular portions); thus

$$s_s = \frac{T}{2A_m t} = \frac{20,000(12)}{2(235)(0.06)} = 8510 \text{ psi} \qquad Ans$$

PROBLEMS

★7.47. A 0.2-in.-thick steel plate is formed into a hollow rectangular section having outside dimensions of 10 by 20 in. Determine the shear flow and maximum shearing stress due to a torque of 60,000 lb-ft.

7.48. A 0.2-in.-thick aluminum plate is formed into a hollow elliptical section having major and minor axes of 15 and 8 in., respectively. Determine the shear flow and maximum shearing stress due to a torque of 35,000 lb-ft.

7.49. A member which is to resist a torque of 40,000 lb-ft is to have a hollow rectangular cross section. The member is to be formed from flat plate and is to have outside dimensions of 8 by 10 in. Based on an allowable shearing stress of 12,000 psi, determine the required plate thickness.

7.50. A member which is to resist a torque of 50,000 lb-ft is to be formed from 0.15-in.-thick aluminum plate and is to have a cross section in the form of a hollow square. Based on an allowable stress of 10,000 psi, determine the required outside dimensions.

★7.51. Based on an allowable stress of 12,000 psi, determine the allowable torque for a member having the cross section shown in Fig. P7.51.

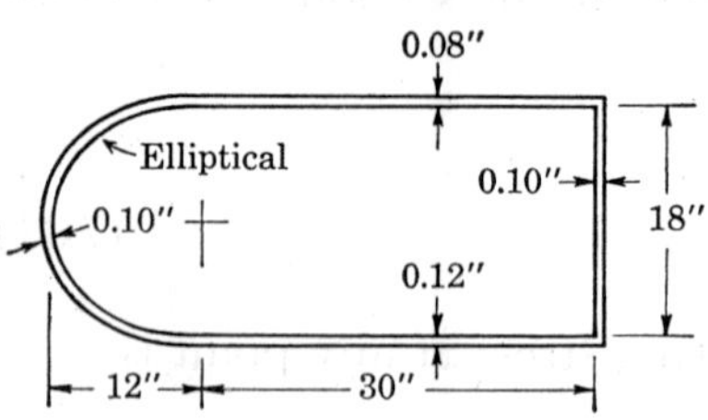

FIG. P7.51

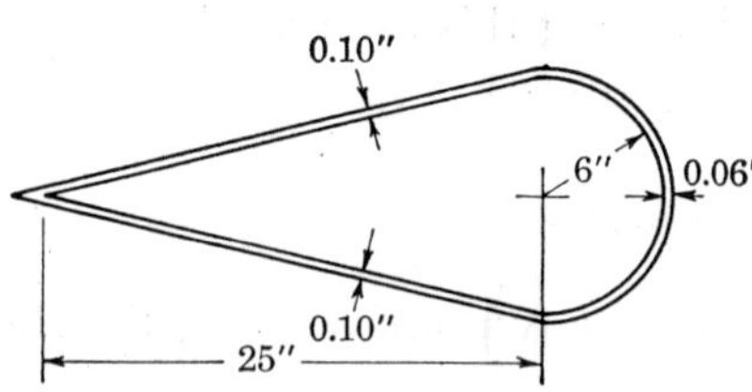

FIG. P7.52

7.52. Based on an allowable stress of 10,000 psi, determine the allowable torque for a member having the cross section shown in Fig. P7.52.

7.9 SHEAR CENTER OF BEAMS

The derivations of the equations for beam stresses in Chap. 5 were based on a number of assumptions. Among these was the assumption that any cross section of a beam must be symmetrical with respect to the plane of loading. When this assumption is met, the beam will bend without twisting. However, it is sometimes necessary, for practical considerations, to orient a beam in such a way

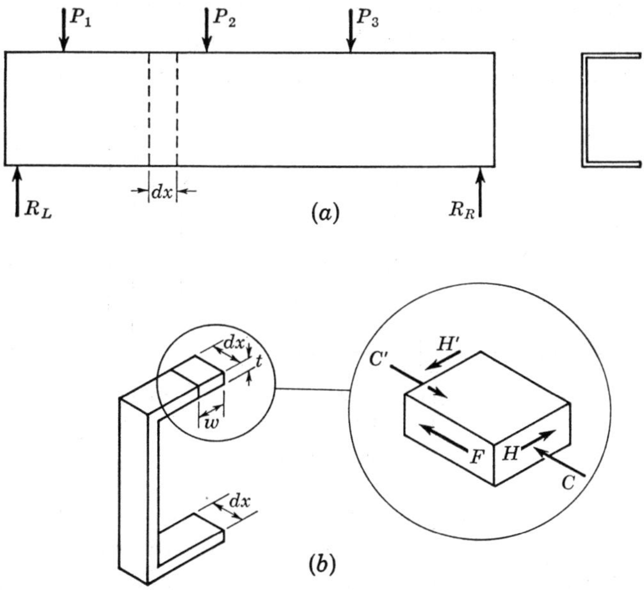

FIG. 7.11

that the plane of loading does not coincide with the longitudinal plane of symmetry of the beam. In such cases, twisting can be avoided by causing the plane of loading to act through a point in the plane of any cross section known as the *shear center*. For any beam with a cross section having an axis of symmetry, the shear center will be located on that axis. Of course, for a beam cross section having two or more axes of symmetry, the shear center will be at the intersection of the axes (i.e., at the centroid of the cross section).

For beams made of relatively thin material, the shear center can be determined using the concept of shear flow discussed in the previous article. Shown in Fig. 7.11*a* is a beam having a cross section without a vertical axis of symmetry and subjected to a number of vertical loads acting in a plane which passes through the shear center of the section. From a typical dx length, a small element is taken from the upper flange and a free-body diagram drawn as shown in Fig. 7.11*b*. The element is acted on by the compressive forces C and C' due to bending and by the longitudinal shear force F. Since, as was shown in Art. 5.5, there must be equal shearing stresses on mutually perpendicular planes, shear forces H and H' must also act on the element. Thus it is seen that transverse shearing stresses exist in the flange and that the transverse

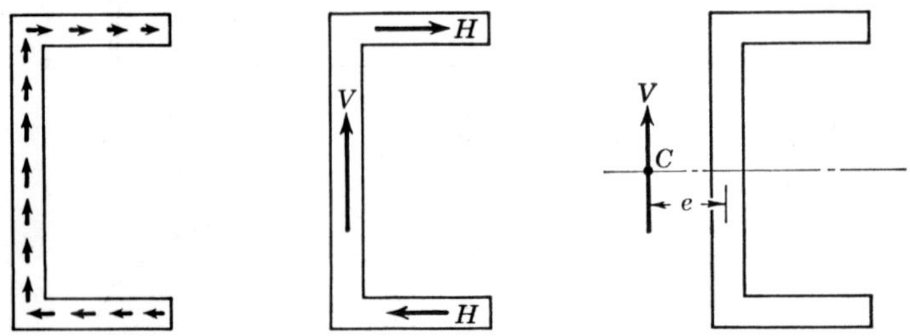

(*a*) Shear flow (*b*) Internal shear forces (*c*) Resultant internal shear force

FIG. 7.12

shearing stress at a given point is numerically equal to the longitudinal shearing stress at the point. The transverse shearing stress at any point on a given section is thus defined by the equation $s_s = VQ/It$, and the shear flow (the shear force per linear inch) by the equation $q = VQ/I$. Thus it should be noted that the shear flow in beams is not constant. The shear flow for the beam of Fig. 7.11 acts as shown in Fig. 7.12*a*; the internal shear forces on the flanges and in the web therefore act as shown in Fig. 7.12*b*. As shown in Fig. 7.12*c* the resultant internal shear force V (the resultant of the three forces shown in Fig. 7.12*b*) would act through point C on the axis of symmetry of the cross section. Point C represents the shear center (since unless the external resultant shear force also acts through this point, it and the resultant internal shear force would constitute a torsional couple). Hence, the shear center can be located merely by summing moments of the shear forces in the various parts of the cross section with respect to any convenient point on the axis of symmetry.

Example 7.6. A beam having the cross section shown in Fig. 7.13*a* is made of a material having a uniform thickness of 0.25 in. Determine the shear center and the shear-flow distribution at a section where the vertical shear is 3000 lb.

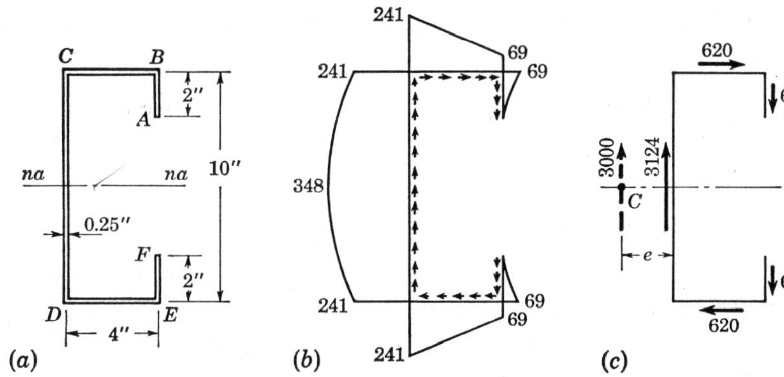

FIG. 7.13

Determining the moment of inertia about the neutral axis (neglecting the half thickness of the various elements),

	a	I_o	d	ad^2
AB and *EF*	1.00	0.33	4	16.0
BC and *DE*	2.00	—	5	50.0
CD	2.50	20.83	0	0.0
Total		21.16		66.0

$I = 21.2 + 66.0 = 87.2 \text{ in.}^4$

The shear flow at B is a function of the first moment of the area between A and B; thus

$$q = \frac{VQ}{I} = \frac{3000}{87.2}\,0.5(4) = \frac{3000}{87.2}\,2 = 69 \text{ lb/in.}$$

The shear flow at C is a function of the first moment of the area between A and C

$$q = \frac{VQ}{I} = \frac{3000}{87.2}\,[2 + 1(5)] = \frac{3000}{87.2}\,7 = 241 \text{ lb/in.}$$

The shear flow at the neutral axis is

$$q = \frac{VQ}{I} = \frac{3000}{87.2}[7 + 1.25(2.5)] = \frac{3000}{87.2} 10.12 = 348 \text{ lb/in.}$$

The values of shear flow at D and E are known from symmetry; the shear-flow distribution is as shown in Fig. 7.13*b* (*Ans*).

Although the shear-flow distribution in elements AB and EF is parabolic, it may be assumed to be linear for the purpose of computing the shear forces in the elements.

$$V_{AB} = V_{EF} = \frac{69(2)}{2} = 69 \text{ lb}$$

$$V_{BC} = V_{DE} = \frac{241 + 69}{2} 4 = 620 \text{ lb}$$

$$V_{CD} = 241(10) + \tfrac{2}{3}(348 - 241)(10) = 3124 \text{ lb}$$

The shear forces on the various elements act as shown in Fig. 7.13*c*. A check of vertical forces reveals a discrepancy of 14 lb or 0.5 per cent due to neglecting the half thickness of the members and assuming a linear-stress distribution for elements AB and EF.

The shear center is located by noting that the moment of the resultant vertical shear (3000 lb) must be equal to the moment of the shear forces on the various elements. Thus, summing moments about the center of element CD,

$$3000(e) = 620(10) + 2(69)(4)$$
$$e = 2.25 \text{ in.} \qquad \textit{Ans}$$

PROBLEMS

★**7.53.** A beam having the cross section shown in Fig. P7.53, with $h = 8$ in., is made of material having a uniform thickness of 0.10 in. Determine the shear center and the shear-flow distribution at a section where the vertical shear is 2000 lb.

7.54. A beam having the cross section shown in Fig. P7.54, with $h = 12$ in., is made of material having a uniform thickness of 0.20 in. Determine the shear center and the shear-flow distribution at a section where the vertical shear is 3000 lb.

7.55. A beam having the cross section shown in Fig. P7.55, with $h = 10$ in., is made of a material having a uniform thickness of 0.15 in. Determine the shear center and the shear-flow distribution at a section where the vertical shear is 3000 lb.

7.56. A beam having the cross section shown in Fig. P7.56, with $h = 16$ in., is made of a material having a uniform thickness of 0.20 in. Determine the shear center and the shear-flow distribution at a section where the vertical shear is 5000 lb.

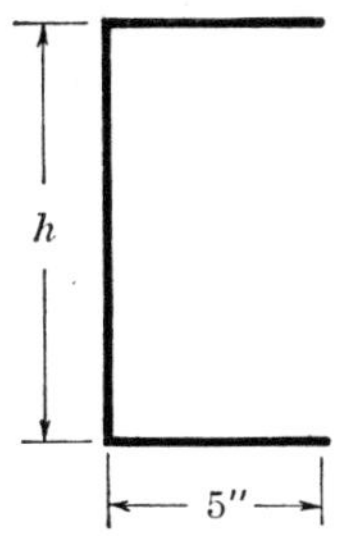

FIG. P7.53, P7.54

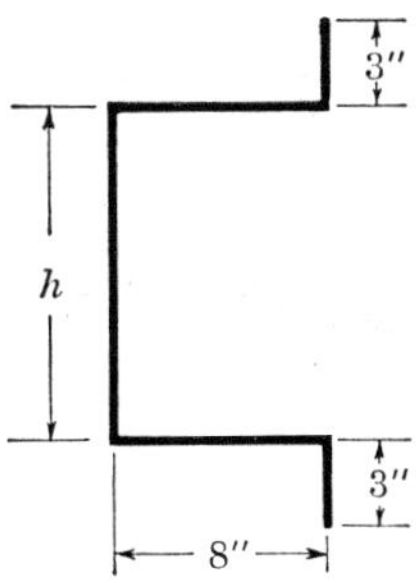

FIG. P7.55, P7.56

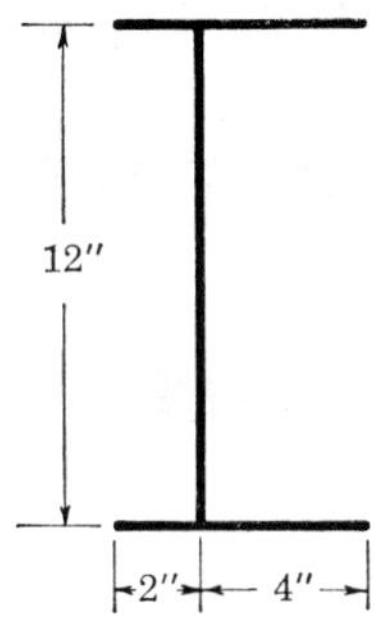

FIG. P7.57, P7.58

★**7.57.** A beam having the cross section shown in Fig. P7.57 is made by welding three flat plates together. The web plate is 0.10 in. thick, while each flange plate has a thickness of 0.15 in. Determine the shear center and the shear-flow distribution at a section where the vertical shear is 1500 lb.

7.58. A beam having the cross section shown in Fig. P7.58 is made by welding three flat plates together. The web plate is 0.15 in. thick, while each flange plate has a thickness of 0.20 in. Determine the shear center and the shear-flow distribution at a section where the vertical shear is 4000 lb.

7.59. For any channel section made of material of thickness t, having a web height of h and equal flanges of width b, show that the shear center is located a distance $b^2h^2t/4I$ from the back of the web.

8

RIVETED AND WELDED CONNECTIONS

8.1 INTRODUCTION

Among the most critical considerations in the design of machines and structures is that of adequately fastening the various members into a complete unit. Many engineers tend to lose sight of the importance of providing proper connections. As a result, the load-carrying capacity of a structure or machine is often substantially less than that of its major components (and, therefore, substantially less than the designer had planned).

Basically, members may be joined by the use of mechanical devices such as nails, pins, bolts, and rivets; by an adhesion process such as gluing or brazing; or by a fusion process such as welding. The "exact" stress conditions in all but the simplest of connections is difficult, if not impossible, to determine. For this reason, certain simplifying assumptions are normally made in the design of any type of connection. The use of such assumptions is justified by and based on past experience with actual structures, as well as on the results of controlled tests. Essentially, allowable stresses are determined by applying an appropriate factor of safety to the statistical average of the failure stresses of a large number of test connections. Thus, the inelastic properties of materials, rather than elastic properties, are the basis of connection analysis and design.

The remainder of this chapter is divided into two parts: Riveted Connections and Welded Connections. These two types of connections are by far the most commonly used for metallic materials.

Riveted Connections

8.2 METHOD OF FABRICATION

The basic method of forming a riveted joint is illustrated in Fig. 8.1. First, suitable holes are drilled or punched in the materials being

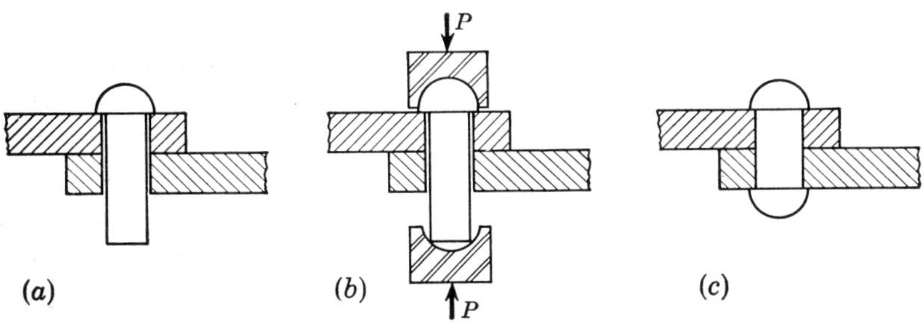

FIG. 8.1 Final forming of rivet.

joined, and a rivet of slightly smaller diameter is inserted, as shown in Fig. 8.1*a*. Then, as shown in Fig. 8.1*b*, sufficient force or pressure is applied through appropriate dies to deform the rivet shank. In deforming, the rivet material completely fills the hole and forms a second head, as shown in Fig. 8.1*c*. The process of thus deforming rivets is referred to as *driving;* rivets may be either *hot-driven* or *cold-driven*. Whether a rivet is driven hot or cold depends primarily on its composition and size. Most steel rivets are hot-driven (at red heat temperatures); many nonferrous rivets are driven cold (at room temperature) or even at refrigerated temperatures.

8.3 BEHAVIOR UNDER LOAD

There are a number of different ways in which even the simplest riveted joint can fail. These methods of failure are listed below and illustrated in Fig. 8.2 with simple tension joints.

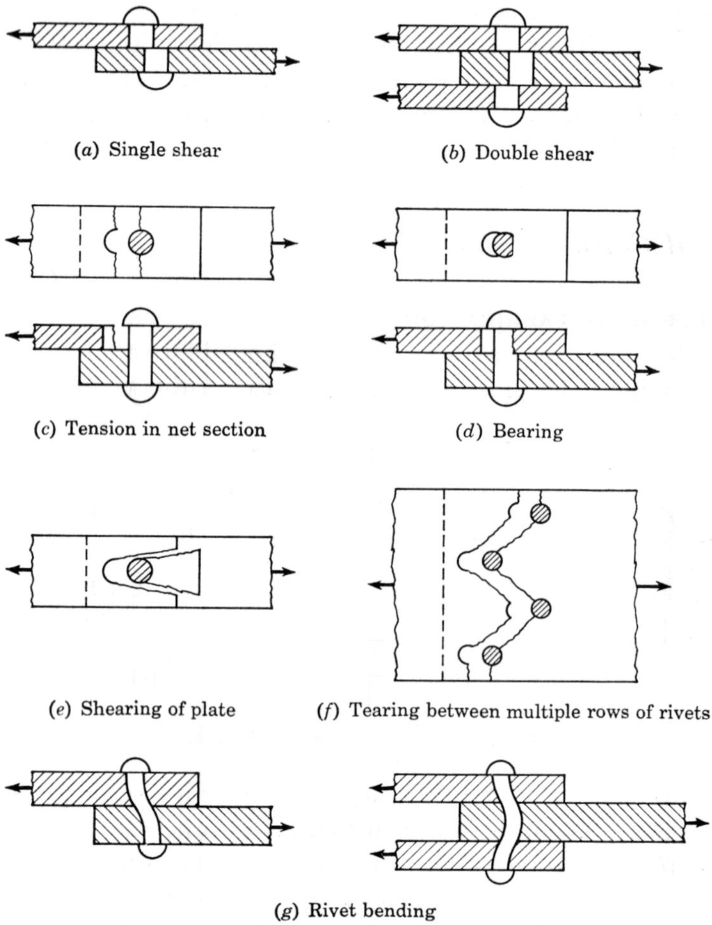

FIG. 8.2 Basic types of failure of riveted joints.

1. *Rivet Shear.* Perhaps the most obvious manner of failure is illustrated in Fig. 8.2*a* and *b*. The rivet in Fig. 8.2*a* is said to be in *single shear*, since failure occurs along only one plane, while the rivet in Fig. 8.2*b* is said to be in *double shear*.

2. *Tension in Net Section.* The materials being joined have minimum cross-sectional areas at sections through the rivet holes. As shown in Fig. 8.2*c* tearing of material would occur at such a "net" section.

3. *Bearing of Plate against Rivet.* Failure can also result from *bearing*, that is, the crushing of the plate by the rivet or the crushing of the rivet shank by the plate. This type of failure is illustrated in Fig. 8.2*d*. Such elongation of rivet holes or crushing of rivets results in a loose connection as well as in excessive deformation.

4. *Shearing of Plate.* If rivet holes are made too near the edge of the plate material, a shear failure of the plate, as shown in Fig. 8.2*e*, may occur.

5. *Tearing between Multiple Rows of Rivets.* When a connection has more than a single row of rivets, failure can occur by a tearing of plate material between adjacent rows of rivet holes. Such a zigzag tearing of plate is illustrated in Fig. 8.2*f*. It occurs when adjacent rows of rivets are too closely spaced.

6. *Bending of Rivets.* If a rivet is relatively long in comparison with its diameter, it may bend excessively prior to failing in shear or bearing. This type of failure, resulting in excessive joint deformation, is illustrated in Fig. 8.2*g*.

Of the possible types of failure listed above, only the first three (rivet shear, bearing, and tension in net section) are used as a basis for design computations. If proper specifications are followed (limiting edge distance, proximity of rivet rows, and rivet length-diameter ratio), the load-carrying capacity of a connection will not be limited by any of the last three types.

8.4 DETERMINATION OF ALLOWABLE STRESSES

Owing to the interaction of the various components, the determination of the actual stresses in a riveted joint is extremely difficult, if not impossible. As a prime example, substantial stresses are normally developed in connections due to the cooling of rivets from the red-heat condition to the temperature of the materials being joined. Since they are prevented from contracting to any considerable degree, high tensile stresses (in some cases up to the yield point) are induced in the rivets. The rivets, in turn, squeeze the materials being joined together and thus greatly increase the friction potential between them. Such a friction force actually contributes significantly to the strength of the connection; in fact, for a well-designed connection, the allowable load may be resisted completely by the friction force between the materials being joined. However, since the magnitude of such friction forces cannot be computed with any degree of certainty, they are neglected in standard practice. Instead, connection design is based on allowable stresses determined from tests, to failure, of actual connections. These test connections are so designed that failure will occur in a particular manner (that is, plate tension, bearing, or rivet shear). Allowable stresses are then determined by dividing the failure loads from a large number of tests by the appropriate area and applying a factor of safety. Since there is no way of knowing the actual bearing-stress distribution of a rivet, for computation purposes bearing stress is assumed to be uniformly distributed over an arbitrarily defined area. For purposes of design and analysis, the bearing area of a rivet is defined as the *product* of the effective *rivet diameter* and the *plate thickness.*

8.5 PRESSURE-VESSEL CONNECTIONS

Illustrated in Fig. 8.3 are two views of several basic types of riveted connections which are used in the fabrication of pressure vessels. The single-riveted *lap joint* gets its name from the fact that plates being joined are lapped over each other and connected by a row of rivets as shown in Fig. 8.3*a*. A butt joint is formed by butting the *main plates* (the plates being joined) and connecting them through auxiliary plates known as *splice plates.* If a single row of rivets is used on each side of the juncture, as shown in Fig. 8.3*b*, the connection is referred to as a single-riveted butt joint. If two rows of rivets are used on each side, as shown in Fig. 8.3*c*, the connection is a double-riveted butt joint. Normally the spacing of rivets in the outer row is increased in order to provide a larger net section (and therefore greater strength) for the main plate. One splice plate of a double-riveted butt joint is normally narrower than the other in order to make calking more effective and the vessel more fluid-tight. When more rivets are required than can be properly spaced in two rows, as when connecting extra-thick plates, triple-riveted or even quadruple-riveted connections may be used.

The center-to-center spacing of rivets in a particular row is known as the rivet *pitch*. For double-riveted joints, having different rivet spacing for its two rows, the larger spacing is called

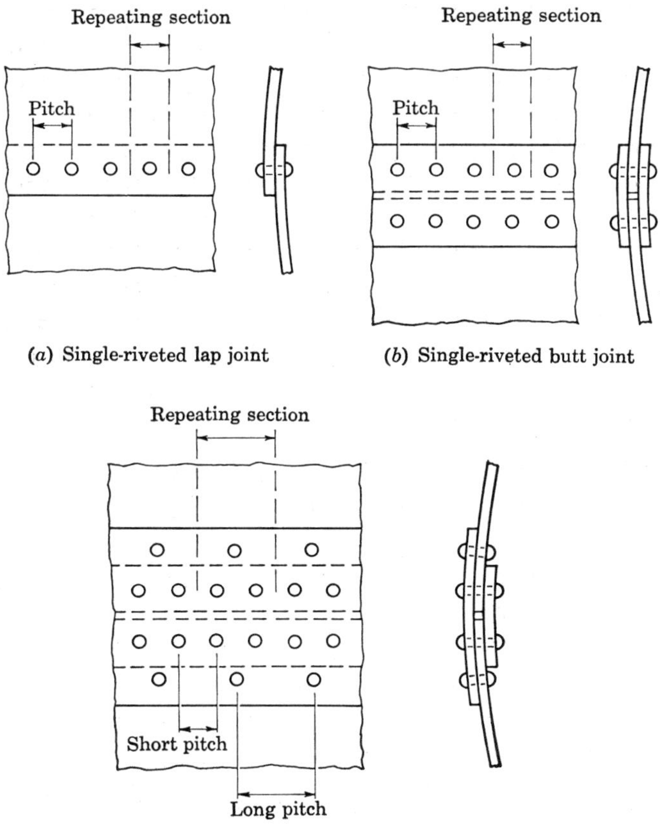

(*a*) Single-riveted lap joint

(*b*) Single-riveted butt joint

(*c*) Double-riveted butt joint

FIG. 8.3 Riveted connections for pressure vessels.

the *long pitch* and the smaller the *short pitch*. Generally, in such a case, the long pitch is made twice as large as the short. The rows of a triple-riveted joint may have three different spacings, referred to as long pitch, intermediate pitch, and short pitch. Most pressure-vessel connections utilize a large number of rivets arranged in some regular pattern. For computation purposes such a connection can be considered as a number of short, identical, parallel joints, or *repeating sections*. As shown in Fig. 8.3, a repeating section is the length comprising a single complete rivet pattern.

Most pressure vessels are designed in accordance with specifications of the ASME Boiler and Pressure Vessel Code.† In general, these specifications require that rivet holes be "match drilled" through the pieces being joined and that such holes be made no more than $\frac{1}{16}$ in. larger than the diameters of the undriven rivets. For strength computations, a driven rivet is assumed to have an effective diameter equal to the diameter of the rivet hole. Allowable stresses are dependent on the operating temperatures of the vessels.

The design of riveted joints is, to some extent, a trial-and-error process. One method of judging the effectiveness of a particular design is to evaluate the *efficiency* of the connection. The efficiency is the *ratio* of the computed *strength of the joint* to the computed *strength of the main plate material.* Joint efficiency may be expressed as a decimal fraction or in percentage form.

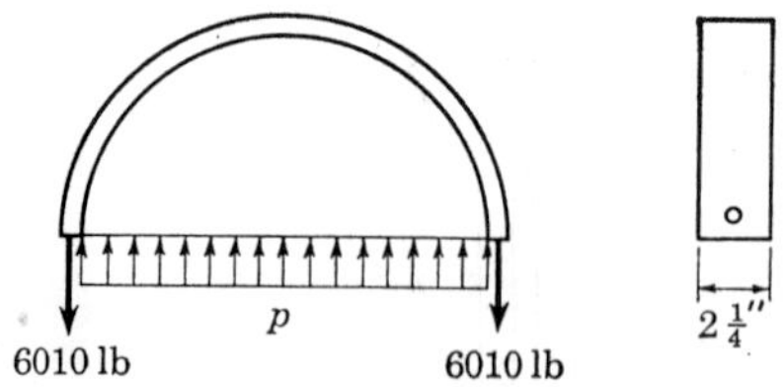

FIG. 8.4

Example 8.1. A cylindrical vessel made from $\frac{3}{8}$-in. plate and having a longitudinal single-riveted lap joint (see Fig. 8.3*a*) has an inside diameter of 40 in. The connection is made with $\frac{3}{4}$-in.-diam rivets (in $\frac{13}{16}$-in.-diam holes) at a pitch of 2.25 in. Allowable stresses are as follows: shearing = 11,600 psi; bearing = 22,000 psi; tension = 13,750 psi. Determine the allowable internal pressure for the vessel.

Determining the allowable load *per repeating section:*
Allowable load in shear

$$P_s = (sA)_s = 11{,}600\left[\frac{\pi}{4}\left(\frac{13}{16}\right)^2\right] = 6010 \text{ lb}$$

Allowable load in bearing

$$P_b = (sA)_b = 22{,}000(\tfrac{13}{16}\ \tfrac{3}{8}) = 6700 \text{ lb}$$

† American Society of Mechanical Engineers, "Boiler and Pressure Vessel Code," Sec. I, Power Boilers, Sec. VIII, Unfired Pressure Vessels, 1959.

Allowable load in tension

$$P_t = (sA)_t = 13{,}750[\tfrac{3}{8}(2.25 - \tfrac{13}{16})] = 7410 \text{ lb}$$

Therefore, the allowable load is 6010 lb as limited by rivet shear.

To find the allowable internal pressure, a free-body diagram of a half ring as shown in Fig. 8.4 is used. From the equations of equilibrium

$$p[(40)(2.25)] = 2(6010)$$
$$p = 134 \text{ psi} \qquad \textit{Ans}$$

Example 8.2. A single-riveted butt joint, similar to Fig. 8.2*b*, is used to connect two ½-in. plates. Splice plates are each ⅜ in. thick; ¾-in. rivets (in 13/16-in. holes) are used at a pitch of 2.5 in. Allowable stresses are as follows: shearing = 11,600 psi; bearing = 22,000 psi; tension = 13,750 psi. Determine the allowable load per linear inch of connection.

Since the combined thickness of the two splice plates is greater than either main plate, the allowable load (in tension or bearing) will not be limited by the splice-plate strength.

Determining the allowable load per repeating section:

Allowable load in shear

$$P_s = 11{,}600\left[2\,\frac{\pi}{4}\left(\frac{13}{16}\right)^2\right] = 12{,}020 \text{ lb}$$

Allowable load in bearing

$$P_b = 22{,}000(\tfrac{13}{16}\,\tfrac{1}{2}) = 8940 \text{ lb}$$

Allowable load in tension

$$P_t = 13{,}750[(2.5 - \tfrac{13}{16})\tfrac{1}{2}] = 11{,}600 \text{ lb}$$

Therefore, the allowable load per repeating section is 8940 lb as limited by bearing. The allowable load per inch of connection is

$$\frac{8940}{2.5} = 3580 \text{ lb/in.} \qquad \textit{Ans}$$

Example 8.3. Determine the efficiency of a double-riveted butt joint, similar to the one shown as Fig. 8.3*c*, which is used to connect two ⅝-in. plates. Splice plates are each ½ in. thick; rivets = 1 in. diam; rivet holes = 1 1/16 in. diam; short pitch = 3 in.; long pitch = 6 in. Allowable stresses are as follows: shearing = 11,600 psi; bearing = 22,000 psi; tension = 13,750 psi.

The determination of the allowable load is simplified by making the following preliminary computations:

Allowable load for one rivet in single shear

$$R = 11{,}600\left[\frac{\pi}{4}\left(\frac{17}{16}\right)^2\right] = 10{,}280 \text{ lb}$$

Allowable load for one rivet bearing on a main plate

$$R = 22{,}000(\tfrac{5}{8}\ \tfrac{17}{16}) = 14{,}610 \text{ lb}$$

Allowable load for one rivet bearing on one splice plate

$$R = 22{,}000(\tfrac{1}{2}\ \tfrac{17}{16}) = 11{,}690 \text{ lb}$$

Determining the allowable load for the 6-in. repeating section:

Allowable load for rivets

The allowable load for each of the two inner row rivets is 14,610 lb. The load is governed by main plate bearing, since each rivet is in double shear and bears on two splice plates (in other words, 14,610 lb is less than twice 10,280 lb or twice 11,690 lb). The allowable load for the rivet in the outer row is 10,280 lb, as limited by (single) shear. The allowable load for the three rivets of a repeating section is therefore

$$P = 2(14{,}610) + 10{,}280 = 39{,}500 \text{ lb}$$

Allowable load for tension in main plates

At a section through the outer row of rivets, the allowable load for tension in main plates is

$$P = 13{,}750[\tfrac{5}{8}(6 - \tfrac{17}{16})] = 42{,}400 \text{ lb}$$

In order for a main plate to tear at a section through the inner row of rivets, *the rivet at the outer row must also fail.* Therefore, the allowable load for main plate tension at this section is

$$P = 13{,}750[\tfrac{5}{8}(6 - \tfrac{34}{16})] + 10{,}280 \qquad \text{or}$$
$$P = 33{,}300 + 10{,}280 = 43{,}580 \text{ lb}$$

Allowable load for tension in splice plates

The allowable load for tension in splice plates is governed by the section through the inner row of rivets:

$$P = 13{,}750(2)[\tfrac{1}{2}(6 - \tfrac{34}{16})] = 53{,}300 \text{ lb}$$

Therefore, the allowable load per repeating section is 39,500 lb and

$$\text{Efficiency} = \frac{39{,}500}{13{,}750[\frac{5}{8}(6)]} = 0.77 \qquad \textit{Ans}$$

PROBLEMS

NOTE: For all problems in this group, use Boiler and Pressure Vessel Code Specifications with the following allowable stresses: shearing = 11,600 psi; bearing = 22,000 psi; tension = 13,750 psi.

***8.1.** A cylindrical pressure vessel having an inside diameter of 15 in. is made from $\frac{3}{8}$-in. plate material with a longitudinal double-riveted lap joint. The joint is made with $\frac{5}{8}$-in. rivets (in $\frac{11}{16}$-in.-diam holes); the pitch of each row of rivets is 2 in. Determine the allowable internal pressure.

8.2. A cylindrical pressure vessel having an inside diameter of 20 in. is made from $\frac{5}{8}$-in. plate material with a longitudinal double-riveted lap joint. The joint is made with $\frac{3}{4}$-in. rivets (in $\frac{13}{16}$-in.-diam holes); the pitch of each row of rivets is 2.5 in. Determine the allowable internal pressure.

8.3. A cylindrical pressure vessel having an inside diameter of 30 in. is made from $\frac{3}{4}$-in. plate material with a longitudinal double-riveted lap joint. The joint is made with $\frac{7}{8}$-in. rivets (in $\frac{15}{16}$-in.-diam holes); the pitch of each row of rivets is 3 in. Determine the allowable internal pressure.

***8.4.** A single-riveted butt joint is used to connect two $\frac{1}{2}$-in. plates. Splice plates are each $\frac{5}{16}$ in. thick; $\frac{5}{8}$-in. rivets (in $\frac{11}{16}$-in.-diam holes) are used at a pitch of 2.5 in. Determine the joint efficiency.

8.5. A single-riveted butt joint is used to connect two $\frac{5}{8}$-in. plates. Splice plates are each $\frac{3}{8}$ in. thick; $\frac{3}{4}$-in. rivets (in $\frac{13}{16}$-in.-diam holes) are used at a pitch of 2.25 in. Determine the joint efficiency.

8.6. A single-riveted butt joint is used to connect two $\frac{3}{4}$-in. plates. Splice plates are each $\frac{1}{2}$ in. thick; $\frac{7}{8}$-in. rivets (in $\frac{15}{16}$-in.-diam holes) are used at a pitch of 3.5 in. Determine the joint efficiency.

***8.7.** A single-riveted butt joint is to be used to connect $\frac{3}{8}$-in. main plates. Each splice plate is to be $\frac{1}{4}$ in. thick. Using $\frac{5}{8}$-in. rivets (in $\frac{11}{16}$-in.-diam holes), determine the rivet pitch which will give the maximum joint efficiency.

8.8. A single-riveted butt joint is to be used to connect $\frac{1}{2}$-in. main plates. Each splice plate is to be $\frac{5}{16}$ in. thick. Using $\frac{3}{4}$-in. rivets (in $\frac{13}{16}$-in.-diam holes), determine the rivet pitch which will give the maximum joint efficiency.

8.9. A single-riveted butt joint is to be used to connect $\frac{5}{8}$-in. main plates. Each splice plate is to be $\frac{3}{8}$ in. thick. Using $\frac{7}{8}$-in. rivets (in $\frac{15}{16}$-in.-diam holes), determine the rivet pitch which will give the maximum joint efficiency.

★8.10. A double-riveted butt joint, similar to the one illustrated in Fig. 8.3*c*, is used to connect two $\frac{3}{8}$-in. plates. Splice plates are each $\frac{1}{4}$ in. thick. Rivets = $\frac{5}{8}$ in. diam; rivet holes = $\frac{11}{16}$ in. diam; short pitch = 3 in.; long pitch = 6 in. Determine the joint efficiency.

8.11. A double-riveted butt joint, similar to the one illustrated in Fig. 8.3*c*, is used to connect two $\frac{5}{8}$-in. plates. Splice plates are each $\frac{3}{8}$ in. thick. Rivets = $\frac{3}{4}$ in. diam; rivet holes = $\frac{13}{16}$ in. diam; short pitch = 2.25 in.; long pitch = 4.5 in. Determine the joint efficiency.

8.12. A double-riveted butt joint, similar to the one illustrated in Fig. 8.3*c*, is used to connect two $\frac{3}{4}$-in. plates. Splice plates are each $\frac{1}{2}$ in. thick. Rivets = $\frac{7}{8}$ in. diam; rivet holes = $\frac{15}{16}$ in. diam; short pitch = 2.5 in.; long pitch = 5 in. Determine the joint efficiency.

8.13. A cylindrical pressure vessel made from $\frac{7}{8}$-in. plate material and having a longitudinal triple-riveted butt joint has an inside diameter of 7.5 ft. Determine the allowable internal pressure if the efficiency of the joint is 0.88.

8.14. A cylindrical pressure vessel made from 1-in. plate material and having a longitudinal quadruple-riveted butt joint has an inside diameter of 10 ft. If the efficiency of the joint is 93 per cent, determine the allowable internal pressure.

8.6 STRUCTURAL CONNECTIONS

Because of the method of fabricating connections for most structures, the areas used in computing connection strength are significantly different from those specified by the Boiler Code (and discussed in the previous article). Rivet holes for structural connections are normally punched instead of being drilled; also, the holes are located by measurements on the separate pieces instead of by matching the holes of one piece to the other. Since punched holes are likely to be somewhat rough and uneven and to compensate for the lack of perfect hole alignment, computations of joint strength are made on the following basis: (1) The effective diameters of rivets in bearing and in shear are assumed to be the same as for the undriven rivets, and (2) rivet-hole diameters are assumed to be $\frac{1}{8}$ in. larger than the corresponding rivet sizes. Actually, the holes are punched only $\frac{1}{16}$ in. larger; the additional $\frac{1}{16}$ in. is an allowance for material damage due to punching.

The specifications most commonly used for structural connections are those of the AISC.† The specifications include different allowable bearing stresses for two different bearing conditions. As shown in Fig. 8.5, the stress distribution along the depth of a plate depends on whether the plate is "enclosed" or "unenclosed." For the lap joint of Fig. 8.5*a*, both plates are *unenclosed;* for the joint of Fig. 8.5*b*, plate B is enclosed while plates A and C are not. Because of the more uniform stress distribution, higher bearing

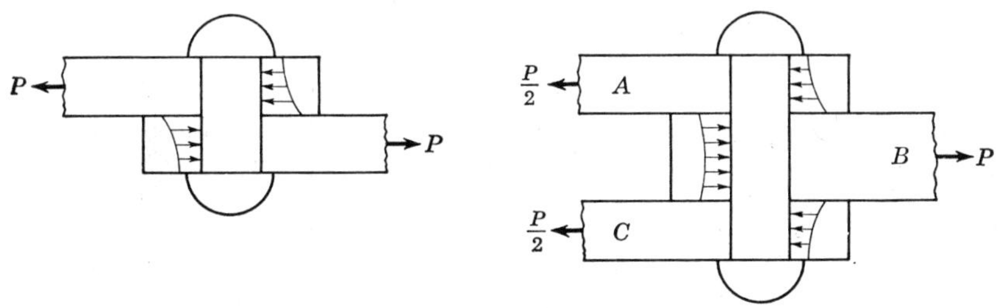

FIG. 8.5 Simplified bearing-stress pattern.

stresses are allowed on enclosed plates. The AISC allowable stresses for the low-carbon structural steel (ASTM A-7) and rivets (ASTM A-141) normally used are:

Tension on net section	20,000 psi
Shearing on rivets	15,000 psi
Bearing, enclosed	40,000 psi
Bearing, unenclosed	32,000 psi

Example 8.4. A roof truss member consisting of two 4- by 3- by $\frac{1}{4}$-in. angles is connected to a $\frac{7}{16}$-in. gusset plate with four $\frac{3}{4}$-in. rivets as shown in Fig. 8.6. Using AISC allowable stresses, determine the allowable axial load P.

Preliminary computations are made to determine the allowable load per rivet:

† "Steel Construction Manual," 5th ed., American Institute of Steel Construction, New York, 1961.

Allowable load for one rivet in double shear

$$R = 2\left[15{,}000\,\frac{\pi}{4}\left(\frac{3}{4}\right)^2\right] = 13{,}250 \text{ lb}$$

Allowable load for one rivet bearing (unenclosed) on the two angles

$$R = 2(32{,}000)(\tfrac{1}{4}\ \tfrac{3}{4}) = 12{,}000 \text{ lb}$$

Allowable load for one rivet bearing (enclosed) on the gusset plate

$$R = 40{,}000(\tfrac{7}{16}\ \tfrac{3}{4}) = 13{,}120 \text{ lb}$$

The allowable load for the rivets is

$$P = 4(12{,}000) = 48{,}000 \text{ lb}$$

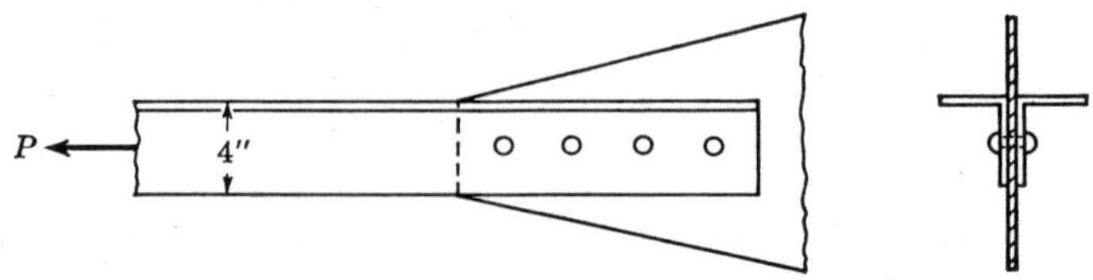

FIG. 8.6

The allowable load for tension in the angles is governed by their net section at the extreme left rivet hole. Obtaining the gross area of the angles from Table A.8,

$$P = 20{,}000(2)(1.69 - \tfrac{1}{4}\ \tfrac{7}{8}) = 58{,}800 \text{ lb}$$

The allowable load, as limited by rivet strength, is 48,000 lb (*Ans*).

Example 8.5. A built-up beam is composed of a 24WF160 and two 16- by ⅞-in. cover plates as shown in Fig. 8.7. At a section of the beam where the shear is 175,000 lb, determine the required longitudinal pitch for ¾-in. rivets used to connect the cover plates to the wide flange. Use AISC specifications.

The purpose of the rivets is to make the plates and wide flange act as a single flexural member; this is accomplished by the resistance of the rivets to horizontal shear. Determining the allowable load for one rivet in shear,

$$R_s = 15{,}000\,\frac{\pi}{4}\left(\frac{3}{4}\right)^2 = 6630 \text{ lb}$$

Since the flanges of the wide flange are thicker than the cover plates, the allowable load for a rivet in bearing is governed by the cover-plate thickness and

$$R_b = 32{,}000(\tfrac{7}{8}\ \tfrac{3}{4}) = 21{,}000 \text{ lb}$$

If the beam were monolithic instead of built-up and of the same cross section, the horizontal shearing stress at either plane where a cover plate adjoins the wide flange would be $s_s = VQ/Iw$; the horizontal shear force

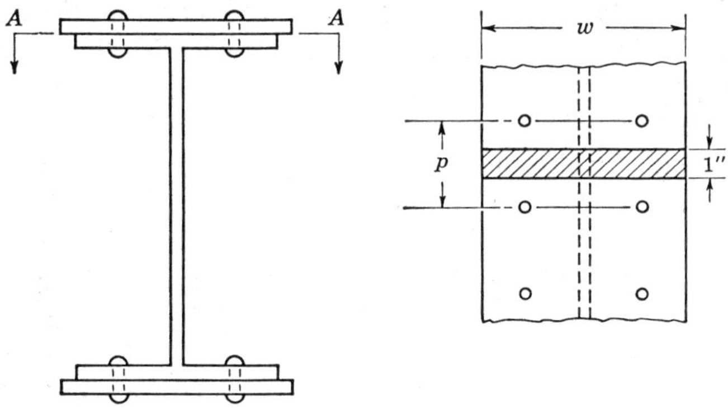

Cross section　　　Plan view *A-A*

FIG. 8.7

per inch of length would be the product of the shearing stress and the shaded area indicated on the plan view of Fig. 8.7. Thus

$$H = s_s(w)(1) = \frac{VQ}{Iw}w = \frac{VQ}{I}$$

Evaluating I, the composite moment of inertia,

WF section (from table)		= 5110
Cover plates	= 2[(16)(0.875)(12.80)²]	= 4590
Total I		9700 in.⁴

Evaluating Q, the first moment of the area of a cover plate,

$$Q = (16)\tfrac{7}{8}(12.80) = 179.2 \text{ in.}^3$$

Therefore, the horizontal shear force per inch of length is

$$H = \frac{VQ}{I} = \frac{175{,}000(179.2)}{9700} = 3230 \text{ lb/in.}$$

The horizontal shearing force to be resisted by a pair of rivets is the product of H and a distance equal to the longitudinal rivet pitch p. Therefore

$$2(6630) = 3230(p)$$
$$p = 4.10 \text{ in.} \qquad \textit{Ans}$$

Unfinished and high-strength bolts are sometimes used for connections instead of rivets. The strength computations for connections utilizing unfinished bolts are identical with those for riveted joints *except for* the use of appreciably *lower allowable stresses.* The use of high-strength bolts for structural connections is a comparatively recent development. These bolts, which have a tensile strength of about twice that of structural steel, are installed in punched holes and tightened to the degree that their shanks are stressed to about 60,000 psi. Because of the greatly increased friction potential between the materials being joined, the use of properly installed high-strength bolts results in very tight connections. There are good reasons to believe that these joints are significantly stronger than similar riveted joints; however, under present practices, strength computations for high-strength bolt connections are identical in all respects (including allowable stresses) with those for structural-steel riveted joints.

PROBLEMS

***8.15.** An 8- by ⅜-in. tension member is to be spliced with two 8- by ¼-in. plates as shown in Fig. P8.15. Using AISC specifications, determine the allowable axial load P and the number of ¾-in. rivets required on each side of the joint.

8.16. An 8- by ½-in. tension member is to be spliced with two 8- by ¼-in. plates as shown in Fig. P8.16. Using AISC specifications, determine the allowable axial load P and the number of ¾-in. rivets required on each side of the joint.

8.17. An 8- by ½-in. tension member is to be spliced with two 5⁄16-in. plates as shown in Fig. P8.17. Using AISC specifications, determine the allowable axial load P and the number of ¾-in. rivets required on each side of the joint.

***8.18.** A building truss member consisting of two 8-in. 11.5-lb channel sections is to be connected to a ½-in. gusset plate as shown in Fig.

P8.18. Using AISC specifications, determine the allowable axial load P and the number of $\frac{3}{4}$-in. rivets required for the connection.

8.19. A building truss member consisting of two 8-in. 13.75-lb channel sections is to be connected to a $\frac{1}{2}$-in. gusset plate as shown in Fig. P8.19. Using AISC specifications, determine the allowable axial load P and the number of $\frac{3}{4}$-in. rivets required for the connection.

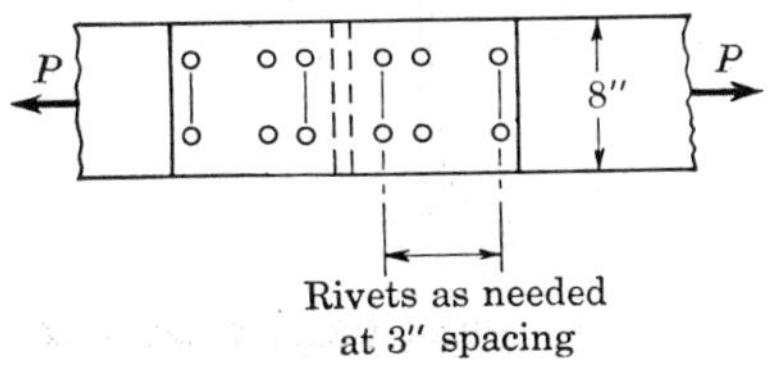

FIG. P8.15, P8.16, P8.17

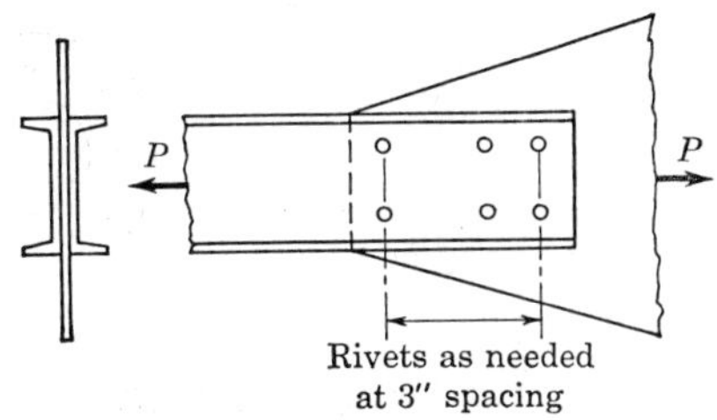

FIG. P8.18, P8.19, P8.20

8.20. A building truss member consisting of two 8-in. 18.75-lb channel sections is to be connected to a $\frac{1}{2}$-in. gusset plate as shown in Fig. P8.20. Using AISC specifications, determine the allowable axial load P and the number of $\frac{7}{8}$-in. rivets required for the connection.

***8.21.** The flanged coupling shown in Fig. P8.21 is made with eight $\frac{3}{4}$-in.-diam unfinished bolts evenly spaced about a circle 9.5 in. in diameter. Each flange thickness is 1.00 in. Allowable stresses are as follows: shearing = 10,000 psi; unenclosed bearing = 20,000 psi. Compute the allowable torque for the coupling.

8.22. The flanged coupling shown in Fig. P8.22 is made with twelve $\frac{7}{8}$-in.-diam unfinished bolts evenly spaced about a circle 14 in. in diameter. Each flange thickness is 1.125 in. Allowable stresses are as follows: shearing = 10,000 psi; unenclosed bearing = 20,000 psi. Compute the allowable torque for the coupling.

8.23. The flanged coupling shown in Fig. P8.23 is to be made with 1-in.-diam unfinished bolts spaced evenly about a circle of 15 in. diameter. Each flange thickness is to be 1.25 in. Allowable stresses are as follows: shearing = 10,000 psi; unenclosed bearing = 20,000 psi. If the coupling is to be loaded with a torque of 900,000 lb-in., determine the required number of bolts.

***8.24.** At a section where the shear is 125 kips, a riveted plate girder has the cross section shown in Fig. P8.24. Using AISC specifications, determine the required longitudinal pitch for $\frac{7}{8}$-in. rivets used to connect the cover plates to the angles.

8.25. At a section where the shear is 100 kips, a riveted plate girder has the cross section shown in Fig. P8.25. Using AISC specifications, determine the required longitudinal pitch for ¾-in. rivets used to connect the cover plates to the angles.

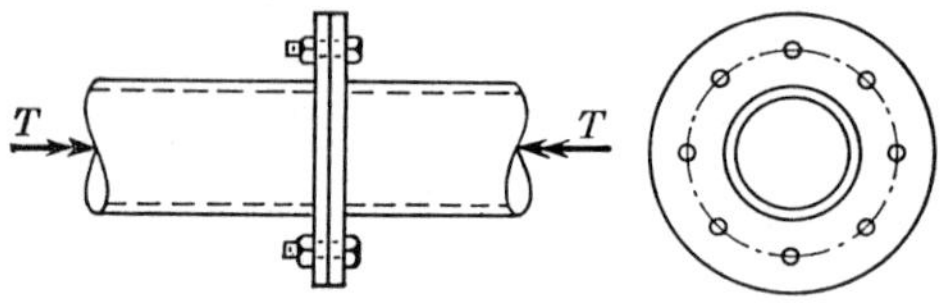

FIG. P8.21, P8.22, P8.23

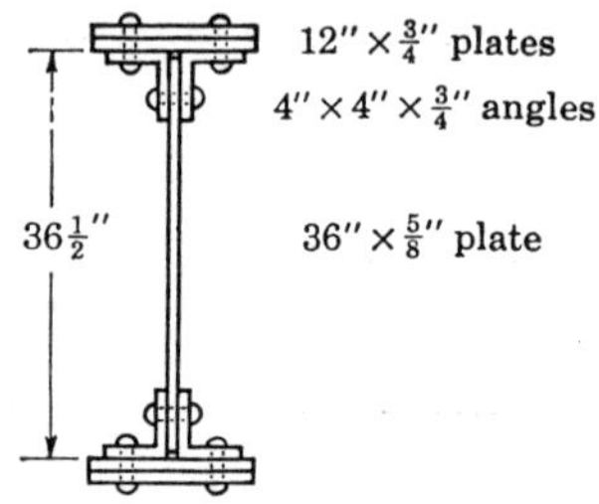

FIG. P8.24, P8.25, P8.26, P8.27

★**8.26.** At a section where the shear is 125 kips, a riveted plate girder has the cross section shown in Fig. P8.26. Using AISC specifications, determine the required longitudinal pitch for ⅞-in. rivets used to fasten the angles to the web plate.

8.27. At a section where the shear is 100 kips, a riveted plate girder has the cross section shown in Fig. P8.27. Using AISC specifications, determine the required longitudinal pitch for ¾-in. rivets used to fasten the angles to the web plate.

8.7 ECCENTRICALLY LOADED CONNECTIONS

While most riveted connections are for practical purposes centrically loaded, connections which are to resist eccentric loading are frequently necessary. Common examples of eccentrically loaded connections are structural brackets and moment-resisting beam connections.

The internal force on any rivet of an eccentrically loaded connection can be determined by resolving the eccentric load, as shown in Fig. 8.8, into a centric load (a load passing through the center of gravity of the rivet group) and a torsion couple. If the rivets are all of the same size, the internal force on each rivet due to the centric load is $R_C = P/n$, where n is the number of rivets. If *it is assumed that the plates being fastened are rigid* and therefore that all the deformation occurs in the rivets, the force on any rivet due to torsion can be determined from a modified version of the equation $s_s = Tr/J$. Referring to Fig. 8.9, let A represent the area of one

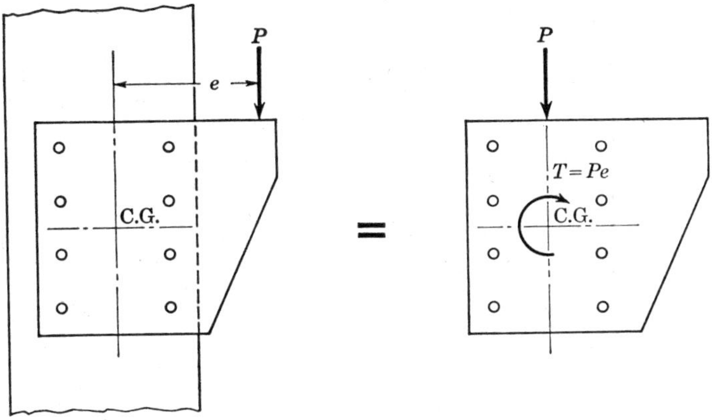

FIG. 8.8

rivet and r represent the distance from the center of gravity of the rivet group to the center of a particular rivet. The force due to torsion R_T acting perpendicular to the distance r would be $R_T = s_s A = (Tr/J)(A)$, where J represents the polar moment of inertia of the rivet areas. Neglecting the moment of inertia of each rivet about its own centroidal axis, $J = \Sigma Ar^2 = \Sigma A(x^2 + y^2)$. Therefore

$$R_T = \frac{Tr(A)}{\Sigma A(x^2 + y^2)} = \frac{Tr}{\Sigma(x^2 + y^2)}$$

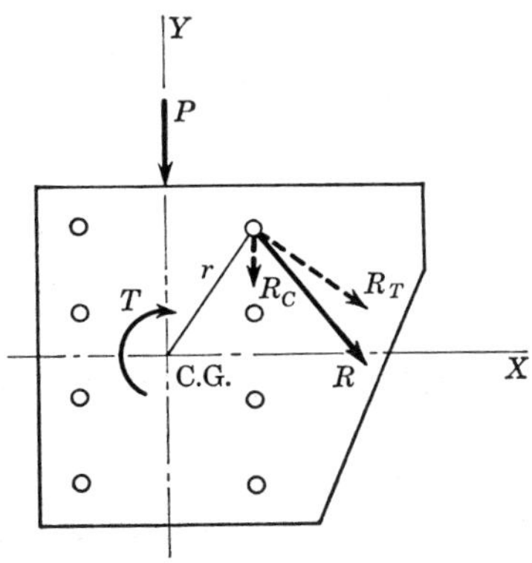

FIG. 8.9

The resultant force on any rivet is the *vector sum* of the force due to the centric load and the force due to torsion, or

$$R = R_C \nrightarrow R_T$$

$$R = \frac{P}{n} \nrightarrow \frac{Tr}{\Sigma(x^2 + y^2)} \tag{8.1}$$

Example 8.6. A 10,000-lb load is to be supported by a 1-in.-thick bracket fastened to a 1.25-in.-thick main plate with six rivets as shown in Fig. 8.10*a*. Using AISC allowable stresses, determine the required

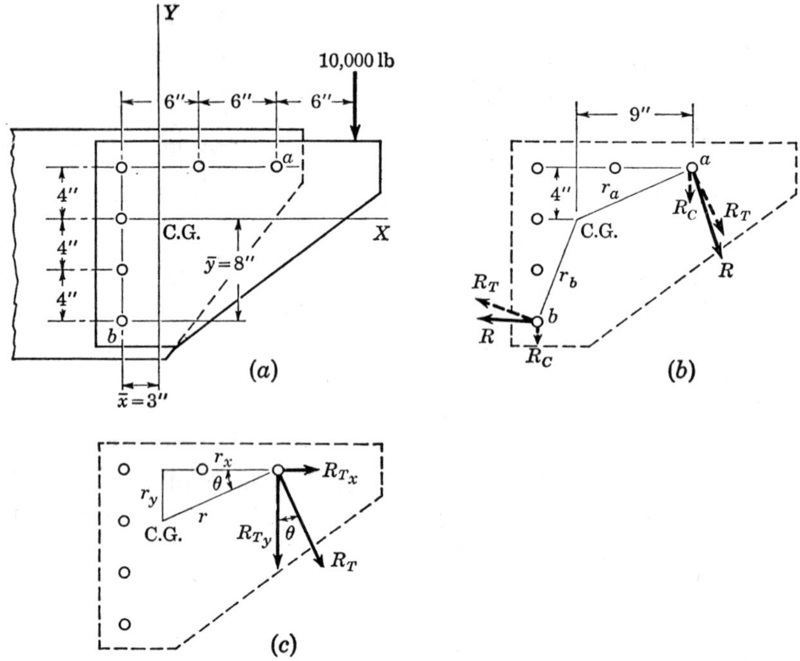

FIG. 8.10

size of rivets. Rivet sizes available are from ½ to 1¼ in. in ⅛-in. increments.

Locating the center of gravity of the rivet group with respect to rivet *b*,

$$\bar{x} = \frac{4(0) + 1(6) + 1(12)}{6} = 3 \text{ in.}$$

$$y = \frac{1(0) + 1(4) + 1(8) + 3(12)}{6} = 8 \text{ in.}$$

Evaluating $\Sigma(x^2 + y^2)$ for the rivet group

$$\Sigma x^2 = 4(3)^2 + 1(3)^2 + 1(9)^2 = 126$$
$$\Sigma y^2 = 3(4)^2 + 1(0)^2 + 1(4)^2 + 1(8)^2 = 128$$
$$\Sigma(x^2 + y^2) = 126 + 128 = 254$$

Since the R_T force on any rivet is a function of its distance from the center of gravity of the rivet group, either rivet *a* or rivet *b* will be the most highly stressed. As shown in Fig. 8.10*b*, rivet *a* will be the most highly stressed (not only because it will have the larger R_T but also because the vertical component of its R_T will act in the same direction as its R_C).

The vector summation of the R_C and R_T forces on rivet *a* can be performed most easily by resolving R_T into its horizontal and vertical components. Since R_T acts normal to its radius, its vertical and horizontal components can be determined by using the horizontal and vertical components of the radial distance. As shown in Fig. 8.10*c*,

$$R_{T_x} = R_T \sin\theta = R_T \frac{r_y}{r} = \frac{Tr}{\Sigma(x^2 + y^2)} \frac{r_y}{r} = \frac{Tr_y}{\Sigma(x^2 + y^2)}$$

Similarly,

$$R_{T_y} = \frac{Tr_x}{\Sigma(x^2 + y^2)}$$

Finding the resultant load on rivet *a*,

$$R = R_C \nrightarrow R_T = R_C \nrightarrow R_{T_x} \nrightarrow R_{T_y}$$

$$R_C = \frac{P}{n} = \frac{10{,}000}{6} = 1670 \text{ lb}\downarrow$$

$$R_{T_x} = \frac{Tr_y}{\Sigma(x^2 + y^2)} = \frac{[10{,}000(15)](4)}{254} = 2360 \text{ lb} \rightarrow$$

$$R_{T_y} = \frac{Tr_x}{\Sigma(x^2 + y^2)} = \frac{[10{,}000(15)](9)}{254} = 5320 \text{ lb}\downarrow$$

Therefore, $R_x = 2360$ and $R_y = 1670 + 5320 = 6990$:

$$R = \sqrt{(2360)^2 + (6990)^2} = 7380 \text{ lb}$$

Finding the required diameter as limited by shear,

$$A = \frac{R}{s_s} \qquad \frac{\pi}{4} d^2 = \frac{7380}{15{,}000}$$

$$d = 0.79 \text{ in.}$$

Finding the required diameter as limited by bearing,

$$A = \frac{R}{s_b} \qquad d = \frac{7380}{32{,}000}$$

$$d = 0.23 \text{ in.}$$

Therefore, use ⅞-in. rivets (*Ans*).

PROBLEMS

★**8.28.** If the eccentric load P on the riveted bracket shown in Fig. P8.28 is 30 kips, determine the resultant force on the most highly stressed rivet.

8.29. A load $P = 20$ kips is to be supported by a $\frac{3}{4}$-in.-thick bracket fastened to a $\frac{7}{8}$-in.-thick main plate as shown in Fig. P8.29. Using AISC allowable stresses, determine the required size of rivets. Rivet sizes available are from $\frac{1}{2}$ to $1\frac{1}{4}$ in. in $\frac{1}{8}$-in. increments.

★**8.30.** If the eccentric load P on the riveted bracket shown in Fig. P8.30 is 40 kips, determine the resultant force on the most highly stressed rivet.

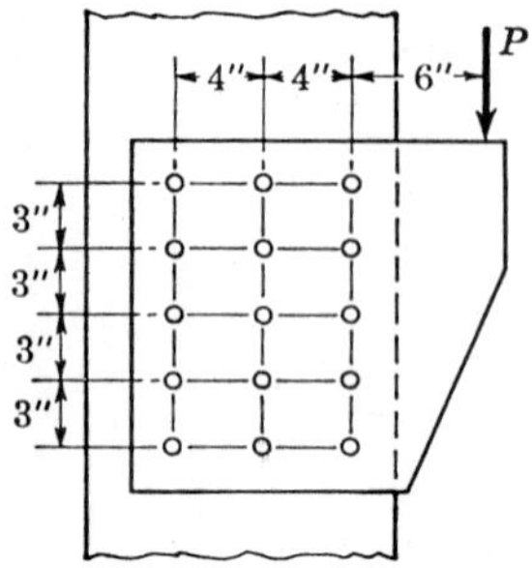

FIG. P8.28, P8.29

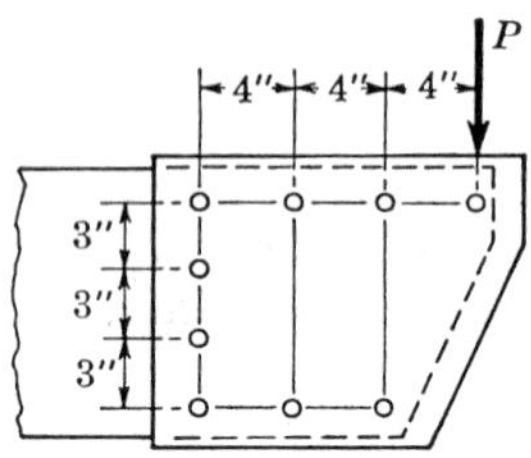

FIG. P8.30, P8.31

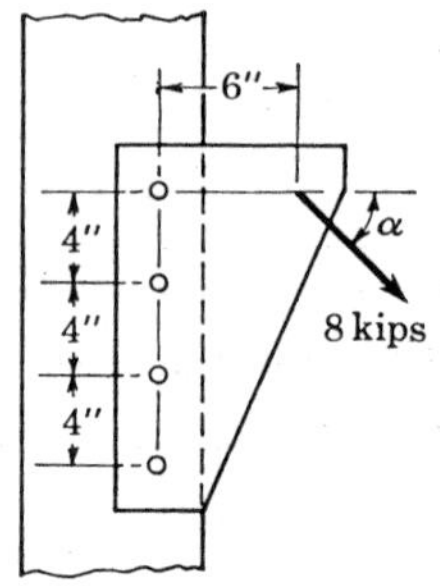

FIG. P8.32, P8.33

8.31. A $\frac{3}{4}$-in.-thick bracket is fastened to a $\frac{7}{8}$-in.-thick main plate with $\frac{3}{4}$-in. rivets as shown in Fig. P8.31. Using AISC specifications, determine the allowable value of the eccentric load P.

★**8.32.** If the angle α of the 8-kip load is 30°, determine the resultant force on the most highly stressed rivet of the bracket shown in Fig. P8.32.

8.33. A $\frac{3}{4}$-in.-thick bracket plate is fastened to a $\frac{7}{8}$-in.-thick main plate as shown in Fig. P8.33. Using AISC allowable stresses, determine the required size of rivets when the angle α is 60°. Rivet sizes available are from $\frac{1}{2}$ to $1\frac{1}{4}$ in. in $\frac{1}{8}$-in. increments.

Welded Connections

8.8 WELDING PROCESSES

Welding is the fusing together of metallic materials; the materials may be joined to each other directly or through a filler (welding) material. Direct welding is accomplished by applying the proper amount of pressure and heat to the juncture of the pieces being joined. Most structural welding is of the type which uses a filler material applied in the form of molten metal. With this method the materials to be joined are heated at the juncture to the melting point while the molten welding material is applied. On cooling, the pieces and the welding material solidify as a unit to form a continuous mass. The heat for most structural welds is furnished either by an electric arc or by the burning of a mixture of oxygen and acetylene gas. In the arc welding process, the *welding rod* is used as an expendable electrode; the welding rod acts as one pole of the arc as well as a source of filler material. In the gas welding process, the welding rod is gradually flame melted and fed into the juncture as filler material is required.

8.9 TYPES OF WELDS; WELD STRENGTH

The two basic types of structural welds are *butt welds* and *fillet welds*. Several of the more common types of butt welds are illustrated in

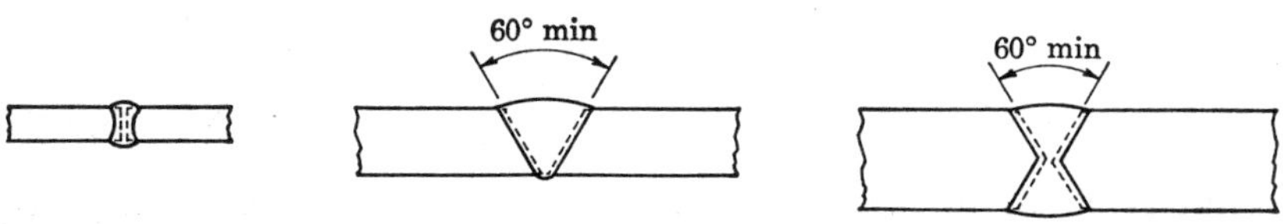

(*a*) Square butt (*b*) Single-vee butt (*c*) Double-vee butt

FIG. 8.11

Fig. 8.11. The simplest type, the square butt, is used to join only relatively thin materials (about $\frac{5}{16}$-in. thickness or less). When thicker materials are joined, a vee or double-vee butt joint may be used. With these types, the members being joined are beveled prior to welding in order to facilitate full penetration of the weld. Specifications generally require the use of welding materials which are as strong as or stronger than the materials being joined; therefore, strength computations for full penetration butt joints are not normally necessary.

Fillet welds connect materials by joining surfaces which are at approximate right angles to each other. As shown in Fig. 8.12, the cross section of a fillet weld is approximately triangular. The size of a fillet weld is designated by its nominal leg dimensions. A weld having unequal legs is designated by the dimensions of both legs (3/8- by 1/2-in. weld, for example). The size of a weld having equal legs is designated by the dimension of one leg only (1/4-in. weld, for example). Almost all fillet welds are of the equal-leg type.

The effective cross-sectional area of a fillet weld is assumed to be triangular (as indicated by the broken lines in Fig. 8.12); the *throat* of a weld is the perpendicular distance from its root to its effective hypotenuse. Regardless of the direction of the applied force, the strength of a fillet weld is considered to be governed by

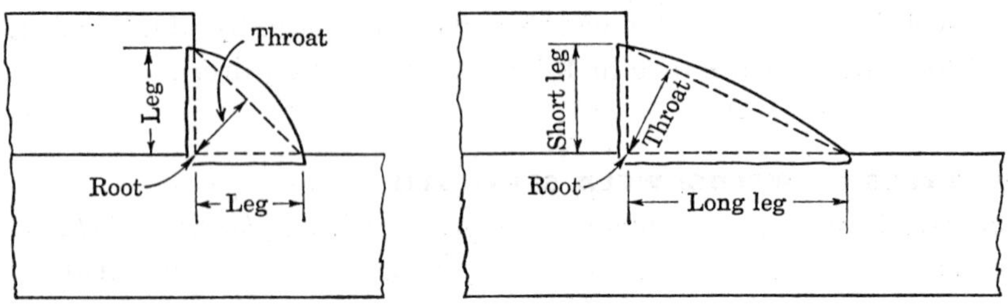

FIG. 8.12 Fillet welds.

the *shearing stress on its throat area.* For an equal-leg weld, the throat is equal to $0.707w$, where w is the weld size. Therefore, the allowable load per unit length of weld is

$$F = s_s(0.707w)$$

where s_s is the allowable shearing stress. For AISC specifications the allowable shearing stress is 13,600 psi and the allowable load per linear inch of weld is

$$F = 13{,}600(0.707w) = 9600w \qquad \text{lb/in.}$$

Thus, for example, the allowable load per linear inch is 3000 lb for a 5/16-in. weld, 3600 lb for a 3/8-in. weld, and 4200 lb for a 7/16-in. weld. It is apparent that the allowable stress of 13,600 psi was specified (instead of, for example, 13,000 or 14,000 psi) in order to have more convenient allowable weld loads for standard-size welds.

8.10 CENTRICALLY LOADED CONNECTIONS

Welded connections are generally easier and simpler to design than equivalent riveted connections. For example, welded butt joints can be made without splice plates. With the use of fillet welds, most truss and beam connections can be made more directly (i.e., with fewer or no connection pieces). Generally, the design process for fillet welds is one of selecting a weld size and using the principles of statics to determine the length and location of the welds required. As a rule, it is more economical, from the standpoint of labor and material required, to use the smallest practical fillet weld size. Specifications normally set both maximum and minimum limits on weld sizes, depending on the thicknesses of the materials being joined. AISC standard sizes for fillet welds start at $\frac{3}{16}$ in. and vary by intervals of $\frac{1}{16}$ in.

Example 8.7. A roof truss member consisting of two 4- by 3- by $\frac{1}{4}$-in. angles is to be connected to a gusset plate with $\frac{3}{16}$-in. fillet welds as shown

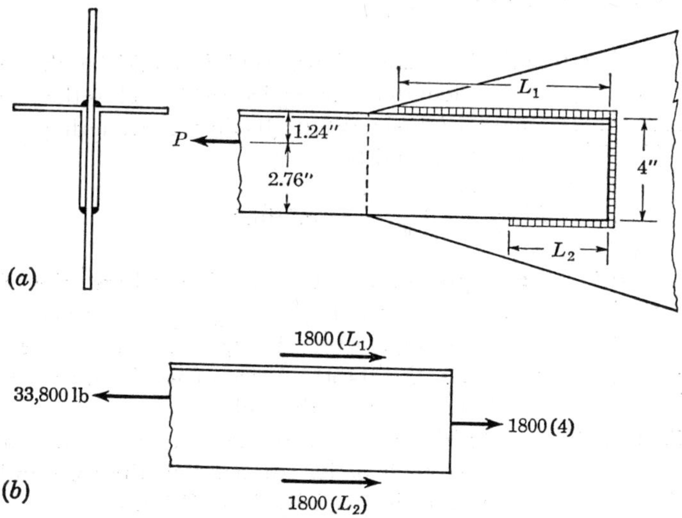

FIG. 8.13

in Fig. 8.13*a*. Using AISC allowable stresses, determine the lengths of side welds required to support the allowable load of the member.

The allowable load for the member is

$$P = sA = 20{,}000[(2)(1.69)] = 67{,}600 \text{ lb}$$

The allowable load per inch of $\frac{3}{16}$-in. weld is

$$F = 9600w = 9600(\tfrac{3}{16}) = 1800 \text{ lb/in.}$$

The weld lengths L_1 and L_2 should be in such proportion that the resultant of the weld forces will be collinear with the axial load. Since the connection is symmetrically loaded, the welding for each angle will be identically stressed. Summing moments of the forces acting on one angle, as shown in Fig. 8.13*b*, about any point along weld L_2,

$$+1800L_1(4) + 1800(4)(2) - 33{,}800(2.76) = 0$$
$$L_1 = 11.0 \text{ in.} \qquad Ans$$

Summing moments about weld L_1

$$+1800L_2(4) + 1800(4)(2) - 33{,}800(1.24) = 0$$
$$L_2 = 3.8 \text{ in.} \qquad Ans$$

As a check, the total length of weld on each angle should be

$$\frac{33{,}800}{1800} = 18.8 \text{ in.}$$

$$4.0 + 11.0 + 3.8 = 18.8 \qquad Check$$

NOTE: For practical considerations, weld L_1 could be changed to a shorter length of larger size weld. Thus the equivalent length of $\frac{1}{4}$-in. weld (having a strength per inch of 2400 lb) would be

$$L_1' = \frac{11.0(1800)}{2400} = 8.2 \text{ in.}$$

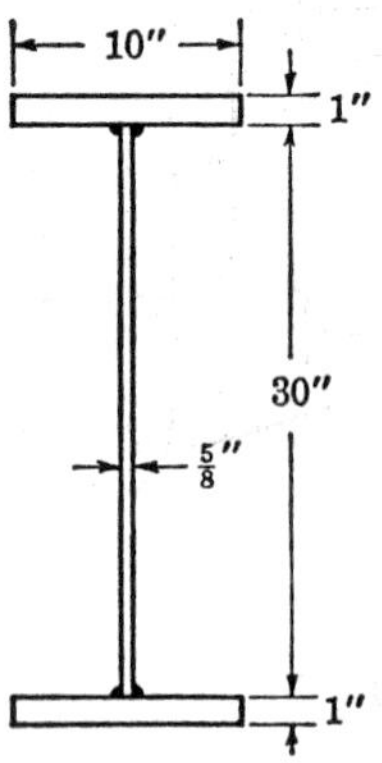

FIG. 8.14

Example 8.8. At a section where the shear is 180,000 lb, a welded plate girder has the cross section shown in Fig. 8.14. Using AISC allowable stresses, determine the standard size of weld required to fasten the flange plates to the web.

As shown in Example 8.5, the horizontal shear force per inch of length between the web and either flange plate is given by $H = VQ/I$. Computing I, the composite moment of inertia,

$$\frac{(\frac{5}{8})(30)^3}{12} = 1410$$
$$2[(10)(1)(15.5)^2] = 4810$$
$$I = \overline{6220} \text{ in.}^4$$

Computing Q, the first moment of the area of a flange plate,

$$Q = 10(1)(15.5) = 155 \text{ in.}^3$$

$$H = \frac{VQ}{I} = \frac{180{,}000(155)}{6220} = 4490 \text{ lb/in.}$$

Since each flange plate is connected by means of a weld on each side of the web plate, the force per inch of weld F is 4490/2 or 2245 lb/in. The required weld size is

$$w = \frac{F}{9600} = \frac{2245}{9600} = 0.23 \text{ in.}$$

For standard size, use ¼-in. weld (*Ans*).

PROBLEMS

NOTE: Use AISC allowable stresses for all problems in this group.

★8.34. A 4- by ½-in. bar is to be connected to a gusset plate with ¼-in. fillet welds as shown in Fig. P8.34. Determine the connection length L required to support the allowable load of the bar.

8.35. A 4- by ⅜-in. bar is to be welded to a gusset plate as shown in Fig. P8.35. If L is to be 4 in., determine the standard weld size required to support the allowable load of the bar.

8.36. A 4- by ⅝-in. bar is fastened to a gusset plate with ⅜-in. welds as shown in Fig. P8.36. If L is 4.5 in., determine the allowable axial load P.

★8.37. A ½-in.-thick bar is to be connected to a gusset plate with $^3/_{16}$-in. fillet welds as shown in Fig. P8.37. Determine the connection length L required to support the allowable axial load of the member.

8.38. A ¾-in.-thick bar is to be connected to a gusset plate with ¼-in. fillet welds as shown in Fig. P8.38. Determine the connection length L required to support the allowable load of the member.

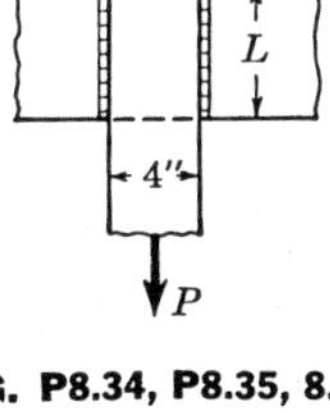

FIG. P8.34, P8.35, 8.36

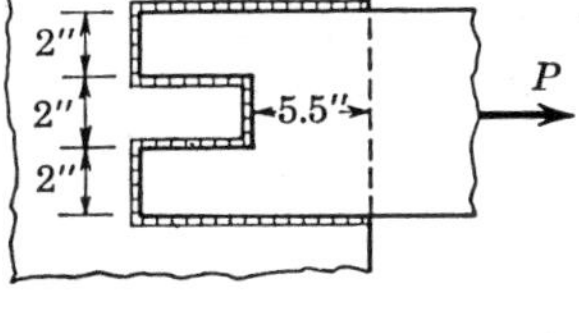

FIG. P8.37, P8.38, 8.39

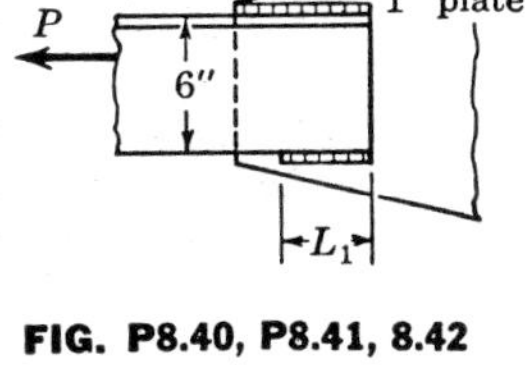

FIG. P8.40, P8.41, 8.42

8.39. A 1-in.-thick bar is to be connected to a gusset plate with $\frac{7}{16}$-in. fillet welds as shown in Fig. P8.39. Determine the connection length L required to support the allowable load of the member.

***8.40.** A building truss member consisting of two 6- by 4- by $\frac{3}{8}$-in. angles is to be connected to a gusset plate with fillet welds as shown in Fig. P8.40. Weld sizes to be used for lengths L_1 and L_2 are $\frac{1}{4}$ and $\frac{1}{2}$ in., respectively. Determine the weld lengths L_1 and L_2 required to support the allowable axial load of the member.

8.41. A building truss member consisting of two 6- by 4- by $\frac{5}{8}$-in. angles is to be connected to a gusset plate with fillet welds as shown in Fig. P8.41. Weld sizes to be used for lengths L_1 and L_2 are $\frac{7}{16}$ and $\frac{3}{4}$ in., respectively. Determine the weld lengths L_1 and L_2 required to support the allowable axial load of the member.

8.42. A building truss member consisting of two 6- by 4- by $\frac{7}{8}$-in. angles is to be connected to a gusset plate with fillet welds as shown in Fig. P8.42. Weld sizes to be used for lengths L_1 and L_2 are $\frac{5}{8}$ and $\frac{3}{4}$ in., respectively. Determine the weld lengths L_1 and L_2 required to support the allowable axial load of the member.

***8.43.** At a section where the shear is 150 kips, a welded plate girder has the cross section shown in Fig. P8.43. Determine the standard size of fillet weld required to fasten the flange plates to the web.

8.44. At a section where the shear is 250 kips, a welded plate girder has the cross section shown in Fig. P8.44. Determine the standard size of fillet weld required to fasten the flange plates to the web.

8.45. At a section where the shear is 325 kips, a welded plate girder has the cross section shown in Fig. P8.45. Determine the standard size of fillet weld required to fasten the flange plates to the web.

***8.46.** A built-up beam is composed of a 21WF142 and two 12- by 1-in. cover plates as shown in Fig. P8.46. At a section where the shear

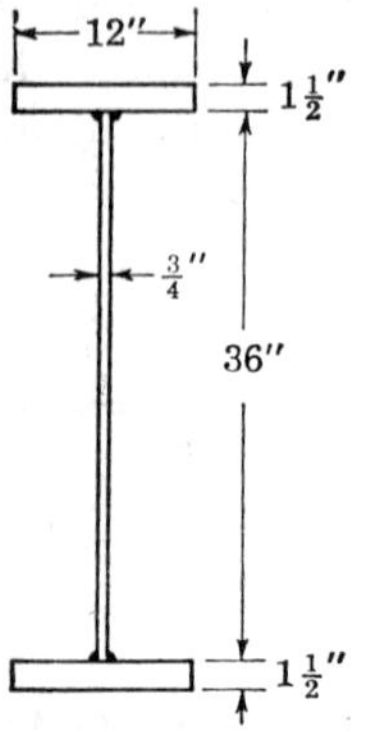

FIG. P8.43, P8.44, 8.45

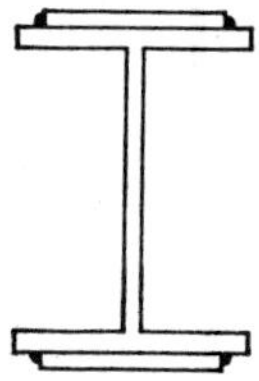

FIG. P8.46, P8.47, 8.48

is 180 kips, determine the standard size of fillet weld required to fasten the cover plates to the wide flange.

8.47. A built-up beam is composed of a 27WF177 and two 12- by 1-in. cover plates as shown in Fig. P8.47. At a section where the shear is 250 kips, determine the standard size of fillet weld required to fasten the cover plates to the wide flange.

8.48. A built-up beam is composed of a 27WF145 and two 12- by ¾-in. cover plates as shown in Fig. P8.48. At a section where the shear is 200 kips, determine the standard size of fillet weld required to fasten the cover plates to the wide flange.

8.11 ECCENTRICALLY LOADED CONNECTIONS

The strength computations for welded connections which are eccentrically loaded are quite similar to those for the equivalent riveted joints. As shown in Fig. 8.15, the eccentric load can be resolved into a centric load passing through the center of gravity of the weld material and a torsion couple. The force per inch of weld at any point is found as the resultant of the force due to centric load and the force due to torsion, or $F = F_C \nrightarrow F_T$.

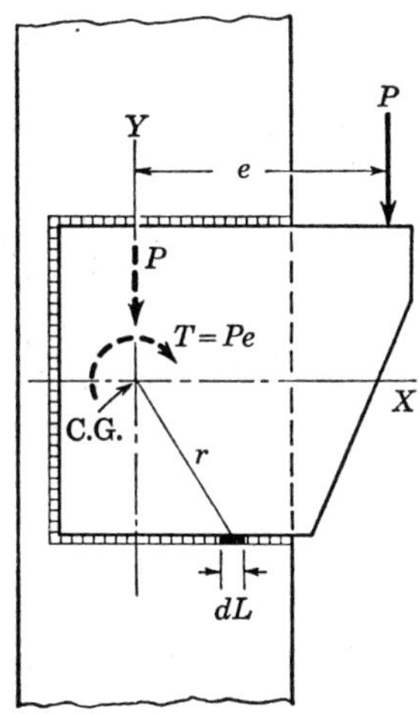

FIG. 8.15

If all the welds are of the same size, $F_C = P/\Sigma L$, where ΣL represents the total length of weld. If *it is assumed that the plates being fastened are rigid* and that all deformation, therefore, occurs in the weld material, F_T can be determined from a modified version of the equation $s_s = Tr/J$. Letting t represent the effective weld thickness, F_T at a distance r from the center of gravity would be

$$F_T = s_s A = s_s(1)(t) = \frac{Tr}{J}t$$

Since J is equal to $\int r^2\,dA$, it can be expressed as $\int r^2 t\,dL$. Therefore

$$F_T = \frac{Trt}{\int r^2 t\,dL} = \frac{Tr}{\int r^2\,dL} = \frac{Tr}{J_L}$$

where J_L represents the polar moment of inertia of the *weld lengths.*

The resultant force per inch at any point on a weld is

$$F = F_C \nrightarrow F_T \qquad \text{or}$$

$$F = \frac{P}{\Sigma L} \nrightarrow \frac{Tr}{J_L} \tag{8.2}$$

(Moments of inertia of lines are computed in much the same way as are moments of inertia of areas. The moment of inertia of any straight line about a centroidal axis normal to its length is $L^3/12$; the transfer term used for the moment of inertia of lines is Ld^2.)

Example 8.9. An eccentrically loaded bracket is to be fastened to a main plate with three lengths of welds as shown in Fig. 8.16*a*. Using AISC allowable stresses, determine the required size of welds.

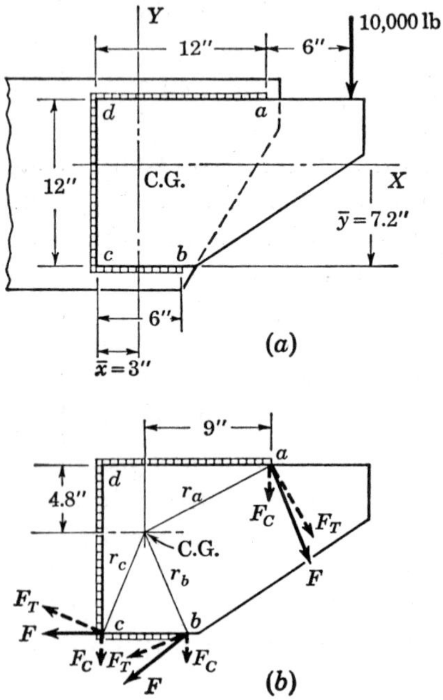

FIG. 8.16

Locating the center of gravity of the welds with respect to point *c*,

$$\bar{x} = \frac{6(3) + 12(0) + 12(6)}{30} = 3.0 \text{ in.}$$

$$\bar{y} = \frac{6(0) + 12(6) + 12(12)}{30} = 7.2 \text{ in.}$$

Computing I_x with respect to the horizontal axis through the center of gravity,

For the 6-in. weld: $0 + 6(7.2)^2 = 311$

For the vertical 12-in. weld: $\frac{(12)^3}{12} + 12(1.2)^2 = 161$

For the horizontal 12-in. weld: $0 + 12(4.8)^2 = 276$

$$I_x = \overline{748} \text{ in.}^3$$

Computing I_y with respect to the vertical axis through the center of gravity,

For the 6-in. weld: $\frac{(6)^3}{12} = 18$

For the vertical 12-in. weld: $0 + 12(3)^2 = 108$

For the horizontal 12-in. weld: $\frac{(12)^3}{12} + 12(3)^2 = 252$

$$I_y = \overline{378} \text{ in.}^3$$

Therefore,

$$J_L = I_x + I_y = 748 + 378 = 1126 \text{ in.}^3$$

Since F_T at any point is a function of its r distance, the highest stress in the welding material will occur at one of the three points shown as a, b, and c in Fig. 8.16*b*. As can be seen, point a will be the most highly stressed, since it has the largest F_T and since the vertical component of this F_T is in the same direction as F_C.

The vector addition of the F_C and F_T forces at point a will be performed after resolving F_T into its vertical and horizontal components. By the procedure used in Example 8.6 it can be shown that

$$F_{T_x} = \frac{Tr_y}{J_L} \quad \text{and} \quad F_{T_y} = \frac{Tr_x}{J_L}$$

$$F_C = \frac{P}{\Sigma L} = \frac{10{,}000}{30} = 333 \text{ lb}\downarrow$$

$$F_{T_x} = \frac{Tr_y}{J_L} = \frac{[10{,}000(15)](4.8)}{1126} = 639 \text{ lb} \rightarrow$$

$$F_{T_y} = \frac{Tr_x}{J_L} = \frac{[10{,}000(15)](9)}{1126} = 1199 \text{ lb}\downarrow$$

Therefore,

$$F_x = 639 \text{ lb} \quad \text{and} \quad F_y = 333 + 1199 = 1532 \text{ lb}$$

and

$$F = \sqrt{(639)^2 + (1532)^2} = 1660 \text{ lb}$$

The required weld size is

$$w = \frac{F}{9600} = \frac{1660}{9600} = 0.173 \text{ in.}$$

For standard size, use $\frac{3}{16}$-in. welds (*Ans.*).

PROBLEMS

NOTE: Use AISC allowable stresses for all problems in this group.

★8.49. For the bracket shown in Fig. P8.49, let $L = 6$ in. and $P = 16$ kips. Determine the standard size of fillet weld required.

8.50. For the bracket shown in Fig. P8.50, let $L = 12$ in. and $P = 32$ kips. Determine the standard size of fillet weld required.

★8.51. For the bracket shown in Fig. P8.51, let $P = 15$ kips. Determine the standard size of fillet weld required.

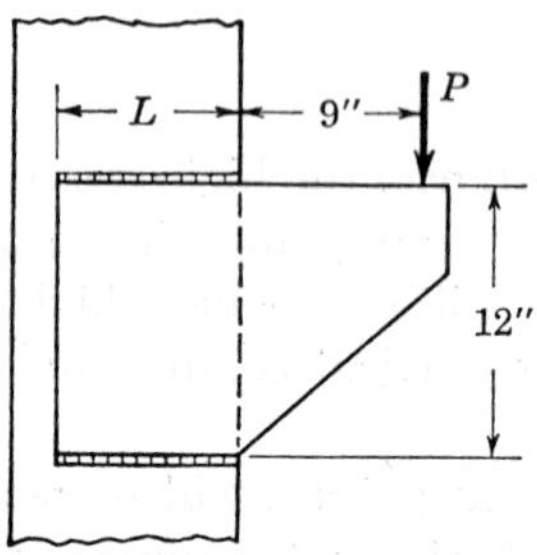

FIG. P8.49, P8.50

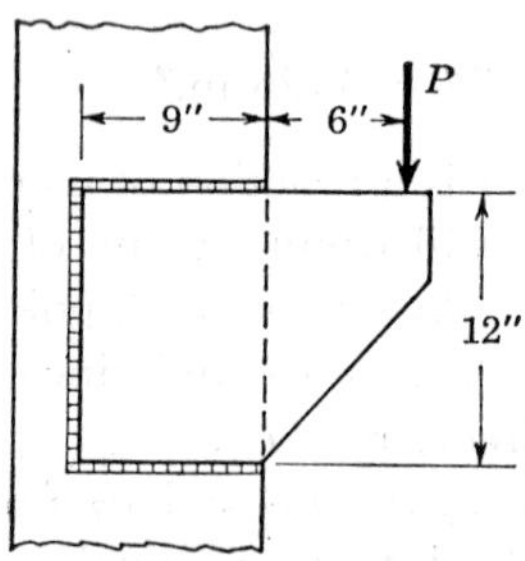

FIG. P8.51, P8.52

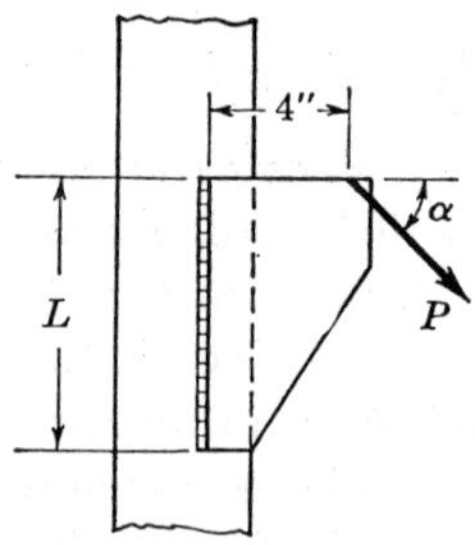

FIG. P8.53, P8.54

8.52. The bracket connection shown in Fig. P8.52 is made with $\frac{5}{8}$-in. fillet welds. Determine the allowable eccentric load P.

★8.53. For the bracket shown in Fig. P8.53, let $L = 8$ in., $P = 2500$ lb, and $\alpha = 30°$. Determine the standard size of fillet weld required.

8.54. A bracket is fastened to a column with a 10-in. length L of $\frac{1}{2}$-in. weld as shown in Fig. P8.54. Determine the allowable load P acting at an angle $\alpha = 60°$.

9

STRESSES FROM COMBINED LOADINGS

9.1 INTRODUCTION

Certain of the previous chapters of this text deal primarily with the stresses due to the three basic types of loading (i.e. axial, bending, and torsion). For simplicity of presentation and since most structural members are subjected essentially to only a single type of loading, stress determinations were treated separately for each type of loading. Many members, however, are stressed by some *combination* of these different loadings. If the deformation of such a member is not excessive, the net stress at any point can be determined as the algebraic sum of the stresses caused by the different loadings.

Two examples of combined loading are shown in Figs. 9.1 and 9.2. Figure 9.1*a* shows a member fixed at one end and supporting a concentrated load Q at the other. As can be seen in Fig. 9.1*b*, the normal stresses on any section x distance from the free end are dependent on both bending moment ($M = Q_y x$) and axial load ($P = Q_x$).

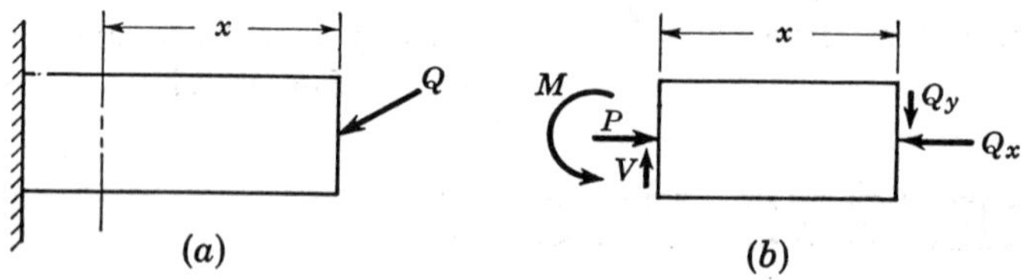

FIG. 9.1

Shown in Fig. 9.2*a* is a bent pipe fixed at one end and acted on by a vertical load Q at the other. As shown in Fig. 9.2*b*, the shearing stresses on the section indicated are dependent on both vertical shear ($V = Q$) and torsion ($T = Qd$).

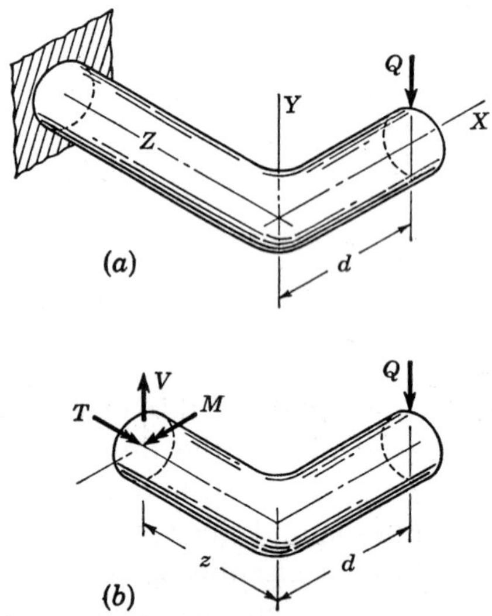

FIG. 9.2

9.2 AXIAL LOADING COMBINED WITH BENDING

Perhaps the most common occurrence of combined stresses is in members which must simultaneously support both bending and axial loading. The normal stresses at any section of such a member can be determined simply by adding algebraically the axial and bending stresses on the section. If the assumptions for both uniform axial stress and the flexure formula are satisfied, the net stress at any point will be

$$s = \pm \frac{P}{A} \pm \frac{My}{I} \tag{9.1a}$$

The maximum tension and compression stresses at any section are given by

$$s = \pm \frac{P}{A} \pm \frac{Mc}{I} \tag{9.1b}$$

In the above equations, it is convenient (and traditional) to regard tension and compression stresses as plus and minus, respectively.

When a member is subjected to both axial load and bending, it should be noted that a certain amount of eccentricity of the axial load will be caused by the bending. This type of eccentricity is illustrated in Fig. 9.3. It can be seen that, at the section shown, a secondary moment equal to $Q_x d$ (or Pd) will result from the bending (caused by the y component of Q). In most structures such secondary moments are negligible. When they are not negligible, they are computed by one of a number of methods, such as, for example, one of those presented in Chaps. 11, 12, and 13. In the problems of this chapter, unless otherwise specified, the stresses due to secondary moments can be neglected.

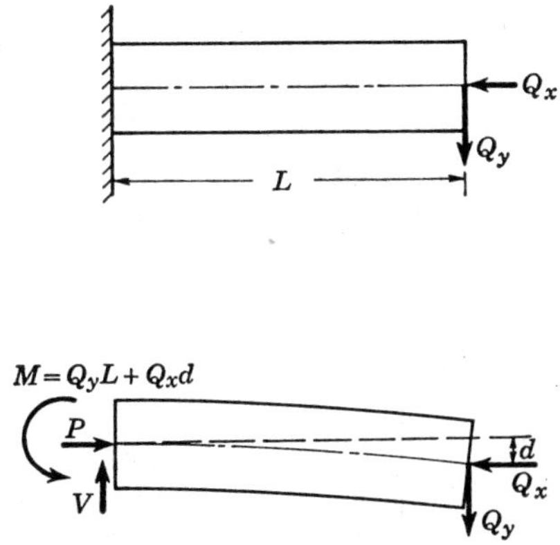

FIG. 9.3

Example 9.1. For the beam shown in Fig. 9.4*a*, determine the maximum tensile and compressive stresses and the normal stress distribution at the section of maximum moment.

Resolving the 10,000-lb force into its horizontal and vertical components, the components of the reactions are determined to be as shown in Fig. 9.4*b*. The moment and axial force diagrams for the beam are drawn as shown in Fig. 9.4*c* and *d*. The moment of inertia of the cross section is computed:

$$I = \frac{4(12)^3}{12} = 576 \text{ in.}^4$$

By examination of the axial force and moment diagrams it is seen that the maximum compressive stress occurs at the top of the section just to the left of the point of application of the 10,000-lb load. Therefore,

$$s = -\frac{P}{A} - \frac{Mc}{I} = -\frac{8660}{4(12)} - \frac{[(14{,}600)(12)](6)}{576}$$

$$s = -180 - 1825 = -2005 \text{ psi} \qquad \textit{Ans}$$

The maximum tensile stress occurs at the bottom of the same section.

$$s = -\frac{P}{A} + \frac{Mc}{I} = -180 + 1825 = +1645 \text{ psi} \qquad \textit{Ans}$$

The stress distribution at the section of maximum moment is drawn as shown in Fig. 9.4*e*. With the use of similar triangles, the neutral axis at this particular section is located as shown.

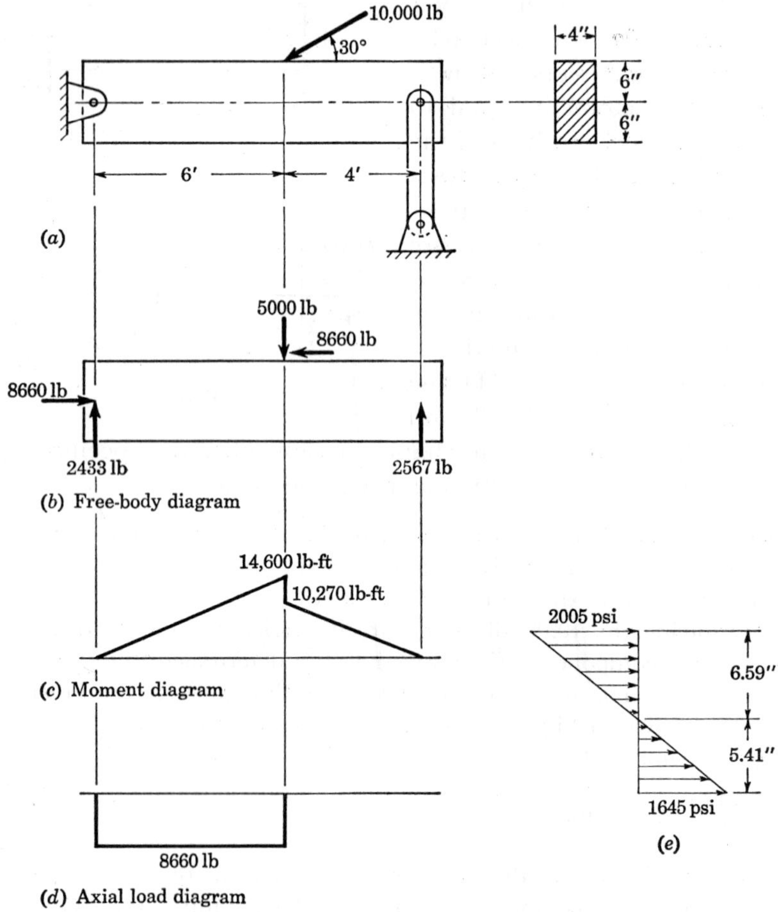

FIG. 9.4

Example 9.2. For the gray cast-iron C-clamp shown in Fig. 9.5*a*, determine the allowable load as limited by normal stress on the section indicated. Allowable stresses are as follows: tension = 6000 psi; compression = 20,000 psi.

From the free-body diagram shown in Fig. 9.5*b*, it is seen that the top part of the clamp must be acted on by a tension force P and a bending moment M (about the centroidal axis of the cross section). The

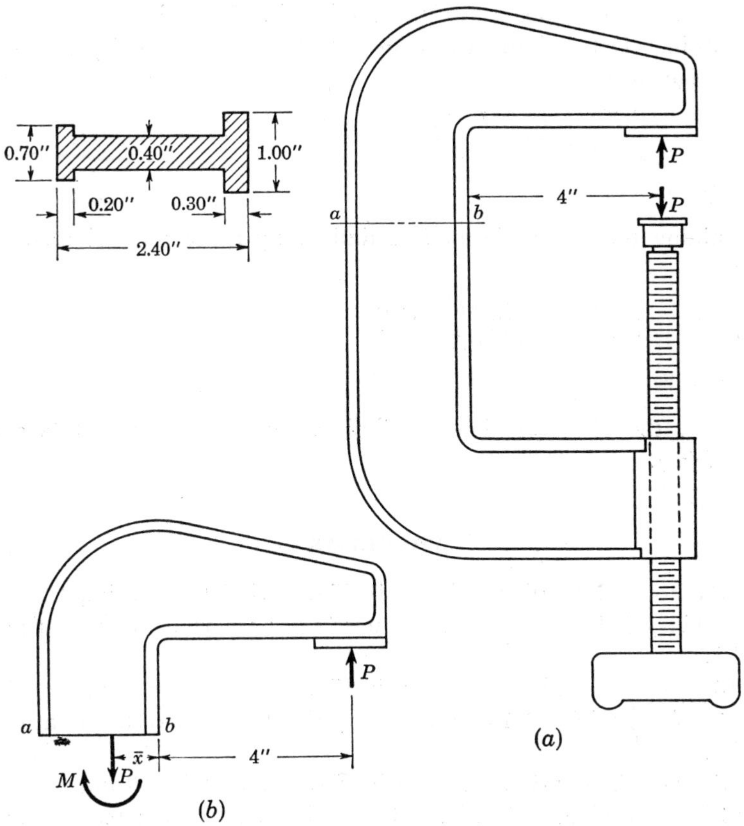

FIG. 9.5

location of the centroidal axis and the centroidal moment of inertia are determined using Table E9.2.

TABLE E9.2

	a	x	ax	I_o	d	ad^2
0.2 by 0.7	0.14	2.30	0.322	0.000	1.20	0.202
0.4 by 1.9	0.76	1.25	0.950	0.229	0.15	0.017
0.3 by 1.0	0.30	0.15	0.045	0.002	0.95	0.271
Total	1.20		1.317	0.231		0.490

$$\bar{x} = \frac{1.317}{1.20} = 1.10 \text{ in.} \qquad I = 0.231 + 0.490 = 0.721 \text{ in.}^4$$

The bending moment at the section indicated is $P(4 + 1.10)$ or $5.10P$ lb-in. Determining the allowable load as limited by the tensile stress at face b,

$$+6000 = +\frac{P}{1.20} + \frac{(5.10P)(1.10)}{0.721}$$

$P = 697$ lb

Determining the allowable load as limited by the compressive stress at face a,

$$-20{,}000 = +\frac{P}{1.20} - \frac{(5.10P)(1.30)}{0.721}$$

$P = 2420$ lb

Therefore, the allowable load as limited by tensile stress at face b is 697 lb (*Ans*).

PROBLEMS

***9.1.** For the cantilever shown in Fig. P9.1, let $h = 3$ in., $t = 2$ in., $Q_1 = 2$ kips, and $Q_2 = 4$ kips. Determine the normal stresses at points a and b.

9.2. For the cantilever shown in Fig. P9.2, let $h = 6$ in., $t = 4$ in., $Q_1 = 15$ kips, and $Q_2 = 10$ kips. Determine the normal stresses at points a and b.

9.3. For the cantilever shown in Fig. P9.3, let $h = 9$ in., $t = 4$ in., $Q_1 = 20$ kips, and $Q_2 = 30$ kips. Determine the normal stresses at points a and b.

***9.4.** The inclined member shown in Fig. P9.4 is 6 ft long and 6 by 8 in. in cross section. Based on an allowable normal stress of 1600 psi, determine the allowable vertical load Q.

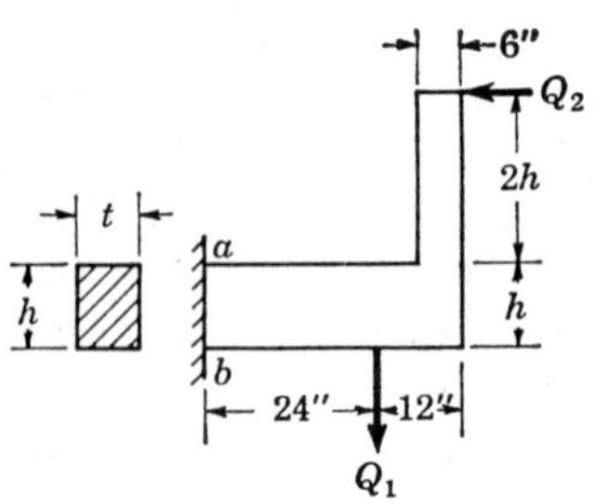

FIG. P9.1, P9.2, P9.3

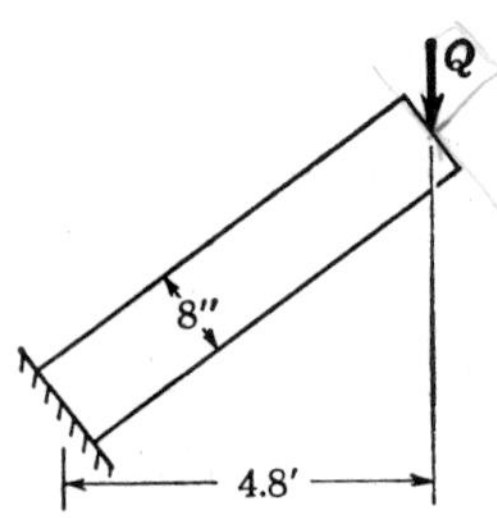

FIG. P9.4, P9.5, P9.6

9.5. The inclined member shown in Fig. P9.5 is 5.2 ft long and 4 by 8 in. in cross section. Based on an allowable normal stress of 1800 psi, determine the allowable vertical load Q.

9.6. The inclined member shown in Fig. P9.6 is 8 ft long and 6 by 8 in. in cross section. Based on an allowable normal stress of 2000 psi, determine the allowable vertical load Q.

★**9.7.** For the beam shown in Fig. P9.7, let $Q = 1000$ lb, $\theta = 45°$, and $d = 6$ in. Determine the maximum tensile and compressive stresses.

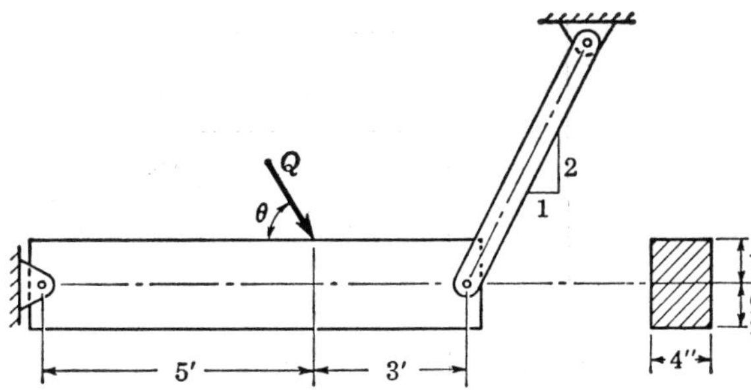

FIG. P9.7, P9.8, P9.9

9.8. For the beam shown in Fig. P9.8, let $Q = 2000$ lb, $\theta = 60°$, and $d = 9$ in. Determine the maximum tensile and compressive stresses.

9.9. For the beam shown in Fig. P9.9, let $\theta = 90°$ and $d = 12$ in. Based on an allowable normal stress of 2000 psi, determine the allowable (vertical) load Q.

★**9.10.** The bent steel rod shown in Fig. P9.10 has a diameter $d = 2$ in. Compare the maximum tensile stresses at the three sections indicated due to the load $Q = 6000$ lb.

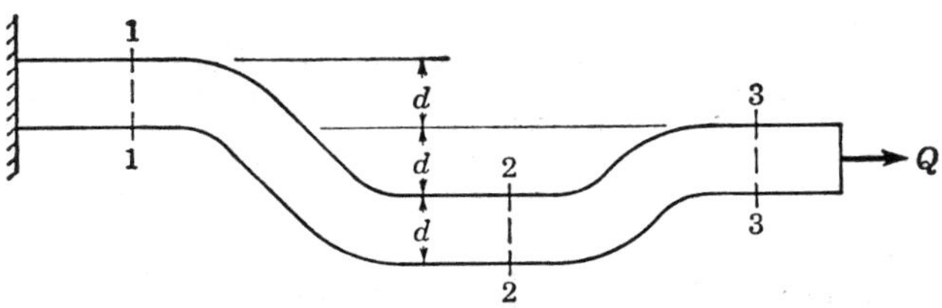

FIG. P9.10, P9.11, P9.12

9.11. The bent steel rod shown in Fig. P9.11 has a diameter $d = 4$ in. Compare the maximum tensile stresses at the three sections indicated due to the load $Q = 25{,}000$ lb.

9.12. The bent steel rod shown in Fig. P9.12 has a diameter $d = 3$ in. Compare the maximum tensile stresses at the three sections indicated due to the load $Q = 12{,}000$ lb.

★9.13. A chain hoist is to be supported by the structure shown in Fig. P9.13. Determine the allowable load Q as limited by normal stress in the vertical member. For the cross section shown let $h = b = 8$ in. Allowable stresses are as follows: tension = 20,000 psi; compression = 18,000 psi.

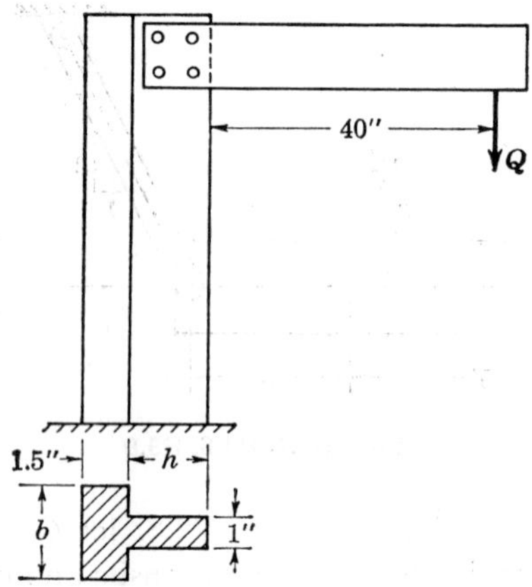

FIG. P9.13, P9.14, P9.15

9.14. A chain hoist is to be supported by the structure shown in Fig. P9.14. Determine the allowable load Q as limited by normal stress in the vertical member. For the cross section shown let $h = 11$ in. and $b = 9$ in. Allowable stresses are as follows: tension = 20,000 psi; compression = 18,000 psi.

9.15. A chain hoist is to be supported by the structure shown in Fig. P9.15. Determine the allowable load Q as limited by normal stress in the vertical member. For the cross section shown let $h = 14$ in. and $b = 10$ in. Allowable stresses are as follows: tension = 20,000 psi; compression = 18,000 psi.

★9.16. The frame of a riveting machine is to withstand a load $Q = 120$ kips as shown in Fig. P9.16. Determine the maximum tensile and compressive stresses at the section indicated where $b_1 = 6$ in. and $b_2 = 10$ in.

9.17. The frame of a riveting machine is to withstand a load $Q = 100$ kips as shown in Fig. P9.17. Determine the maximum tensile and compressive stresses at the section indicated where $b_1 = 5$ in. and $b_2 = 9$ in.

9.18. The frame of a riveting machine is to withstand a load $Q = 80$ kips as shown in Fig. P9.18. Determine the maximum tensile and compressive stresses at the section indicated where $b_1 = 4$ in. and $b_2 = 8$ in.

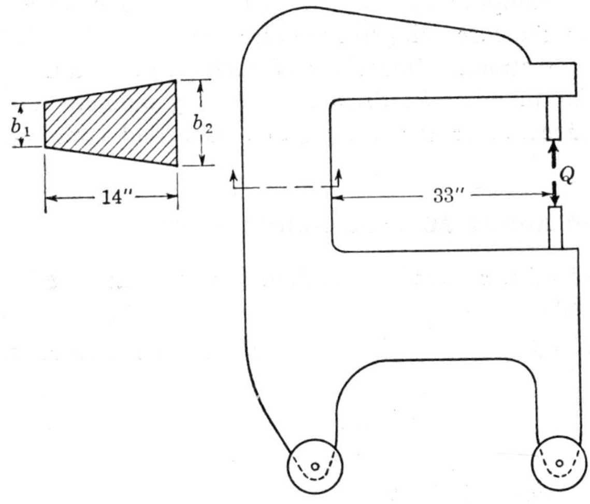

FIG. P9.16, P9.17, P9.18

★9.19. For the three-hinged structure shown in Fig. P9.19, let $h = 6$ ft. The cross-sectional dimensions of each member are $d = 8$ in. and $t = 4$ in. For member ab, draw the bending moment and axial force diagrams and compute the maximum tensile and compressive stresses.

9.20. For the three-hinged structure shown in Fig. P9.20, let $h = 6$ ft. The cross-sectional dimensions of each member are $d = 8$ in. and $t = 4$ in. For member bc, draw the bending moment and axial force diagrams and compute the maximum tensile and compressive stresses.

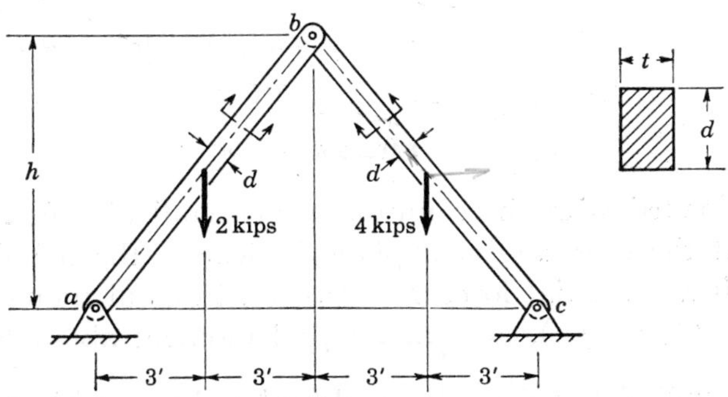

FIG. P9.19, P9.20, P9.21, P9.22

9.21. For the three-hinged structure shown in Fig. P9.21, let $h = 8$ ft. The cross-sectional dimensions of each member are $d = 8$ in. and $t = 6$ in. For member ab, draw the bending moment and axial force diagrams and compute the maximum tensile and compressive stresses.

9.22. For the three-hinged structure shown in Fig. P9.22, let $h = 8$ ft. The cross-sectional dimensions of each member are $d = 8$ in. and $t = 6$ in. For member bc, draw the bending moment and axial force diagrams and compute the maximum tensile and compressive stresses.

9.3 BENDING ABOUT BOTH PRINCIPAL AXES

The method of the previous article can be extended to apply to members which are subjected to bending about both cross-sectional principal axes.† Figure 9.6 shows a member fixed at one

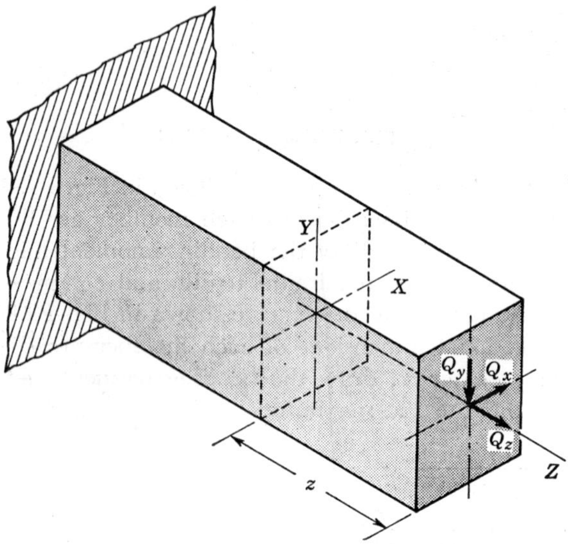

FIG. 9.6

end and loaded with three mutually perpendicular forces at the other. At the cross-sectional plane indicated, the member must have an internal axial force $P(=Q_z)$, an internal moment about the y axis $M_y(= Q_x z)$, and an internal moment about the x axis

† The principal axes of an area are defined as the mutually perpendicular axes about which the area has maximum and minimum moments of inertia. Any axis of symmetry of an area is a principal axis.

$M_x(= Q_y z)$. The net normal stress at any point on the cross section of such a member will be equal to the algebraic sum of the stresses due to the internal axial force and the two bending moments, or

$$s = \pm \frac{P}{A} \pm \frac{M_x y}{I_x} \pm \frac{M_y x}{I_y} \tag{9.2}$$

Example 9.3. For the beam shown in Fig. 9.7*a* determine the maximum tensile and compressive stresses and the neutral axis of the cross section *abcd*.

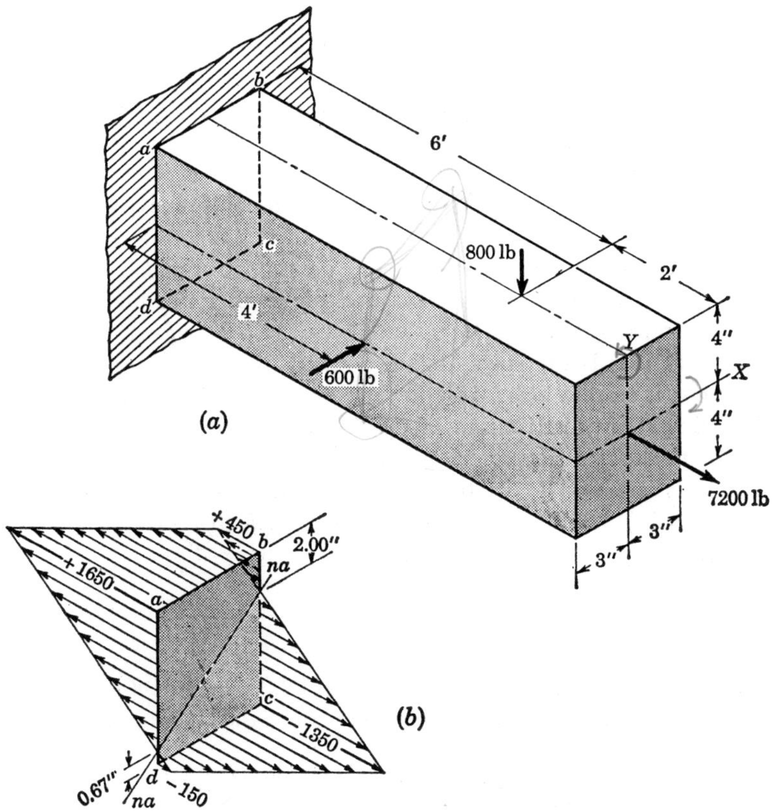

FIG. 9.7

Computing the moments of inertia of the cross-sectional area about its principal axes,

$$I_x = \frac{6(8)^3}{12} = 256 \text{ in.}^4 \qquad I_y = \frac{8(6)^3}{12} = 144 \text{ in.}^4$$

The maximum values of both M_x and M_y are at section *abcd*; the critical points for normal stress are the four corners of this section. Evaluating the terms of Eq. (9.2) separately,

$$\frac{P}{A} = \frac{7200}{(6)(8)} = 150 \text{ psi}$$
$$\frac{M_x y}{I_x} = \frac{[(800)(6)(12)](4)}{256} = 900 \text{ psi}$$
$$\frac{M_y x}{I_y} = \frac{[(600)(4)(12)](3)}{144} = 600 \text{ psi}$$

Determining the net stress at each of the four corners:

At a, $s = +150 + 900 + 600 = +1650$ psi
At b, $s = +150 + 900 - 600 = +450$ psi
At c, $s = +150 - 900 - 600 = -1350$ psi
At d, $s = +150 - 900 + 600 = -150$ psi

Maximum tensile stress = 1650 psi; maximum compressive stress = 1350 psi (*Ans*).

Since the flexure formula is based on the assumption that a cross-sectional plane before bending remains a plane after bending, the normal stress distribution of section *abcd* can be drawn as shown in Fig. 9.7*b*. The neutral axis is the straight line defined by the intersection of these two planes; it can be located by finding any two of its points. From the straight-line stress distribution of the front edge of the section (line *ad*) and with the use of similar triangles, the vertical distance from *d* to the neutral axis is found to be 0.67 in.; from the stress distribution of the back edge (line *bc*) the vertical distance from *b* to the neutral axis is 2 in. (*Ans*).

PROBLEMS

***9.23.** For the cantilever shown in Fig. P9.23, let $h = 8$ in., $b = 6$ in., $Q_x = 500$ lb, and $Q_{yz} = 2000$ lb. Determine the maximum tensile and compressive stresses.

9.24. For the cantilever shown in Fig. P9.24, let $h = 12$ in., $b = 8$ in., $Q_x = 1$ kip, and $Q_{yz} = 6$ kips. Determine the maximum tensile and compressive stresses.

9.25. For the cantilever shown in Fig. P9.25, let $h = 12$ in., $b = 6$ in., $Q_x = 1$ kip, and $Q_{yz} = 4$ kips. Determine the maximum tensile and compressive stresses.

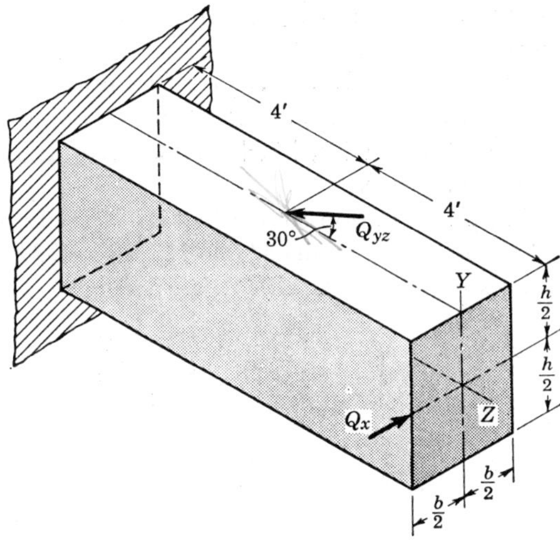

FIG. P9.23, P9.24, P9.25

★**9.26.** The block shown in Fig. P9.26 is acted on by a force $Q = 12$ kips. The line of action of the force passes through point O and a point having (x,y,z) coordinates of $(+1,+2,+2)$. Determine the maximum tensile and compressive stresses.

9.27. The block shown in Fig. P9.27 is acted on by a force $Q = 26$ kips. The line of action of the force passes through point O and a point

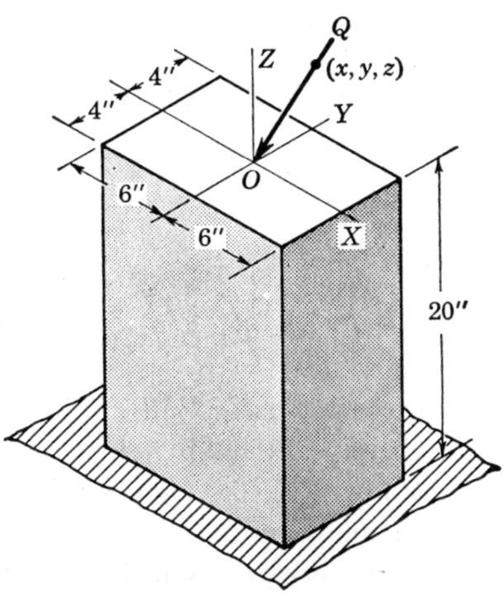

FIG. P9.26, P9.27, P9.28

having (x,y,z) coordinates of $(+3,+4,+12)$. Determine the maximum tensile and compressive stresses.

9.28. The block shown in Fig. P9.28 is acted on by a force $Q = 14$ kips. The line of action of the force passes through point O and a point having (x,y,z) coordinates of $(+2,+3,+6)$. Determine the maximum tensile and compressive stresses.

***9.29.** The column foundation shown in Fig. P9.29 is acted on by a centric load $P = 18$ kips and moments $M_x = 5$ kip-ft and $M_y = 1.5$ kip-ft. Compute the soil pressure on the footing at corners a, b, c, and d.

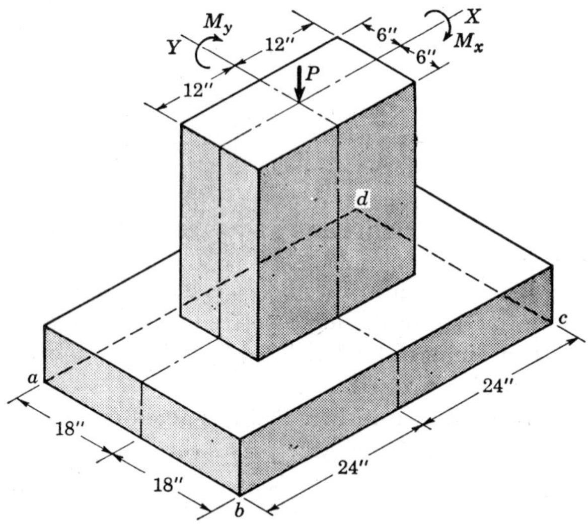

FIG. P9.29, P9.30, P9.31

9.30. The column foundation shown in Fig. P9.30 is acted on by a centric load $P = 24$ kips and moments $M_x = 4$ kip-ft and $M_y = 7$ kip-ft. Compute the soil pressure on the footing at corners a, b, c, and d.

9.31. The column foundation shown in Fig. P9.31 is acted on by a centric load $P = 36$ kips and moments $M_x = 1.5$ kip-ft and $M_y = 13$ kip-ft. Compute the soil pressure on the footing at corners a, b, c, and d.

***9.32.** Boom *abc* is supported by a ball-in-socket at *a* and by cables *db* and *eb*. It is loaded with a force having components $Q_y = 600$ lb and $Q_z = 500$ lb as shown in Fig. P9.32. Determine the maximum tensile and compressive stresses in the boom.

9.33. Boom *abc* is supported by a ball-in-socket at *a* and by cables *db* and *eb*. It is loaded with a force having components $Q_y = 1400$ lb and $Q_z = 200$ lb as shown in Fig. P9.33. Determine the maximum tensile and compressive stresses in the boom.

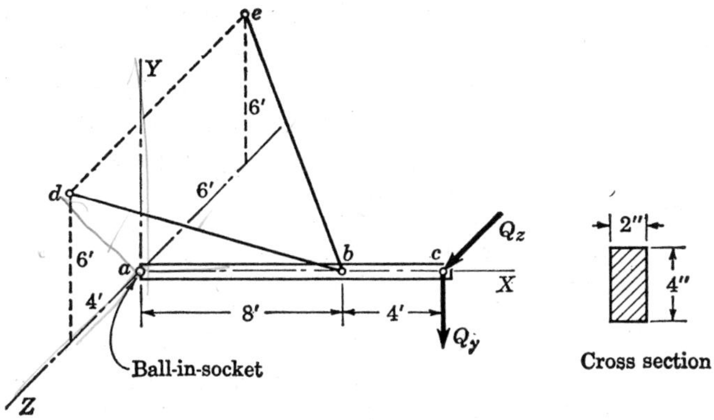

FIG. P9.32, P9.33, P9.34

9.34. Boom *abc* is supported by a ball-in-socket at *a* and by cables *db* and *eb*. It is loaded with a force having components $Q_y = 700$ lb and $Q_z = 400$ lb as shown in Fig. P9.34. Determine the maximum tensile and compressive stresses in the boom.

9.4 ECCENTRICALLY LOADED MEMBERS

Eccentric loading of members is simply a special case of axial loading combined with bending. The normal stresses in such members can be determined directly by the methods of the previous article. Traditionally, however, the stresses in eccentrically loaded mem-

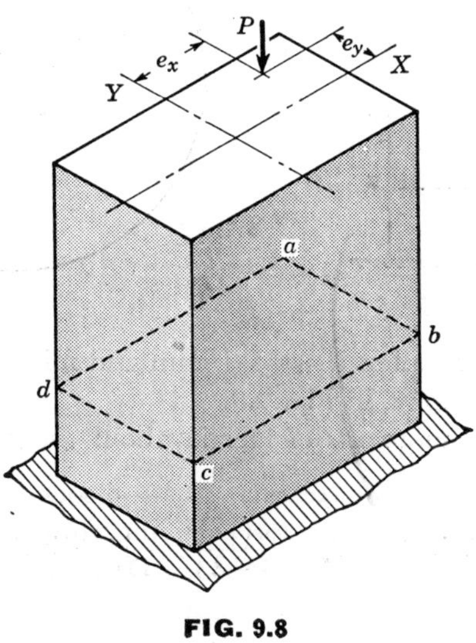

FIG. 9.8

bers are determined from a modified version of Eq. (9.2). This modification will be made using a prismatic member supporting an eccentric load P, as shown in Fig. 9.8. At any cross section, such as section *abcd*, the moment about the x axis will be Pe_y while the moment about the y axis will be Pe_x. Therefore, Eq. (9.2) adapted for eccentric loading becomes

$$s = \pm\frac{P}{A} \pm \frac{Pe_y y}{I_x} \pm \frac{Pe_x x}{I_y} \tag{9.3}$$

It may be noted that the stress distribution will be the same for any cross section of an eccentrically loaded member (except in the immediate vicinity of the point of load application).

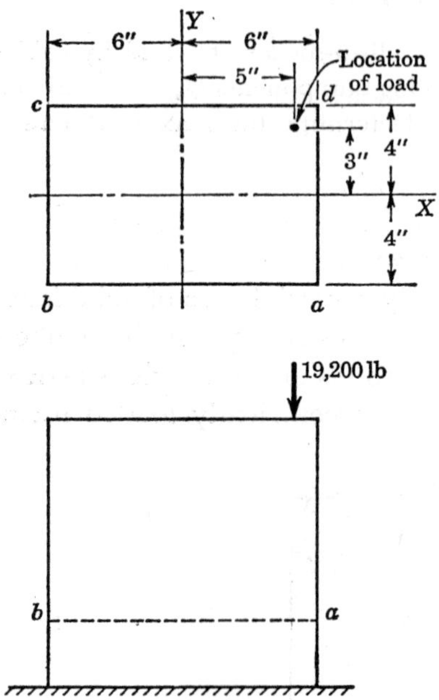

FIG. 9.9

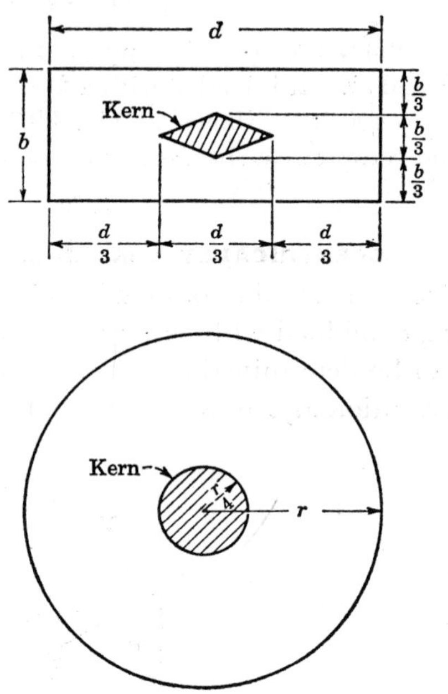

FIG. 9.10

Example 9.4. Determine the maximum and minimum normal stresses for the timber post shown in Fig. 9.9.

Computing the moments of inertia about the principal axes of the cross section,

$$I_x = \frac{(12)(8)^3}{12} = 512 \text{ in.}^4 \qquad I_y = \frac{8(12)^3}{12} = 1152 \text{ in.}^4$$

The extreme values of normal stress occur at two of the four corners of any typical cross section. Evaluating the terms of Eq. (9.3) separately,

$$\frac{P}{A} = \frac{19{,}200}{8(12)} = 200 \text{ psi}$$

$$\frac{Pe_y y}{I_x} = \frac{(19{,}200)(3)(4)}{512} = 450 \text{ psi}$$

$$\frac{Pe_x x}{I_y} = \frac{(19{,}200)(5)(6)}{1152} = 500 \text{ psi}$$

The net stress at the four corners of the typical cross section shown is

At a, $s = -200 + 450 - 500 = -250$ psi
At b, $s = -200 + 450 + 500 = +750$ psi
At c, $s = -200 - 450 + 500 = -150$ psi
At d, $s = -200 - 450 - 500 = -1150$ psi

Maximum tensile stress = 750 psi; maximum compressive stress = 1150 psi (*Ans*).

It can be seen from Example 9.4 that a compressive load can cause significant tensile stress if it is applied sufficiently off center. Some commonly used construction materials have practically no tensile strength (plain concrete and masonry brick for example). The load eccentricities for compression members made of such materials are generally limited to the extent that no tensile stress will result. For any particular cross section, the area defined by these eccentricity limits is called the *kern* (of the section). Shown in Fig. 9.10 are the kerns for members of rectangular and circular cross sections. By way of amplification, no tensile stress will result if a compressive load is applied within the shaded area of either cross section shown. (Similarly no compressive stress will result if a tensile load is applied within the shaded area.)

PROBLEMS

***9.35.** A 14WF34 member is to support a tensile load $P = 100$ kips as shown in Fig. P9.35. If allowable stresses are tension = 20,000 psi and compression = 18,000 psi, determine the allowable eccentricity e.

9.36. An 8WF17 member is to support a tensile load $P = 20$ kips as shown in Fig. P9.36. If the allowable stresses are tension = 20,000 psi and compression = 18,000 psi, determine the allowable eccentricity e.

9.37. For the 21WF112 member shown in Fig. P9.37, determine the combination of load P and eccentricity e which will cause maximum and minimum tensile stresses of 20,000 and 10,000 psi.

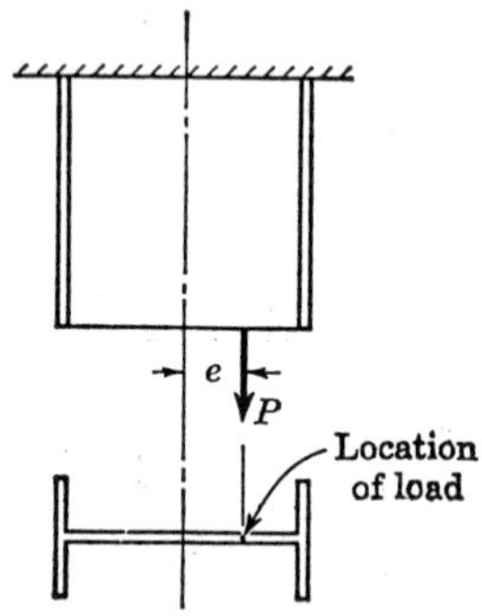

FIG. P9.35, P9.36, P9.37

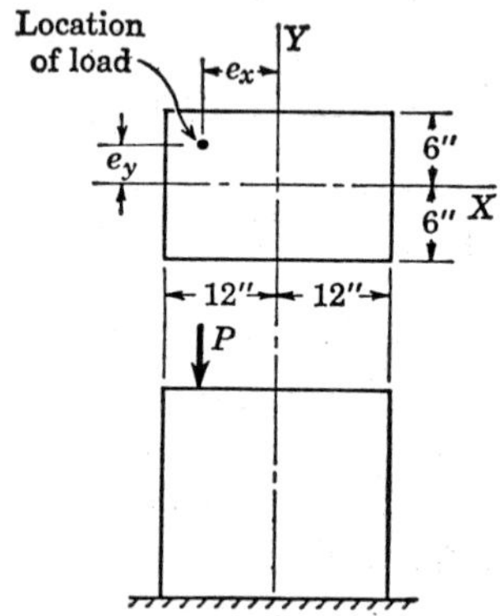

FIG. P9.38, P9.39, P9.40

★**9.38.** For the post shown in Fig. P9.38, let $P = 300$ kips, $e_x = 8$ in., and $e_y = 4$ in. For any typical cross section, determine the stress at each corner and the location of the neutral axis.

9.39. For the post shown in Fig. P9.39, let $P = 400$ kips, $e_x = 5$ in., and $e_y = 2$ in. For any typical cross section, determine the stress at each corner and the location of the neutral axis.

9.40. For the post shown in Fig. P9.40, let $P = 600$ kips, $e_x = 1$ in., and $e_y = 3$ in. For any typical cross section, determine the stress at each corner and the location of the neutral axis.

★**9.41.** For the hollow post shown in Fig. P9.41, let $a = 2$ in. and determine the allowable load P. Allowable stresses are as follows: tension = 4000 psi; compression = 10,000 psi.

9.42. For the hollow post shown in Fig. P9.42, let $a = 3$ in. and determine the allowable load P. Allowable stresses are as follows: tension = 4000 psi; compression = 10,000 psi.

9.43. For the hollow post shown in Fig. P9.43, let $a = 1$ in. and determine the allowable load P. Allowable stresses are as follows: tension = 4000 psi; compression = 10,000 psi.

★**9.44.** For the post shown in Fig. P9.44, e_x is to be 4 in. Determine the combination of P and e_y which will cause the maximum tensile stress to be 4000 psi when the maximum compressive stress is 10,000 psi.

9.45. For the post shown in Fig. P9.45, e_y is to be 3 in. Determine the combination of P and e_x which will cause the maximum tensile stress to be 4000 psi when the maximum compressive stress is 10,000 psi.

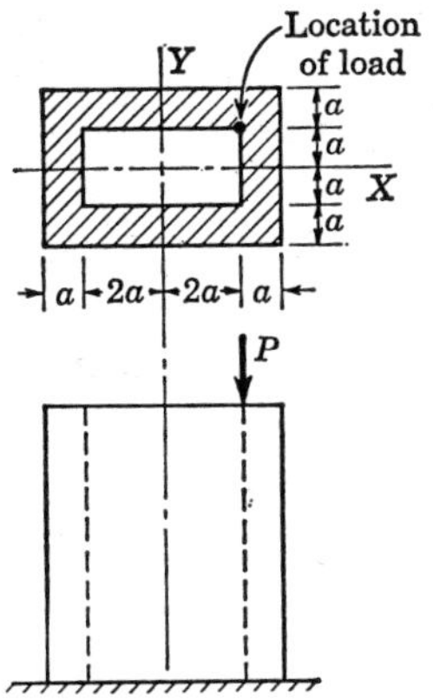

FIG. P9.41, P9.42, P9.43

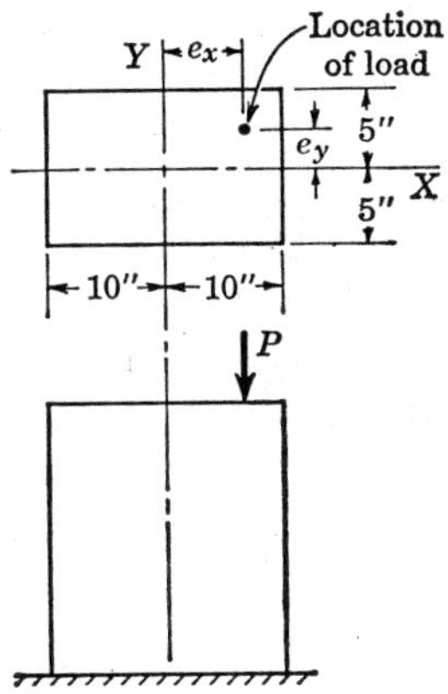

FIG. P9.44, P9.45, P9.46

9.46. For the post shown in Fig. P9.46, e_x is to be 5 in. Determine the combination of P and e_y which will cause the maximum tensile stress to be 4000 psi when the maximum compressive stress is 10,000 psi.

9.47. Verify the limits of the kern areas shown in Fig. 9.10.

9.48. For the cross section of Prob. 9.41 let $a = 2$ in. and determine the kern. Illustrate answer with a sketch similar to Fig. 9.10.

9.49. Determine the kern of a 12WF120 section. Illustrate answer with a sketch similar to Fig. 9.10.

10

PRINCIPAL STRESSES

10.1 INTRODUCTION

The stress theory and computations of the previous chapters have been largely limited to the special case of stresses acting on cross-sectional planes. Thus the formulas derived apply only to stresses which act in the directions of the axes of a member. As a matter of convenience such stresses are said to act in the x and y directions and have been designated as s_x or s_y for normal stresses and s_{xy} or s_{yx} for shearing stresses. Since the stresses are not necessarily the largest stresses which exist at any particular point, and since a material can fail owing to excessive stress in *any* direction, it should be apparent that a more general analysis of stress is necessary. This chapter will be concerned with the determination of stresses, both normal and shearing, which act in other than the x and y directions. Of greatest importance, of course, are the maximum and minimum stresses which exist at any particular point. The maximum and minimum normal stresses at any point are known as the *principal stresses*; the planes on which these stresses act are known as the *principal planes*. The planes on which the maximum shearing stress at any point acts are known as the *planes of maximum shear*. It will be seen that, once the state of stress in the x and y directions has been determined completely, it is a relatively simple matter to determine the principal stresses, the maximum shearing stress, and the planes on which they act.

10.2 GENERAL STRESS CONDITION AT A POINT; STRESSES ON INCLINED PLANES

For any point of a member subjected to biaxial stress, equations for the normal and shearing stresses on any plane can be determined using the x and y stress condition and the principles of statics. Shown in Fig. 10.1*a* is the stress condition of a differential element from any biaxially stressed member. The element is a parallelepiped of dimensions dx by dy by dt. To find the stresses on any inclined plane making an angle θ with the vertical, a section is passed through the element as shown. The stress condition of the portion of the element to the left of the section is as shown in Fig. 10.1*b*. In the general case, the force on the inclined plane of this wedge-shaped piece will be such that both normal stress s_n and shearing stress s_s will occur. These stresses can be evaluated in terms of θ with the use of the free-body diagram shown in Fig. 10.1*c*. Note that the area of the inclined plane is designated as A and that the areas of the vertical and horizontal faces are $A \cos \theta$ and $A \sin \theta$, respectively. The various forces acting on the wedge are expressed as the products of corresponding stresses and areas. The normal stress s_n is most easily evaluated by summing forces in the N direction indicated and resolving the various forces into N and T components as shown in Fig. 10.1*d*. Thus, $\Sigma F_N = 0$:

$$+s_nA - s_xA \cos^2 \theta - s_yA \sin^2 \theta + s_{xy}A \sin \theta \cos \theta + s_{yx}A \sin \theta \cos \theta = 0$$

Since $s_{yx} = s_{xy}$ and dividing the equation by A,

$$s_n = s_x \cos^2 \theta + s_y \sin^2 \theta - 2s_{xy} \sin \theta \cos \theta$$

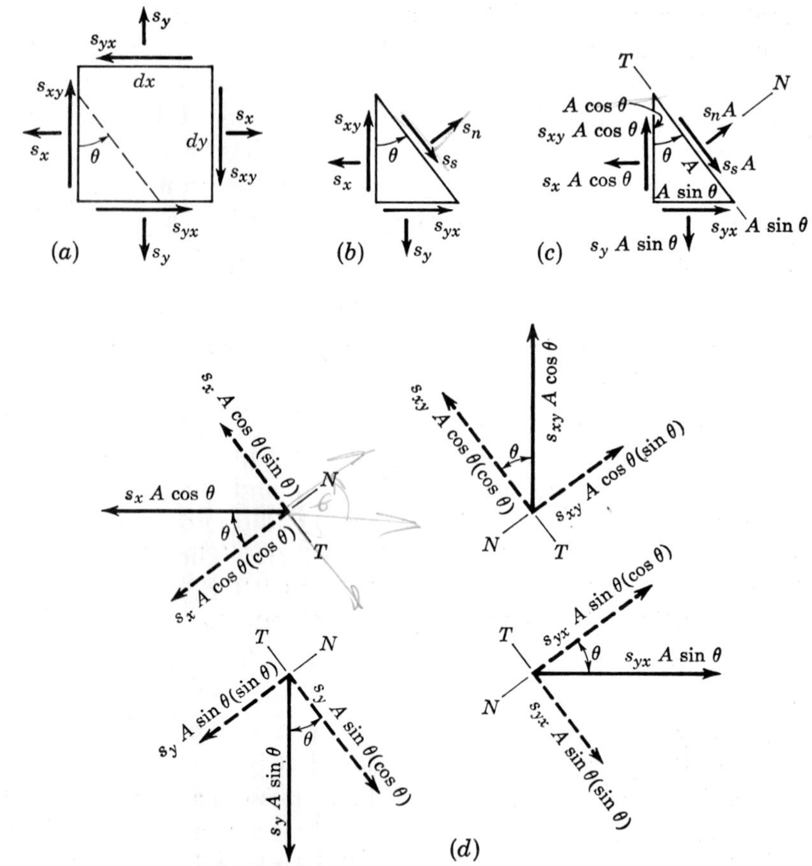

FIG. 10.1

A more useful form of the equation is obtained by using the trigonometric identities $\cos^2\theta = (1 + \cos 2\theta)/2$, $\sin^2\theta = (1 - \cos 2\theta)/2$, and $2\sin\theta\cos\theta = \sin 2\theta$ and collecting terms. Thus

$$s_n = \frac{s_x + s_y}{2} + \frac{s_x - s_y}{2}\cos 2\theta - s_{xy}\sin 2\theta \tag{10.1}$$

The shearing stress s_s is found by summing forces in the T direction. $\Sigma F_T = 0$:

$$s_sA - s_xA\cos\theta\sin\theta + s_yA\sin\theta\cos\theta - s_{xy}A\cos^2\theta + s_{yx}A\sin^2\theta = 0$$

Using the identities $2 \sin \theta \cos \theta = \sin 2\theta$ and $\cos^2 \theta - \sin^2 \theta = \cos 2\theta$, and collecting terms,

$$s_s = \frac{s_x - s_y}{2} \sin 2\theta + s_{xy} \cos 2\theta \tag{10.2}$$

It should be noted that signs of the values in Eqs. (10.1) and (10.2) are dependent on the directions used in the derivation. Thus, positive values of s_x and s_y represent tension, and negative values compression. Vertical shearing stress s_{xy} is positive when acting upward on the left face; the angle θ to the inclined plane is positive when measured counterclockwise from the vertical.

PROBLEMS

10.1–10.9. The stress condition at a point of a loaded member is as shown in Figs. P10.1 to P10.9. All stresses are in pounds per square inch. Using Eqs. (10.1) and (10.2), determine the normal and shearing stresses on the plane indicated. Illustrate answers with a sketch. (*10.1, *10.4, *10.7)

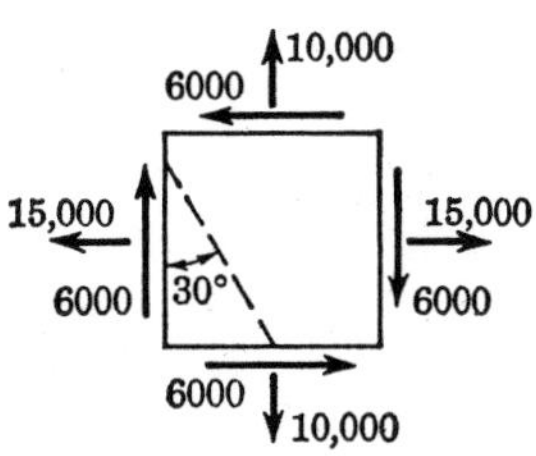

FIG. P10.1

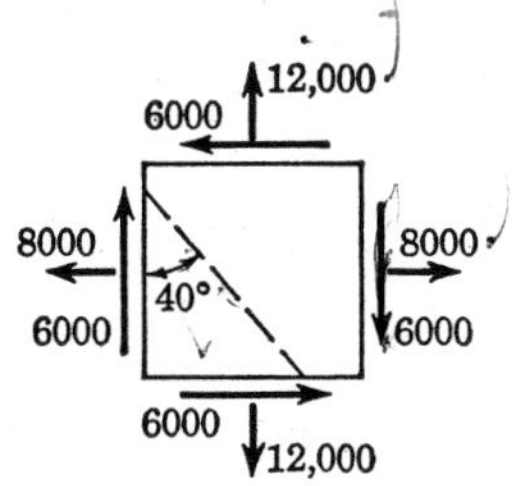

FIG. P10.2

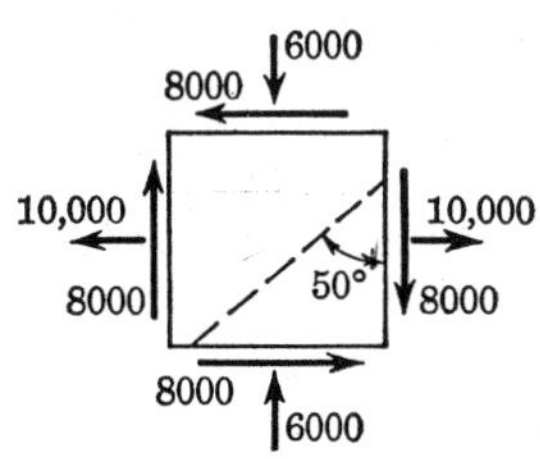

FIG. P10.3

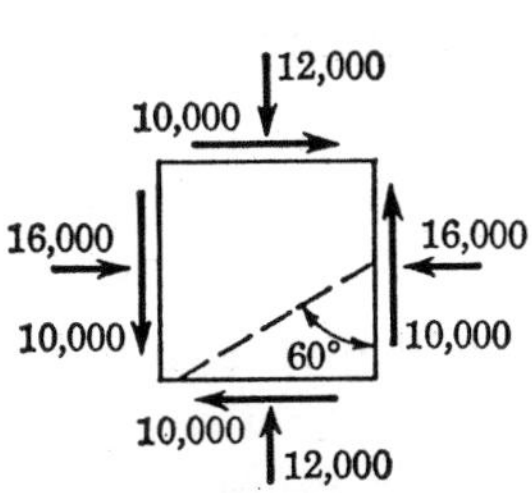

FIG. P10.4

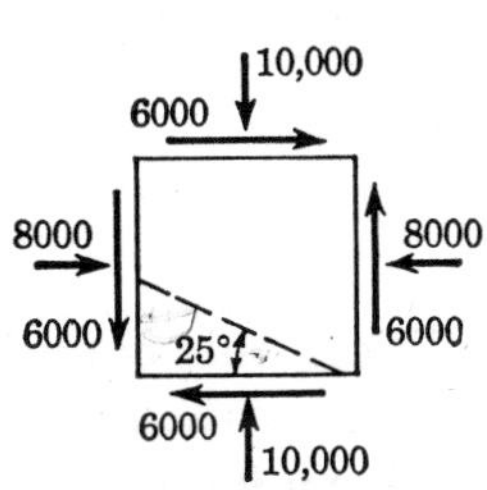

FIG. P10.5

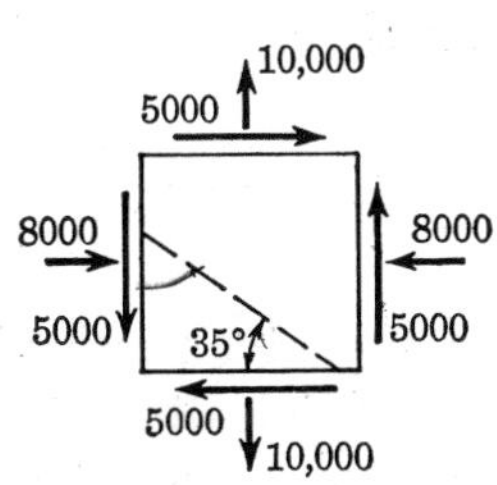

FIG. P10.6, P10.10

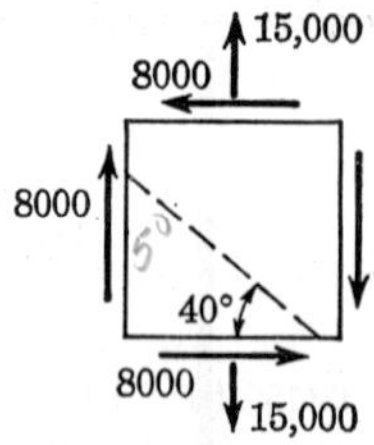

FIG. P10.7, P10.11

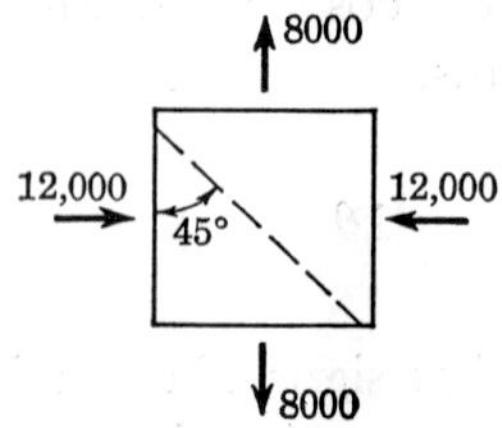

FIG. P10.8, P10.12

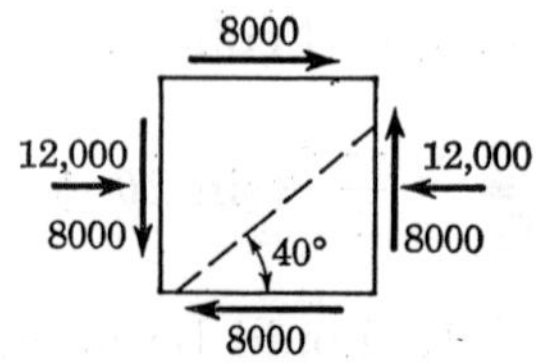

FIG. P10.9, P10.13

10.10–10.13. The stress condition at a point of a loaded member is as shown in Figs. P10.10 to P10.13. All stresses are in pounds per square inch. Using a free-body diagram of a portion of the element and the principles of statics, determine the normal and shearing stresses on the plane indicated.

10.3 MAXIMUM STRESS EQUATIONS

As has been shown, the stresses at any point of a loaded member vary with the inclination of the plane on which they act. Since the normal stress as expressed by Eq. (10.1) is a function of θ, the planes of maximum or minimum normal stress can be determined by setting the derivative $ds_n/d\theta$ equal to zero. Thus

$$\frac{ds_n}{d\theta} = 0 + \frac{s_x - s_y}{2}(-\sin 2\theta)(2) - s_{xy}(\cos 2\theta)(2) = 0$$

and

$$\frac{\sin 2\theta}{\cos 2\theta} = -\frac{2s_{xy}}{s_x - s_y}$$

Therefore, the equation for the principal planes can be expressed as

$$\tan 2\theta_n = -\frac{2s_{xy}}{s_x - s_y} \tag{10.3}$$

For any given stress condition, this equation will give two values of 2θ which differ by 180° or two values of θ which differ by 90°. Therefore, the principal planes at any particular point must be mutually perpendicular. It is significant that principal planes are also planes of zero shearing stress. This fact can be verified by setting s_s in Eq. (10.2) equal to zero:

$$0 = [(s_x - s_y)/2] \sin 2\theta + s_{xy} \cos 2\theta$$

Thus $(\sin 2\theta)/(\cos 2\theta) = -2s_{xy}/(s_x - s_y)$, which defines the same planes as Eq. (10.3).

The angle θ_n, as defined by Eq. (10.3), can be substituted in Eq. (10.1) to obtain an equation for principal stresses in terms of the stresses in the x and y directions. This substitution can be facilitated by expressing Eq. (10.3) as $\tan 2\theta = -s_{xy}/[(s_x - s_y)/2]$ and using the trigonometric identities $\cos \theta = 1/\sqrt{1 + \tan^2 \theta}$ and $\sin \theta = \tan \theta \cos \theta$. Therefore

$$\cos 2\theta = \frac{1}{\sqrt{1 + \left[\dfrac{-s_{xy}}{(s_x - s_y)/2}\right]^2}} = \frac{(s_x - s_y)/2}{\pm \sqrt{[(s_x - s_y)/2]^2 + s_{xy}^2}}$$

and

$$\sin 2\theta = \frac{-s_{xy}}{(s_x - s_y)/2} \frac{(s_x - s_y)/2}{\pm \sqrt{[(s_x - s_y)/2]^2 + s_{xy}^2}} \qquad \textbf{or}$$

$$\sin 2\theta = \frac{-s_{xy}}{\pm \sqrt{[(s_x - s_y)/2]^2 + s_{xy}^2}}$$

Substituting for $\cos 2\theta$ and $\sin 2\theta$ in Eq. (10.1),

$$s_{n,\max,\min} = \frac{s_x + s_y}{2} + \frac{[(s_x - s_y)/2][(s_x - s_y)/2]}{\pm \sqrt{[(s_x - s_y)/2]^2 + s_{xy}^2}} - \frac{s_{xy}(-s_{xy})}{\pm \sqrt{[(s_x - s_y)/2]^2 + s_{xy}^2}}$$

$$s_{n,\max,\min} = \frac{s_x + s_y}{2} + \frac{[(s_x - s_y)/2]^2 + (s_{xy})^2}{\pm \sqrt{[(s_x - s_y)/2]^2 + s_{xy}^2}}$$

Therefore

$$s_{n,\max,\min} = \frac{s_x + s_y}{2} \pm \sqrt{\left(\frac{s_x - s_y}{2}\right)^2 + s_{xy}^2}$$

The maximum and minimum normal stresses are known as the major and minor principal stresses; they will be designated as p and q, respectively. Therefore, the equation for principal stresses can be expressed as

$$p = \frac{s_x + s_y}{2} + \sqrt{\left(\frac{s_x - s_y}{2}\right)^2 + s_{xy}^2} \tag{10.4a}$$

$$q = \frac{s_x + s_y}{2} - \sqrt{\left(\frac{s_x - s_y}{2}\right)^2 + s_{xy}^2} \tag{10.4b}$$

Since the shearing stress as expressed by Eq. (10.2) is a function of θ, the planes of maximum shear can be determined by setting the derivative $ds_s/d\theta$ equal to zero. Thus

$$\frac{ds_s}{d\theta} = \frac{s_x - s_y}{2}(\cos 2\theta)(2) + s_{xy}(-\sin 2\theta)(2) = 0$$

and

$$\frac{\sin 2\theta}{\cos 2\theta} = \frac{s_x - s_y}{2s_{xy}}$$

Therefore, the equation for the planes of maximum shear can be written as

$$\tan 2\theta_s = \frac{s_x - s_y}{2s_{xy}} \tag{10.5}$$

It may be noted that, for any given stress condition, this equation will give two values of θ which differ by 90°. As was shown in Art. 5.5, the shearing stresses on any two mutually perpendicular planes must be equal; therefore the maximum shearing stress will occur on both of the planes defined by the above equation.

The angle θ_s, as defined by Eq. (10.5), can be substituted in Eq. (10.2) to obtain an equation for maximum shearing stress in terms of the stresses in the x and y directions. Using the identities $\cos\theta = 1/\sqrt{1 + \tan^2\theta}$ and $\sin\theta = \tan\theta\cos\theta$,

$$\cos 2\theta = \frac{1}{\sqrt{1 + [(s_x - s_y)/2s_{xy}]^2}} = \frac{s_{xy}}{\sqrt{s_{xy}^2 + [(s_x - s_y)/2]^2}}$$

and

$$\sin 2\theta = \frac{(s_x - s_y)/2}{s_{xy}} \frac{s_{xy}}{\sqrt{s_{xy}^2 + [(s_x - s_y)/2]^2}} \qquad \text{or}$$

$$\sin 2\theta = \frac{(s_x - s_y)/2}{\sqrt{s_{xy}^2 + [(s_x - s_y)/2]^2}}$$

Substituting for sin 2θ and cos 2θ in Eq. (10.2),

$$s_{s,\max} = \frac{[(s_x - s_y)/2][(s_x - s_y)/2]}{\sqrt{s_{xy}^2 + [(s_x - s_y)/2]^2}} + \frac{s_{xy}(s_{xy})}{\sqrt{s_{xy}^2 + [(s_x - s_y)/2]^2}}$$

Therefore

$$s_{s,\max} = \sqrt{\left(\frac{s_x - s_y}{2}\right)^2 + s_{xy}^2} \tag{10.6}$$

It may be noted that the maximum shearing stress, as defined in Eq. (10.6), is equivalent to the radical term of the equations for principal stresses. Thus, Eqs. (10.4*a*) and (10.4*b*) could be written as

$$p = \frac{s_x + s_y}{2} + s_{s,\max} \qquad \text{and}$$

$$q = \frac{s_x + s_y}{2} - s_{s,\max}$$

When these two equations are subtracted, it can be seen that at any point $s_{s,\max} = (p - q)/2$; when the two equations are added, it can be noted that $p + q = s_x + s_y$.

A comparison of Eqs. (10.3) and (10.5) reveals an interesting relationship between the maximum shear planes and the principal planes. Since tan $2\theta_n$ is the negative reciprocal of tan $2\theta_s$, $2\theta_n$ differs from $2\theta_s$ by 90°. Therefore, at any point, the maximum shear planes are at 45° with the principal planes.

The principal stresses and principal planes for a given stress condition can be determined by Eqs. (10.4) and (10.3). After the principal stresses have been determined, it is sometimes important to know which of the principal planes is the major principal plane (the plane on which p acts) and which the minor. One way of making this determination is to substitute one value of θ_n in Eq. (10.1); the value of s_n thus computed will serve to identify the corresponding plane. A better way of making the determination is by using the following rule:

If s_x is algebraically larger than s_y, the plane on which s_x acts will make an angle less than 45° with the major principal plane; conversely, if s_x is algebraically smaller than s_y, the plane on which s_x acts will make an angle less than 45° with the minor principal plane. This rule will be verified by the process of deductive reasoning for the case of s_x being larger than s_y. As shown by the top row of elements in Fig. 10.2, any general biaxial stress condition (element c) can be considered as a combination of pure shear (element b) and biaxial normal stress (element a). The principal planes for each of these stress conditions are as indicated by the bottom row of elements in Fig. 10.2. As shown by element a',

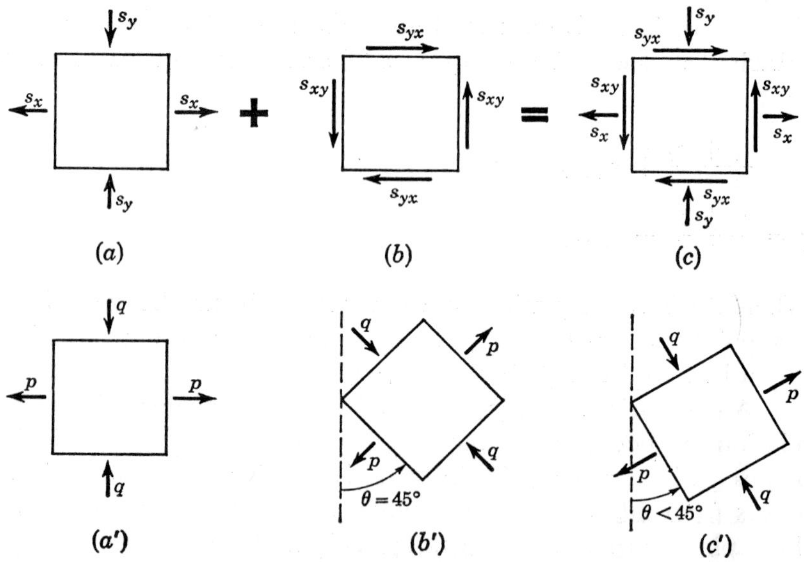

FIG. 10.2 Orientation of principal planes for biaxial normal stress, pure shear, and general biaxial stress.

since the faces of element a have no shearing stress, they are principal planes. As can be easily proved, the principal planes for an element in pure shear are at 45° to the x and y directions; thus the principal planes for the stress condition of element b are as shown by element b'. Since the major principal planes for normal stress alone (element a') and for shearing stress alone (element b') are 45° apart, the major principal plane for the general case of biaxial stress must be at some intermediate position (element c'). Thus it can be seen that the plane on which s_x acts must be less than 45° from the major principal plane. It could similarly be shown that, if s_x were smaller than s_y, the plane on which it acted would be less than 45° from the minor principal plane.

Example 10.1. The state of stress at a point, in pounds per square inch, on a loaded member is as shown in Fig. 10.3*a*. Determine the principal stresses and planes, the maximum shearing stress, and planes of maximum shear.

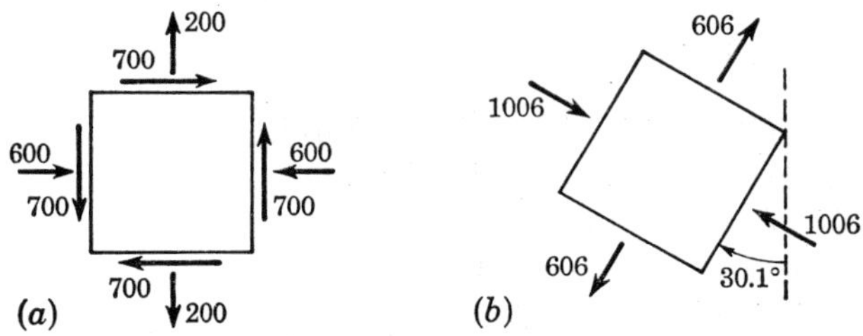

FIG. 10.3

According to the sign convention used in the derivation, $s_x = -600$ psi, $s_y = +200$ psi, and $s_{xy} = -700$ psi.

To find the principal stresses

$$p = \frac{s_x + s_y}{2} + \sqrt{\left(\frac{s_x - s_y}{2}\right)^2 + s_{xy}^2}$$

$$p = \frac{-600 + 200}{2} + \sqrt{\left(\frac{-600 - 200}{2}\right)^2 + (-700)^2}$$

$$p = -200 + 806 = +606 \text{ psi} \qquad Ans$$

$$q = -200 - 806 = -1006 \text{ psi} \qquad Ans$$

To find the principal planes

$$\tan 2\theta_n = -\frac{2s_{xy}}{s_x - s_y} = -\frac{2(-700)}{-600 - 200} = -1.75$$

$$2\theta_n = -60.2° \text{ and } -240.2°$$

$$\theta_n = -30.1° \text{ and } -120.1°$$

Since θ_n is negative, it is measured clockwise from the plane acted on by s_x. Since s_x is smaller than s_y, the plane defined by $\theta_n = -30.1°$ is the *minor* principal plane. The principal stresses, therefore, act as shown on the inclined element of Fig. 10.3*b* (*Ans*).

The maximum shearing stress is equal to the radical term of the principal stress equation. Therefore

$$s_{s,\max} = 806 \text{ psi} \qquad Ans$$

Since the maximum shear planes are at 45° to the principal planes,

$$\theta_s = -30.1 + 45 = +14.9° \qquad \text{and}$$
$$\theta_s = -120.1 + 45 = -75.1° \qquad \textit{Ans}$$

Example 10.2. Determine the principal stresses and principal planes at point A on the member shown in Fig. 10.4*a*.

Determining the stresses in x and y directions at point A,

$$I = \frac{6(8)^3}{12} = 256 \text{ in.}^4$$

$$M = +19{,}200(12) - 9600(2) = 211{,}000 \text{ lb-in.}$$

$$s_y = \pm\frac{P}{A} \pm \frac{Mx}{I} = +\frac{9600}{6(8)} - \frac{211{,}000(1.5)}{256}$$

$$s_y = +200 - 1236 = -1036 \text{ psi}$$

$$s_{xy} = \frac{VQ}{It} = \frac{19{,}200[(6)(2.5)(2.75)]}{256(6)} = 516 \text{ psi}$$

The state of stress in the x and y directions is as shown in Fig. 10.4*b*. Note that, since there is no normal stress in the x direction, $s_x = 0$. From Fig. 10.4*a* it can be seen that shearing stress must act to the left at the

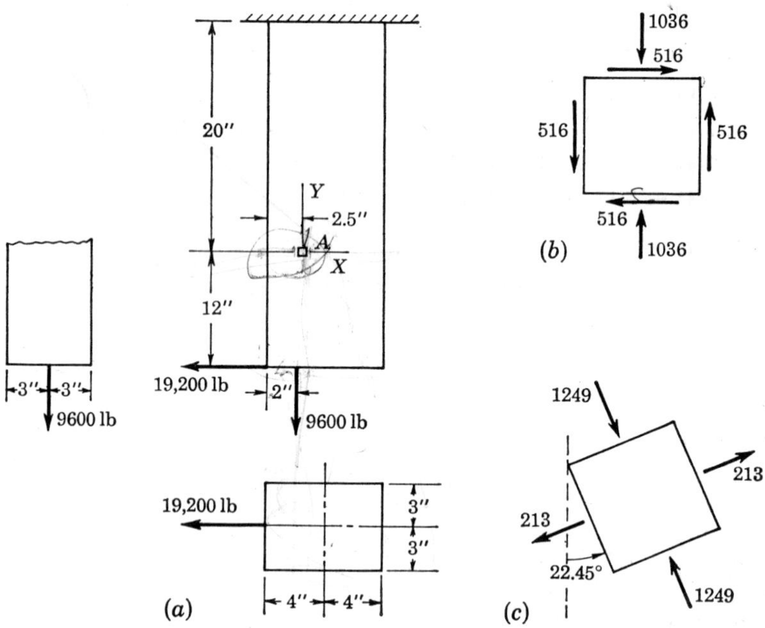

FIG. 10.4

bottom of element A; therefore, the mutually perpendicular shearing stresses must act as shown and s_{xy} is minus.

To find the principal stresses and planes

$$p = \frac{0 + (-1036)}{2} + \sqrt{\left(\frac{0 - (-1036)}{2}\right)^2 + (-516)^2}$$

$$p = -518 + 731 = +213 \text{ psi} \qquad Ans$$
$$q = -518 - 731 = -1249 \text{ psi} \qquad Ans$$
$$\tan 2\theta_n = -\frac{2(-516)}{0 - (-1036)} = +0.996$$
$$2\theta_n = +44.9° \text{ and } +224.9°$$
$$\theta_n = +22.45° \text{ and } +112.45°$$

The principal stresses act on the principal planes as shown in Fig. 10.4*c* (*Ans*).

Example 10.3. For the cylindrical member shown in Fig. 10.5*a*, determine the principal stresses and planes at points A and B.

Determining the moment of inertia and polar moment of inertia of the cross section,

$$I = \frac{\pi(3)^4}{4} = \frac{\pi(81)}{4} = 63.6 \text{ in.}^4 \qquad J = \frac{\pi(3)^4}{2} = 2(63.6) = 127.2 \text{ in.}^4$$

At point A both s_x and s_y are zero; shearing stress s_{xy} results from both beam shear and torsion. The shearing stress due to beam action acts down on the right of the element at A; conversely, the shearing stress due to torsion acts up on the right. Thus

$$s_{xy} = +\frac{VQ}{It} - \frac{Tr}{J}$$
$$s_{xy} = \frac{+15{,}000[(4(3)/3\pi)(9\pi/2)]}{63.6(6)} - \frac{150{,}000(3)}{127.2}$$
$$s_{xy} = +710 - 3540 = -2830 \text{ psi}$$

The state of stress at point A in the x and y directions is as shown in Fig. 10.5*b*, and

$$p = \frac{0 + 0}{2} + \sqrt{(0)^2 + (-2830)^2} = +2830 \text{ psi}$$

$$q = 0 - 2830 = -2830 \text{ psi}$$
$$\tan 2\theta_n = -\frac{2(-2830)}{0} = \infty$$
$$2\theta_n = 90° \text{ and } 270°$$
$$\theta_n = 45° \text{ and } 135°$$

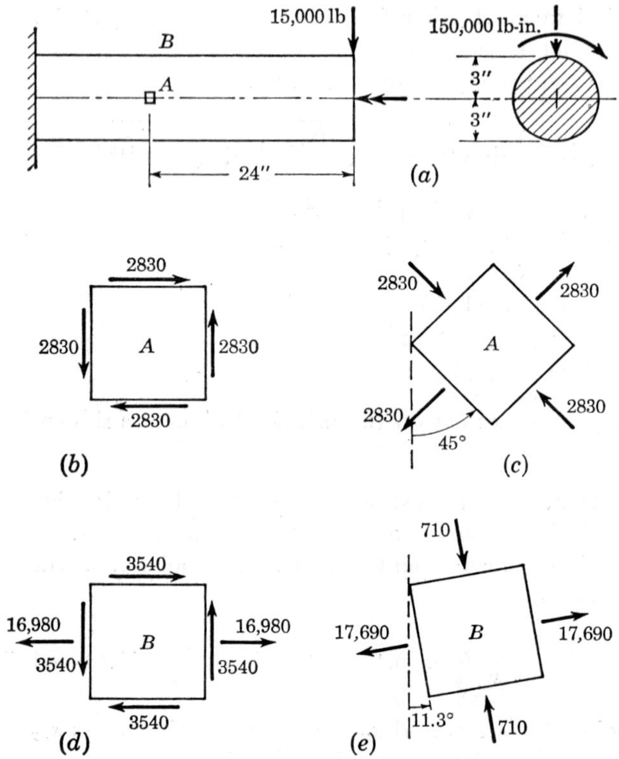

FIG. 10.5

The principal stresses act as shown on the inclined element of Fig. 10.5*c* (*Ans*). The major and minor principal planes are identified by visualizing a free-body diagram of half of the element in Fig. 10.5*b*.

For convenience of notation, in determining the principal stresses and planes at point *B*, the *xy* plane will be taken as being tangent to the top of the member. Therefore,

$$s_y = 0$$

$$s_x = \frac{Mc}{I} = +\frac{[15{,}000(24)](3)}{63.6} = +16{,}980 \text{ psi}$$

$$s_{xy} = \frac{Tr}{J} = -3540 \text{ psi}$$

The state of stress at point *B* (in the new *x* and *y* directions) is as shown in Fig. 10.5*d*, and

$$p = \frac{16{,}980 + 0}{2} + \sqrt{\left(\frac{16{,}980 - 0}{2}\right)^2 + (-3540)^2}$$

$$p = 8490 + 9200 = +17{,}690 \text{ psi}$$
$$q = 8490 - 9200 = -710 \text{ psi}$$
$$\tan 2\theta_n = -\frac{2(-3540)}{16{,}980 - 0} = +0.417$$
$$2\theta_n = +22.6° \text{ and } +202.6°$$
$$\theta_n = +11.3° \text{ and } +101.3°$$

The principal stresses act as shown on the inclined element of Fig. 10.5*e* (*Ans*).

PROBLEMS

10.14–10.22. The state of stress at a point of a loaded member is as shown in Figs. P10.14 to P10.22. All stresses are in pounds per square inch. Determine the maximum shearing stress and the principal stresses and planes. With an inclined element, show how the principal stresses act on the principal planes. (*10.14, *10.17, *10.20)

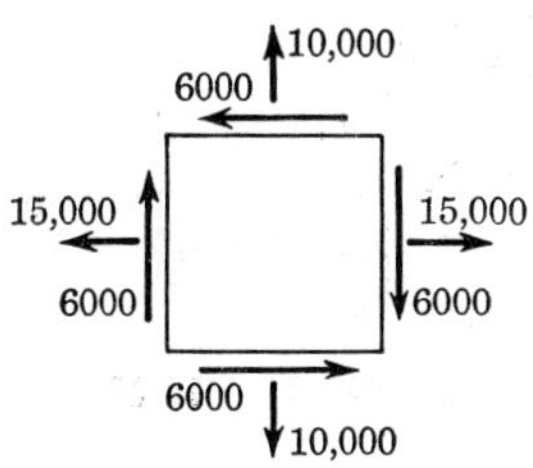

FIG. P10.14

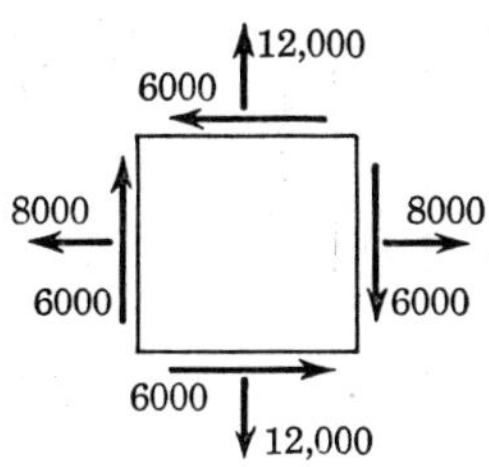

FIG. P10.15

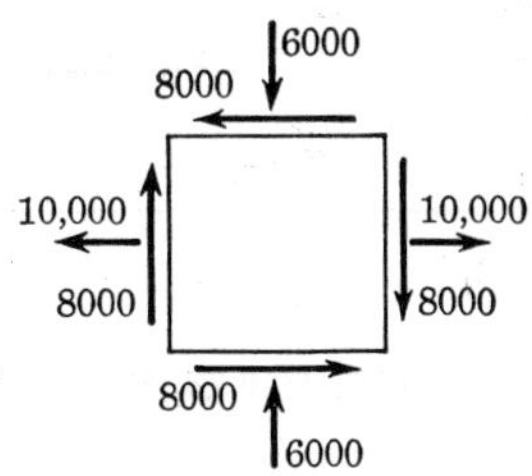

FIG. P10.16

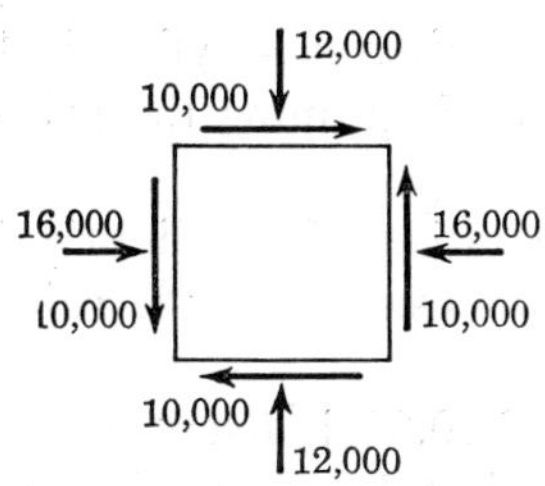

FIG. P10.17

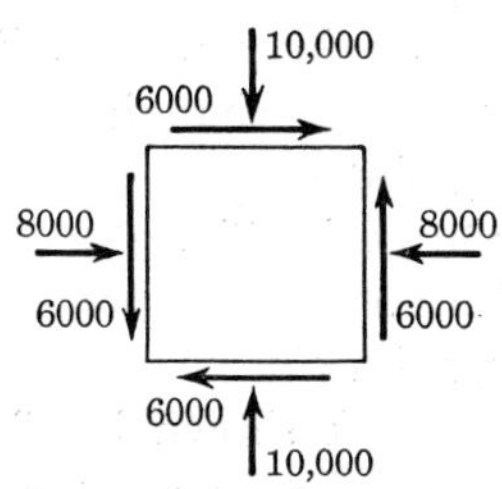

FIG. P10.18

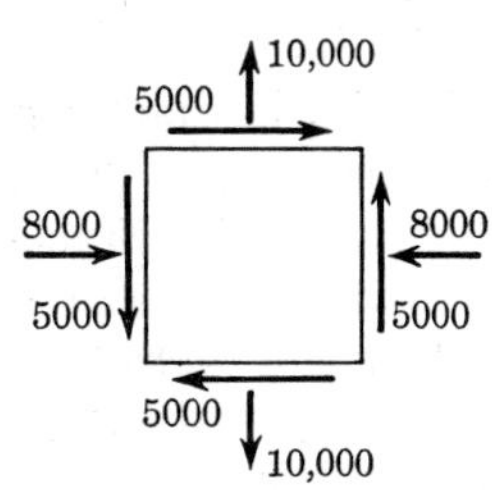

FIG. P10.19

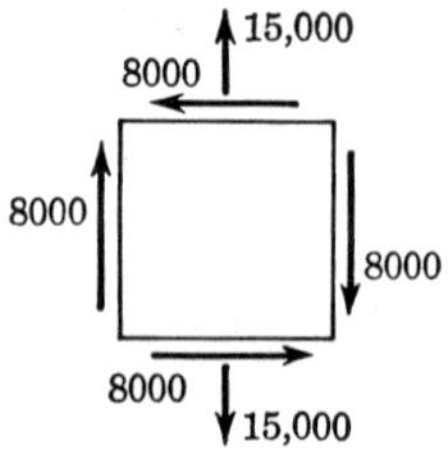

FIG. P10.20

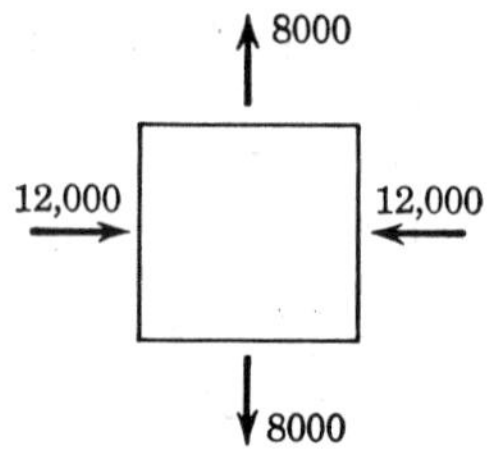

FIG. P10.21

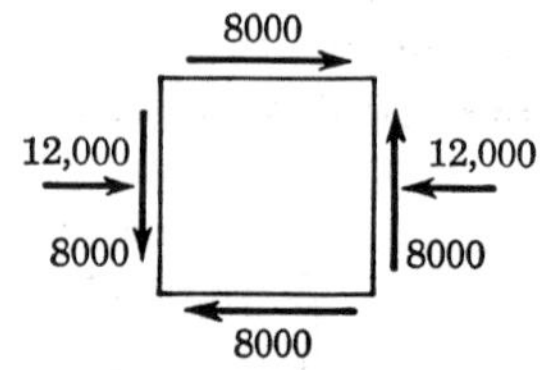

FIG. P10.22

★**10.23.** Point A is located 2 ft from the left reaction and 2 in. from the top of the beam shown in Fig. P10.23. Determine the principal stresses and planes at this point.

10.24. Point B is located 2 in. from the bottom at the midspan of the beam shown in Fig. P10.24. Determine the principal stresses and planes at this point.

10.25. Point C is located 2 ft from the right reaction and 2 in. from the bottom of the beam shown in Fig. P10.25. Determine the principal stresses and planes at this point.

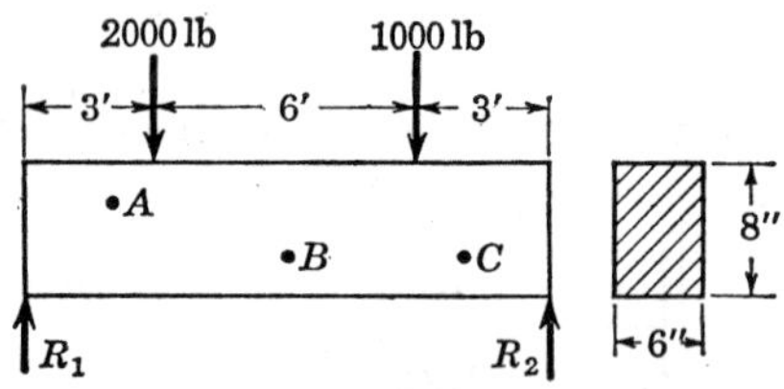

FIG. P10.23, P10.24, P10.25

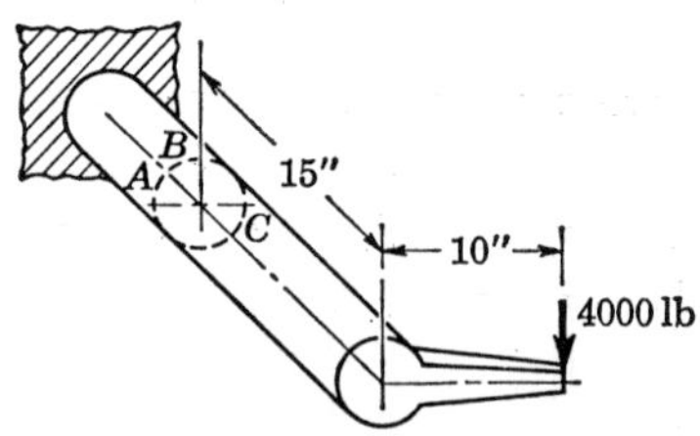

FIG. P10.26, P10.27, P10.28

★**10.26.** The bracket shown in Fig. P10.26 consists of a cylindrical shaft 4 in. in diameter with an overhanging arm. Determine the principal stresses and planes at point A.

10.27. The bracket shown in Fig. P10.27 consists of a cylindrical shaft 4 in. in diameter with an overhanging arm. Determine the principal stresses and planes at point B.

10.28. The bracket shown in Fig. P10.28 consists of a cylindrical shaft 4 in. in diameter with an overhanging arm. Determine the principal stresses and planes at point C.

★**10.29.** The 8-in.-diam shaft shown in Fig. P10.29 is fixed at one end and subjected to a torque and concentrated load at the other. Determine the principal stresses and planes at point A.

10.30. The 8-in.-diam shaft shown in Fig. P10.30 is fixed at one end and subjected to a torque and concentrated load at the other. Determine the principal stresses and planes at point B.

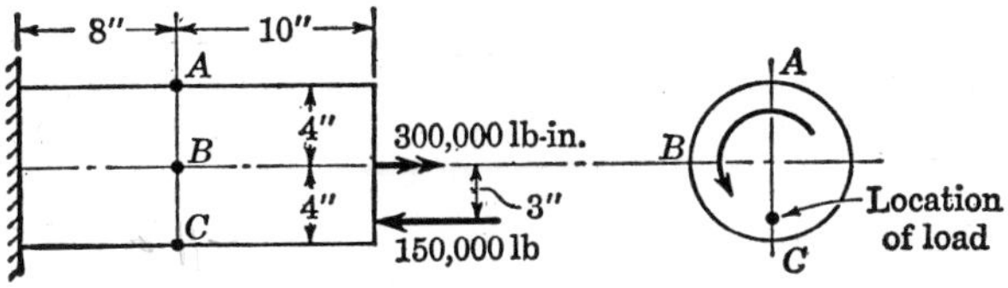

FIG. P10.29, P10.30, P10.31

10.31. The 8-in.-diam cylinder shown in Fig. P10.31 is fixed at one end and subjected to a torque and concentrated load at the other. Determine the principal stresses and planes at point C.

★**10.32.** The tank shown in Fig. P10.32, having inside and outside diameters of 8 and 9 in., is subjected to an internal pressure of 400 psi while supporting an end torque $T = 300$ kip-in. and an end centric load $Q = 60$ kips. Determine the principal stresses and planes at point A on the front surface.

10.33. The tank shown in Fig. P10.33, having inside and outside diameters of 8 and 9 in., is subjected to an internal pressure of 600 psi while supporting an end torque $T = 400$ kip-in. and an end centric load $Q = 50$ kips. Determine the principal stresses and planes at point A on the front surface.

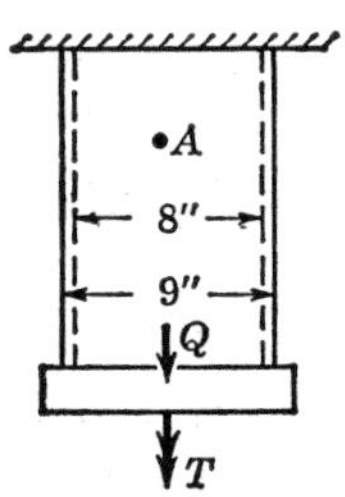

FIG. P10.32, P10.33, P10.34

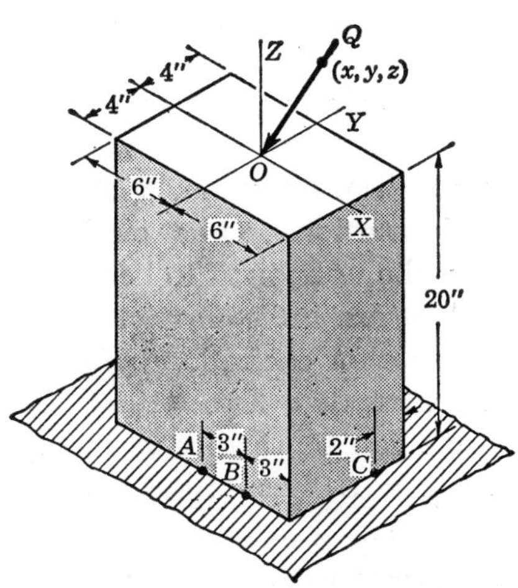

FIG. P10.35, P10.36, P10.37

10.34. The tank shown in Fig. P10.34, having inside and outside diameters of 8 and 9 in., is subjected to an internal pressure of 800 psi while supporting an end torque $T = 500$ kip-in. and an end centric load $Q = 30$ kips. Determine the principal stresses and planes at point A on the front surface.

★**10.35.** The block shown in Fig. P10.35 is acted on by force $Q = 12$ kips. The line of action of the force passes through point O and a point having (x,y,z) coordinates of $(+1,+2,+2)$. Determine the principal stresses and planes at point A.

10.36. The block shown in Fig. P10.36 is acted on by force $Q = 26$ kips. The line of action of the force passes through point O and a point having (x,y,z) coordinates of $(+3,+4,+12)$. Determine the principal stresses and planes at point B.

10.37. The block shown in Fig. P10.37 is acted on by force $Q = 14$ kips. The line of action of the force passes through point O and a point having (x,y,z) coordinates of $(+2,+3,+6)$. Determine the principal stresses and planes at point C.

10.4 MOHR'S CIRCLE FOR STRESS

The equations previously derived in this chapter may be solved graphically by a method known as Mohr's circle.† The basis of this method is a "circle of stress" which graphically represents Eqs. (10.1) and (10.2) on mutually perpendicular s_n and s_s axes; each point on the circle is the plot of s_n and s_s for the same value of 2θ. Thus, once the circle has been constructed, the normal and shearing stresses on any plane can be obtained by merely scaling the coordinates of the appropriate point.

Once it is understood, Mohr's circle can also be used for semigraphical solutions. The circle is sketched, and its geometry solved for the unknowns to be determined. The computations for this method are quite simple, consisting essentially of solving one right triangle for each plane of unknown stress.

The fact that Eqs. (10.1) and (10.2) may be represented by a circle can be readily verified. Letting $(s_x + s_y)/2 = A$ and $(s_x - s_y)/2 = B$, Eqs. (10.1) and (10.2) can be written as

† Developed in the latter part of the nineteenth century by the German engineer and professor Otto Mohr (1835–1918).

$$s_n - A = B \cos 2\theta - s_{xy} \sin 2\theta \tag{a}$$
$$s_s = B \sin 2\theta + s_{xy} \cos 2\theta \tag{b}$$

Squaring both sides of each equation,

$$(s_n - A)^2 = B^2 \cos^2 2\theta - 2Bs_{xy} \sin 2\theta \cos 2\theta + s_{xy}{}^2 \sin^2 2\theta \tag{a'}$$
$$s_s{}^2 = B^2 \sin^2 2\theta + 2Bs_{xy} \sin 2\theta \cos 2\theta + s_{xy}{}^2 \cos^2 2\theta \tag{b'}$$

Adding Eqs. (a') and (b') gives

$$(s_n - A)^2 + s_s{}^2 = B^2 + s_{xy}{}^2 \tag{c}$$

But $B^2 + s_{xy}{}^2 = [(s_x - s_y)/2]^2 + s_{xy}{}^2 = s^2_{s,\max}$, therefore

$$(s_n - A)^2 + s_s{}^2 = s^2_{s,\max} \tag{d}$$

which is the equation of a circle of radius $s_{s,\max}$, the center of which is located $(s_x + s_y)/2$ to the right of the origin (on the s_n axis).

In using Mohr's circle it is convenient to regard shearing stresses on mutually perpendicular planes as opposite in sign (although, of course, numerically equal). The sign convention traditionally used is as follows: Shearing stress on any element is positive if it tends to rotate the element clockwise. This sign convention is illustrated in Fig. 10.6. Note that it is consistent with the sign convention (based on the direction of s_{xy}) used in the previous article.

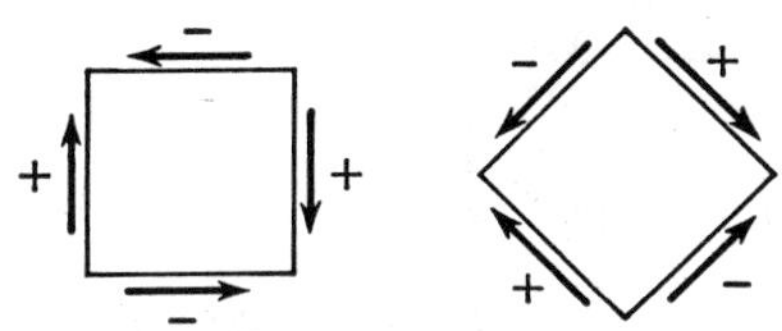

FIG. 10.6 Sign convention for shearing stress with Mohr's circle.

One method of constructing and using Mohr's circle will be explained with respect to the general biaxial stress condition shown as Fig. 10.7a. The stresses on mutually perpendicular faces of the element are plotted as points with respect to s_n and s_s axes as shown in Fig. 10.7b. These points are labeled V for stresses on the vertical face and H for stresses on the horizontal face. Since the circle is based on equations expressed in terms of 2θ, and since the horizontal and vertical planes are 90° apart, points V and H must be at the ends of radii which are 180° apart. Since s_{xy} and s_{yx} are equal in magnitude but opposite in sign, the midpoint of diameter VH (and the center of the stress circle C) must be on the s_n axis.

After the circle has been drawn, the stresses on any plane making an angle θ with the vertical face can be determined by drawing a radius at 2θ from radius CV. For example, as shown in Fig. 10.7*c*, the stresses on a plane $\theta°$ from the vertical face are obtained by measuring the coordinates of point G on the circle. Similarly the stresses on any plane making an angle θ with the horizontal face can be obtained by drawing a radius at 2θ from radius CH.

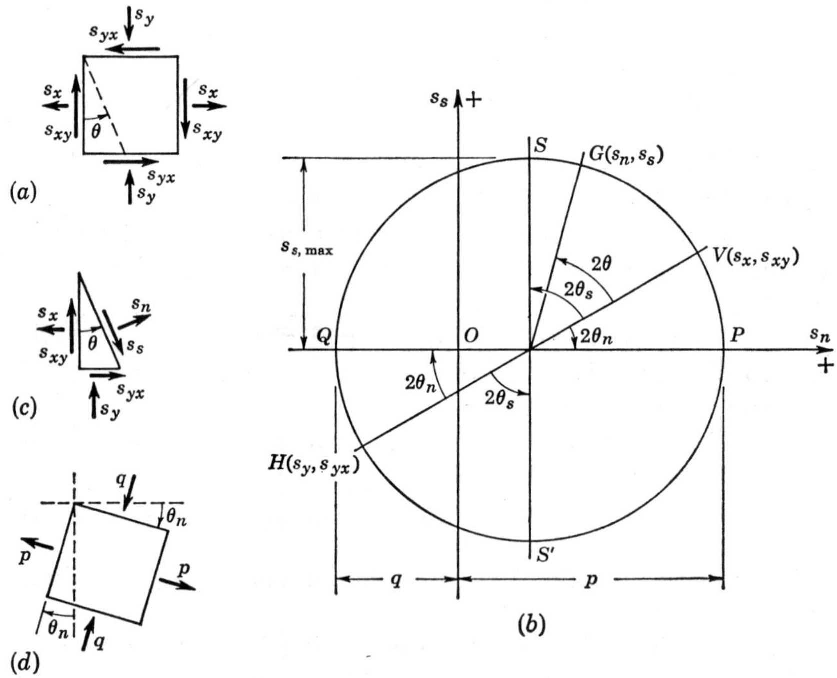

FIG. 10.7

It is apparent that the maximum and minimum s_n coordinates on the circle are those of the points labeled P and Q; the principal stresses p and q are therefore equal to these coordinates. It may be noted that the s_s coordinates of points P and Q are zero, thus indicating zero shearing stress on the principal planes. The direction of the principal planes can be defined by measuring $2\theta_n$ from either radius CV or CH. Thus in Fig. 10.7*d*, the major principal plane is shown at a clockwise angle θ_n from the vertical face. As an alternate procedure, the minor principal plane could be located, as shown, by a clockwise angle θ_n from the horizontal face.

It should also be apparent that the largest s_s coordinates on the circle are those of the points labeled S and S'. Thus the maximum shearing stress is represented by a vertical radius of the stress circle; the mutually perpendicular planes of maximum shear can be defined by measuring the angles $2\theta_s$ from radii CV and CH.

As a summary, the method of solving stress problems by Mohr's circle can be outlined as follows:

1. On s_n and s_s axes plot points V and H having coordinates of (s_x,s_{xy}) and (s_y,s_{yx}), respectively.

2. Draw diameter VH, and label its intersection with the s_n axis as C.

3. Using point C as the center, draw Mohr's circle through points V and H.

4. Label the right and left ends of the horizontal diameter as P and Q, respectively. The coordinates of these points represent the stresses on the major and minor principal planes.

5. Label the upper and lower ends of the vertical diameter as S and S', respectively. The coordinates of these points represent the stresses on the planes of maximum shear.

6. Since radius CV represents the direction of the vertical face, the angle from radius CV to any other radius represents double the angle from the vertical face to any other plane. Similarly the angle from radius CH to any other radius represents double the angle from the horizontal face to any other plane. Therefore, to find the stress on any particular plane, draw a radius representing the direction of the plane and measure the s_n and s_s coordinates of its extremity.

Example 10.4. The state of stress, in pounds per square inch, at a point in a loaded member is as shown in Fig. 10.8*a*. Using Mohr's circle graphically, determine the stresses on (*a*) the principal planes, (*b*) the planes of maximum shear, (*c*) a plane at a counterclockwise angle of 40° with the vertical face of the element.

Noting that $s_x = -600$ psi, $s_{xy} = -700$ psi, $s_y = +200$ psi, and $s_{yx} = +700$ psi, Mohr's circle is drawn as shown in Fig. 10.8*b*; all answers are scaled from the circle.

(*a*) $p = +605$ psi at $2\theta_n = 60°$ ↷ from CH

$q = -1005$ psi at $2\theta_n = 60°$ ↷ from CV

The principal stresses act as shown on the inclined element of Fig. 10.8*c* (*Ans*).

(*b*) $s_{s,\max} = +805$ psi and $s_n = -200$ psi at $2\theta = 30°$ ↶ from *CH*
$s_{s,\max} = -805$ psi and $s_n = -200$ psi at $2\theta = 30°$ ↶ from *CV*

The stresses on the planes of maximum shear act as shown on the inclined element of Fig. 10.8*d* (*Ans*).

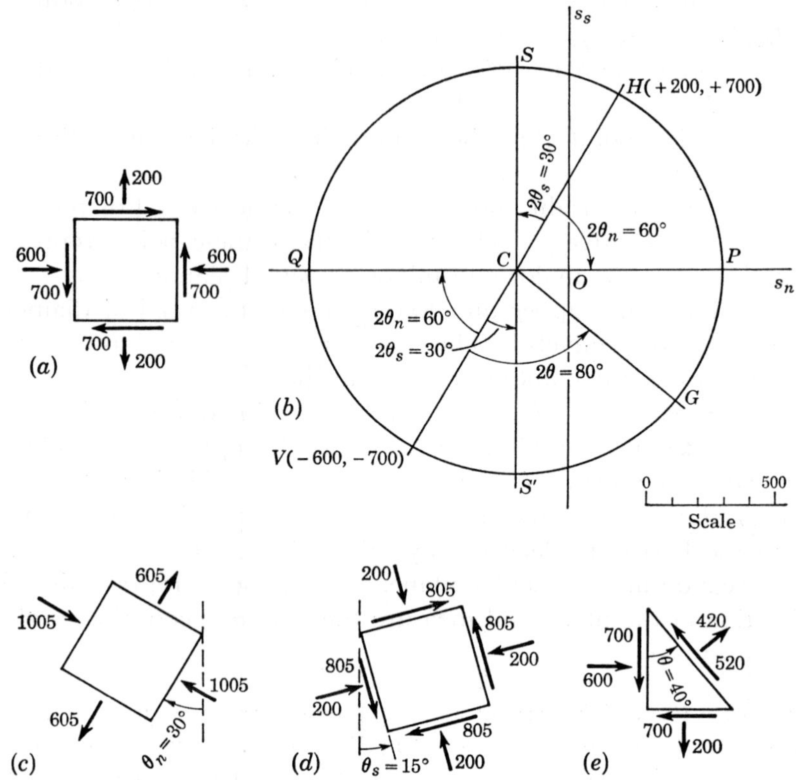

FIG. 10.8

(*c*) To find the stresses on the plane making a counterclockwise angle of 40° with the vertical face, radius *CG* is drawn 80° counterclockwise from radius *CV* and $s_n = +420$ psi, $s_s = -520$ psi. These stresses act as shown on the inclined plane of Fig. 10.8*e* (*Ans*).

Example 10.5. The state of stress, in pounds per square inch, at a point of a loaded member is as shown in Fig. 10.9*a*. Using Mohr's circle semigraphically, determine the maximum shearing stress and the principal stresses and planes.

Noting that $s_x = -5600$ psi, $s_{xy} = +4800$ psi, $s_y = -18{,}400$ psi, and $s_{yx} = -4800$ psi, Mohr's circle is sketched as shown in Fig. 10.9*b*.

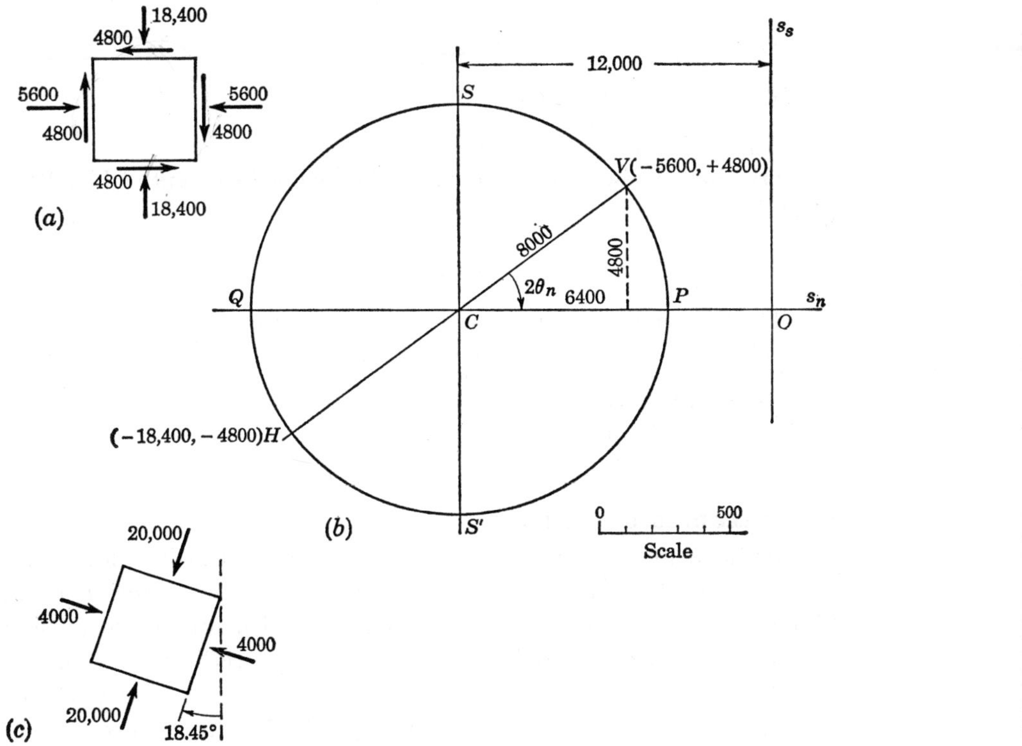

FIG. 10.9

The s_n coordinate of the center of the circle, point C, is equal to

$$\frac{-5600 - 18{,}400}{2} \quad \text{or} \quad -12{,}000$$

The radius of the circle is

$$\sqrt{(4800)^2 + (6400)^2} = 8000 \text{ psi}$$

Therefore,

$s_{s,\max} = 8000$ psi *Ans*

The principal stresses are

$$p = -12{,}000 + 8000 = -4000 \text{ psi}$$
$$q = -12{,}000 - 8000 = -20{,}000 \text{ psi}$$

To find the principal planes

$$\tan 2\theta_n = \frac{4800}{6400} = 0.75$$
$$2\theta_n = 36.9^\circ \curvearrowright \text{ from } CV$$
$$\theta_n = 18.45^\circ$$

The principal stresses act as shown on the inclined element of Fig. 10.9*c* (*Ans*).

PROBLEMS

The following problems may be assigned to be worked using Mohr's circle either graphically or semigraphically:

10.1–10.9 (at the end of Art. 10.2)
10.14–10.37 (at the end of Art. 10.3)

10.5 THEORIES OF FAILURE

As has been previously discussed, strength computations for members are based on the mechanical properties of the materials used. Since these properties are usually determined from simple tension or compression tests, their use in accurately predicting failure in members subjected to uniaxial stress is both simple and straightforward. The problem of accurately predicting the failure stresses for members subjected to biaxial or triaxial stress is, of course, much more complex. In fact, the problem is so complex, in such cases, that a large number of different theories for predicting failure have been formulated. Since each of these theories is based on certain more or less logical assumptions, their relative accuracy must ultimately be judged on the basis of test results. Such test results indicate that any one of the accepted theories will be accurate for certain materials under certain loading conditions; however, they also indicate that none of the theories will be accurate for all materials under all conditions of load.†

† Joseph Marin, *Trans. ASCE*, vol. 101, p. 1162, 1936.

Perhaps the greatest influence on engineering of theories of failure is derived from their use in the determination of design criteria for specifications. However, for unusual design problems not fully covered by specifications, the theories must be considered by the individual engineer.

Several of the best known theories of failure will be discussed in separate paragraphs below. For the sake of simplicity, they will be expressed in equation form as they apply to a biaxial stress condition; these equations can be easily extended to apply also to a triaxial condition of stress. In these equations, $s_{L(t)}$ and $s_{L(c)}$ will be used to designate the limiting strength of the material in tension and compression, respectively. Since ductile materials usually fail as the result of excessive deformation and brittle materials fail by fracture, the limiting strength for these two classes of materials is normally measured by different mechanical properties. For ductile materials the limiting strength is the yield point (or yield strength) as determined from simple tension or compression tests; for brittle materials the breaking strength is the limiting strength.

Maximum Normal-stress Theory. This theory predicts that failure will occur when the maximum or minimum principal stress at any point in a member reaches the limiting strength of the material. In equation form, failure is predicted when

$$p = s_{L(t)} \qquad \text{or} \qquad q = s_{L(c)} \tag{10.7}$$

Since this theory ignores the possibility of failure due to shearing stress, it is generally unsafe for use with ductile materials. However, for brittle materials, which are relatively strong in shear, this theory is generally accurate. The maximum normal-stress theory is frequently referred to as Rankine's theory.

Maximum Shearing-stress Theory. This theory predicts that failure will occur when the maximum shearing stress at any point in a member reaches the limiting shearing strength of the material. In other words, failure is predicted when $s_{s,\max} = s_{L(s)}$. Since tests to determine the limiting strength in shear $s_{L(s)}$ directly are difficult to perform, the theory is generally expressed in terms of normal stress. Since $s_{s,\max} = (p - q)/2$, and since the maximum shearing stress in a simple tension or compression test is equal to $\frac{1}{2}(P/A)$ or one-half the maximum normal stress, failure is predicted when

$$p - q = s_{L(t)} \quad \text{or} \quad s_{L(c)}$$

It should be noted that the above equation is misleading for the case of principal stresses of like sign (i.e., both tension or both compression). For this case, the equation would predict that regardless of the magnitude of the principal stresses, failure would not occur so long as their difference in magnitude was less than the limiting stress. This apparent fallacy arises from the fact that, when p and q are of like sign, the maximum shearing stress occurs on planes which are normal to one of the principal planes. Thus for the element shown in Fig. 10.10, the maximum shearing stress will occur on planes *aegc* and *dhfb* and will be equal to $p/2$. Similarly, if p and q were both compressive, q would be the arithmetically larger and the maximum shearing stress would be equal to $q/2$ (occurring on planes *ehcb* and *dgfa*). Thus the maximum shearing-stress theory can be expressed as

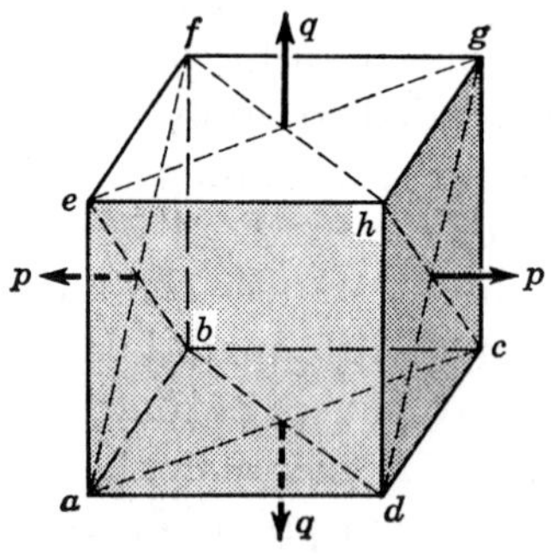

FIG. 10.10

$$p - q = s_{L(t)} \quad \text{or} \quad s_{L(c)} \tag{10.8a}$$

for principal stresses of unlike signs and

$$p = s_{L(t)} \qquad \text{or} \qquad q = s_{L(c)} \tag{10.8b}$$

for principal stresses of like signs. This theory is generally safe, though somewhat conservative, and is used as the basis of many design specifications. It should be noted that Eq. (10.8*b*) is identical with Eq. (10.7); thus, for a biaxial stress condition, the maximum shearing-stress theory as expressed by Eqs. (10.8*a*) and (10.8*b*) is equivalent to a "maximum normal- and shearing-stress" theory. The maximum shearing-stress theory is frequently referred to as Guest's theory.

Maximum Normal-strain Theory. This theory predicts that failure will occur when the normal strain at any point in a member reaches the limiting value of strain as determined from a simple tension or compression test. Noting that the principal strains can be expressed as

$$\epsilon_{\max} = \frac{p}{E} - \frac{\mu q}{E} \qquad \text{and} \qquad \epsilon_{\min} = \frac{q}{E} - \frac{\mu p}{E}$$

the theory can be written as

$$\frac{p}{E} - \frac{\mu q}{E} = \frac{s_{L(t)}}{E} \qquad \text{or} \qquad \frac{q}{E} - \frac{\mu p}{E} = \frac{s_{L(c)}}{E}$$

In simpler form the theory can be expressed as

$$p - \mu q = s_{L(t)} \qquad \text{or} \qquad q - \mu p = s_{L(c)} \tag{10.9}$$

Although this theory certainly appears to be quite logical, it does not predict values which are generally consistent with test results. It has been proved to be generally unsafe for ductile materials, especially when the principal stresses are of like sign. The maximum normal-strain theory is frequently referred to as Saint Venant's theory.

Maximum Strain-energy Theory. This theory predicts that failure will occur when the strain energy per unit volume at any point in a member reaches the limiting strain energy per unit volume of the material (from a simple tension or compression test). If strain energy is expressed in terms of normal stress, as explained in a later chapter, the theory can be written as

$$p^2 - 2\mu pq + q^2 = s^2_{L(t)} \qquad \text{or} \qquad s^2_{L(c)} \tag{10.10}$$

Although this theory is generally safe, it has been largely replaced by the maximum distortion-energy theory which generally gives better agreement with test results.

Maximum Distortion-energy Theory. This theory is a modification of the maximum strain-energy theory described above. It considers that the strain energy of an element results from two effects of stress and strain: (1) change of volume of the element and (2) relative change of dimensions of the elements (i.e., distortion). The theory assumes that the energy associated with volume change has no effect on strength and that failure is dependent on the distortion energy alone. Thus it predicts that failure will occur when the distortion strain energy per unit volume at any point in a member reaches the limiting distortion energy per unit volume of the material (as determined from a simple tension or compression test). In equation form, the theory can be expressed as

$$p^2 - pq + q^2 = s_{L(t)}^2 \quad \text{or} \quad s_{L(c)}^2 \tag{10.11}$$

The maximum distortion-energy theory is becoming increasingly popular because it is generally safe and yet is less conservative than the maximum shearing-stress theory and the maximum strain-energy theory. It is referred to by a wide variety of titles such as Hencky-von Mises, Huber-Hencky-von Mises, maximum shearing distortion, and maximum shearing-strain energy.

11

ELASTIC CURVE: DOUBLE INTEGRATION

11.1 THE ELASTIC CURVE

The edge view of the neutral surface of a beam is known as its elastic curve. Whereas the elastic curve (of a straight beam) will be a straight line, after bending, it will be a smooth curve the equation of which can be readily determined. Many techniques have been developed for using this equation to solve for the slopes and deflections of beams and to make possible the determination of the reactions of beams which are statically indeterminate. It should be emphasized that all these techniques require a thorough understanding of the meaning and a clear visualization of the elastic curve.

In general, for a structure to serve its intended purpose it must be relatively rigid; that is, its maximum deflection must be strictly limited. For example, excessive deflection of a beam in a ceiling would cause cracking or fracture of the plaster underneath. In cases of machinery, excessive deflection of the individual members may prevent the machinery from operating properly or render it useless. Where a structure is to be used by the public (such as a bridge or cantilever floor), the deflections may be limited for purely psychological reasons. Although the structure may be safe, any deflection that could be felt would tend to discourage its use.

The slopes at certain points on elastic curves are useful mainly as tools for determining the reactions and deflections of statically indeterminate structures.

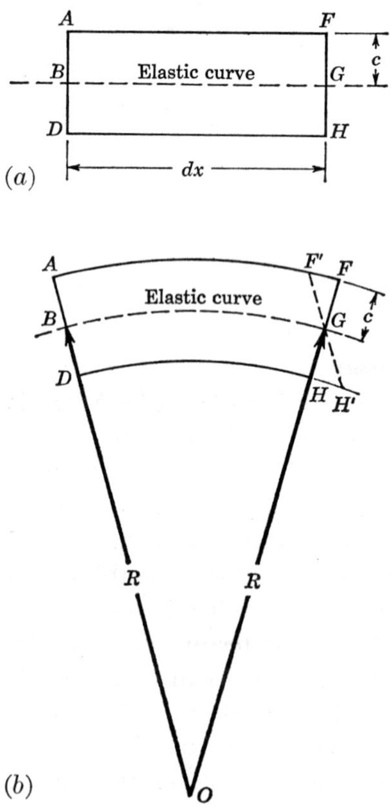

FIG. 11.1

11.2 CURVATURE DUE TO BENDING

The relationship between the curvature at a point on the elastic curve of a beam and the bending moment at that section is basic to the derivation of the equation of the elastic curve. This relationship will be derived with respect to Fig. 11.1, which shows an elemental length of beam before and after bending. Since the ends of the length in Fig. 11.1*a* are defined by parallel lines *ABD* and *FGH*, lines *AF*, *BG*, and *DH* are each *dx* in length. If the beam is subjected to negative bending moment, the elemental length will be bent into the shape shown in Fig. 11.1*b*. Line *BG*, being part of the neutral surface, remains unstressed and *dx* in length. If line *F'GH'* is constructed parallel to line *ABD*, the elongation δ of fiber *AF* is seen to be equal to the distance *FF'* and its unit strain ϵ to FF'/AF' or FF'/dx. It can be observed that sector *F'GF* is similar to sector *BOG*, since they have homologous sides parallel. Therefore, $GF/OG = FF'/BG$ or $c/R = FF'/dx = \epsilon$. If the proportional limit is not exceeded, ϵ can be expressed as s/E, so that $s/E = c/R$. Since, from the flexure formula, the stress at the top is Mc/I, it follows that $Mc/EI = c/R$ and

$$\frac{1}{R} = \frac{M}{EI} \tag{11.1}$$

where R is the radius of curvature and $1/R$ is referred to as the curvature. Thus it should be noted that the curvature at a particular point on the elastic curve is a function of the bending moment at that section of the beam, the moment of inertia of the cross section, and the modulus of elasticity of the material.

11.3 EQUATION OF THE ELASTIC CURVE

The general equation of the elastic curve can be derived by finding the relation between the curvature $1/R$ and the rate of change in

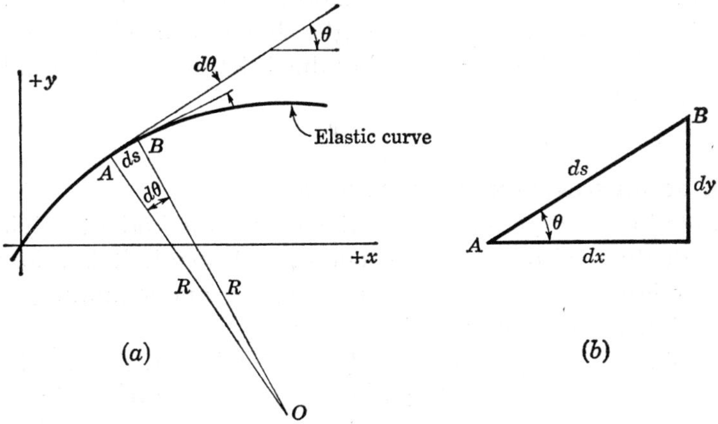

FIG. 11.2

slope of the curve d^2y/dx^2 and then equating this rate of change to the equivalent of the curvature M/EI. Figure 11.2*a* shows the elastic curve of part of a bent beam, for which the differential length AB subtends the differential angle $d\theta$. As can be seen, $d\theta$ represents the change in slope between points A and B. Figure 11.2*b* is an enlarged diagram of the arc AB showing the slope angle θ. The slope of the curve, which is also the slope of the tangent, is dy/dx; the rate of change of slope with respect to the length s is $d\theta/ds$. Therefore, the rate of change of slope with respect to s is $[d(dy/dx)]/ds$ and since the arc length ds is equal to $R\,d\theta$, the rate of change of slope $d\theta/ds$ is also equal to $1/R$. Since the curvature of beams is generally very small, ds differs only slightly from dx, so that $[d(dy/dx)]/ds$ can be taken as equal to $[d(dy/dx)]/dx$ or d^2y/dx^2. Thus it follows that, for the assumptions made, $d^2y/dx^2 = 1/R$.† Since $1/R = M/EI$, then $d^2y/dx^2 = M/EI$, which is the general equation of the elastic curve. This differential equation is more conveniently written in the form

$$EI\frac{d^2y}{dx^2} = M \tag{11.2}$$

where M is the bending moment in terms of x, E is the Young's modulus of the material, I is the moment of inertia of the beam cross section, and d^2y/dx^2 is the rate of change of the slope with respect to x. If the differential equation is integrated twice, an equation in terms of x and y defining the shape of the elastic curve is obtained. The technique for doing this is known as the double-integration method.

11.4 THE DOUBLE-INTEGRATION METHOD

The double-integration method is the oldest method of solving problems dealing with the elastic curve. Although it is seldom used for design purposes, it is most useful for understanding the behavior of structural members in bending.

The solution of a problem by the double-integration method is a multistep operation which can generally be facilitated by following a definite procedure. Steps for such a procedure are listed below. The student will find it advantageous to refer to this list whenever he experiences difficulty in the solution of a "double-integration problem."

† If the assumption of $dx = ds$ is not made, the expression for curvature becomes $\dfrac{1}{R} = \dfrac{d^2y/dx^2}{[1 + (dy/dx)^2]^{3/2}}$, the derivation of which can be found in almost any calculus book. It can be noted that, for small slopes, the denominator of this expression approaches unity.

Steps for the solution of problems by double integration

1. Draw a neat line diagram of the loaded beam (including reactions).
2. Sketch the elastic curve.
3. Select a *convenient* coordinate origin (including positive x and y directions).
4. Write the moment equations, and show their limits.
5. Set $EI\, d^2y/dx^2$ equal to M, and integrate twice.
6. Using the known conditions of the elastic curve, solve for the constants of integration and other unknowns (if any).
7. Solve the applicable equations for the slopes or deflections required.

When the origin for a particular problem is selected, frequently the solution is greatly simplified by taking the x distance as positive

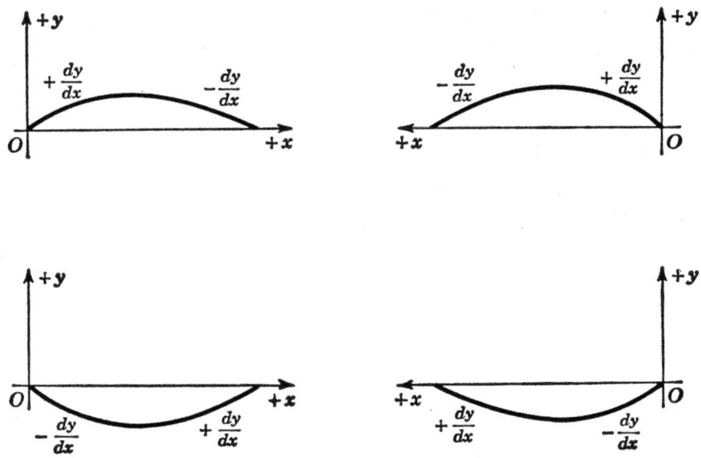

FIG. 11.3 Sign convention for slopes.

toward the left (instead of to the right as is traditional). This procedure does not, of course, affect the sign of the bending moment or the deflection. It might, however, cause some confusion regarding the sign of the slope. Therefore, it should be remembered that the slope is positive if the tangent to the curve is up in the positive direction of x. This sign convention is illustrated in Fig. 11.3. For beams of constant EI, the differential equation of the elastic curve can be solved for slope dy/dx and deflection y by the following steps:

$$EI \frac{d(dy/dx)}{dx} = M \tag{a}$$

$$EI \int d\frac{dy}{dx} = \int M\,dx$$

$$EI \frac{dy}{dx} = \left(\int M\,dx\right) + C_1 \tag{b}$$

$$EI \int dy = \int \left[\left(\int M\,dx\right) + C_1\right] dx$$

$$EIy = \left(\iint M\,dx\,dx\right) + C_1x + C_2 \tag{c}$$

For simplicity of notation, d^2y/dx^2 and dy/dx are frequently designated as y'' and y', respectively.

In working with problems dealing with beam deflections, care must be taken to avoid errors due to inconsistency of units. The units usually assigned are:

Bending moment M	lb-ft
Young's modulus E	psi
Moment of inertia I	in.^{4}
Horizontal distance x	ft
Curvature $1/R$ or d^2y/dx^2	in.$^{-1}$
Slope dy/dx	in./in.
Deflection y	in.

With these units the "unit equations" of Eqs. (a), (b), and (c) above would be inconsistent as shown below:

$$\left(\frac{\text{lb}}{\text{in.}^2}\right)(\text{in.}^4)\left(\frac{1}{\text{in.}}\right) \neq \text{lb-ft} \qquad \text{for Eq. } (a)$$

$$\left(\frac{\text{lb}}{\text{in.}^2}\right)(\text{in.}^4)\left(\frac{\text{in.}}{\text{in.}}\right) \neq \text{lb-ft}^2 \qquad \text{for Eq. } (b)$$

$$\left(\frac{\text{lb}}{\text{in.}^2}\right)(\text{in.}^4)(\text{in.}) \neq \text{lb-ft}^3 \qquad \text{for Eq. } (c)$$

The units can be made consistent by multiplying the right-hand sides of Eqs. (a), (b), and (c) by 12, 144, and 1728, respectively.

SEVERO GARZA
EG 131-
LAB A

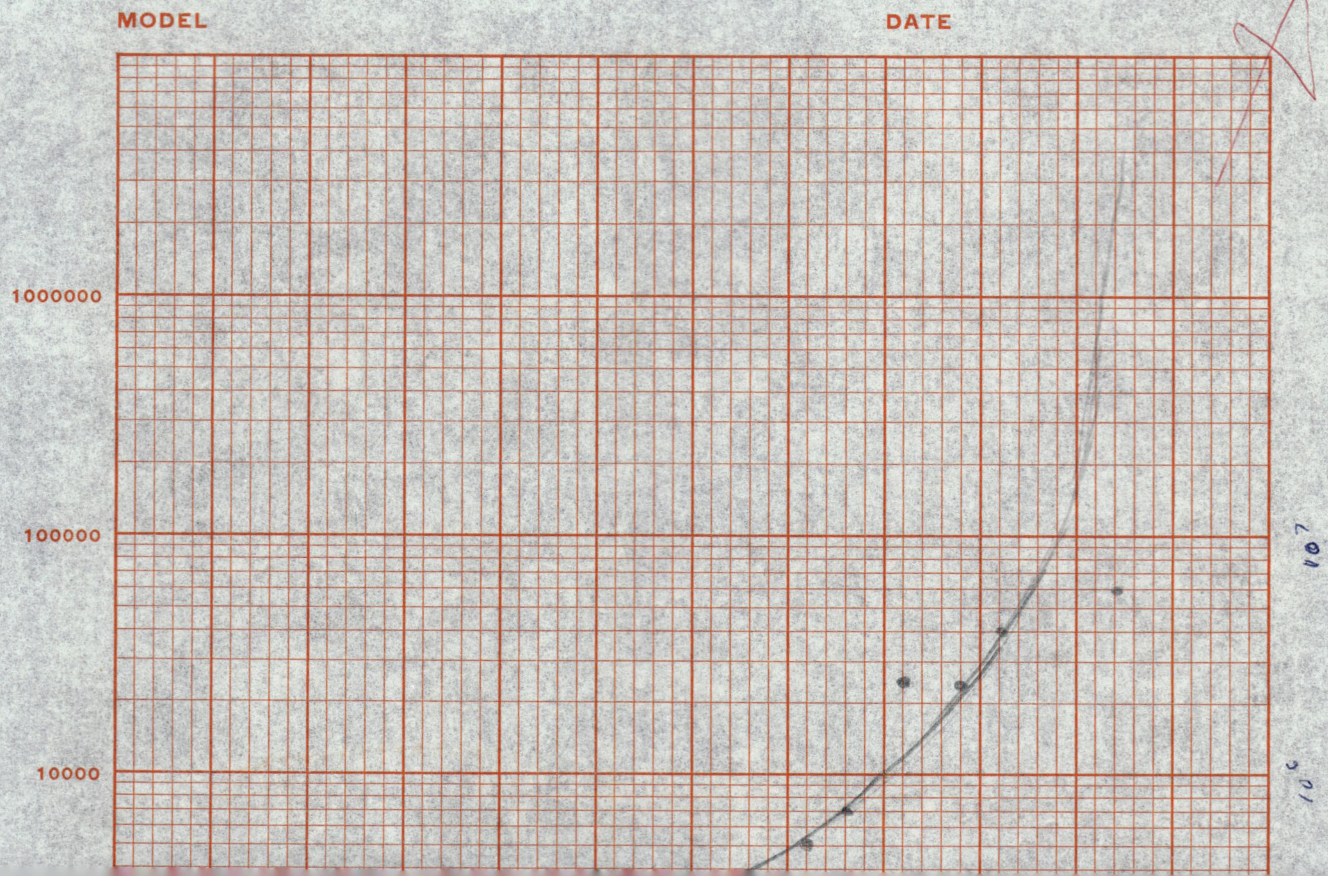

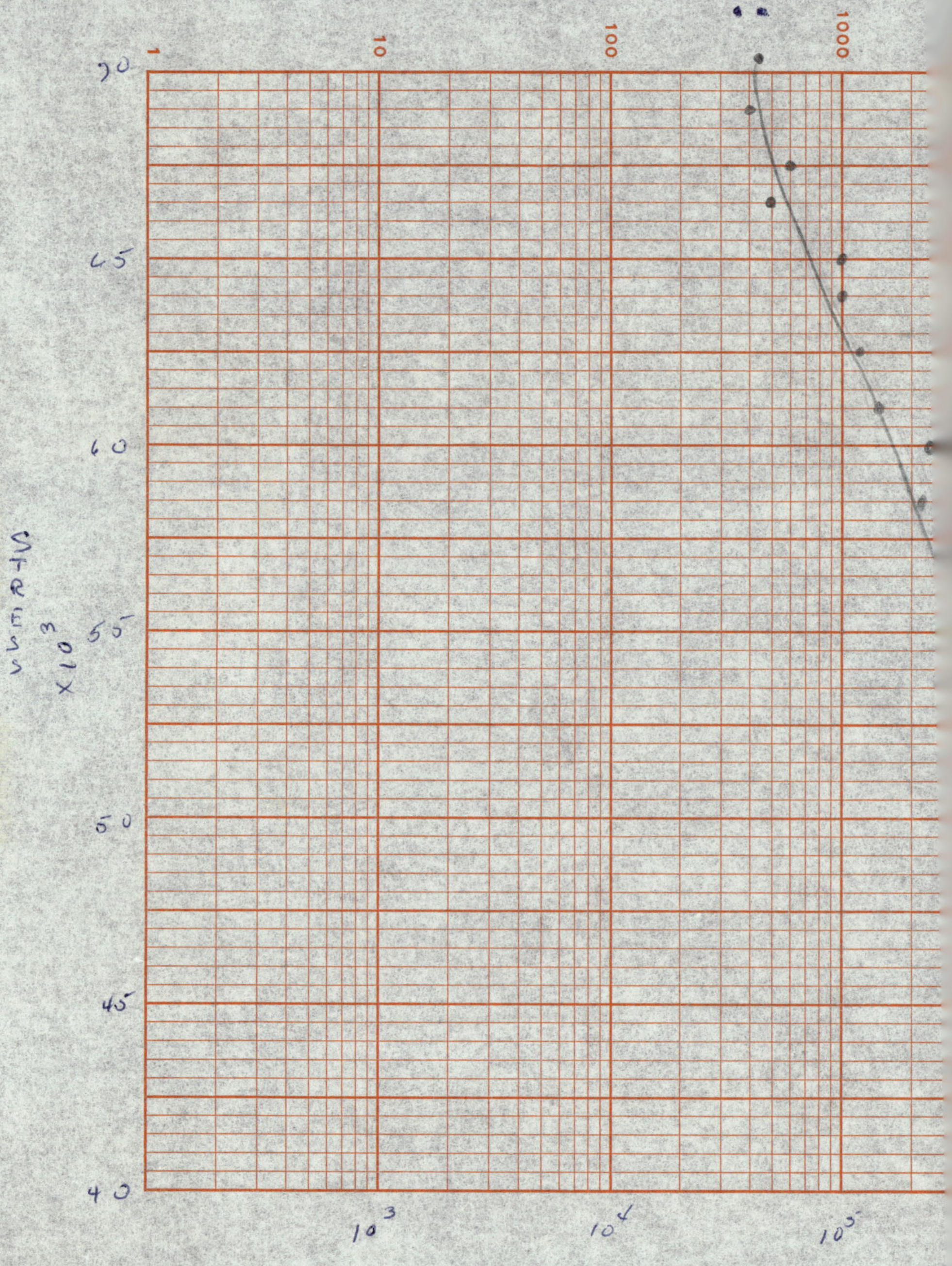
K+E SEMI-LOGARITHMIC
7 CYCLES X 60 DIVISIONS
KEUFFEL & ESSER CO.
46 6463
MADE IN U.S.A.
1
10
100
1000
70
65
60
55
50
45
40
10^3
10^4
10^5
× 10^3

Four types of elastic-curve problems are presented in the next four articles. The problems are separated into types only for teaching purposes. *The basic method of solution*, as discussed in this article, *is the same for all elastic-curve problems.*

11.5 SYMMETRICALLY LOADED BEAMS

A special case which is frequently encountered or assumed in engineering design is that of beams with symmetrical loading. Symmetrical loading means that the reactions as well as the applied loading are symmetrical. Since the elastic curve for such a beam is symmetrical, equations for moment, slope, and deflection need be written for only half of the beam.

Example 11.1. For the beam shown in Fig. 11.4, $E = 30{,}000{,}000$ psi and $I = 15$ in.^{4} Determine the maximum deflection and the slope angle at each reaction.

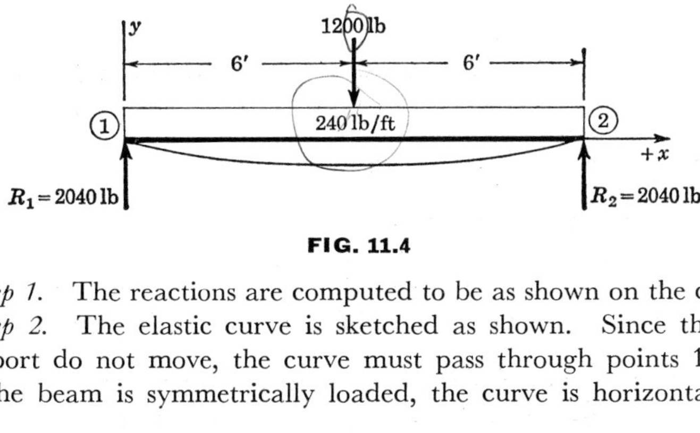

FIG. 11.4

Step 1. The reactions are computed to be as shown on the diagram.

Step 2. The elastic curve is sketched as shown. Since the points of support do not move, the curve must pass through points 1 and 2; since the beam is symmetrically loaded, the curve is horizontal at the center.

Step 3. The origin is selected at point 1 with x positive to the right.

Step 4. The bending-moment equation for half of the beam is $M_x = +2040x - 120x^2$ (valid between the limits of $x = 0$ and $x = 6$).

Step 5. The moment equation is set equal to $EI\, d^2y/dx^2$ and the equation integrated twice:

$$EI\frac{d^2y}{dx^2} = +2040x - 120x^2 \tag{a}$$

$$EI\frac{dy}{dx} = +\frac{2040x^2}{2} - \frac{120x^3}{3} + C_1 \tag{b}$$

$$EIy = +\frac{1020x^3}{3} - \frac{40x^4}{4} + C_1x + C_2 \tag{c}$$

Step 6. Equations (*b*) and (*c*) in Step 5 cannot be utilized until the unknown constants of integration have been evaluated. These two constants can be evaluated by using *two known conditions* of *the elastic curve.* From the sketch of the elastic curve it can be seen that the curve passes through the point $x = 0$, $y = 0$. Also because of symmetry the slope must be zero at $x = 6$. Using these two known conditions,

At $x = 0$, $y = 0$ [in Eq. (*c*)]
$$EI(0) = +340(0)^3 - 10(0)^4 + C_1(0) + C_2$$

Therefore,

$$C_2 = 0$$

At $x = 6$, $dy/dx = 0$ [in Eq. (*b*)]
$$EI(0) = +1020(6)^2 - 40(6)^3 + C_1$$

Therefore,

$$C_1 = -28{,}080$$

With the constants of integration evaluated, Eqs. (*b*) and (*c*) can now be written as

$$EI\frac{dy}{dx} = +1020x^2 - 40x^3 - 28{,}080 \qquad [b]$$

$$EIy = +340x^3 - 10x^4 - 28{,}080x \qquad [c]$$

These equations can now be used to evaluate the slope and deflection at any point on the elastic curve within the limits of the equations ($x = 0$ to $x = 6$).

Step 7. From the sketch of the elastic curve it can be seen that the maximum deflection occurs at $x = 6$. Substituting in Eq. [*c*],

$$EIy_{\max} = +340(6)^3 - 10(6)^4 - 28{,}080(6) = -108{,}000 \text{ lb-ft}^3$$

$$y_{\max} = \frac{-108{,}000(1728)}{30{,}000{,}000(15)} = -0.415 \text{ in.} \qquad \textit{Ans}$$

To find the slope at the left reaction, let $x = 0$ in Eq. [*b*].

$$EI\left(\frac{dy}{dx}\right)_1 = +1020(0)^2 - 40(0)^3 - 28{,}080 = -28{,}080 \text{ lb-ft}^2$$

$$\left(\frac{dy}{dx}\right)_1 = \frac{-28{,}080(144)}{30{,}000{,}000(15)} = -0.00899 \text{ in./in.}$$

Since, for very small angles, the angle in radians is equal to the slope (tangent) of the angle,

$\theta_1 = -0.00899$ radian *Ans*

From symmetry

$\theta_2 = +0.00899$ radian *Ans*

PROBLEMS

★11.1. Compute the maximum deflection of a simply supported beam 12 ft long due to a midspan concentrated load of 1200 lb. $E = 10{,}000{,}000$ psi; $I = 40$ in.^{4}

11.2. Compute the maximum deflection of a simply supported beam 10 ft long due to a midspan concentrated load of 4800 lb. $E = 30{,}000{,}000$ psi; $I = 50$ in.^{4}

★11.3. A uniform load of 2400 lb/ft acts on the entire span of a simply supported beam 10 ft long. $E = 30{,}000{,}000$ psi; $I = 70$ in.^{4} Compute (*a*) the maximum deflection, (*b*) the slope angle at each support.

11.4. A uniform load of 96 lb/ft acts on the entire span of a simply supported beam 12 ft long. $E = 1{,}600{,}000$ psi; $I = 500$ in.^{4} Compute (*a*) the maximum deflection, (*b*) the slope angle at each support.

★11.5. For the beam shown in Fig. P11.5, let $P = 4$ kips, $L = 24$ ft, $E = 10{,}000{,}000$ psi, $I = 300$ in.^{4} Compute (*a*) the maximum deflection, (*b*) the deflection 6 ft from each reaction.

11.6. For the beam shown in Fig. P11.6, let $P = 10$ kips, $L = 20$ ft, $E = 30{,}000{,}000$ psi, $I = 1000$ in.^{4} Compute (*a*) the maximum deflection, (*b*) the deflection 5 ft from each reaction.

★11.7. For the beam shown in Fig. P11.7, let $w = 720$ lb/ft, $L = 20$ ft, $E = 10{,}000{,}000$ psi, $I = 100$ in.^{4} Compute (*a*) the maximum deflection, (*b*) the slope angle at each reaction.

11.8. For the beam shown in Fig. P11.8, let $w = 600$ lb/ft, $L = 16$ ft, $E = 10{,}000{,}000$ psi, $I = 80$ in.^{4} Compute (*a*) the maximum deflection, (*b*) the slope angle at each reaction.

★11.9. A 14-ft span of a continuous beam has *positive bending moments* of 1200 lb-ft at each end. $E = 1{,}600{,}000$ psi; $I = 25$ in.^{4} Compute the maximum deflection.

11.10. A 22-ft span of a continuous beam has *negative bending moments* of 240 kip-ft at each end. $E = 30{,}000{,}000$ psi; $I = 1200$ in.^{4} Compute the maximum deflection.

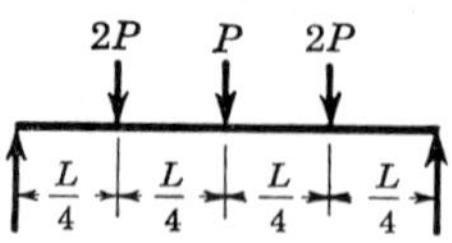

FIG. P11.5, P11.6

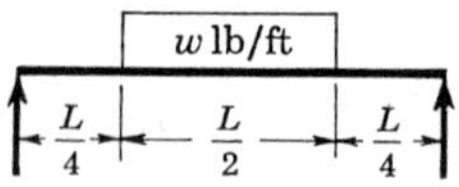

FIG. P11.7, P11.8

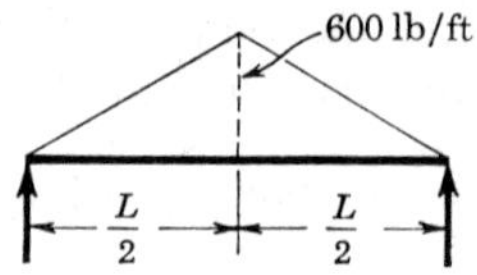

FIG. P11.13, P11.14

11.11. A simply supported beam of length L ft is loaded with a uniform load of w lb/ft over the entire span. Show that the maximum deflection is $5wL^4/384EI$ and that the slope angle at either reaction is $wL^3/24EI$.

11.12. A simply supported beam of length L is loaded at midspan with a concentrated load P. Show that the maximum deflection is $PL^3/48EI$ and that the slope angle at either reaction is $PL^2/16EI$.

★11.13. For the beam shown in Fig. P11.13, with triangular loading, let $L = 20$ ft, $EI = 1{,}800{,}000{,}000$ lb-in.^{2} Compute the maximum deflection.

11.14. For the beam shown in Fig. P11.14, with triangular loading, let $L = 12$ ft, $EI = 360{,}000{,}000$ lb-in.^{2} Compute the maximum deflection.

11.6 CANTILEVER BEAMS

The procedure for obtaining the slope and deflection equations of cantilever beams can usually be simplified by selecting the origin at the fixed end. Taking the origin at the fixed end generally results in constants of integration which have values of zero.

Example 11.2. For the beam shown in Fig. 11.5, $E = 30{,}000{,}000$ psi (or 30,000 ksi), $I = 200$ in.^{4} Determine the maximum deflection, the deflection at midspan, and the slope angle at the free end.

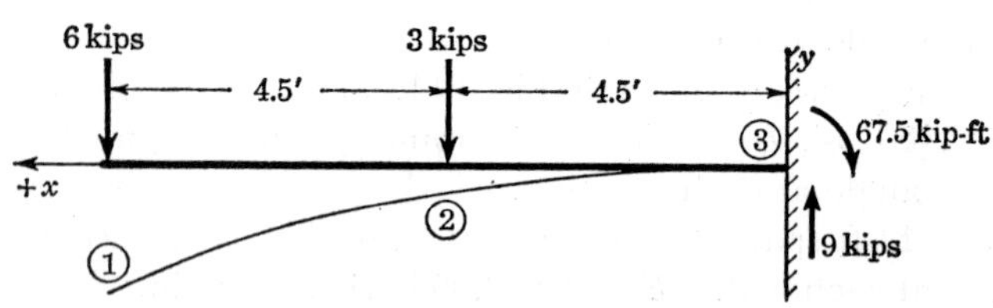

FIG. 11.5

The reactions are a vertical force *and a couple* as shown. The moment equations (with their limits) are

$0 \leqq x \leqq 4.5$	$4.5 \leqq x \leqq 9$
(a) $EIy'' = -67.5 + 9x$	(b) $EIy'' = -67.5 + 9x - 3(x - 4.5)$
(c) $EIy' = -67.5x + \dfrac{9x^2}{2} + C_1$	(d) $EIy' = -67.5x + \dfrac{9x^2}{2} - \dfrac{3(x-4.5)^2}{2} + C_2$
(e) $EIy = \dfrac{-67.5x^2}{2} + \dfrac{4.5x^3}{3} + C_1x + C_3$	(f) $EIy = -\dfrac{67.5x^2}{2} + \dfrac{4.5x^3}{3} - \dfrac{1.5(x-4.5)^3}{3} + C_2x + C_4$

For the two sets of equations there are four constants of integration to be evaluated, thus requiring four known conditions of the elastic curve. These are $x = 0$, $y = 0$ in Eq. (e); $x = 0$, $y' = 0$ in Eq. (c); for $x = 4.5$ the slope y' obtained from Eqs. (c) and (d) is equal; for $x = 4.5$ the deflection y obtained from Eqs. (e) and (f) is equal.

The constants are evaluated as follows:

For $x = 4.5$, $y'(c) = y'(d)$	therefore $C_1 = C_2$
For $x = 4.5$, $y(e) = y(f)$	therefore $C_3 = C_4$
For $x = 0$, $y'(c) = 0$	therefore $C_1 = 0$ and $C_2 = C_1 = 0$
For $x = 0$, $y(e) = 0$	therefore $C_3 = 0$ and $C_4 = C_3 = 0$

It can be noted that $C_2 = C_1$ and $C_4 = C_3$ because the expression $(x - 4.5)$ becomes zero when x is 4.5. It should also be noted that, if Eq. (b) had been simplified prior to integration, C_2 would *not* have been equal to C_1, C_4 would *not* have been equal to C_3, and C_2 and C_4 would have values other than zero.

With the constants of integration evaluated, Eqs. (c), (d), (e), and (f) can now be written in a more usable form.

$$[c]\quad EIy' = -67.5x + 4.5x^2$$
$$[d]\quad EIy' = -67.5x + 4.5x^2 - 1.5(x - 4.5)^2$$
$$[e]\quad EIy = -33.75x^2 + 1.5x^3$$
$$[f]\quad EIy = -33.75x^2 + 1.5x^3 - 0.5(x - 4.5)^3$$

From the elastic curve, the maximum deflection occurs at $x = 9$. Using Eq. [f],

$$EIy_{\max} = -33.75(9)^2 + 1.5(9)^3 - 0.5(9 - 4.5)^3 = -1686 \text{ kip-ft}^3$$
$$y_{\max} = \frac{-1686(1728)}{30{,}000(200)} = -0.486 \text{ in.} \qquad \textit{Ans}$$

For the deflection at midspan either Eq. [*e*] or [*f*] can be used with $x = 4.5$. Using Eq. [*e*],

$$EIy_2 = -33.75(4.5)^2 + 1.5(4.5)^3 = -546 \text{ kip-ft}^3$$

$$y_2 = \frac{-546(1728)}{30{,}000(200)} = -0.157 \text{ in.} \qquad Ans$$

To find the slope at the end, $x = 9$ is used in Eq. [*d*]:

$$EIy_1' = -67.5(9) + 4.5(9)^2 - 1.5(9 - 4.5)^2 = -274 \text{ kip-ft}^2$$

$$y_1' = \frac{-274(144)}{30{,}000(200)} = -0.00658 \text{ in./in.}$$

$$\theta_1 = -0.00658 \text{ radian} \qquad Ans$$

PROBLEMS

★**11.15.** A 10-ft cantilever beam fixed at the left end supports a concentrated 900-lb load at the right end. $E = 1{,}600{,}000$ psi; $I = 300$ in.^{4} Compute the deflection at (*a*) the right end, (*b*) midspan.

11.16. A 6-ft cantilever beam fixed at the left end supports a concentrated 20-kip load at the right end. $E = 30{,}000{,}000$ psi; $I = 400$ in.^{4} Compute the deflection at (*a*) the right end, (*b*) midspan.

11.17. A 12-ft cantilever beam fixed at the right end has a uniform load of 120 lb/ft on its entire span. $E = 1{,}600{,}000$ psi; $I = 200$ in.^{4} Compute the deflection and slope angle at the free end.

11.18. An 8-ft cantilever beam fixed at the right end has a uniform load of 3000 lb/ft on its entire span. $E = 30{,}000{,}000$ psi; $I = 300$ in.^{4} Compute the deflection and slope angle at the free end.

★**11.19.** For the beam shown in Fig. P11.19, let $L = 10$ ft, $E = 1{,}800{,}000$ psi, $I = 600$ in.^{4} Compute the midspan and maximum deflections.

11.20. For the beam shown in Fig. P11.20, let $L = 6$ ft, $E = 1{,}600{,}000$ psi, $I = 200$ in.^{4} Compute the midspan and maximum deflections.

11.21. For the beam shown in Fig. P11.21, let $a = 3$ ft, $b = 7$ ft, $EI = 700{,}000{,}000$ lb-in.^{2} Compute the midspan and maximum deflections.

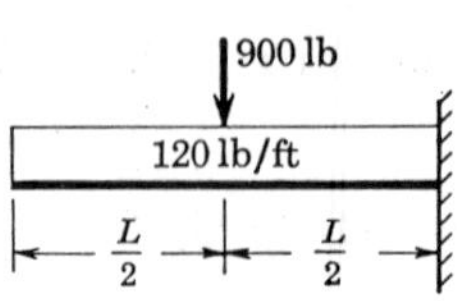

FIG. P11.19, P11.20

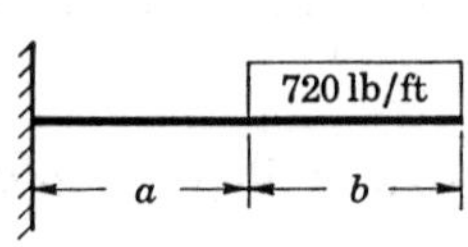

FIG. P11.21, P11.22

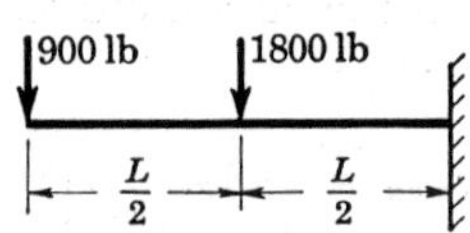

FIG. P11.23, P11.24

11.22. For the beam shown in Fig. P11.22, let $a = 8$ ft, $b = 4$ ft, $EI = 2{,}100{,}000{,}000$ lb-in.^{2} Compute the midspan and maximum deflections.

★**11.23.** For the beam shown in Fig. P11.23, let $L = 15$ ft, $E = 30{,}000{,}000$ psi, $I = 60$ in.^{4} Compute the midspan and maximum deflections.

11.24. For the beam shown in Fig. P11.24, let $L = 9$ ft, $E = 10{,}000{,}000$ psi, $I = 50$ in.^{4} Compute the midspan and maximum deflections.

11.25. For the beam shown in Fig. P11.25, let $a = 4$ ft, $b = 6$ ft, $EI = 500{,}000{,}000$ lb-in.^{2} Compute the maximum deflection.

★**11.26.** For the beam shown in Fig. P11.26, let $a = 7$ ft, $b = 5$ ft, $EI = 900{,}000{,}000$ lb-in.^{2} Compute the maximum deflection.

11.27. For the beam shown in Fig. P11.27, let $a = 7$ ft, $b = 3$ ft, $EI = 1{,}500{,}000{,}000$ lb-in.^{2} Compute the slope and deflection at the free end.

11.28. For the beam shown in Fig. P11.28, let $a = 8$ ft, $b = 4$ ft, $EI = 550{,}000{,}000$ lb-in.^{2} Compute the slope and deflection at the free end.

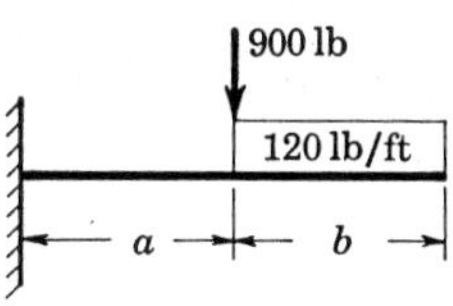

FIG. P11.25, P11.26

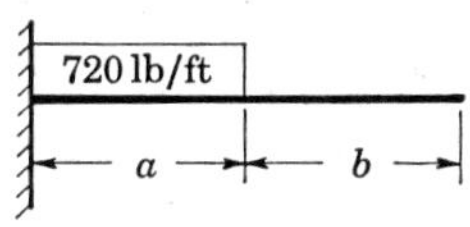

FIG. P11.27, P11.28

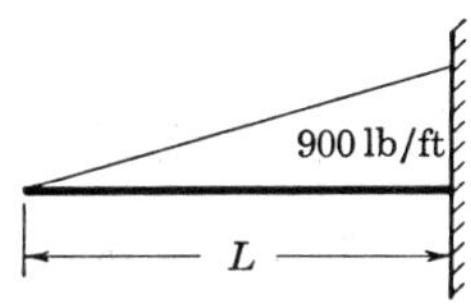

FIG. P11.33, P11.34

★**11.29.** A 10-ft cantilever beam fixed at the right end supports a 15-kip concentrated load 5 ft from the free end. $E = 30{,}000{,}000$ psi; $I = 300$ in.^{4} Compute the maximum deflection.

★**11.30.** An 11-ft cantilever beam is fixed at the right end and loaded with a clockwise 600 lb-ft couple at the left end. $EI = 32{,}000{,}000$ lb-in.^{2} Compute the slope and deflection at the free end.

11.31. A cantilever beam of length L is loaded with a concentrated load P at the free end. Show that the slope at the free end is $PL^2/2EI$; show that the maximum deflection is $PL^3/3EI$.

11.32. A cantilever beam of length L ft is loaded with a uniform load of w lb/ft over the entire span. Show that the slope angle at the free end is $wL^3/6EI$; show that the maximum deflection is $wL^4/8EI$.

11.33. For the cantilever beam shown in Fig. P11.33, with uniformly increasing loading, let $L = 10$ ft, $E = 2{,}000{,}000$ psi, $I = 700$ in.^{4} Compute the maximum deflection.

★**11.34.** For the cantilever beam shown in Fig. P11.34, with uniformly increasing loading, let $L = 6$ ft, $E = 1{,}800{,}000$ psi, $I = 200$ in.^{4} Compute the maximum deflection.

11.7 SIMPLY SUPPORTED BEAMS WITH UNSYMMETRICAL LOADING

The general case of a simply supported beam with unsymmetrical loading is the one most frequently encountered in engineering design. For this type of beam the computation of slope and deflection equations can generally be simplified by a judicious selection of the origin. Three general rules to be used as a guide in the selection of such an origin follow: (1) Select the origin at the "least loaded end," that is, the end farthest from the greatest number of applied loads. *Generally* the least loaded end has the smallest reaction. (2) When a uniform load extends over part of the beam select the origin at the end farthest from the uniform load. (3) When the beam has a single overhang, select the origin at the opposite end.

Example 11.3. Compute the maximum deflection for the beam shown in Fig. 11.6. $E = 10{,}000{,}000$ psi, $I = 12$ in.^{4}

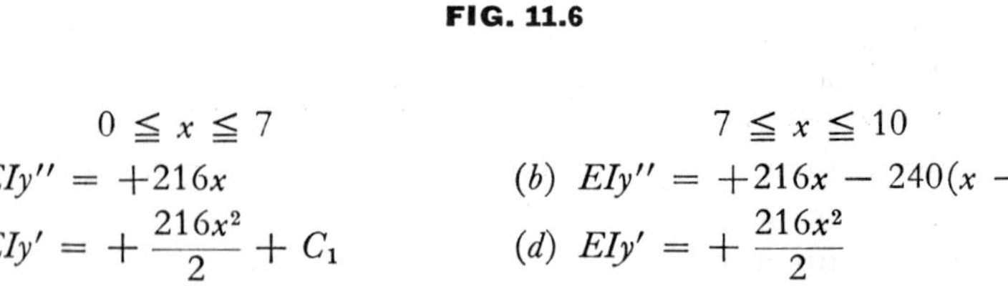

FIG. 11.6

$0 \leqq x \leqq 7$

(a) $EIy'' = +216x$

(c) $EIy' = +\dfrac{216x^2}{2} + C_1$

(e) $EIy = +\dfrac{108x^3}{3} + C_1x + C_3$

$7 \leqq x \leqq 10$

(b) $EIy'' = +216x - 240(x-7)^2$

(d) $EIy' = +\dfrac{216x^2}{2} - \dfrac{240(x-7)^3}{3} + C_2$

(f) $EIy = +\dfrac{108x^3}{3} - \dfrac{80(x-7)^4}{4} + C_2x + C_4$

At $x = 7,\ y'(c) = y'(d)$ therefore $C_1 = C_2$
At $x = 7,\ y(e) = y(f)$ therefore $C_3 = C_4$
At $x = 0,\ y(e) = 0$ therefore $C_3 = 0$ and $C_4 = 0$
At $x = 10,\ y(f) = 0$ therefore $C_2 = -3438$ and $C_1 = -3438$

Note that, although the location of the point of zero slope was not known, enough known conditions of the elastic curve were available to solve for all the constants of integration.

$[c]\ EIy' = +108x^2 - 3438$ $[d]\ EIy' = +108x^2 - 80(x-7)^3 - 3438$
$[e]\ EIy = +36x^3 - 3438x$ $[f]\ EIy = +36x^3 - 20(x-7)^4 - 3438x$

From the elastic curve it can be seen that the maximum deflection occurs where the slope is zero. This appears to be at some point between $x = 0$ and $x = 7$. Setting y' in Eq. [c] equal to zero,

$$EI(0) = +108x^2 - 3438$$

from which

$$x = \pm 5.64 \text{ ft}$$

The -5.64 value is invalid because it is outside the limits of Eq. [c]. Using the value $+5.64$ for x in Eq. [e],

$$EIy_{max} = +36(5.64)^3 - 3438(5.64) = -12{,}930 \text{ lb-ft}^3$$

$$y_{max} = \frac{-12{,}930(1728)}{10{,}000{,}000(12)} = -0.186 \text{ in.} \qquad Ans$$

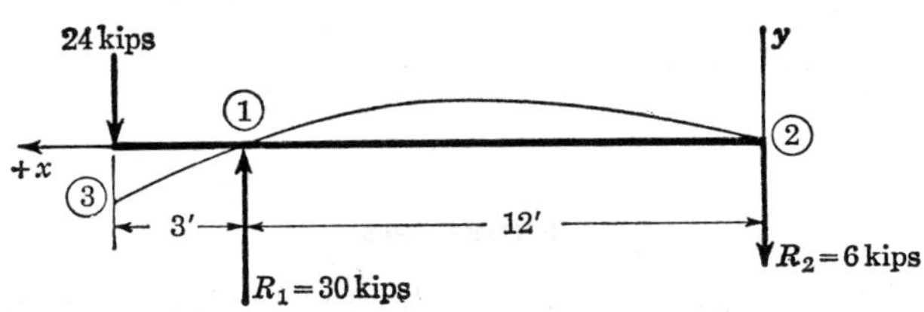

FIG. 11.7

Example 11.4. For the beam shown in Fig. 11.7, $E = 30{,}000$ ksi, $I = 240$ in.^{4} Determine the maximum upward and downward deflections.

$0 \leqq x \leqq 12$	$12 \leqq x \leqq 15$
$(a)\ EIy'' = -6x$	$(b)\ EIy'' = -6x + 30(x - 12)$
$(c)\ EIy' = \dfrac{-6x^2}{2} + C_1$	$(d)\ EIy' = \dfrac{-6x^2}{2} + \dfrac{30}{2}(x - 12)^2 + C_2$
$(e)\ EIy = \dfrac{-3x^3}{3} + C_1x + C_3$	$(f)\ EIy = \dfrac{-3x^3}{3} + \dfrac{15}{3}(x - 12)^3 + C_2x + C_4$

At $x = 12$, $y'(c) = y'(d)$	therefore $C_1 = C_2$
At $x = 12$, $y(e) = y(f)$	therefore $C_3 = C_4$
At $x = 0$, $y(e) = 0$	therefore $C_3 = 0 = C_4$
At $x = 12$, $y(e) = 0$	therefore $C_1 = +144 = C_2$
$[c]\ EIy' = -3x^2 + 144$	$[d]\ EIy' = -3x^2 + 15(x - 12)^2 + 144$
$[e]\ EIy = -x^3 + 144x$	$[f]\ EIy = -x^3 + 5(x - 12)^3 + 144x$

The maximum upward deflection occurs where the slope is equal to zero, that is, where $y'[c] = 0$.

$$EI(0) = -3x^2 + 144$$
$$x = \pm 6.93 \text{ ft}$$

using $x = +6.93$ in Eq. [*e*],

$$EIy_{\max} = -(6.93)^3 + 144(6.93) = +665 \text{ kip-ft}^3$$
$$y_{\max} = \frac{+665(1728)}{30{,}000(240)} = +0.160 \text{ in.} \qquad \textit{Ans}$$

The maximum downward deflection occurs at point 3.

$$EIy_3 = -1(15)^3 + 5(15 - 12)^3 + 144(15) = -1080 \text{ kip-ft}^3$$
$$y_3 = \frac{-1080(1728)}{30{,}000(240)} = -0.259 \text{ in.} \qquad \textit{Ans}$$

PROBLEMS

★**11.35.** For the beam shown in Fig. P11.35, let $a = 8$ ft, $b = 4$ ft, $E = 30{,}000{,}000$ psi, $I = 80$ in.^{4} Compute the maximum and midspan deflections.

11.36. For the beam shown in Fig. P11.36, let $a = 6$ ft, $b = 3$ ft, $E = 10{,}000{,}000$ psi, $I = 60$ in.^{4} Compute the maximum and midspan deflections.

11.37. For the beam shown in Fig. P11.37, let $a = 12$ ft, $b = 6$ ft, $E = 30{,}000{,}000$ psi, $I = 100$ in.^{4} Compute the maximum and midspan deflections.

★11.38. For the beam shown in Fig. P11.38, let $a = 4$ ft, $b = 8$ ft, $EI = 400{,}000{,}000$ lb-in.^{2} Compute the maximum deflection.

11.39. For the beam shown in Fig. P11.39, let $a = 6$ ft, $b = 12$ ft, $EI = 2{,}400{,}000{,}000$ lb-in.^{2} Compute the maximum deflection.

11.40. For the beam shown in Fig. P11.40, let $a = 3$ ft, $b = 6$ ft, $EI = 97{,}500{,}000$ lb-in.^{2} Compute the maximum deflection.

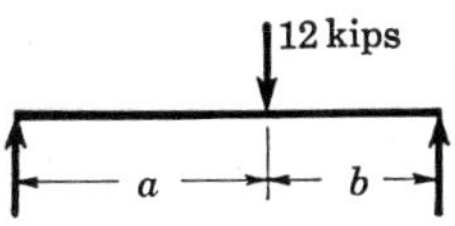

FIG. P11.35, P11.36, P11.37

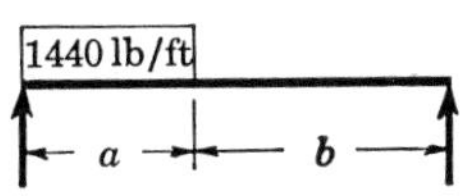

FIG. P11.38, P11.39, P11.40

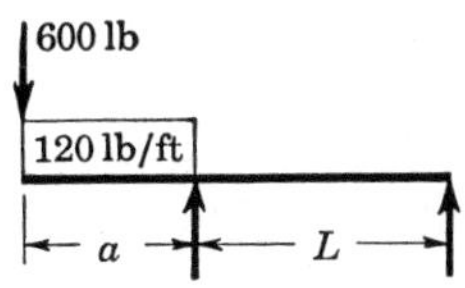

FIG. P11.47, P11.48

★11.41. A 10-ft simply supported beam is loaded with a couple at the right reaction. The couple causes a *positive* 6000 lb-ft *bending moment;* $EI = 300{,}000{,}000$ lb-in.^{2} Determine the maximum deflection.

11.42. A 15-ft simply supported beam is loaded with a couple at the left reaction. The couple causes a *negative* 1350 lb-ft *bending moment;* $EI = 300{,}000{,}000$ lb-in.^{2} Determine the maximum deflection.

★11.43. A 12-ft simply supported steel beam is loaded with couples at each end. The couple at the right reaction causes a *positive* 12,000 lb-ft *bending moment;* the couple at the left reaction causes a *positive* 19,200 lb-ft *bending moment;* $I = 50$ in.^{4} Compute the maximum deflection.

11.44. A 12-ft simply supported aluminum beam is loaded with couples at each end. The couple at the right reaction causes a *negative* 19,200 lb-ft *bending moment;* the couple at the left reaction causes a *positive* 12,000 lb-ft *bending moment;* $I = 50$ in.^{4} Compute the maximum upward and downward deflections.

11.45. A 16-ft timber beam, 10 ft between supports, has a 6-ft overhang at the right end. A uniform load of 9 kips/ft is applied to the overhang. $E = 1{,}800{,}000$ psi; $I = 600$ in.^{4} Compute the maximum upward and downward deflections.

★11.46. A 12-ft steel beam, 9 ft between supports, has a 3-ft overhang at the right end. $I = 50$ in.^{4} Determine the maximum upward and downward deflections due to a 6-kip load applied at the end of the overhang.

11.47. For the beam shown in Fig. P11.47, let $a = 4$ ft, $L = 10$ ft, $EI = 360{,}000{,}000$ lb-in.^{2} Determine the maximum upward and downward deflections.

★11.48. For the beam shown in Fig. P11.48, let $a = 6$ ft, $L = 12$ ft, $EI = 120{,}000{,}000$ lb-in.^{2} Determine the maximum upward and downward deflections.

11.49. For the steel beam shown in Fig. P11.49, let $a = 3$ ft, $b = 9$ ft, $c = 5$ ft, $I = 150$ in.^{4} Compute the maximum upward and downward deflections.

11.50. For the steel beam shown in Fig. P11.50, let $a = 4$ ft, $b = 8$ ft, $c = 7$ ft, $I = 200$ in.^{4} Compute the maximum upward and downward deflections.

11.51. A simply supported beam of length L is loaded with a couple M at the right end. Show that the maximum deflection occurs at $0.577L$ from the left end and is equal to $ML^2/(9\sqrt{3}\,EI)$.

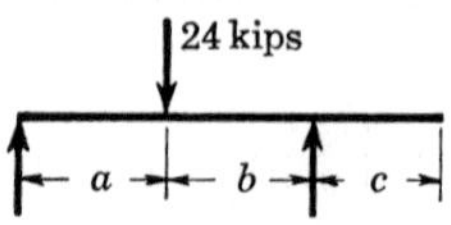

FIG. P11.49, P11.50

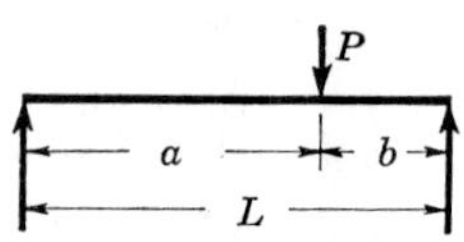

FIG. P11.52

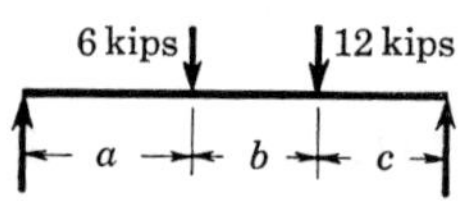

FIG. P11.53, P11.54

11.52. A simply supported beam of length L is loaded with an off-center load P as shown in Fig. P11.52. If a is greater than b, show that the maximum deflection occurs at $\sqrt{(L^2 - b^2)/3}$ from the left end and is equal to $Pb(L^2 - b^2)^{3/2}/(9\sqrt{3}\,LEI)$.

11.53. For the beam shown in Fig. P11.53, let $a = 6$ ft, $b = c = 3$ ft, $EI = 4{,}500{,}000{,}000$ lb-in.^{2} Compute the midspan deflection.

★11.54. For the beam shown in Fig. P11.54, let $a = 4$ ft, $b = 5$ ft, $c = 3$ ft, $EI = 1{,}400{,}000{,}000$ lb-in.^{2} Compute the midspan deflection.

11.8 RESTRAINED BEAMS

Very few beams in real structures are actually simply supported; they are restrained to some degree by their supports or connections. Where this restraint is small, it is good engineering practice to neglect it; where the restraint is almost complete, the beam is usually assumed to be fixed at that end. A fixed end condition means that the end of the beam is prevented from any rotation. Restraint at the end of a beam, except for a cantilever, causes it to be statically indeterminate. This restraint causes a reaction couple which is in addition to the force reaction. With this additional unknown there are not enough equilibrium equations to solve for reactions. However, with the use of the equation of the elastic curve in addition to the equilibrium equations, the reactions can be determined.

Example 11.5. Compute the reactions of the beam shown in Fig. 11.8. The reactions cannot be determined by statics and are therefore labeled as unknowns. Since the beam is fixed at both ends, the elastic

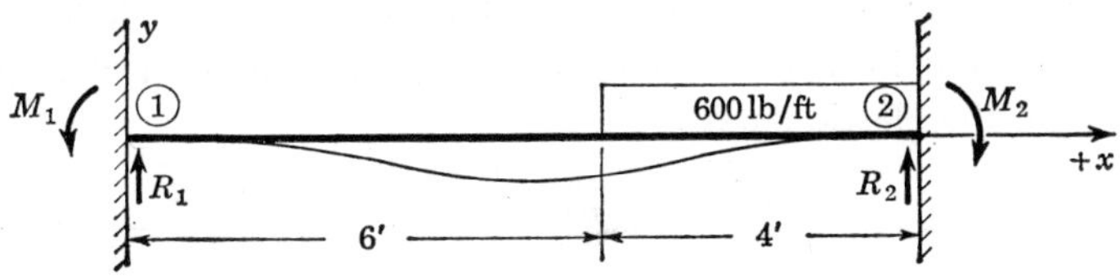

FIG. 11.8

curve is sketched with the *ends horizontal. Note that* M_1 *and* M_2 *must have the directions shown in order to keep the ends of the curve horizontal.*

$0 \leqq x \leqq 6$

(a) $EIy'' = -M_1 + R_1x$

(c) $EIy' = -M_1x + \dfrac{R_1x^2}{2} + C_1$

(e) $EIy = \dfrac{-M_1x^2}{2} + \dfrac{R_1x^3}{6} + C_1x + C_3$

$6 \leqq x \leqq 10$

(b) $EIy'' = -M_1 + R_1x - 300(x-6)^2$

(d) $EIy' = -M_1x + \dfrac{R_1x^2}{2} - \dfrac{300(x-6)^3}{3} + C_2$

(f) $EIy = -\dfrac{M_1x^2}{2} + \dfrac{R_1x^3}{6} - \dfrac{100(x-6)^4}{4} + C_2x + C_4$

The above equations contain six unknown constants: four constants of integration and two reactions. In order to evaluate these six unknowns, six known conditions of the elastic curve are needed. These six conditions and the evaluation of the unknowns are as follows:

At $x = 6$, $y'(c) = y'(d)$ therefore $C_1 = C_2$
At $x = 6$, $y(e) = y(f)$ therefore $C_3 = C_4$
At $x = 0$, $y'(c) = 0$ therefore $C_1 = 0$ and $C_2 = 0$
At $x = 0$, $y(e) = 0$ therefore $C_3 = 0$ and $C_4 = 0$
At $x = 10$, $y'(d) = 0$ therefore $0 = -10M_1 + 50R_1 - 6400$ (g)
At $x = 10$, $y(f) = 0$ therefore $0 = -50M_1 + 166.7R_1 - 6400$ (h)

Solving Eqs. (*g*) and (*h*) simultaneously gives

$R_1 = 307$ lb and $M_1 = 895$ lb-ft *Ans*

M_2 and R_2 are now found by statics:

$\Sigma F_y = 0$ therefore $R_2 = 2093$ lb *Ans*
$\Sigma M_2 = 0$ therefore $M_2 = 2625$ lb-ft *Ans*

Note that the signs of all reactions are plus, thus indicating that their directions were taken correctly. This also means that both M_1 and M_2 *cause negative bending moment.* The equations for slope and deflection can now be rewritten in a form useful for computing slope or deflection at any point on the elastic curve:

$$[c]\ EIy' = -895x + \frac{307x^2}{2} \qquad [d]\ EIy' = -895x + \frac{307x^2}{2} - 100(x-6)^3$$

$$[e]\ EIy = \frac{-895x^2}{2} + \frac{307x^3}{6} \qquad [f]\ EIy = \frac{-895x^2}{2} + \frac{307x^3}{6} - 25(x-6)^4$$

Example 11.6. For the beam shown in Fig. 11.9, derive the equation for the maximum deflection in terms of the midspan load *P*, the length

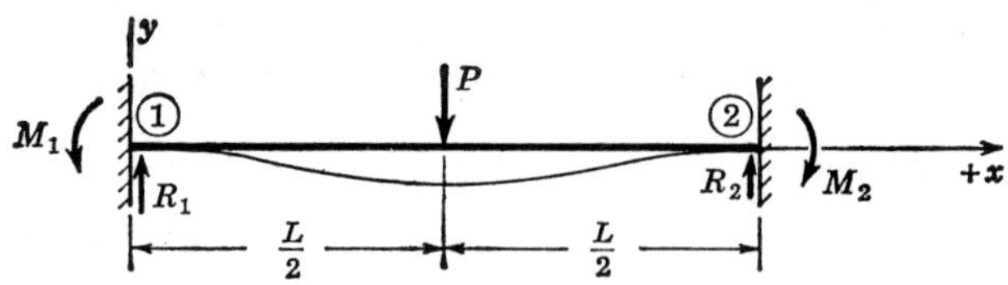

FIG. 11.9

L, and *EI*. From statics and symmetry, $R_1 = R_2 = P/2$ and $M_1 = M_2$. Since the beam is symmetrical, equations for the moment, slope, and deflection are written for half the beam only.

$$0 \leqq x \leqq \frac{L}{2}$$

$$EIy'' = -M_1 + \frac{P}{2}x \tag{a}$$

$$EIy' = -M_1x + \frac{P}{4}x^2 + C_1 \tag{b}$$

$$EIy = -\frac{M_1}{2}x^2 + \frac{P}{12}x^3 + C_1x + C_2 \tag{c}$$

At $x = 0$, $y'(b) = 0$ therefore $C_1 = 0$
At $x = 0$, $y(c) = 0$ therefore $C_2 = 0$

At $x = \frac{L}{2}$, $y'(b) = 0$ therefore $M_1 = \frac{PL}{8}$

$$EIy' = -\frac{PL}{8}x + \frac{P}{4}x^2 \quad [b]$$

$$EIy = -\frac{PL}{16}x^2 + \frac{P}{12}x^3 \quad [c]$$

The maximum deflection occurs at $x = L/2$.

$$EIy_{max} = -\frac{PL}{16}\left(\frac{L}{2}\right)^2 + \frac{P}{12}\left(\frac{L}{2}\right)^3 = -\frac{PL^3}{192}$$

$$y_{max} = -\frac{PL^3}{192EI} \quad \textit{Ans}$$

Example 11.7. The beam in Fig. 11.10 is supported on rollers at end 1 and is fixed at end 2. $E = 2{,}000{,}000$ psi; $I = 600$ in.^{4} Compute the reactions and the maximum deflection.

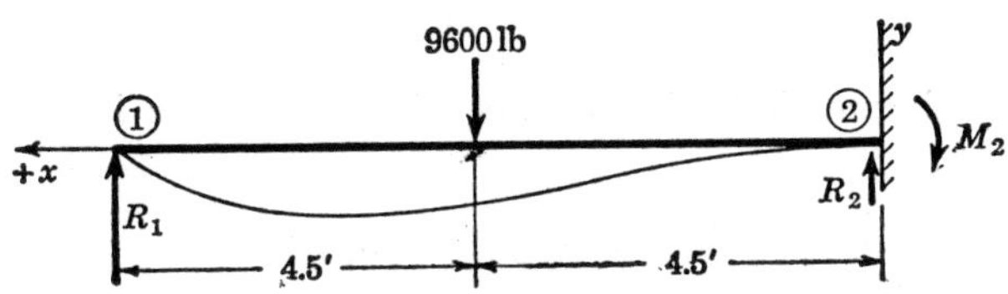

FIG. 11.10

$0 \leqq x \leqq 4.5$

(a) $EIy'' = -M_2 + R_2x$

(c) $EIy' = -M_2x + \frac{R_2x^2}{2} + C_1$

(e) $EIy = \frac{-M_2x^2}{2} + \frac{R_2x^3}{6} + C_1x + C_3$

$4.5 \leqq x \leqq 9$

(b) $EIy'' = -M_2 + R_2x - 9600(x - 4.5)$

(d) $EIy' = -M_2x + \frac{R_2x^2}{2} - \frac{9600}{2}(x - 4.5)^2 + C_2$

(f) $EIy = \frac{-M_2x^2}{2} + \frac{R_2x^3}{6} - \frac{4800}{3}(x-4.5)^3 + C_2x + C_4$

At $x = 4.5$, $y'(c) = y'(d)$ therefore $C_1 = C_2$
At $x = 4.5$, $y(e) = y(f)$ therefore $C_3 = C_4$
At $x = 0$, $y'(c) = 0$ therefore $C_1 = 0$ and $C_2 = 0$
At $x = 0$, $y(e) = 0$ therefore $C_3 = 0$ and $C_4 = 0$
At $x = 9$, $y(f) = 0$ therefore $-40.5M_2 + 121.5R_2 - 145{,}800 = 0$ (g)

The number of unknown constants in the above equations exceeds the number of known conditions of the elastic curve by one. However, an equation relating two of the unknowns can be obtained from statics:

$$\Sigma M_1 = 0 \qquad +M_2 - 9R_2 + 43{,}200 = 0 \tag{h}$$

Solving Eqs. (*g*) and (*h*) simultaneously gives

$$R_2 = 6600 \text{ lb}\uparrow \qquad \text{and} \qquad M_2 = 16{,}200 \text{ lb-ft} \circlearrowright \qquad \textit{Ans}$$

From statics

$$\Sigma F_y = 0 \qquad R_1 = 3000 \text{ lb}\uparrow \qquad \textit{Ans}$$

$$[c]\ EIy' = -16{,}200x + \frac{6600}{2}x^2 \qquad [d]\ EIy' = -16{,}200x + \frac{6600}{2}x^2 - 4800(x-4.5)^2$$

$$[e]\ EIy = -\frac{16{,}200}{2}x^2 + \frac{6600}{6}x^3 \qquad [f]\ EIy = -\frac{16{,}200}{2}x^2 + \frac{6600}{6}x^3 - 1600(x-4.5)^3$$

The maximum deflection occurs where the slope is equal to zero. Using Eq. [*d*],

$$EI(0) = -16{,}200x + 3300x^2 - 4800(x-4.5)^2$$

which reduces to

$$x^2 - 18x + 64.8 = 0$$

therefore

$$x = +13.03, +4.97$$

The value of $+13.03$ is outside the limits of Eq. [*d*]. Using $+4.97$ for x in Eq. [*f*],

$$EIy_{\max} = -65{,}000 \text{ lb-ft}^3$$

$$y_{\max} = \frac{-65{,}000(1728)}{2{,}000{,}000(600)} = -0.0936 \text{ in.} \qquad \textit{Ans}$$

PROBLEMS

★**11.55.** A 20-ft steel beam is restrained at both ends; equal loads of 36 kips each are applied 5 ft from each end. $I = 600 \text{ in.}^4$ Compute (*a*) the reactions and (*b*) the maximum deflection.

11.56. An 18-ft timber beam is restrained at both ends; equal loads of 1200 lb each are applied 6 ft from each end. $EI = 320{,}000{,}000$ lb-in.^{2} Compute (*a*) the reactions and (*b*) the maximum deflection.

★**11.57.** For the beam shown in Fig. P11.57, let $L = 16$ ft. (*a*) Compute the reactions. (*b*) Draw the shear and bending moment diagrams.

11.58. For the beam shown in Fig. P11.58, let $L = 24$ ft. (*a*) Compute the reactions. (*b*) Draw the shear and bending moment diagrams.

★**11.59.** A 12-ft timber beam is restrained at both ends; a load of 8100 lb is applied 4 ft from the left end. $E = 1{,}200{,}000$ psi; $I = 450$ in.^{4} Compute (*a*) the reactions and (*b*) the maximum deflection.

11.60. A 12-ft steel beam is restrained at both ends; a load of 64 kips is applied 3 ft from the left end. $I = 400$ in.^{4} Compute (*a*) the reactions and (*b*) the maximum deflection.

11.61. A 24-ft beam is restrained at both ends; a load of 1280 lb is applied 6 ft from the left end. Determine the reactions.

★**11.62.** For the beam shown in Fig. P11.62, let $a = b = 6$ ft, $w =$ 1200 lb/ft. (*a*) Determine the reactions. (*b*) Draw the shear and moment diagrams.

11.63. For the beam shown in Fig. P11.63, let $a = 5$ ft, $b = 7$ ft, $w = 6000$ lb/ft. (*a*) Compute the reactions. (*b*) Draw the shear and moment diagrams.

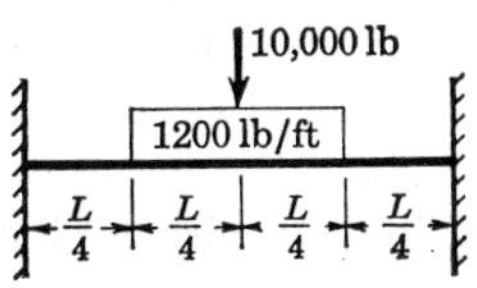

FIG. P11.57, P11.58

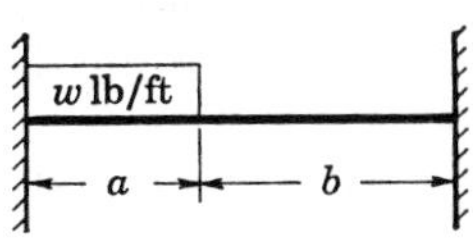

FIG. P11.62, P11.63

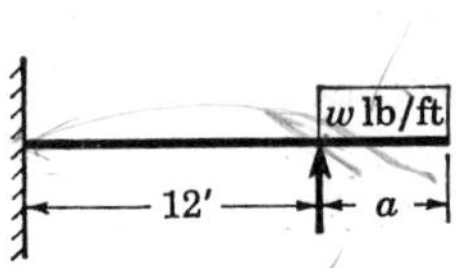

FIG. P11.70, P11.71

★**11.64.** A 10-ft beam is fixed at the left end and supported on rollers at the right end; a uniform load of 600 lb/ft is applied to the left 4 ft of the beam. (*a*) Compute the reactions. (*b*) Draw the shear and moment diagrams.

11.65. An 18-ft beam is fixed at the right end and supported on rollers at the left end; a uniform load of 24 kips/ft is applied to the right 6 ft of the beam. Compute the reactions.

11.66. A 12-ft beam is fixed at the right end and supported on rollers at the left end; a uniform load of 600 lb/ft is applied to the left 4 ft of the beam. Compute the reactions.

11.67. A 12-ft beam is fixed at the left end and supported on rollers at the right end; a uniform load of 24 kips/ft is applied to the right 7 ft of the beam. (*a*) Compute the reactions. (*b*) Draw the shear and moment diagrams.

11.68. A 12-ft beam is fixed at the right end and supported on rollers at the left end; a load of 6 kips is applied 8 ft from the left end. $EI =$ 1,500,000,000 lb-in.^{2} (*a*) Compute the reactions. (*b*) Compute the maximum deflection.

11.69. An 18-ft beam is fixed at the right end and supported on rollers at the left end; a 9-kip load is applied 12 ft from the left end. $EI =$ 240,000,000 lb-in.^{2} (*a*) Compute the reactions. (*b*) Compute the maximum deflection.

★11.70. For the beam shown in Fig. P11.70, let $w =$ 1440 lb/ft, $a =$ 4 ft. Compute the reactions.

11.71. For the beam shown in Fig. P11.71, let $w =$ 720 lb/ft, $a =$ 6 ft. Compute the reactions.

11.72. A beam of length L ft is fixed at both ends and supports a uniform load of w lb/ft over the entire span. (*a*) Show that the moment at each end is $wL^2/12$. (*b*) Show that the maximum deflection is $wL^4/384EI$.

11.73. A beam of length L ft supported at one end by rollers and fixed at the other supports a uniform load of w lb/ft over the entire span. Show that the moment and the vertical reaction at the fixed end are $wL^2/8$ and $5wL/8$.

★11.74. A 10-ft beam is fixed at both ends and loaded with a clockwise couple of 36 kip-ft applied 4 ft from the left end. Compute the reactions.

11.75. A 12-ft beam is fixed at both ends and loaded with a clockwise couple of 1200 lb-ft applied 3 ft from the left end. Compute the reactions.

12

ELASTIC CURVE: AREA MOMENT

12.1 THE AREA-MOMENT METHOD

The area-moment method is frequently used in the solution of design problems dealing with the elastic curve. Its use is especially advantageous for determining the reactions of statically indeterminate beams and the deflections of beams supporting several loads. The basis of this method is the application of certain principles of geometry to the elastic curve. As the name implies, the area-moment method makes use of the area of the moment diagram of a beam to determine certain angles and distances related to its elastic curve.

12.2 THE AREA-MOMENT THEOREMS

Fundamental to the area-moment method are two theorems which can be stated (for beams of constant cross section) as follows:

The First Area-moment Theorem. EI times the difference in the slope angles of the tangents at any two points on the elastic curve is equal to the area of the moment diagram *between the two points.*

The Second Area-moment Theorem. EI times the displacement of any point on the elastic curve from the tangent at a second point on the curve is equal to the *first moment* of the moment diagram area *between the two points* with respect to the *first point.*

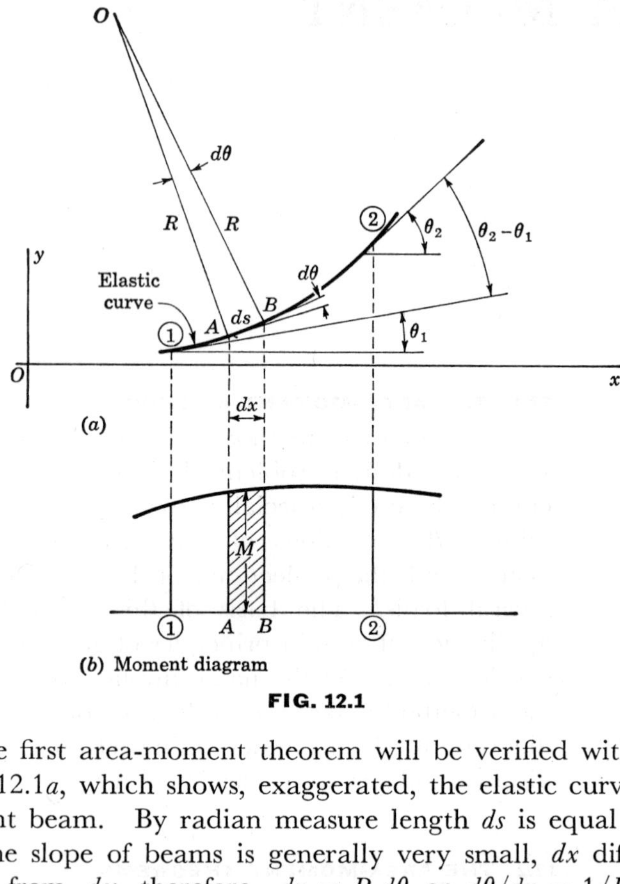

FIG. 12.1

The first area-moment theorem will be verified with respect to Fig. 12.1*a*, which shows, exaggerated, the elastic curve of part of a bent beam. By radian measure length ds is equal to $R\,d\theta$. Since the slope of beams is generally very small, dx differs only slightly from ds; therefore, $dx = R\,d\theta$ or $d\theta/dx = 1/R$. From Eq. (11.1), $1/R$ is equal to M/EI; therefore, $d\theta/dx = M/EI$ and $EI\,d\theta = M\,dx$. Assuming EI constant and integrating both sides of the equation between points 1 and 2 gives

$$EI(\theta_2 - \theta_1) = \int_1^2 M\,dx$$

The expression $(\theta_2 - \theta_1)$ represents the difference in the slope angles of the tangents at points 1 and 2. From Fig. 12.1*b* it can be seen that the term $M\,dx$ represents a differential area under the moment diagram; therefore $\int_1^2 M\,dx$ is equal to A_M, the area of the moment diagram between points 1 and 2. The first area-moment theorem can then be written in equation form as

$$EI\,(\theta_2 - \theta_1) = A_M \tag{12.1}$$

where θ_1 and θ_2 are the slope angles at points 1 and 2 respectively and A_M is the area of the moment diagram between points 1 and 2.

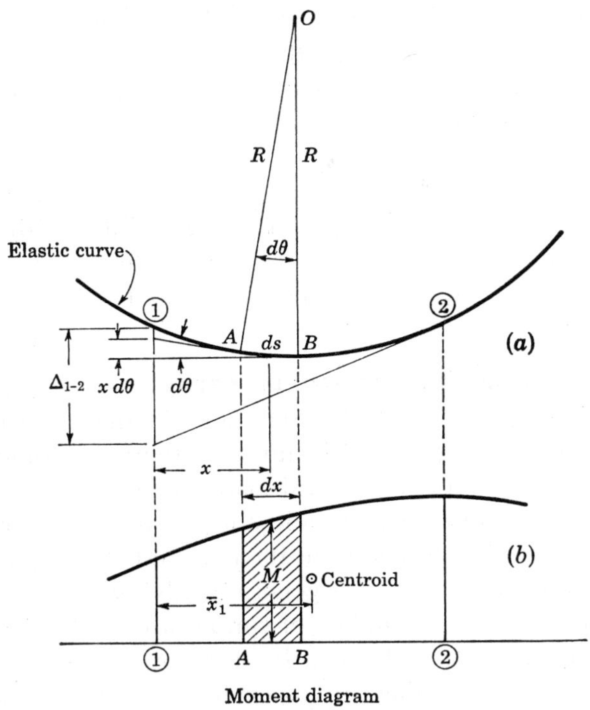

FIG. 12.2

Because of the designation used in the derivation, the signs of θ_2 and θ_1 will be correct when point 2 is to the right of point 1.

The second area-moment theorem will be verified by reference to Fig. 12.2*a*, which shows the elastic curve of a bent beam with tangents at points *A* and *B* extended to intersect a vertical line passing through point 1. The distance *on the vertical line* passing through point 1 intercepted by the tangents at A and B is $x\,d\theta$. The vertical displacement Δ_{1-2} of point 1 from the tangent at point 2 is equal to the summation of the intercepts $x\,d\theta$ between point 1 and the tangent from 2, or $\Delta_{1-2} = \int_1^2 x\,d\theta$. In the verification of the first area-moment theorem, it was shown that $EI\,d\theta = M\,dx$ or $d\theta = M\,dx/EI$. Therefore, $\Delta_{1-2} = \int_1^2 x\,d\theta$ becomes $\Delta_{1-2} = \int_1^2 xM\,dx/EI$ or $EI\Delta_{1-2} = \int_1^2 xM\,dx$. The term $xM\,dx$ represents the moment of a differential area under the moment diagram with respect to point 1. Therefore, $\int_1^2 xM\,dx$ represents the first moment of the area between points 1 and 2 *with respect to point* 1. The second area-moment theorem can then be written in equation form as

$$EI\Delta_{1-2} = A_M\bar{x}_1 \tag{12.2}$$

where Δ_{1-2} is the vertical displacement of point 1 with respect to a tangent at point 2, and $A_M\bar{x}_1$ is the first moment of the area between points 1 and 2 with respect to point 1.

12.3 MOMENT DIAGRAMS AS SIMPLE AREAS

As previously stated, the two area-moment theorems utilize the areas of moment diagrams of beams and the first moments of the areas. If a beam supports several loads, the moment diagram has a more or less complex shape, and thus it would be cumbersome and time consuming to compute its area and the first moment of the area. For this reason, it is generally advisable to represent the moment diagram as a number of simple areas whose values can readily be computed. This can be done by drawing in a logical manner the moment diagram of each external load separately. The diagrams thus obtained will consist of rectangles, triangles, and parabolic segments whose pertinent properties are shown in Table 12.1.

The procedure for drawing moment diagrams as simple areas will be demonstrated by reference to the beam loaded as shown in Fig. 12.3*a*. The complete moment diagram, drawn by the methods covered in Chap. 4, is shown in Fig. 12.3*b*. This complete moment

TABLE 12.1 Properties of Simple Areas

Simple area	Area	$\bar{x}$
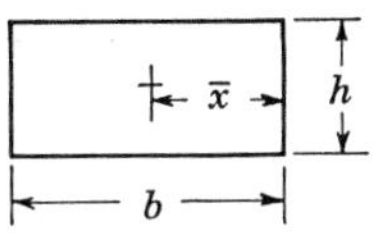	bh	$\frac{1}{2}b$
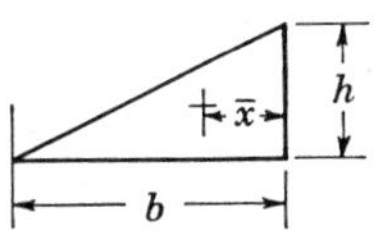	$\frac{1}{2}bh$	$\frac{1}{3}b$
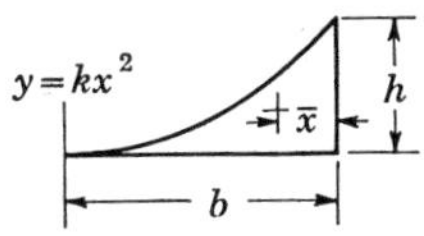	$\frac{1}{3}bh$	$\frac{1}{4}b$
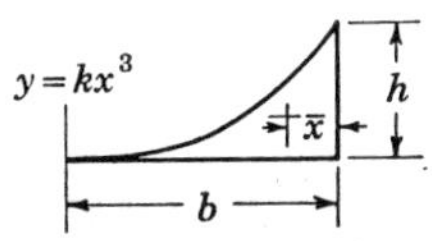	$\frac{1}{4}bh$	$\frac{1}{5}b$

diagram could have been drawn from the left end toward the right, from the right end toward the left, or from each end to a common section of the beam. Similarly, the moment diagram of each external load separately can be drawn from either end. Figure 12.3*c* is the moment diagram as simple areas drawn from left to right. The triangle is obtained by drawing the moment diagram from the left due to the left reaction *as if it were the only load on the beam.* The parabolic area is obtained by drawing the moment diagram from the left due to the uniform load *as if it were the only load on the beam.* It should be understood that, although the beam acted on by any one force would not be in equilibrium, equilibrium would result from the summation of the effects of all the forces. At any section of the beam the algebraic sum of the ordinates of the simple-area moment diagrams should be equal to the ordinate of the complete moment diagram. This statement can easily be checked by comparing the ordinates of Fig. 12.3*b* and *c* at each reaction and at a section 6 ft from the left reaction. Figure 12.3*d* is the moment diagram as simple areas drawn from right to left. The top triangle is due to the right reaction. The moment diagram due to the uniform load is *parabolic under the uniform load* and a straight line to the left of the uniform load. It is represented as the three simple areas indicated, that is, a parabolic segment, a rectangle, and a triangle. It can be seen that the

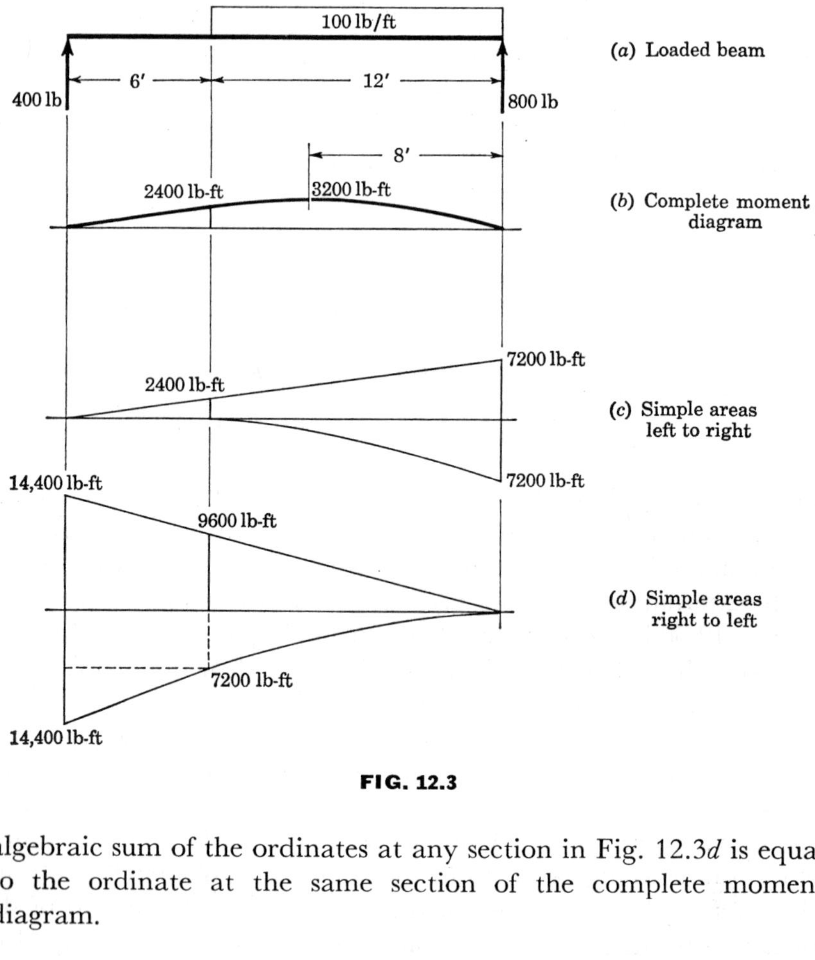

FIG. 12.3

algebraic sum of the ordinates at any section in Fig. 12.3*d* is equal to the ordinate at the same section of the complete moment diagram.

Example 12.1. For the beam loaded as shown in Fig. 12.4, draw the moment diagram as simple areas (*a*) from left to right, (*b*) from right to left, and (*c*) from each end to a section at point 2. (*d*) Compute the first moment of the area of the moment diagram with respect to point 3.

Figure 12.4*a* is the moment diagram as simple areas drawn from the left. Triangle A is due to R_1, triangle B is due to the 48-kip load, triangle C is due to R_2, and the parabolic area D is due to the uniform load.

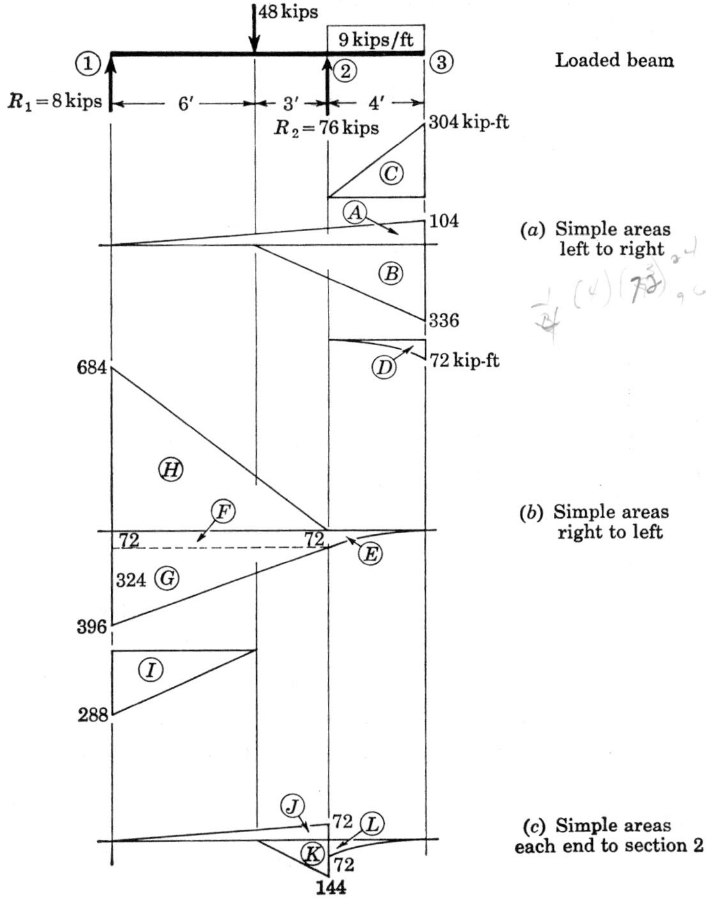

FIG. 12.4

Figure 12.4*b* is the moment diagram as simple areas drawn from the right. Areas E, F, and G are due to the uniform load; triangle H is due to R_2; and triangle I is due to the 48-kip load.

Figure 12.4*c* is the moment diagram as simple areas drawn from each end to a section at point 2. Triangle J is due to R_1, triangle K is due to the 48-kip load, and the parabolic area L is due to the uniform load. It can be observed that area L is the same as area E and that areas J

and K are the left portions of areas A and B, respectively. It can also be noted that at section 2 the ordinate of area L is equal to the algebraic sum of the ordinates of areas J and K. This is a check, since the bending moment at a section must be the same whether computed from the left or from the right.

For the computation of the first moment of the moment diagram area about point 3, Fig. 12.4*c* is used, since it is comprised of a smaller number of simple areas than either Fig. 12.4*a* or *b*. With Table 12.1 used as a reference, the first moment of each simple area is computed. The algebraic sum of these first moments is equal to the first moment of the area of the moment diagram with respect to point 3.

TABLE E12.1

Figure	Area	$\bar{x}_3$	$A\bar{x}_3$	
J	$\frac{(+72)(9)}{2}$	$(4 + 3)$	$+2268$	
K	$\frac{(-144)(3)}{2}$	$(4 + 1)$	-1080	
L	$\frac{(-72)(4)}{3}$	(3)	$-\ 288$	
Total			$+\ 900$ k-ft^3	*Ans*

12.4 GEOMETRY APPLIED TO THE ELASTIC CURVE

The area-moment method is actually a method of applying geometry of small angles to the elastic curve. By "small angle" is meant any angle small enough so that *for practical purposes* the sine, tangent, and the angle in radians are equal. For slide-rule computations, the term small angles includes all angles up to at least 6° (0.10 radian).

It should be remembered that the two area-moment theorems deal with the *tangents* at various points on the elastic curve; they do not, in general, provide a means of obtaining directly the deflection or slope at a point. The basis of the solution of the area-moment problems is the relating of the geometry of elastic curves to properly selected tangents used as references; the area-moment theorems are merely tools used in the solution.

The solution of elastic-curve problems by the area-moment method can be facilitated by following the procedure outlined below.

Steps used in the solution of problems by area moment

1. Draw a neat line diagram of the loaded beam (including reactions).
2. Sketch the elastic curve.
3. Plan the geometry needed for the solution of the problem, and select a tangent to be used as a reference.
4. Draw the moment diagram as simple areas in a manner consistent with the selection of the reference tangent.
5. Solve the geometry of the problem for the unknowns using the two area-moment theorems.

An understanding of the signs of the quantities obtained from the area-moment equations is of the utmost importance. A posi-

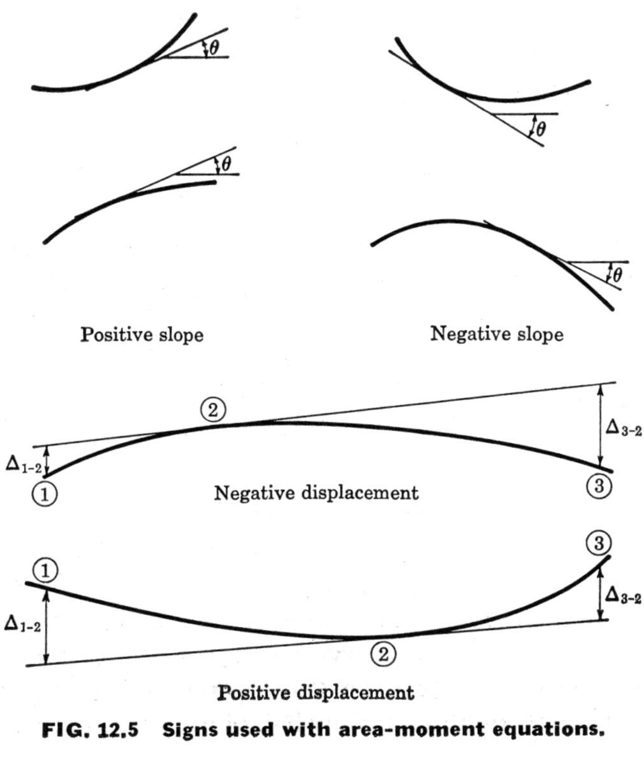

FIG. 12.5 Signs used with area-moment equations.

tive displacement value means that the point on the elastic curve is above the reference tangent. The slope angle at a point is positive if the tangent to the elastic curve at that point slopes up to the right. The signs of the quantities used in or obtained from the area-moment equations are illustrated in Fig. 12.5.

In working with the area-moment theorems as expressed by Eqs. (12.1) and (12.2), care must be taken to avoid errors due to inconsistency of units. The units usually assigned to the quantities involved are:

Bending moment M	lb-ft
Modulus of elasticity E	psi
Moment of inertia I	in.^{4}
Horizontal distance x	ft
Slope angle θ	radian or in./in.
Vertical displacement Δ	in.

With these units the "unit equations" of the area-moment equations would be

$$\left(\frac{\text{lb}}{\text{in.}^2}\right)(\text{in.}^4)\left(\frac{\text{in.}}{\text{in.}}\right) \neq \text{lb-ft}^2 \qquad \text{for Eq. (12.1)}$$

$$\left(\frac{\text{lb}}{\text{in.}^2}\right)(\text{in.}^4)(\text{in.}) \neq \text{lb-ft}^3 \qquad \text{for Eq. (12.2)}$$

The units can be made consistent by multiplying the right-hand side of Eqs. (12.1) and (12.2) by 144 and 1728, respectively.

Five types of elastic-curve problems are presented in the next five articles. The problems are separated into types only for teaching purposes; *the basic method of solution*, as discussed in this article, is the *same for all elastic curve problems*.

12.5 SYMMETRICALLY LOADED BEAMS

Beams with symmetrical loading constitute a special case which is frequently encountered in engineering. The term *symmetrical loading* means that the reactions as well as the applied loads are symmetrical. Since the elastic curve for such a beam is symmetrical, any deflection or slope can be found by applying geometry and the area-moment equations to only half of the elastic curve. Therefore, for a symmetrical beam, only half of the moment diagram need be drawn.

Example 12.2. For the beam shown in Fig. 12.6, $E = 30{,}000{,}000$ psi and $I = 15$ in.^{4} Find the maximum deflection and the slope angle at each reaction.

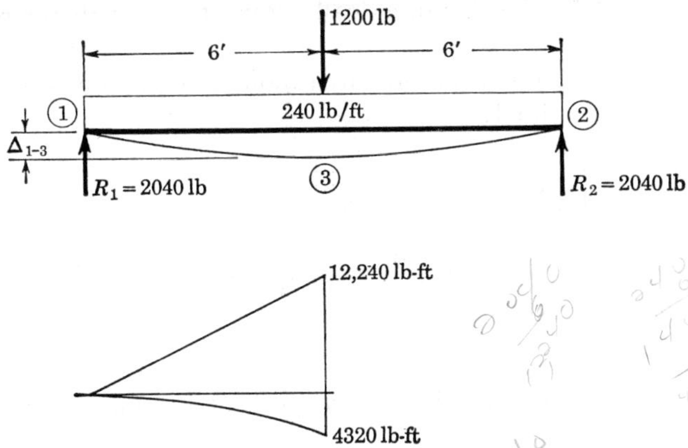

FIG. 12.6

Step 1. The reactions are computed to be as shown on the load diagram.

Step 2. The elastic curve is sketched as shown. Since the points of support do not move, the curve passes through points 1 and 2. Because of symmetry, the curve is horizontal at the center.

Step 3. The maximum deflection occurs at midspan (point 3). Since the tangent at point 3 is horizontal, the vertical displacement of point 1 from the tangent at point 3 is numerically equal to the deflection at point 3. Therefore the tangent at point 3 is selected as the reference tangent.

The slope angle at point 3 is known to be zero. Therefore the expression $EI(\theta_3 - \theta_1)$ will give θ_1 directly if set equal to the area of the moment diagram between the two points.

Step 4. Since only the left half of the elastic curve is being used, drawing the left half of the moment diagram from left to right gives a minimum number of simple areas.

Step 5. The geometry of the problem is solved as follows: Using the second area-moment equation in order to determine the displacement of point 1 from the tangent at point 3,

$$EI\Delta_{1-3} = A_M\bar{x}_1$$

$$EI\Delta_{1-3} = +\frac{12{,}240(6)}{2}\,4 - \frac{4320(6)}{3}\,4.5$$

$$EI\Delta_{1-3} = +146{,}900 - 38{,}900 = +108{,}000 \text{ lb-ft}^3$$

$$\Delta_{1-3} = \frac{+\,108{,}000(1728)}{30{,}000{,}000(15)} = +0.415 \text{ in.}$$

The maximum deflection is numerically equal to Δ_{1-3}; therefore $y_{max} = -0.415$ in. *Ans*

As planned in Step 3, the first area-moment equation is used to find the slope angle at point 1.

$$EI(\theta_3 - \theta_1) = A_M$$

$$EI(0 - \theta_1) = +\frac{12{,}240(6)}{2} - \frac{4320(6)}{3}$$

$$EI(-\theta_1) = +36{,}720 - 8640 = +28{,}080 \text{ lb-ft}^2$$

$$\theta_1 = \frac{-28{,}080(144)}{30{,}000{,}000(15)} = -0.00899 \text{ radian} \qquad \textit{Ans}$$

Because of symmetry, θ_2 is numerically equal to θ_1. Therefore

$$\theta_2 = +0.00899 \text{ radian} \qquad \textit{Ans}$$

PROBLEMS

★12.1. A uniform load of 2400 lb/ft acts on the entire span of a simply supported beam 10 ft long. $E = 30{,}000{,}000$ psi; $I = 70$ in.^{4} Compute (*a*) the maximum deflection, (*b*) the slope angle at each support.

12.2. A uniform load of 96 lb/ft acts on the entire span of a simply supported beam 12 ft long. $E = 1{,}600{,}000$ psi; $I = 500$ in.^{4} Compute (*a*) the maximum deflection, (*b*) the slope angle at each support.

★12.3. For the beam shown in Fig. P12.3, let $P = 4$ kips, $L = 24$ ft, $E = 10{,}000{,}000$ psi, $I = 300$ in.^{4} Compute (*a*) the maximum deflection, (*b*) the deflection 6 ft from each reaction.

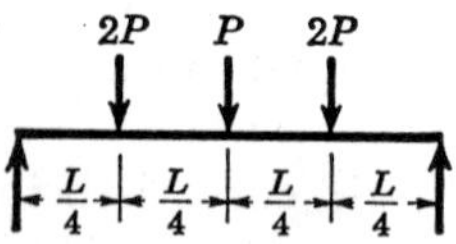

FIG. P12.3, P12.4

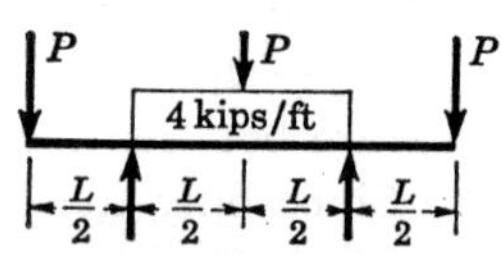

FIG. P12.5, P12.6

12.4. For the beam shown in Fig. P12.4, let $P = 10$ kips, $L = 20$ ft, $E = 30{,}000{,}000$ psi, $I = 1000$ in.^{4} Compute (*a*) the maximum deflection, (*b*) the deflection 5 ft from each reaction.

★**12.5.** For the steel beam shown in Fig. P12.5, let $L = 12$ ft, $P = 10$ kips, $I = 150$ in.^{4} Determine the midspan deflection and the deflection at the end of each overhang.

12.6. For the steel beam shown in Fig. P12.6, let $L = 6$ ft, $P = 12$ kips, $I = 80$ in.^{4} Determine the midspan deflection and the deflection at the end of each overhang.

★**12.7.** A 16-ft span of a continuous beam is loaded with a uniform load of 200 lb/ft over its entire length and has negative bending moments of 1200 lb-ft at each end. $E = 1{,}600{,}000$ psi; $I = 200$ in.^{4} Determine the maximum deflection.

12.8. A 24-ft span of a continuous beam is loaded with a uniform load of 6 kips/ft over its entire length and has *negative bending moments* of 240 kip-ft at each end. $E = 30{,}000{,}000$ psi; $I = 1200$ in.^{4} Determine the maximum deflection.

12.9. A simply supported beam of length L ft is loaded with a uniform load of w lb/ft over the entire span. Show that the maximum deflection is $5wL^4/384EI$ and that the slope angle at either reaction is $wL^3/24EI$.

12.10. A simply supported beam of length L is loaded at midspan with a concentrated load P. Show that the maximum deflection is $PL^3/48EI$ and that the slope angle at either reaction is $PL^2/16EI$.

★**12.11.** For the beam shown in Fig. P12.11, with triangular loading, let $L = 20$ ft, $EI = 1{,}800{,}000{,}000$ lb-in.^{2} Compute the maximum deflection.

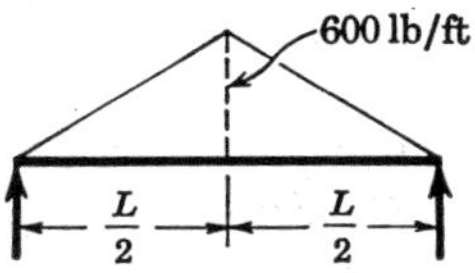

FIG. P12.11, P12.12

12.12. For the beam shown in Fig. P12.12, with triangular loading, let $L = 12$ ft, $EI = 360{,}000{,}000$ lb-in.^{2} Compute the maximum deflection.

12.6 CANTILEVER BEAMS

The slope or deflection at any point on the elastic curve of a cantilever beam can be easily found by selecting a reference tangent at the fixed end. Using such a tangent simplifies the planning and allows the deflections and slopes to be found directly.

Example 12.3. For the beam shown in Fig. 12.7, $E = 30{,}000{,}000$ psi (or 30,000 ksi), $I = 200$ in.[4] Determine the maximum deflection, the deflection at midspan, and the slope angle at the free end.

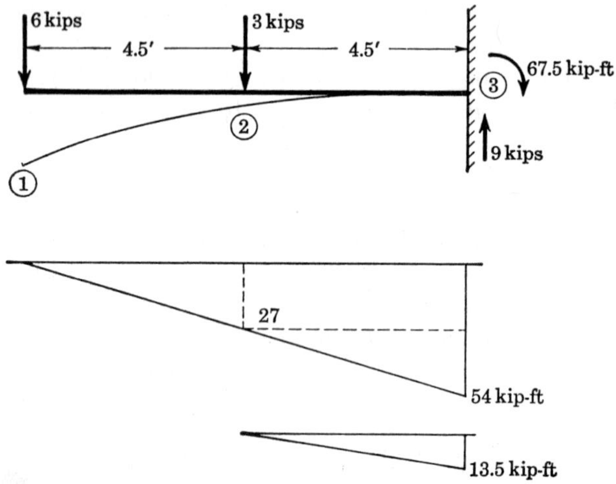

FIG. 12.7

The reaction consists of a vertical force *and a couple* as shown. The moment diagram is drawn as simple areas from the left. To find the maximum deflection

$$EI\Delta_{1-3} = A_M\bar{x}_1$$

$$EI\Delta_{1-3} = -\frac{54(9)}{2}6 - \frac{13.5(4.5)}{2}7.5$$

$$EI\Delta_{1-3} = -1458 - 228 = -1686 \text{ kip-ft}^3$$

$$y_{max} = \Delta_{1-3} = \frac{-1686(1728)}{30{,}000(200)} = -0.486 \text{ in.} \qquad \textit{Ans}$$

To find the deflection at point 2

$$EI\Delta_{2-3} = A_M\bar{x}_2$$

$$EI\Delta_{2-3} = -27(4.5)(2.25) - \frac{27(4.5)}{2}3 - \frac{13.5(4.5)}{2}3$$

$$EI\Delta_{2-3} = -273.2 - 182.1 - 91.2 = -546 \text{ kip-ft}^3$$

$$y_2 = \Delta_{2-3} = \frac{-546(1728)}{30{,}000(200)} = -0.157 \text{ in.} \qquad \textit{Ans}$$

To find the slope angle at the free end

$$EI(\theta_3 - \theta_1) = A_M$$

$$EI(0 - \theta_1) = -\frac{54(9)}{2} - \frac{13.5(4.5)}{2}$$

$$-EI\theta_1 = -243 - 30.4 = -273 \text{ kip-ft}^2$$

$$\theta_1 = \frac{273(144)}{30{,}000(200)} = +0.00656 \text{ radian} \qquad \textit{Ans}$$

PROBLEMS

★12.13. For the beam shown in Fig. P12.13, let $L = 15$ ft, $E = 30{,}000{,}000$ psi, $I = 60$ in.^{4} Compute the midspan and maximum deflections.

12.14. For the beam shown in Fig. P12.14, let $L = 9$ ft, $E = 10{,}000{,}000$ psi, $I = 50$ in.^{4} Compute the midspan and maximum deflections.

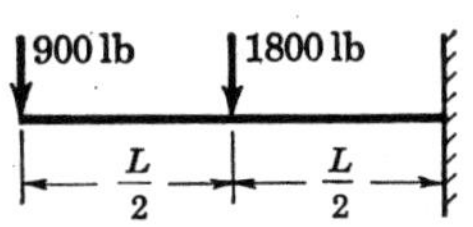

FIG. P12.13, P12.14

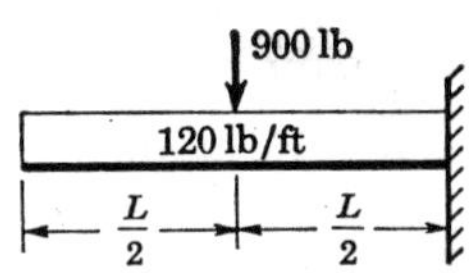

FIG. P12.17, P12.18

★12.15. A 10-ft cantilever beam fixed at the right end supports a 15-kip concentrated load 5 ft from the free end. $E = 30{,}000{,}000$ psi; $I = 300$ in.^{4} Compute the maximum deflection.

12.16. A 13-ft cantilever beam fixed at the right end supports a 6-kip load at 7 ft from the free end. $E = 30{,}000$ ksi; $I = 80$ in.^{4} Determine the maximum deflection.

★12.17. For the beam shown in Fig. P12.17, let $L = 10$ ft, $E = 1{,}800{,}000$ psi, $I = 600$ in.^{4} Compute the slope and deflection at the free end.

12.18. For the beam shown in Fig. P12.18, let $L = 6$ ft, $E = 1{,}600{,}000$ psi, $I = 200$ in.^{4} Compute the slope and deflection at the free end.

12.19. For the beam shown in Fig. P12.19, let $a = 4$ ft, $b = 6$ ft, $EI = 500{,}000{,}000$ lb-in.^{2} Compute the maximum deflection.

★12.20. For the beam shown in Fig. P12.20, let $a = 7$ ft, $b = 5$ ft, $EI = 900{,}000{,}000$ lb-in.^{2} Compute the maximum deflection.

12.21. For the beam shown in Fig. P12.21, let $L = 12$ ft and $EI = 2{,}000{,}000{,}000$ lb-in.^{2} Compute the maximum deflection.

★**12.22.** For the beam shown in Fig. P12.22, let $L = 6$ ft and $EI = 800{,}000{,}000$ lb-in.^{2} Compute the maximum deflection.

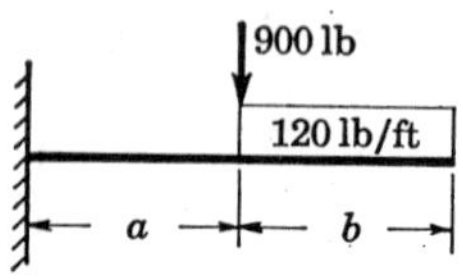

FIG. P12.19, P12.20

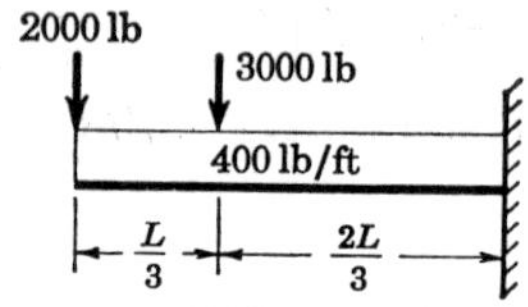

FIG. P12.21, P12.22

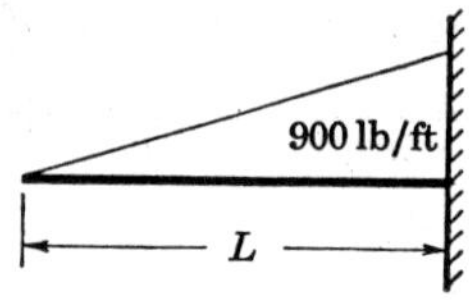

FIG. P12.24, P12.25

★**12.23.** An 11-ft cantilever beam is fixed at the right end and loaded with a clockwise 600 lb-ft couple at the left end. $EI = 32{,}000{,}000$ lb-in.^{2} Compute the slope and deflection at the free end.

12.24. For the cantilever beam shown in Fig. P12.24, with uniformly increasing loading, let $L = 10$ ft, $E = 2{,}000{,}000$ psi, $I = 700$ in.^{4} Compute the maximum deflection.

★**12.25.** For the cantilever beam shown in Fig. P12.25, with uniformly increasing loading, let $L = 6$ ft, $E = 1{,}800{,}000$ psi, $I = 200$ in.^{4} Compute the maximum deflection.

12.7 SIMPLY SUPPORTED BEAMS WITH UNSYMMETRICAL LOADING

The general case of simply supported beams with unsymmetrical loading is the one most frequently encountered in engineering design. In this type of problem the slope of the tangent is not known for any point. Therefore the direction of the reference tangent must be located after it is selected. The reference tangent is selected at one point of support, and its direction located by finding the displacement from the tangent of the second point of support. It is usually advisable to select the reference tangent at the least loaded end, that is, the end farthest from the greatest number of applied loads. Generally the least loaded end has the smallest reaction. A judicious selection of the method of drawing the moment diagram as simple areas will minimize the number of computations required for this type of beam. It is advisable to sketch the moment diagram as simple areas (without computing ordinates) by two or more methods and to adopt the one method best suited to the particular problem. For a beam without an overhang, it is usually advantageous to draw the moment diagram from the end at which the reference tangent is taken. For a beam with a single overhang, it is usually advantageous to draw the moment diagram from each end to the reaction at the overhang.

Example 12.4. Compute the deflection at points 3 and 4 for the beam shown in Fig. 12.8. $E = 30{,}000$ ksi; $I = 100$ in.^{4}

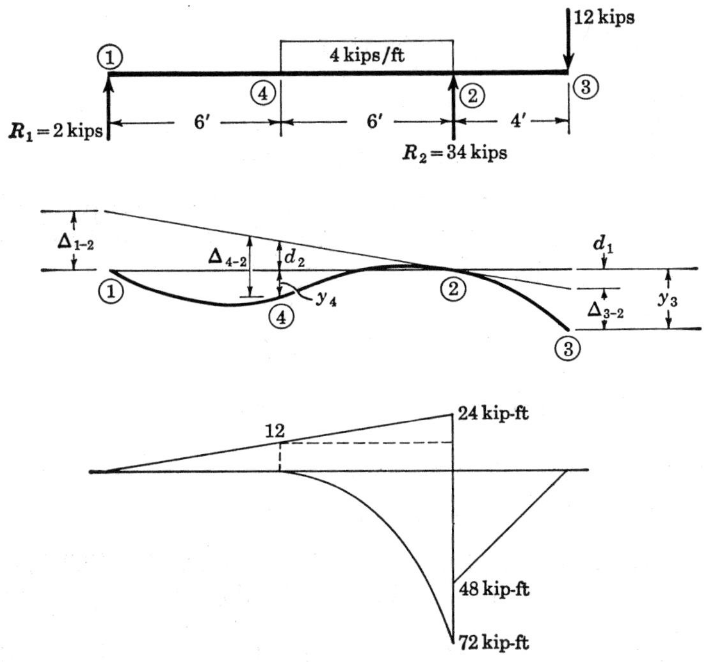

FIG. 12.8

The reference tangent is selected at point 2, and the moment diagram drawn from each end to point 2. The reference tangent is located by finding the displacement of point 1 from the reference tangent.

$$EI\Delta_{1-2} = A_M\bar{x}_1$$

$$EI\Delta_{1-2} = +\frac{24(12)}{2}\,8 - \frac{72(6)}{3}\,10.5$$

$$EI\Delta_{1-2} = +1152 - 1512 = -360 \text{ kip-ft}^3$$

$$\Delta_{1-2} = \frac{-360(1728)}{30{,}000(100)} = -0.207 \text{ in.}$$

This means that point 1 is 0.207 in. *below* the tangent at point 2; therefore the tangent at point 2 slopes *up to the left* as shown. It should be understood that point 1 did not move but that the tangent at point 2 rotated.

It can be seen from the sketch of the elastic curve that the deflection at point 3 is equal to the sum of the distances d_1 and Δ_{3-2}. The distance d_1 is found by similar triangles $d_1/\Delta_{1-2} = 4/12$; therefore $d_1 = 4/12(0.207) = 0.069$ in.

$$EI\Delta_{3-2} = A_M\bar{x}_3$$

$$EI\Delta_{3-2} = -\frac{48(4)}{2}\frac{8}{3} = -256 \text{ kip-ft}^3$$

$$\Delta_{3-2} = \frac{-256(1728)}{30{,}000(100)} = -0.147 \text{ in.}$$

$$y_3 = -(0.069 + 0.147) = -0.216 \text{ in.} \qquad \textit{Ans}$$

The deflection of point 4 is equal to the difference between the distances Δ_{4-2} and d_2. By similar triangles $d_2/\Delta_{1-2} = 6/12$; therefore $d_2 = 6/12(0.207) = 0.104$ in.

$$EI\Delta_{4-2} = A_M\bar{x}_4$$

$$EI\Delta_{4-2} = +12(6)(3) + \frac{12(6)}{2}4 - \frac{72(6)}{3}4.5$$

$$EI\Delta_{4-2} = +216 + 144 - 648 = -288 \text{ kip-ft}^3$$

$$\Delta_{4-2} = \frac{-288(1728)}{30{,}000(100)} = -0.166 \text{ in.}$$

$$y_4 = -(0.166 - 0.104) = -0.062 \text{ in.} \qquad \textit{Ans}$$

Example 12.5. For the beam shown in Fig. 12.9, $E = 10{,}000{,}000$ psi, $I = 12$ in.^{4} Compute the maximum deflection and the deflection at midspan.

The reference tangent is selected at point 1, the least loaded end, and the moment diagram as simple areas drawn from the left.

At the point of maximum deflection the slope angle is zero. The location of this point, point *m*, can be found using the first area-moment equation if the slope at another point is known. The slope angle at point 1 is equal to its tangent $\Delta_{2-1}/10$. Since the tangent at point *m* is horizontal, Δ_{1-m} is numerically equal to the maximum deflection.

$$EI\Delta_{2-1} = A_M\bar{x}_2$$

$$EI\Delta_{2-1} = +\frac{2160(10)}{2}\frac{10}{3} - \frac{2160(3)}{3}\frac{3}{4}$$

$$EI\Delta_{2-1} = +36{,}000 - 1620 = +34{,}380 \text{ lb-ft}^3$$

$$EI\theta_1 = \frac{-EI\Delta_{2-1}}{10} = \frac{-34{,}380}{10} = -3438 \text{ lb-ft}^2$$

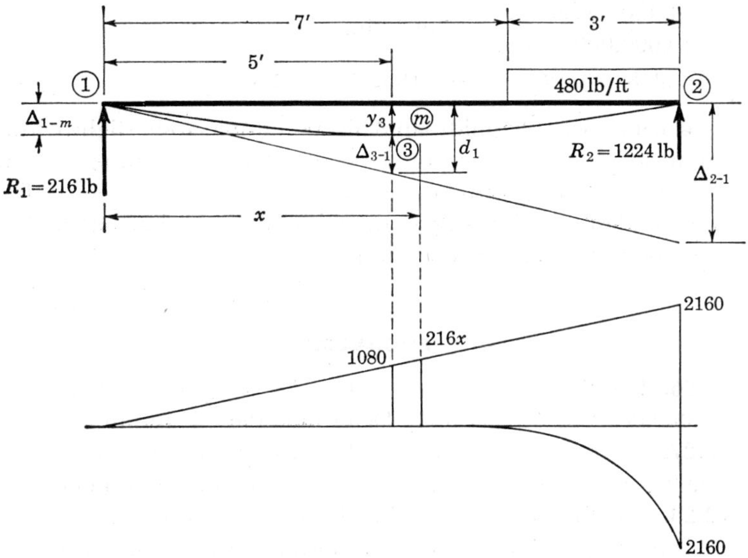

FIG. 12.9

$EI(\theta_m - \theta_1) = A_M$, where A_M must be expressed in terms of x, the unknown distance between points 1 and m.

$$EI\left(0 - \frac{-3438}{EI}\right) = +\frac{(216x)(x)}{2}$$

$$x^2 = 31.8 \qquad x = 5.64 \text{ ft}$$

$$EI\Delta_{1-m} = A_M\bar{x}_1$$

$$EI\Delta_{1-m} = +\frac{[(216)(5.64)](5.64)}{2}\left[\frac{2}{3}\,5.64\right] = +12{,}900 \text{ lb-ft}^3$$

$$\Delta_{1-m} = \frac{+12{,}900(1728)}{10{,}000{,}000(12)} = +0.186 \text{ in.}$$

$$y_{max} = -0.186 \text{ in.} \qquad \textit{Ans}$$

As shown, the deflection at midspan is equal to the difference between distances d_1 and Δ_{3-1}. From similar triangles, $d_1 = \frac{5}{10}\,\Delta_{2-1}$. Since $EI\Delta_{2-1}$ was found to be equal to $+34{,}380$ lb-ft^3, then

$$\Delta_{2-1} = \frac{+34{,}380(1728)}{10{,}000{,}000(12)} = +0.495 \text{ in.}$$

$$d_1 = \tfrac{5}{10}\,0.495 = 0.248 \text{ in.}$$

$$EI\Delta_{3-1} = A_M\bar{x}_3$$

$$EI\Delta_{3-1} = +\frac{1080(5)}{2}\,\frac{5}{3} = +4500 \text{ lb-ft}^3$$

$$\Delta_{3-1} = \frac{+4500(1728)}{10{,}000{,}000(12)} = +0.0648 \text{ in.}$$

$$y_3 = -(0.248 - 0.065) = -0.183 \text{ in.} \qquad \textit{Ans}$$

As can be seen, the midspan deflection differs only slightly from the maximum deflection. This would be true for any loading of a simply supported beam having zero bending moment at each reaction. For this reason many designers investigate the midspan deflection rather than the maximum deflection for such a beam.

PROBLEMS

★**12.26.** For the beam shown in Fig. P12.26, let $a = 8$ ft, $b = 4$ ft, $E = 30{,}000{,}000$ psi, $I = 80$ in.^{4} Determine the maximum deflection.

12.27. For the beam shown in Fig. P12.27, let $a = 6$ ft, $b = 3$ ft, $E = 10{,}000{,}000$ psi, $I = 60$ in.^{4} Determine the maximum deflection.

12.28. For the beam shown in Fig. P12.28, let $a = 12$ ft, $b = 6$ ft, $E = 30{,}000{,}000$ psi, $I = 100$ in.^{4} Determine the maximum deflection.

★**12.29.** For the beam shown in Fig. P12.29, let $a = 4$ ft, $b = 8$ ft, $EI = 400{,}000{,}000$ lb-in.^{2} Compute the maximum deflection.

12.30. For the beam shown in Fig. P12.30, let $a = 6$ ft, $b = 12$ ft, $EI = 2{,}400{,}000{,}000$ lb-in.^{2} Compute the maximum deflection.

12.31. For the beam shown in Fig. P12.31, let $a = 3$ ft, $b = 6$ ft, $EI = 97{,}500{,}000$ lb-in.^{2} Compute the maximum deflection.

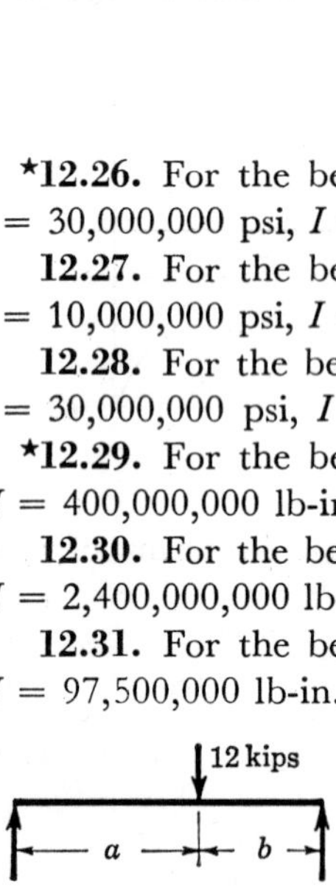

FIG. P12.26, P12.27, P12.28

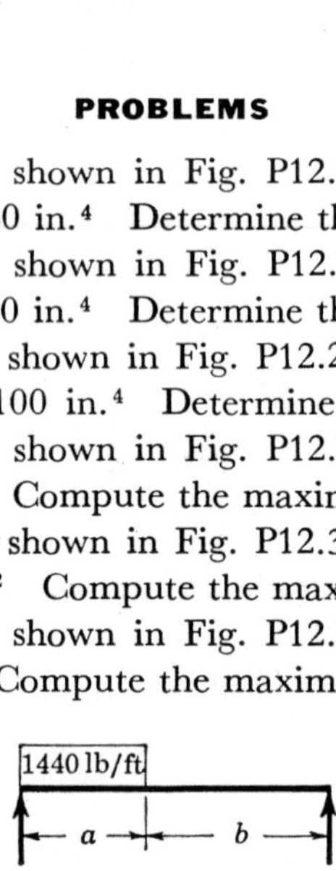

FIG. P12.29, P12.30, P12.31

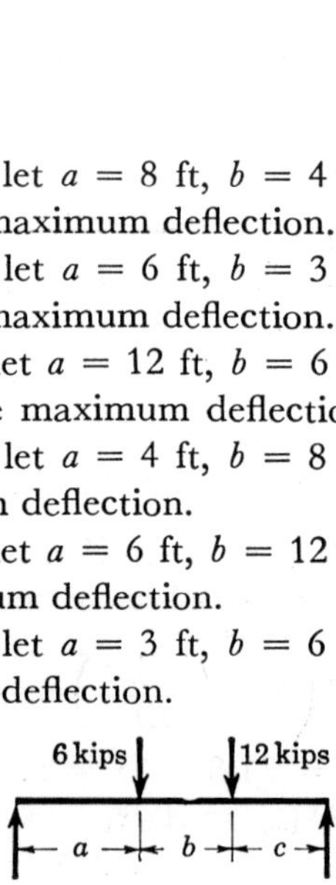

FIG. P12.35, P12.36

★**12.32.** A 10-ft simply supported beam is loaded with a couple at the right reaction. The couple causes a *positive* 6000 lb-ft *bending moment;* $EI = 300{,}000{,}000$ lb-in.^{2} Determine the maximum deflection.

12.33. A 15-ft simply supported beam is loaded with a couple at the left reaction. The couple causes a *negative* 1350 lb-ft *bending moment;* $EI = 300{,}000{,}000$ lb-in.^{2} Determine the maximum deflection.

★**12.34.** A 12-ft simply supported steel beam is loaded with couples at each end. The couple at the right reaction causes a *positive* 12,000 lb-ft *bending moment;* the couple at the left reaction causes a *positive* 19,200 lb-ft *bending moment;* $I = 50$ in.^{4} Compute the maximum deflection.

12.35. For the beam shown in Fig. P12.35, let $a = 6$ ft, $b = 3$ ft, $c = 3$ ft, $EI = 4{,}500{,}000{,}000$ lb-in.^{2} Compute the midspan deflection.

★**12.36.** For the beam shown in Fig. P12.36, let $a = 4$ ft, $b = 5$ ft, $c = 3$ ft, $EI = 1{,}400{,}000{,}000$ lb-in.^{2} Compute the midspan deflection.

★**12.37.** A 12-ft steel beam 9 ft between supports has a 3-ft overhang at the right end. $I = 50$ in.^{4} Determine the maximum upward and downward deflections due to a 6-kip load applied at the end of the overhang.

12.38. A 16-ft beam 10 ft between supports has an overhang at the left end. $E = 1800$ ksi; $I = 600$ in.^{4} Determine the maximum upward and downward deflections due to a uniform load of 9 kips/ft applied to the 6-ft overhang.

12.39. For the beam shown in Fig. P12.39, let $a = 3$ ft and $EI = 2{,}400{,}000{,}000$ lb-in.^{2} Determine the deflection (*a*) under the 12-kip load and (*b*) at the end of the overhang.

★**12.40.** For the beam shown in Fig. P12.40, let $a = 4.5$ ft and $EI = 4{,}000{,}000{,}000$ lb-in.^{2} Determine the deflection (*a*) under the 12-kip load and (*b*) at the end of the overhang.

12.41. A simply supported beam of length L is loaded with a couple M at the right end. Show that the maximum deflection occurs at $0.577L$ from the left end and is equal to $ML^2/(9\sqrt{3}\,EI)$.

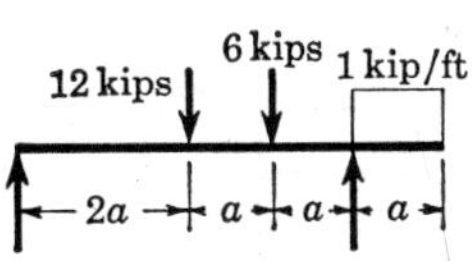

FIG. P12.39, P12.40

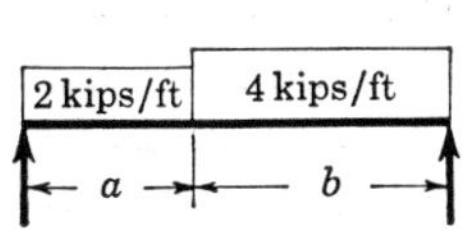

FIG. P12.42, P12.43

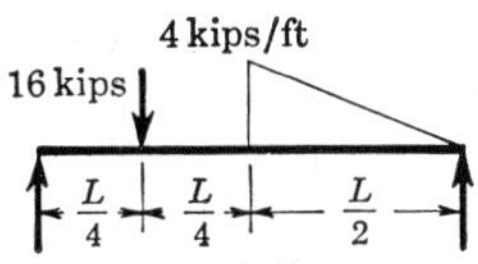

FIG. P12.44, P12.45

12.42. For the steel beam shown in Fig. P12.42, let $a = 4$ ft, $b = 8$ ft, $I = 150$ in.^{4} Determine the midspan deflection.

★**12.43.** For the steel beam shown in Fig. P12.43, let $a = b = 8$ ft, $I = 300$ in.^{4} Determine the midspan deflection.

12.44. For the steel beam shown in Fig. P12.44, let $L = 12$ ft, $I = 100$ in.^{4} Compute the midspan deflection.

12.45. For the steel beam shown in Fig. P12.45, let $L = 20$ ft, $I = 300$ in.^{4} Compute the midspan deflection.

12.46. A 12-ft steel beam 6 ft between supports has a 3-ft overhang at each end. A 6-kip load is applied at the end of the left overhang, and a 12-kip load is applied at the end of the right overhang. $I = 100$ in.^{4} Compute the deflection at each end of the beam.

12.47. An 18-ft steel beam 9 ft between supports has a 4.5-ft overhang at each end. A 15-kip load is applied at the end of the left overhang, and a 3-kip load is applied at the end of the right overhang. $I = 300$ in.^{4} Compute the deflection at each end of the beam.

12.8 RESTRAINED BEAMS

Very few beams in real structures are actually simply supported; they are restrained to some degree by their supports or connections. Where this restraint is small, it is good engineering practice to neglect it; where the restraint is almost complete, the beam is usually assumed to be fixed at that end. A fixed end condition means that the end of the beam is prevented from any rotation. Restraint at the end of a beam, except for a cantilever, causes it to be statically indeterminate. This restraint is caused by an external reaction couple which is in addition to the force reaction. With this additional unknown, there are not enough equilibrium equations to solve for the reactions. However, with the use of the known conditions of the geometry of the elastic curve in addition to the equilibrium equations, the reactions can be found.

Example 12.6. Determine the reactions for the beam shown in Fig. 12.10.

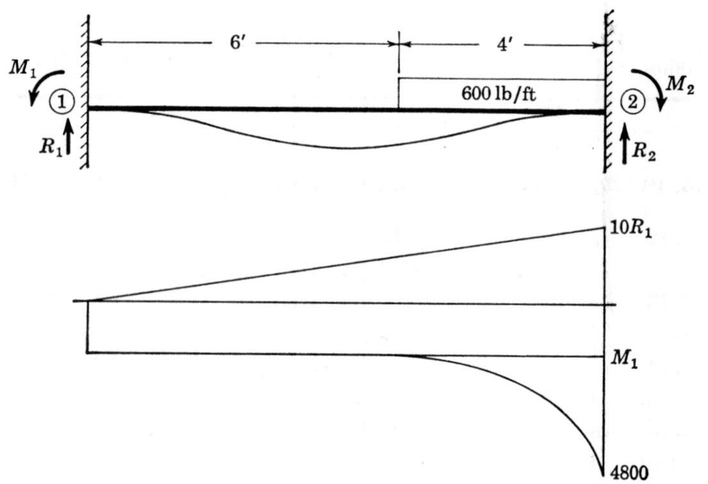

FIG. 12.10

The reactions cannot be determined by statics and are therefore labeled as unknowns. Since the beam is fixed at both ends, the elastic curve is sketched with the ends horizontal. It should be noted that M_1 and M_2 must have the directions shown in order to keep the ends of the curve horizontal.

The reference tangent, selected at point 1, is horizontal. The moment diagram is drawn from the left as shown.

$$EI(\theta_2 - \theta_1) = A_M$$

$$EI(0 - 0) = +\frac{10R_1(10)}{2} - M_1(10) - \frac{4800(4)}{3}$$

$$0 = +50R_1 - 10M_1 - 6400 \qquad (a)$$

$$EI\Delta_{2-1} = A_M\bar{x}_2$$

$$EI(0) = +\frac{10R_1(10)}{2}\frac{10}{3} - M_1(10)(5) - \frac{4800(4)}{3}1$$

$$0 = +\frac{1000R_1}{6} - 50M_1 - 6400 \qquad (b)$$

Solving Eqs. (*a*) and (*b*) simultaneously gives

$R_1 = 307$ lb and $M_1 = 895$ lb-ft *Ans*

M_2 and R_2 are now found by statics.

$\Sigma F_y = 0 \quad R_2 = 2093$ lb *Ans*
$\Sigma M_2 = 0 \quad M_2 = 2625$ lb-ft *Ans*

Note that the signs of all reactions are plus, thus indicating that their directions were taken correctly. This means that both M_1 and M_2 cause *negative bending moments.*

Example 12.7. For the beam shown in Fig. 12.11, derive the equation for the maximum deflection in terms of the uniform load w, the length L, and EI.

From statics and symmetry, $R_1 = R_2 = wL/2$ and $M_1 = M_2$. Since the beam is symmetrical, the problem can be solved by applying geometry to only half of the elastic curve. The reference tangent selected at point 1 is horizontal. The moment diagram is drawn from the left for half of the beam as shown. Because of symmetry the slope angle at point 3 is zero.

$$EI(\theta_3 - \theta_1) = A_M$$

$$EI(0 - 0) = +\frac{1}{2}\frac{wL^2}{4}\frac{L}{2} - M_1\frac{L}{2} - \frac{1}{3}\frac{wL^2}{8}\frac{L}{2}$$

$$0 = +\frac{wL^3}{16} - \frac{M_1L}{2} - \frac{wL^3}{48}$$

$$M_1 = \frac{wL^2}{12}$$

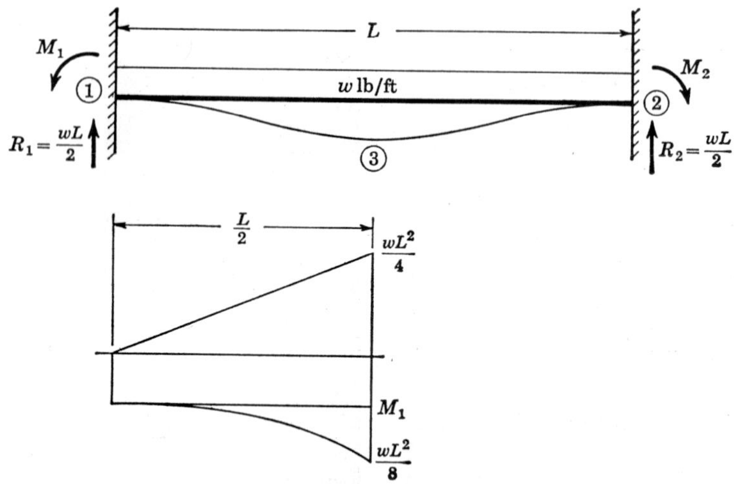

FIG. 12.11

Now that M_1 has been determined, the maximum deflection can be found.

$$EI\Delta_{3-1} = A_M\bar{x}_3$$

$$EI\Delta_{3-1} = +\frac{1}{2}\frac{wL^2}{4}\frac{L}{2}\frac{L}{6} - \frac{wL^2}{12}\frac{L}{2}\frac{L}{4} - \frac{1}{3}\frac{wL^2}{8}\frac{L}{2}\frac{L}{8}$$

$$EI\Delta_{3-1} = -\frac{wL^4}{384}$$

$$y_{max} = \Delta_{3-1} = \frac{wL^4}{384EI} \qquad Ans$$

Example 12.8. The beam in Fig. 12.12 is supported on rollers at point 1 and is fixed at point 2. Determine the reactions.

The elastic curve is sketched, as shown, passing through point 1 and horizontal at point 2. The reference tangent is selected at point 2, and the moment diagram drawn from the left. The problem can be solved by noting that the displacement of point 1 from a tangent at point 2 is equal to zero.

$$EI\Delta_{1-2} = A_M\bar{x}_1$$

$$EI(0) = +\frac{12R_1(12)}{2}8 - 36(12)(6) - \frac{144(12)}{2}8 - \frac{72(6)}{3}10.5$$

$R_1 = 19.12$ kips ↑ *Ans*

$\Sigma M_2 = 0$ $M_2 = 22.56$ kip-ft ⟳ *Ans*

$\Sigma F_y = 0$ $R_2 = 16.88$ kips ↑ *Ans*

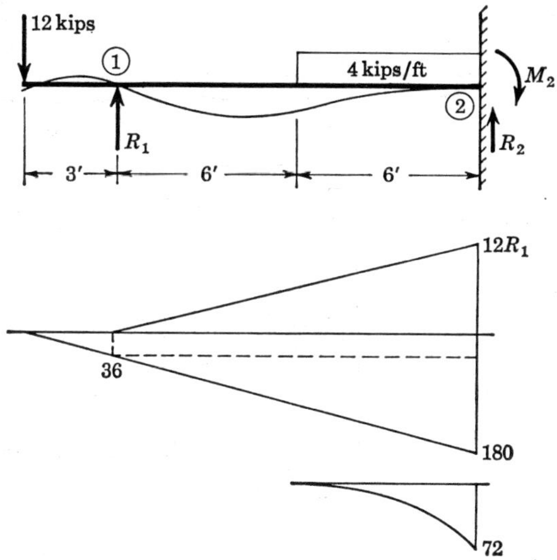

FIG. 12.12

Example 12.9. The beam shown in Fig. 12.13 is supported on rollers at point 1 and is fixed at point 2. $E = 2{,}000{,}000$ psi; $I = 36$ in.^{4} Compute the maximum deflection.

The reactions cannot be determined by statics and are therefore labeled as unknowns. The elastic curve is sketched, and the reference tangent selected at point 2. The moment diagram is drawn from the right.

M_2 and R_2 must be evaluated before the maximum deflection can be found.

$$EI\Delta_{1-2} = A_M \bar{x}_1$$

$$EI(0) = -M_2(9)(4.5) + \frac{9R_2(9)}{2}\,3 - \frac{5400(6)}{2}\,2 - \frac{900(3)}{3}\,\frac{3}{4}$$

$$0 = M_2 - 3R_2 + 816.7 \qquad (a)$$

Equation (a) contains two unknowns, and all the known conditions of the elastic curve have been utilized. However, from statics

$$\Sigma M_1 = 0 \qquad M_2 - 9R_2 + 6300 = 0 \qquad (b)$$

Solving Eqs. (a) and (b) simultaneously gives

$$R_2 = 914 \text{ lb} \qquad M_2 = 1926 \text{ lb-ft}$$

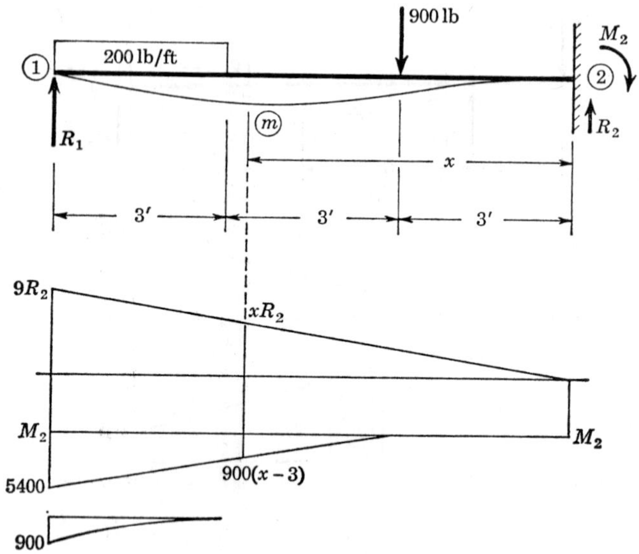

FIG. 12.13

To find the point of maximum deflection

$$EI(\theta_2 - \theta_m) = A_M$$

$$EI(0 - 0) = -1926(x) + \frac{914x(x)}{2} - \frac{900(x-3)(x-3)}{2}$$

$$0 = x^2 + 110.6x - 578.7$$
$$x = 5.0 \text{ ft}$$

To find the maximum deflection

$$EI\Delta_{m-2} = A_M\bar{x}_m$$

$$EI\Delta_{m-2} = -1926(5)(2.5) + \frac{914(5)(5)}{2}\frac{5}{3} - \frac{900(2)(2)}{2}\frac{2}{3}$$

$$EI\Delta_{m-2} = -24{,}075 + 19{,}050 - 1200 = -6225 \text{ lb-ft}^3$$

$$y_{\max} = \Delta_{m-2} = \frac{-6225(1728)}{2{,}000{,}000(36)} = -0.149 \text{ in.} \qquad \textit{Ans}$$

PROBLEMS

★**12.48.** A 20-ft steel beam is restrained at both ends; equal loads of 36 kips each are applied 5 ft from each end. $I = 600$ in.[4] Compute (*a*) the reactions and (*b*) the maximum deflection.

12.49. An 18-ft timber beam is restrained at both ends; equal loads of 1200 lb each are applied 6 ft from each end. $EI = 320{,}000{,}000$ lb-in.^{2} Compute (*a*) the reactions and (*b*) the maximum deflection.

★**12.50.** For the beam shown in Fig. P12.50, let $L = 12$ ft. (*a*) Compute the reactions. (*b*) Draw the complete shear and moment diagrams.

12.51. For the beam shown in Fig. P12.51, let $L = 16$ ft. (*a*) Compute the reactions. (*b*) Draw the complete shear and moment diagrams.

★**12.52.** A 12-ft timber beam is restrained at both ends; a load of 8100 lb is applied 4 ft from the left end. $E = 1{,}200{,}000$ psi; $I = 450$ in.^{4} Compute (*a*) the reactions and (*b*) the maximum deflection.

12.53. A 12-ft steel beam is restrained at both ends; a load of 64 kips is applied 3 ft from the left end. $I = 400$ in.^{4} Compute (*a*) the reactions and (*b*) the maximum deflection.

★**12.54.** For the beam shown in Fig. P12.54, let $L = 12$ ft and compute the reactions.

12.55. For the beam shown in Fig. P12.55, let $L = 16$ ft and compute the reactions.

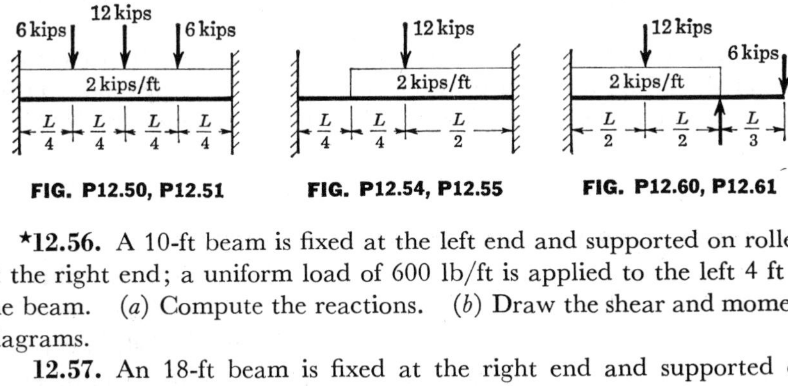

FIG. P12.50, P12.51 **FIG. P12.54, P12.55** **FIG. P12.60, P12.61**

★**12.56.** A 10-ft beam is fixed at the left end and supported on rollers at the right end; a uniform load of 600 lb/ft is applied to the left 4 ft of the beam. (*a*) Compute the reactions. (*b*) Draw the shear and moment diagrams.

12.57. An 18-ft beam is fixed at the right end and supported on rollers at the left end; a uniform load of 24 kips/ft is applied to the right 6 ft of the beam. Compute the reactions.

12.58. A 12-ft beam is fixed at the right end and supported on rollers at the left end; a uniform load of 600 lb/ft is applied to the left 4 ft of the beam. Compute the reactions.

12.59. A 12-ft beam is fixed at the left end and supported on rollers at the right end; a uniform load of 24 kips/ft is applied to the right 7 ft of the beam. (*a*) Compute the reactions. (*b*) Draw the shear and moment diagrams.

12.60. For the beam shown in Fig. P12.60, let $L = 12$ ft and compute the reactions.

★**12.61.** For the beam shown in Fig. P12.61, let $L = 9$ ft and compute the reactions.

12.62. A beam of length L is fixed at each end and supports a midspan concentrated load P. Prove that the moment at each end is $PL/8$ and that the maximum deflection is $PL^3/192EI$.

12.63. A beam of length L ft supported at one end by rollers and fixed at the other supports a uniform load of w lb/ft over the entire span. Show that the moment and the vertical reaction at the fixed end are $wL^2/8$ and $5wL/8$.

12.64. A beam of length L is supported at one end by rollers and fixed at the other. The beam supports a midspan concentrated load P. Show that the moment and the vertical reaction at the fixed end are $3PL/16$ and $11P/16$.

★**12.65.** A 10-ft beam is fixed at both ends and loaded with a clockwise couple of 36 kip-ft applied 4 ft from the left end. Compute the reactions.

12.66. A 12-ft beam is fixed at both ends and loaded with a clockwise couple of 1200 lb-ft applied 3 ft from the left end. Compute the reactions.

12.67. A beam of length L is supported at one end by rollers and fixed at the other; it is loaded with a couple M at the roller end. Prove that the moment and the vertical reaction at the fixed end are $M/2$ and $M/2L$.

12.9 BEAMS WITH VARIABLE MOMENTS OF INERTIA

As has been discussed in Chaps. 5 and 6, the section modulus I/c required at any section of a beam is a function of the bending moment at that section. Generally, the I/c needed at the point of maximum bending moment is used for the entire span, even though a smaller I/c would be sufficient for the greater part of the beam. However, in many cases it is economical to design beams of variable I/c or variable moments of inertia.

The area-moment method is well suited to the solution of elastic-curve problems dealing with beams of variable moments of inertia. For beams of constant E, such problems can be easily solved by using the area of the M/I diagram and adapting the area-moment equations into the following form:†

† The student can verify these equations by making the appropriate modification to the derivations in Art. 12.2.

$$E(\theta_2 - \theta_1) = A_{M/I} \tag{12.3}$$
$$E\Delta_{2-1} = A_{M/I}\bar{x}_2 \tag{12.4}$$

Example 12.10. For the beam loaded as shown in Fig. 12.14, compute the maximum deflection. I for the 6-ft center segment is 60 in.^{4}; I for the two 3-ft end segments is 30 in.^{4}; E = 30,000 ksi.

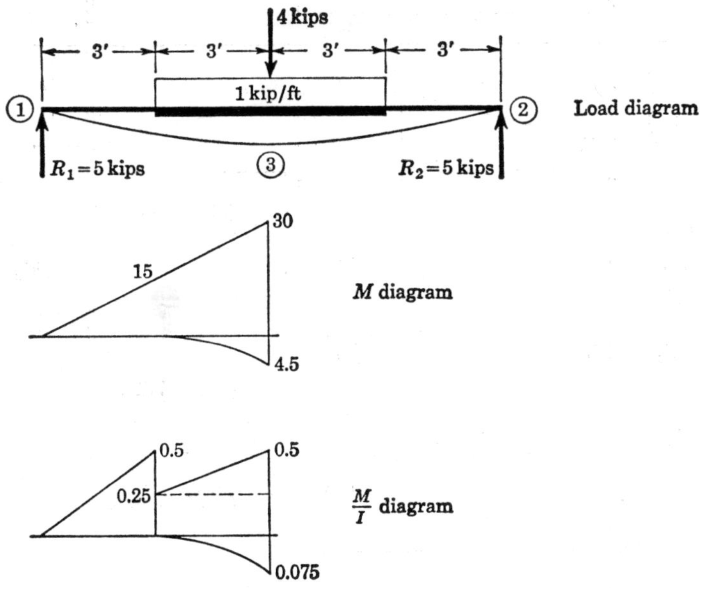

FIG. 12.14

Since the beam is symmetrically loaded, the moment diagram as simple areas is drawn for only half of the beam. Then the critical ordinates of the moment diagram are divided by the appropriate I and the M/I diagram drawn as shown.

$$E\Delta_{1-3} = A_{M/I}\bar{x}_1$$

$$E\Delta_{1-3} = +\frac{0.5(3)}{2}\,2 + \frac{0.25(3)}{2}\,5 + 0.25(3)(4.5) - \frac{0.075(3)}{3}\,5.25$$

$$E\Delta_{1-3} = +1.500 + 1.875 + 3.375 - 0.394 = +6.356 \quad \text{kip-ft}^3/\text{in.}^4$$

$$\Delta_{1-3} = \frac{+6.356(1728)}{30{,}000} = +0.366 \text{ in.}$$

$$y_{\text{max}} = -0.366 \text{ in.} \qquad \textit{Ans}$$

Example 12.11. The beam shown in Fig. 12.15*a* has a constant width and a variable depth. Determine the reactions.

The load diagram and elastic curve are as shown in Fig. 12.15*b*. Since the beam is symmetrical, only half of the elastic curve need be used for the solution of the problem. The moment diagram of the left half of the beam is shown in Fig. 12.15*c*. For practical purposes the tapered portion of the beam can be divided into short segments, each having a

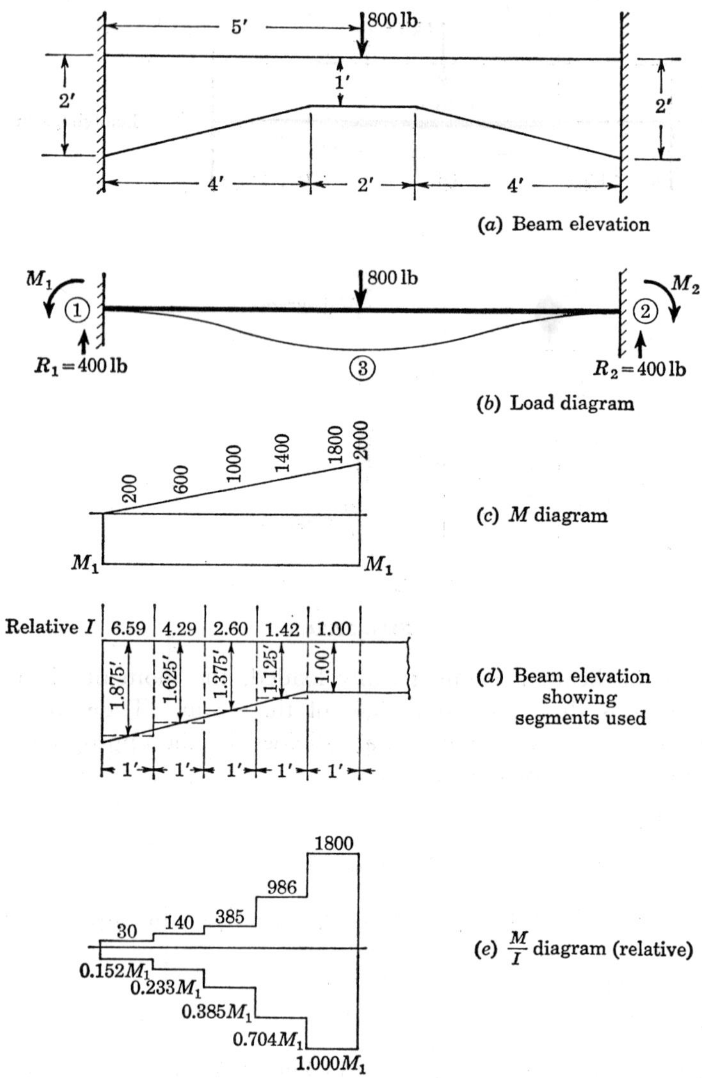

FIG. 12.15

constant depth and therefore a constant moment of inertia. Figure 12.15*d* shows the tapered portion represented as a series of 1-ft segments having different depths. As will be seen, the reactions of the beam are dependent on the *relative* moments of inertia of the different segments, and the absolute values need not be computed. Assigning a value of unity to the I of the center portion of the beam, the relative I of each 1-ft segment is computed. Noting that I varies as the cube of the depth, the values of relative I are computed to be as shown in Fig. 12.15*d*. The M/I diagram, shown in Fig. 12.15*e*, is obtained by dividing the moment at the center of each 1-ft segment by the corresponding relative I.

$$E(\theta_3 - \theta_1) = A_{M/I}$$
$$E(0 - 0) = 30 + 140 + 385 + 986 + 1800 - M_1(0.152 + 0.233 + 0.385 + 0.704 + 1.000)$$
$$0 = +3341 - 2.474M_1$$
$$M_1 = 1350 \text{ lb-ft} \qquad Ans$$

It can now be seen that, if the absolute value of I for the center section had been used, all the terms in the right side of the above equation would have been divided by this value. Since the left side of the equation is equal to zero, the value of M_1 would not have been affected.

PROBLEMS

***12.68.** For the steel beam shown in Fig. P12.68, let $P = 24$ kips, I for the right half of the beam $= 500$ in.^{4}, I for the left half $= 2000$ in.^{4} Compute the deflection at the free end.

12.69. For the aluminum beam shown in Fig. P12.69, let $P = 12$ kips, I for the right half of the beam $= 200$ in.^{4}, I for the left half $= 1000$ in.^{4} Compute the deflection at the free end.

12.70. For the beam shown in Fig. P12.70, let $a = b = 6$ ft, I for the middle 6 ft of the beam $= 200$ in.^{4}, I for the rest of the beam $= 400$ in.^{4} (*a*) Compute the reactions. (*b*) Draw the complete shear and moment diagrams.

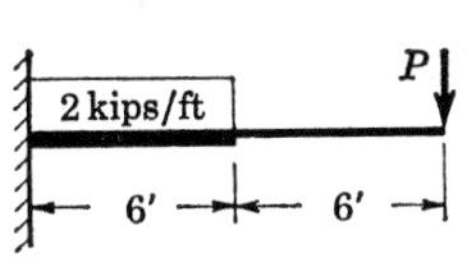

FIG. P12.68, P12.69

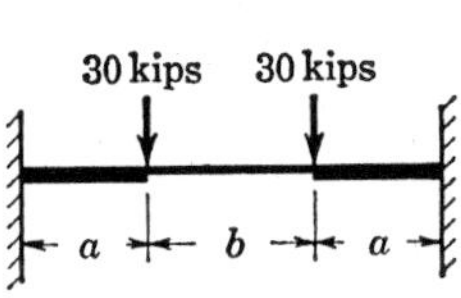

FIG. P12.70, P12.71

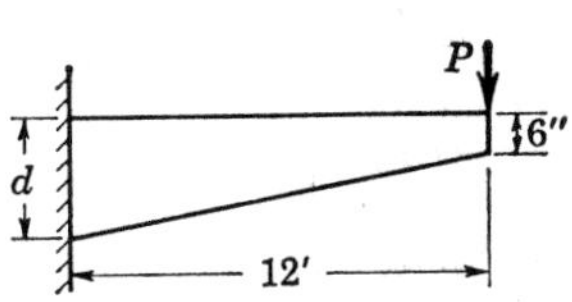

FIG. P12.72, P12.73, P12.74

12.71. For the beam shown in Fig. P12.71, let $a = 4$ ft, $b = 8$ ft, I for the middle 8 ft of the beam $= 100$ in.^{4}, I for the rest of the beam $=$ 300 in.^{4} (*a*) Compute the reactions. (*b*) Draw the complete shear and moment diagrams.

★**12.72.** The beam shown in Fig. P12.72 has a constant 6-in. width and a variable depth. Let $P = 3600$ lb, $E = 1{,}600{,}000$ psi, $d = 18$ in. Using 2-ft segments, compute the deflection at the free end.

12.73. The beam shown in Fig. P12.73 has a constant 6-in. width and a variable depth. Let $P = 1600$ lb, $E = 2{,}000{,}000$ psi, $d = 12$ in. Using 2-ft segments, compute the deflection at the free end.

12.74. The beam shown in Fig. P12.74 has a constant 6-in. width and a variable depth. Let $P = 6400$ lb, $E = 1{,}800{,}000$ psi, $d = 24$ in. Using 2-ft segments, compute the deflection at the free end.

13

ELASTIC CURVE: SUPERPOSITION

13.1 INTRODUCTION

Many elastic-curve problems can be most easily solved using the principle of superposition. As is the case with the methods of the two previous chapters, the method of superposition is applicable to both statically determinate and statically indeterminate beams.

The principle of superposition can be stated as follows: The effect on a body of several loads acting simultaneously is the algebraic sum of the effects of the loads acting separately. The only limitation to this principle is that there must be a linear relationship between loads and effects. For elastic-curve problems, the effects which are normally obtained by superposition are slopes and deflections. However, the principle of superposition is also very useful for determining the reactions of certain types of indeterminate beams.

13.2 DEFLECTIONS

Because it is less susceptible to mistakes, and because the computations required are relatively simple, the method of superposition is frequently used for routine determinations of beam deflections. When this method is used, the deflections (or slopes) due to the individual loads are computed from formulas which have been

TABLE 13.1 Slope and Deflection Formulas for Cantilever Beams

Loading diagram	Slope at end	Deflection at end
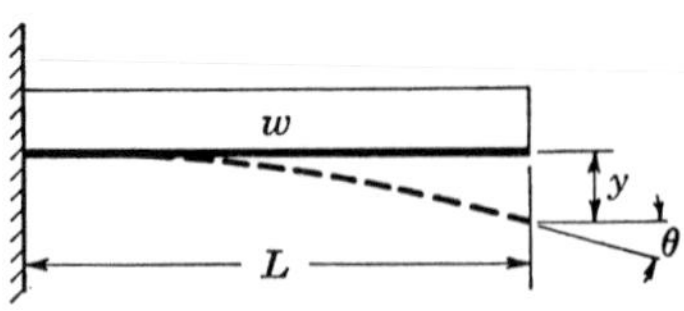	$\theta = \frac{wL^3}{6EI}$	$y = \frac{wL^4}{8EI}$
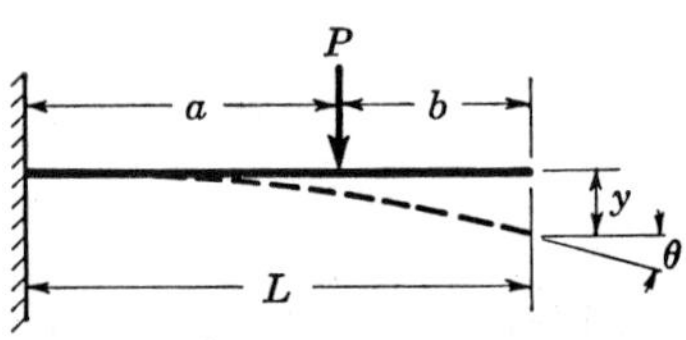	$\theta = \frac{Pa^2}{2EI}$	$y = \frac{Pa^2}{6EI}(3L - a)$
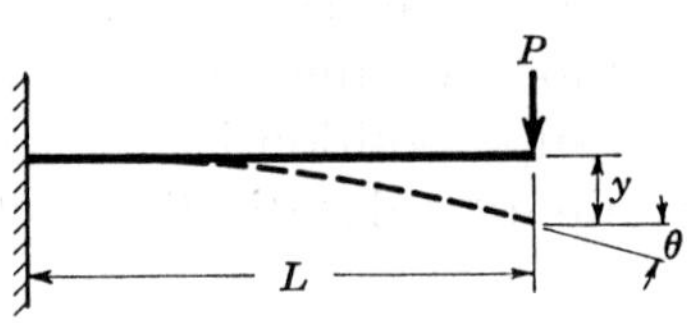	$\theta = \frac{PL^2}{2EI}$	$y = \frac{PL^3}{3EI}$

previously determined. Such formulas for a wide variety of load conditions are available from handbooks and other sources. Of course, the more routine formulas will ordinarily be memorized from frequent usage. The formulas for a few of the more commonly used load conditions are shown in Tables 13.1 and 13.2.

Except for cantilever beams, the method of superposition does not lend itself to finding maximum deflections of beams which are unsymmetrically loaded. However, in most cases, the midspan deflection for such beams will differ only slightly from the maximum

TABLE 13.2 Slope and Deflection Formulas for Simply Supported Beams

Loading diagram	Slope at R_1	Slope at R_2	Deflection at midspan
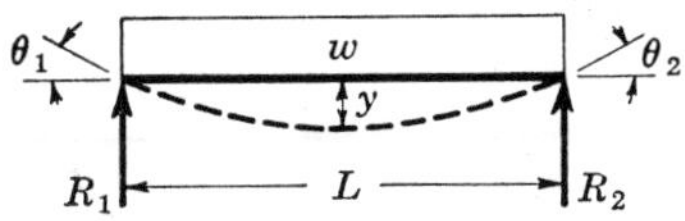	$\theta_1=\dfrac{wL^3}{24EI}$	$\theta_2=\dfrac{wL^3}{24EI}$	$y=\dfrac{5wL^4}{384EI}$
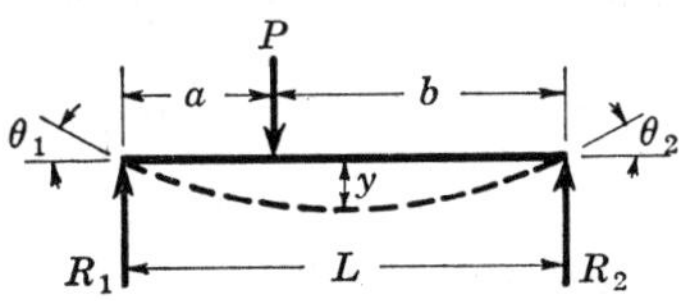	$\theta_1=\dfrac{Pb(L^2-b^2)}{6LEI}$	$\theta_2=\dfrac{Pa(L^2-a^2)}{6LEI}$	$y=\dfrac{Pa(3L^2-4a^2)}{48EI}$ for $a<b$
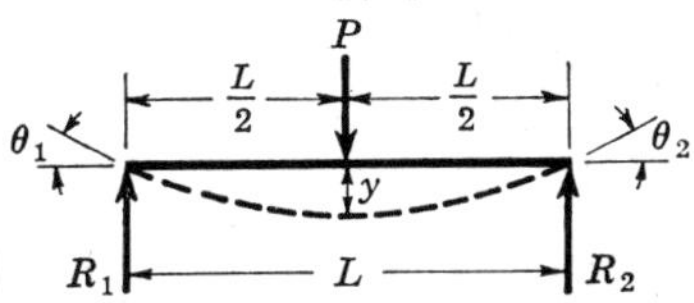	$\theta_1=\dfrac{PL^2}{16EI}$	$\theta_2=\dfrac{PL^2}{16EI}$	$y=\dfrac{PL^3}{48EI}$
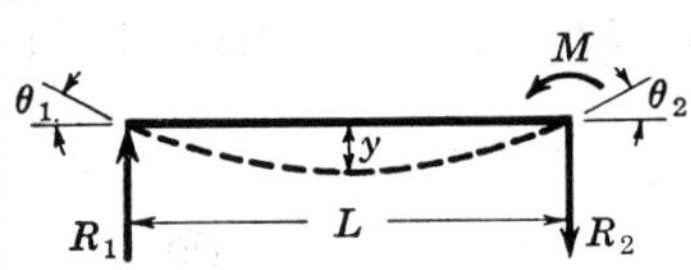	$\theta_1=\dfrac{ML}{6EI}$	$\theta_2=\dfrac{ML}{3EI}$	$y=\dfrac{ML^2}{16EI}$

deflection. Generally a sketch or visualization of the elastic curve will reveal whether or not the midspan deflection is sufficiently close to the maximum. If it is not, the maximum deflection can be determined by one of the other methods.

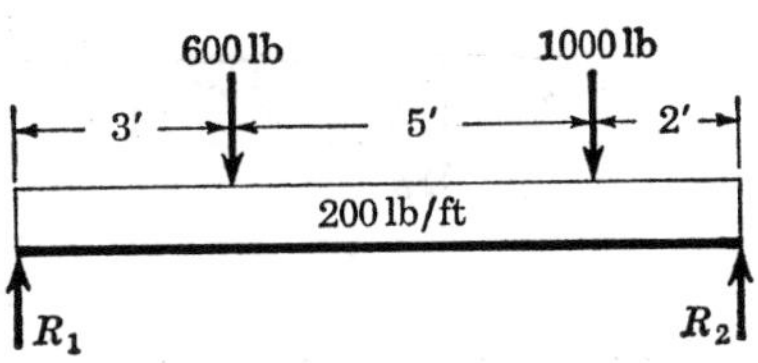

FIG. 13.1

Example 13.1. For the beam shown in Fig. 13.1, E = 30,000,000 psi and I = 15 in.^{4} Determine the midspan deflection and the slope at the left reaction.

Using the formulas of Table 13.2 and noting that a is 3 ft for the 600-lb load and 2 ft for the 1000-lb load, the midspan deflection is determined as follows:

$$y = \frac{5wL^4}{384EI} + \sum \frac{Pa(3L^2 - 4a^2)}{48EI}$$

$$EIy = \frac{5(200)(10)^4}{384} + \frac{600(3)}{48}[3(10)^2 - 4(3)^2] + \frac{1000(2)}{48}[3(10)^2 - 4(2)^2]$$

$$EIy = 26{,}050 + 9900 + 11{,}830 = 47{,}780 \text{ lb-ft}^3$$

$$y = \frac{47{,}780(1728)}{30{,}000{,}000(15)} = 0.183 \text{ in.}\downarrow \qquad Ans$$

Using the formulas of Table 13.2 and noting that b is 7 ft for the 600-lb load and 2 ft for the 1000-lb load, the slope at the left reaction is determined as

$$\theta = \frac{wL^3}{24EI} + \sum \frac{Pb(L^2 - b^2)}{6LEI}$$

$$EI\theta = \frac{200(10)^3}{24} + \frac{600(7)(10^2 - 7^2)}{6(10)} + \frac{1000(2)(10^2 - 2^2)}{6(10)}$$

$$EI\theta = 8330 + 3570 + 3200 = 15{,}100 \text{ lb-ft}^2$$

$$\theta = \frac{15{,}100(144)}{30{,}000{,}000(15)} = 0.00483 \text{ radian} \qquad Ans$$

Example 13.2. For the beam shown in Fig. 13.2*a*, $E = 30{,}000$ ksi and $I = 240$ in.^{4} Determine the deflection at midspan and at the end of the overhang.

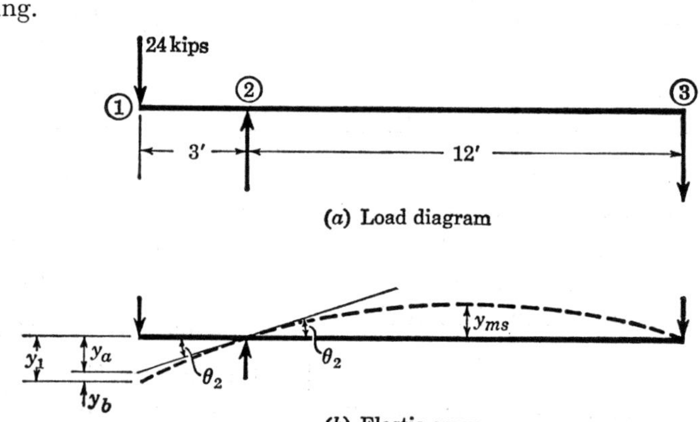

(*a*) Load diagram

(*b*) Elastic curve

FIG. 13.2

Since the overhang load causes a moment at the left reaction of $M = 24(3) = 72$ kip-ft, the midspan deflection can be determined directly from the appropriate formula in Table 13.2. Thus

$$y_{ms} = \frac{ML^2}{16EI} = \frac{72(12)^2(1728)}{16(30{,}000)(240)} = 0.156 \text{ in.}\uparrow \qquad Ans$$

As shown in Fig. 13.2*b*, the deflection at the end of the overhang y_1 can be considered as consisting of two parts. One part y_a due to the rotation of the tangent at the support is equal to $\theta_2(3)$. The other part is the vertical displacement of the end with respect to the tangent at the support; this part y_b is equivalent to the deflection of a cantilever beam of the length of the overhang. Thus, using the equations of Tables 13.1 and 13.2,

$$\theta_2 = \frac{ML}{3EI} = \frac{72(12)}{3EI} = \frac{288}{EI}$$

$$y_a = \frac{288}{EI}3 = \frac{864}{EI}$$

$$y_b = \frac{PL^3}{3EI} = \frac{24(3)^3}{3EI} = \frac{216}{EI}$$

$$y_1 = y_a + y_b = \frac{864}{EI} + \frac{216}{EI} = \frac{1080}{EI}$$

$$y_1 = \frac{1080(1728)}{30{,}000(240)} = 0.259 \text{ in.}\downarrow \qquad \textit{Ans}$$

PROBLEMS

★13.1. A 24-ft simply supported beam is loaded with a 4-kip load at midspan and 8-kip loads at each quarter point. $E = 10{,}000$ ksi, and $I = 300$ in.^{4} Determine the maximum deflection and the slope at each support.

13.2. A 20-ft simply supported beam is loaded with a 15-kip load at midspan, a 25-kip load at the left quarter point, and a 10-kip load at the right quarter point. $E = 30{,}000$ ksi, and $I = 1000$ in.^{4} Determine the midspan deflection and the slope at each support.

★13.3. A 10-ft cantilever beam supports a uniform load of 120 lb/ft and a midspan concentrated load of 900 lb. $E = 1{,}800{,}000$ psi, and $I = 600$ in.^{4} Determine the slope and deflection at the free end.

13.4. For the beam shown in Fig. P13.4, let $L = 12$ ft, and $EI = 2{,}000{,}000{,}000$ lb-in.^{2} Determine the maximum deflection.

★13.5. For the beam shown in Fig. P13.5, let $L = 6$ ft and $EI = 800{,}000{,}000$ lb-in.^{2} Determine the maximum deflection.

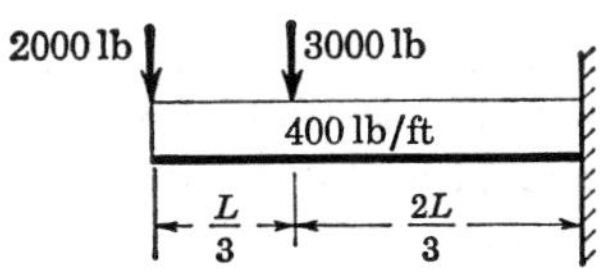

FIG. P13.4, P13.5

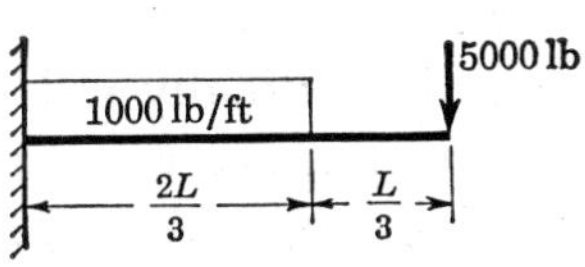

FIG. P13.6, P13.7

13.6. For the beam shown in Fig. P13.6, let $L = 12$ ft, $E = 30,$-000,000 psi, and $I = 300$ in.^{4} Determine the maximum deflection.

13.7. For the beam shown in Fig. P13.7, let $L = 9$ ft, $E = 30,000$-000 psi, and $I = 200$ in.^{4} Determine the maximum deflection.

***13.8.** A 12-ft simply supported steel beam is loaded with couples at each end. The couple at the right reaction causes a positive 12,000 lb-ft bending moment; the couple at the left reaction causes a positive 19,200 lb-ft bending moment; $I = 50$ in.^{4} Determine the midspan deflection.

***13.9.** For the steel beam shown in Fig. P13.9, let $L = 12$ ft, $P = 10$ kips, $I = 150$ in.^{4} Determine the midspan deflection.

13.10. For the steel beam shown in Fig. P13.10, let $L = 6$ ft, $P = 12$ kips, $I = 80$ in.^{4} Determine the midspan deflection.

13.11. A 16-ft span of a continuous beam supports a uniform load of 200 lb/ft over its entire length. The bending moment at the left support is minus 800 lb-ft; at the right support the bending moment is minus 1600 lb-ft. Determine the midspan deflection.

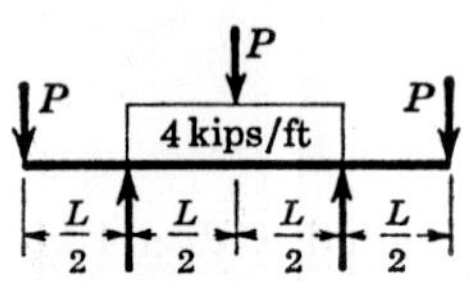

FIG. P13.9, P13.10

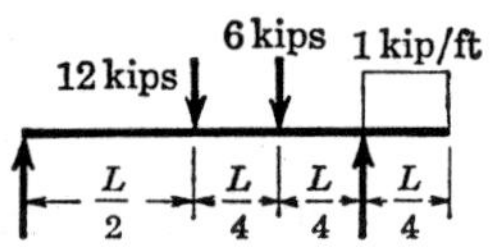

FIG. P13.13, P13.14, P13.15

13.12. A 24-ft span of a continuous beam supports a uniform load of 6 kips/ft over its entire length. The bending moment at the left support is minus 300 kip-ft; at the right support the bending moment is minus 180 kip-ft. Determine the midspan deflection.

13.13. For the beam shown in Fig. P13.13, let $L = 12$ ft and $EI = 2,400,000,000$ lb-in.^{2} Determine the deflection under the 12-kip load.

***13.14.** For the beam shown in Fig. P13.14, let $L = 18$ ft and $EI = 4,000,000,000$ lb-in.^{2} Determine the deflection under the 12-kip load.

13.15. For the beam shown in Fig. P13.15, let $L = 24$ ft and $EI = 6,000,000,000$ lb-in.^{2} Determine the deflection under the 12-kip load.

13.3 RESTRAINED BEAMS

For certain statically indeterminate beams, the principle of superposition provides a rapid and simple method for the determination of reactions as well as deflections and slopes. This is especially true when the appropriate statically indeterminate formulas are available from tables or from other sources. When such formulas are available, the method of using superposition is essentially like that of the previous article. The formulas for a few of the more commonly used restrained-beam load conditions are shown in Table 13.3.

TABLE 13.3 End Moments and Deflection Formulas for Restrained Beams

Loading diagram	End moment at end 1	End moment at end 2	Deflection at midspan
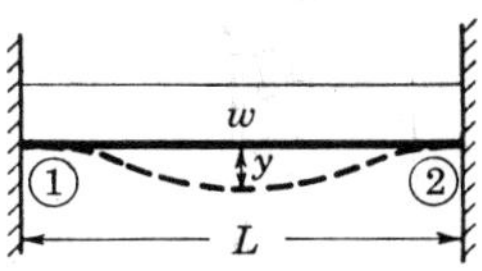	$M_1 = \frac{wL^2}{12}$	$M_2 = \frac{wL^2}{12}$	$y = \frac{wL^4}{384EI}$
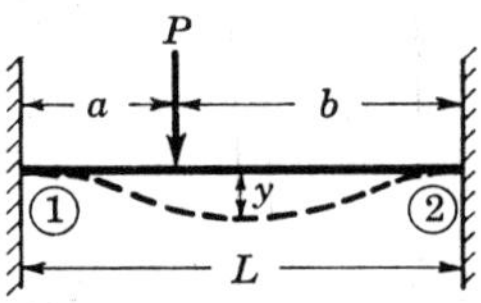	$M_1 = \frac{Pab^2}{L^2}$	$M_2 = \frac{Pa^2b}{L^2}$	$y = \frac{Pa^2(3L-4a)}{48EI}$ for $a < b$
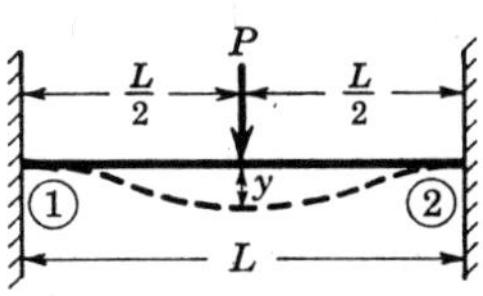	$M_1 = \frac{PL}{8}$	$M_2 = \frac{PL}{8}$	$y = \frac{PL^3}{192EI}$

When the appropriate statically indeterminate formulas are not available, a more general method can be used. By this method, the effect to be determined is obtained by considering the indeterminate load condition as resulting from the superposition of two or more statically determinate load conditions. For any indeterminate beam, one or more of the reaction components are in excess of the number required for equilibrium. Therefore, if certain of these "surplus" components were removed, an indeterminate beam would become statically determinate. These "removable" reaction components are known as *redundants.* The general method consists of first "removing" the redundants and then considering them as separate load conditions of the statically determinate beam.

The values of the redundants must be such that the superposition of the statically determinate load conditions produces the elastic curve of the original statically indeterminate beam.

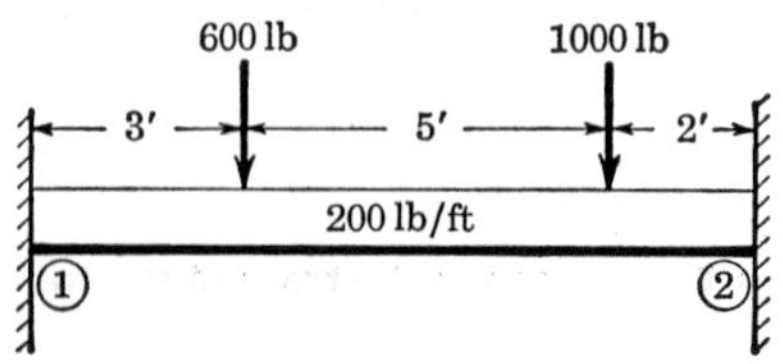

FIG. 13.3

Example 13.3. For the beam shown in Fig. 13.3, $E = 30{,}000{,}000$ psi and $I = 15$ in.^{4} Determine the beam reactions and the midspan deflection.

Using the formulas of Table 13.3,

$$M_1 = \frac{wL^2}{12} + \sum \frac{Pab^2}{L^2}$$

$$M_1 = \frac{200(10)^2}{12} + \frac{600(3)(7)^2}{10^2} + \frac{1000(8)(2)^2}{10^2}$$

$$M_1 = 1667 + 882 + 320 = 2869 \text{ lb-ft} \quad \textit{Ans}$$

$$M_2 = \frac{wL^2}{12} + \sum \frac{Pa^2b}{L^2}$$

$$M_2 = \frac{200(10)^2}{12} + \frac{600(3)^2(7)}{10^2} + \frac{1000(8)^2(2)}{10^2}$$

$$M_2 = 1667 + 378 + 1280 = 3325 \text{ lb-ft} \quad \textit{Ans}$$

From statics,

$$R_1 = 1574 \text{ lb}\uparrow \qquad \text{and} \qquad R_2 = 2026 \text{ lb}\uparrow \qquad \textit{Ans}$$

Using the formulas of Table 13.3 and noting that a must be used as 3 ft for the 600-lb load and 2 ft for the 1000-lb load, the midspan deflection is determined as

$$y = \frac{wL^4}{384EI} + \sum \frac{Pa^2}{48EI}(3L - 4a)$$

$$y = \frac{200(10)^4}{384EI} + \frac{600(3)^2}{48EI}[3(10) - 4(3)] + \frac{1000(2)^2}{48EI}[3(10) - 4(2)]$$

$$EIy = 5210 + 2025 + 1833 = 9068 \text{ lb-ft}^3$$

$$y = \frac{9068(1728)}{30{,}000{,}000(15)} = 0.035 \text{ in.}\downarrow \qquad \textit{Ans}$$

Example 13.4. Determine the reactions of the propped beam shown in Fig. 13.4*a*.

The elastic curve and reactions for the beam are as shown in Fig. 13.4*b*. If any one of the three reaction unknowns is determined from the condition of the elastic curve, the other two can be found from the conditions of equilibrium.

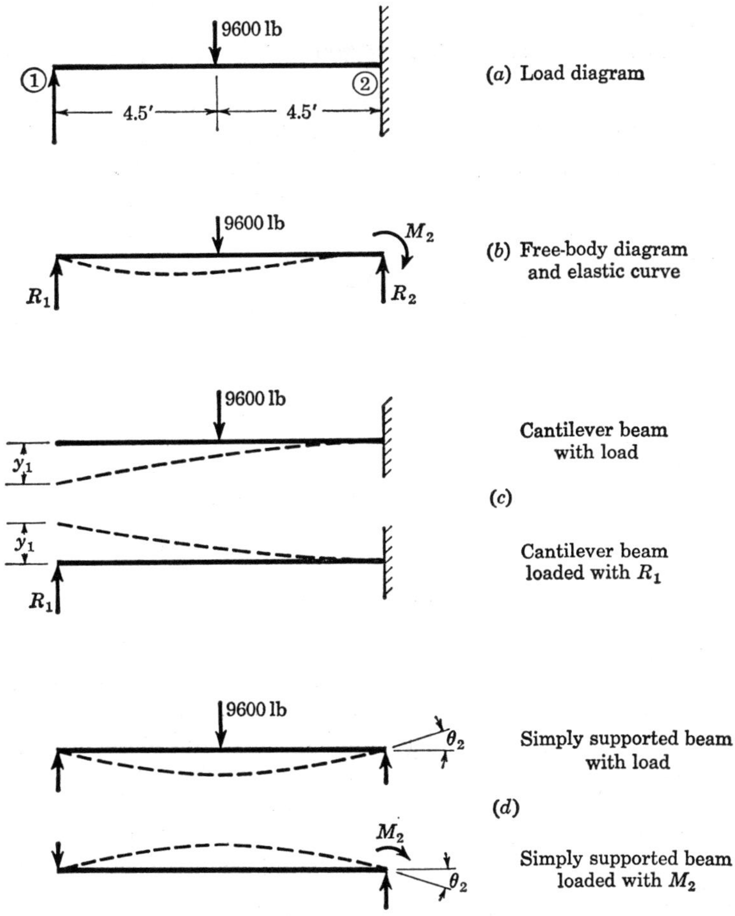

FIG. 13.4

Considering R_1 as the redundant, the two statically determinate load conditions to be superimposed are as shown in Fig. 13.4*c*. In order for this superposition to produce the elastic curve of Fig. 13.4*b*, the value of R_1 must be such that the two end deflections in Fig. 13.4*c* are numerically equal. Therefore, using the formulas of Table 13.1,

$$(y_1)_{\text{redundant}} = (y_1)_{\text{load}}$$

$$\frac{R_1(9)^3}{3EI} = \frac{9600(4.5)^2}{6EI}[3(9) - 4.5]$$

$R_1 = 3000 \text{ lb}\uparrow$ *Ans*

From statics,

$R_2 = 6600 \text{ lb}\uparrow$ and $M_2 = 16{,}200$ lb-ft ⟳ *Ans*

An *alternate solution* can be made by regarding the fixed-end moment M_2 as the redundant and using the two statically determinate load conditions shown in Fig. 13.4*d*. For these two simply supported load conditions to produce the elastic curve of Fig. 13.4*b*, the value of M_2 must be such that the right-end slopes in Fig. 13.4*d* are numerically equal. Therefore, using the formulas of Table 13.2,

$$(\theta_2)_{\text{redundant}} = (\theta_2)_{\text{load}}$$

$$\frac{M_2(9)}{3EI} = \frac{9600(9)^2}{16EI}$$

$M_2 = 16{,}200$ lb-ft ⟳ *Check*

From statics,

$R_1 = 3000 \text{ lb}\uparrow$ and $R_2 = 6600 \text{ lb}\uparrow$ *Check*

PROBLEMS

★**13.16.** For the beam shown in Fig. P13.16, let $L = 12$ ft. Determine the reactions, and draw the complete shear and moment diagrams.

13.17. For the beam shown in Fig. P13.17, let $L = 24$ ft. Determine the reactions, and draw the complete shear and moment diagrams.

13.18. A 12-ft steel beam is restrained at both ends and supports the following loads: 2 kips/ft over the entire span, 12 kips at 3 ft from the left end, and 6 kips at 4 ft from the right end. If $I = 125$ in.^{4}, determine the moment at each support and the midspan deflection.

★**13.19.** A 15-ft steel beam is restrained at both ends and supports the following loads: 2 kips/ft over the entire span, 4 kips at 5 ft from the left end, and 10 kips at 6 ft from the right end. If $I = 200$ in.^{4}, determine the moment at each support and the midspan deflection.

13.20. A 10-ft steel beam is restrained at both ends and supports the following loads: 3 kips/ft over the entire span, 10 kips at 4 ft from the left end, and 8 kips at 3 ft from the right end. If $I = 100$ in.^{4}, determine the moment at each support and the midspan deflection.

13.21. For the steel beam shown in Fig. P13.21, let $L = 10$ ft, $P_1 = 10$ kips, $P_2 = 20$ kips, and $P_3 = 30$ kips. If $I = 300$ in.^{4}, determine the moment at each support and the midspan deflection.

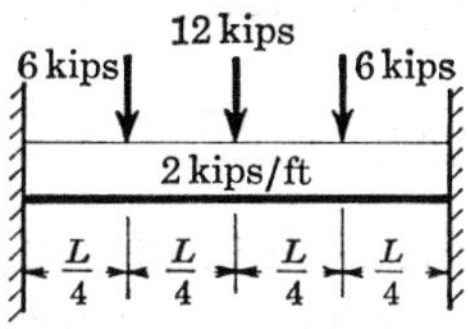

FIG. P13.16, P13.17

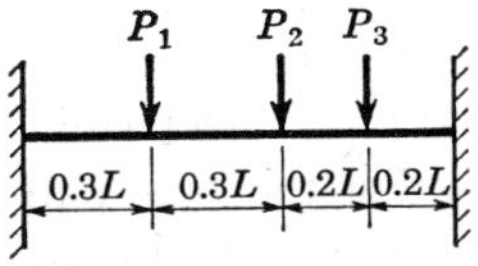

FIG. P13.21, P13.22, P13.23

★**13.22.** For the steel beam shown in Fig. P13.22, let $L = 15$ ft, $P_1 = 25$ kips, $P_2 = 15$ kips, and $P_3 = 12$ kips. If $I = 400$ in.^{4}, determine the moment at each support and the midspan deflection.

13.23. For the steel beam shown in Fig. P13.23, let $L = 20$ ft, $P_1 = 12$ kips, $P_2 = 20$ kips, and $P_3 = 8$ kips. If $I = 400$ in.^{4}, determine the moment at each support and the midspan deflection.

13.24. A 16-ft propped beam supports a uniform load of 2000 lb/ft over its entire span. Determine the reactions.

★**13.25.** A 12-ft propped beam supports a uniform load of 4 kips/ft over its entire span. Determine the reactions.

13.26. A 12-ft propped beam supports a concentrated load of 30 kips at 8 ft from the restrained end. Determine the reactions.

13.27. A 16-ft propped beam supports a concentrated load of 40 kips at 10 ft from the restrained end. Determine the reactions.

13.28. For the propped beam shown in Fig. P13.28, let $L = 12$ ft, $P = 40$ kips, and $w = 4$ kips/ft. Based on an allowable bending stress of 20,000 psi, determine the required section modulus.

★**13.29.** For the propped beam shown in Fig. P13.29, let $L = 15$ ft, $P = 30$ kips, and $w = 3$ kips/ft. Based on an allowable bending stress of 18,000 psi, determine the required section modulus.

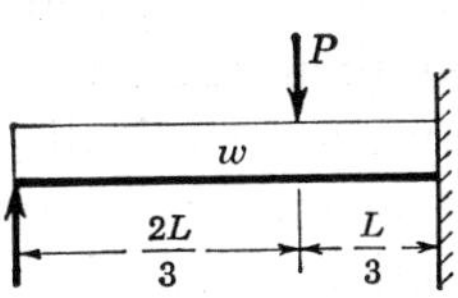

FIG. P13.28, P13.29

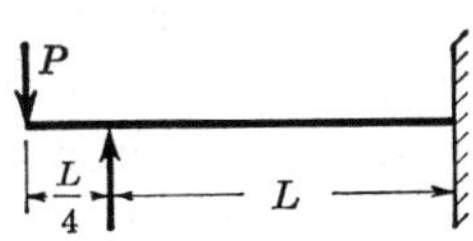

FIG. P13.30, P13.31

13.30. For the propped steel beam shown in Fig. P13.30, let $L = 12$ ft and $P = 24$ kips. If $I = 400$ in.^{4}, determine the deflection at the end of the overhang.

13.31. For the propped steel beam shown in Fig. P13.31, let $L = 16$ ft and $P = 10$ kips. If $I = 100$ in.^{4}, determine the deflection at the end of the overhang.

★13.32. For the propped beam shown in Fig. P13.32, let $L = 12$ ft and $P = 15$ kips. Determine the reactions.

13.33. For the propped beam shown in Fig. P13.33, let $L = 15$ ft and $P = 20$ kips. Determine the reactions.

13.34. For the two-span continuous beam shown in Fig. P13.34, let $L = 10$ ft and $w = 4$ kips/ft. Using R_2 as the redundant, determine the reactions and draw the complete bending-moment diagram.

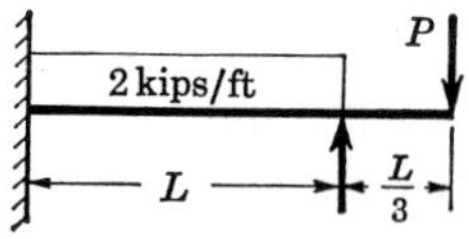

FIG. P13.32, P13.33

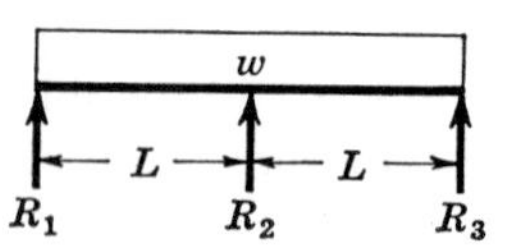

FIG. P13.34, P13.35

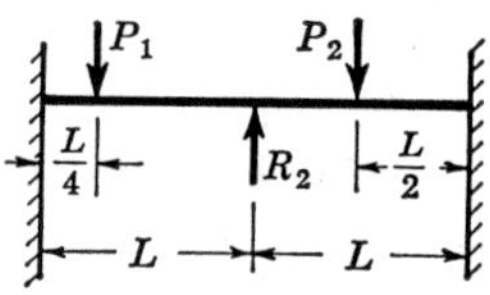

FIG. P13.36, P13.37

★13.35. For the two-span continuous beam shown in Fig. P13.35, let $L = 16$ ft and $w = 3$ kips/ft. Using R_2 as the redundant, determine the reactions and draw the complete bending-moment diagram.

13.36. For the two-span continuous beam shown in Fig. P13.36, let $L = 10$ ft, $P_1 = 5$ kips, and $P_2 = 10$ kips. Using R_2 as the redundant, determine the reactions.

★13.37. For the two-span continuous beam shown in Fig. P13.37, let $L = 16$ ft, $P_1 = 8$ kips, and $P_2 = 12$ kips. Using R_2 as the redundant, determine the reactions.

13.4 MEMBERS HAVING RELATED DEFLECTIONS

Occasionally it is necessary or expedient to connect two or more beams to each other in such a way that their elastic curves are interrelated. The reactions of such beams are statically indeterminate in that the force exerted by one beam on the other is a function of common or related deflections. Although other methods can be used, the method of superposition is especially suited to this type of problem.

Example 13.5. Figure 13.5*a* is a plan view of two steel beams which are in contact at their midspans. The bottom of beam 1-2 rests on the top of beam 3-4; a 30,000-lb vertical load is applied at the point indicated.

Beam 1-2 is pin connected at both ends; beam 3-4 is fixed at both ends. If I for beam 3-4 is twice that of beam 1-2, determine the reactions.

The elastic curve and free-body diagram for each beam are as shown in Fig. 13.5*b* and *c*. The force at the point of contact between the two

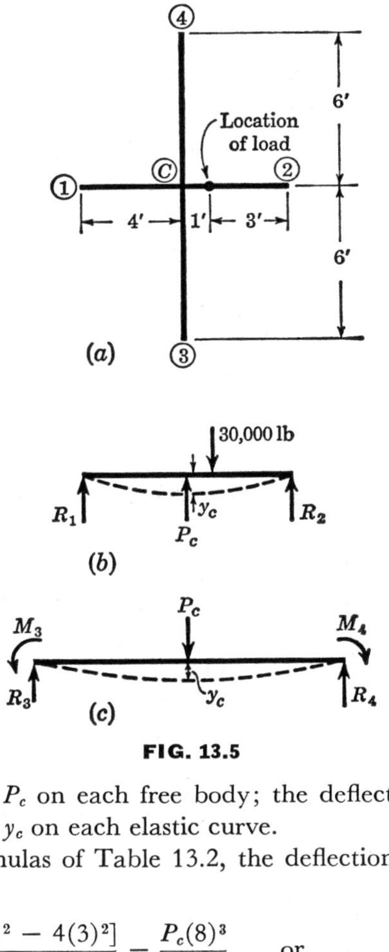

FIG. 13.5

beams is shown as P_c on each free body; the deflection at the point of contact is shown as y_c on each elastic curve.

Using the formulas of Table 13.2, the deflection y_c for beam 1-2 is written as

$$y_c = \frac{30{,}000(3)[3(8)^2 - 4(3)^2]}{48EI} - \frac{P_c(8)^3}{48EI} \qquad \text{or}$$

$$y_c = \frac{14{,}040{,}000 - 512P_c}{48EI}$$

Using the formulas of Table 13.3, the deflection y_c for beam 3-4 is written as

$$y_c = \frac{P_c(12)^3}{192E(2I)} = \frac{1728P_c}{384EI} = \frac{216P_c}{48EI}$$

Since y_c is the same for each beam,

$$\frac{14{,}040{,}000 - 512P_c}{48EI} = \frac{216P_c}{48EI}$$

Therefore

$$P_c = 19{,}290 \text{ lb}$$

From statics and Fig. 13.5*b*

$R_1 = 1605 \text{ lb}\uparrow$ and $R_2 = 9105 \text{ lb}\uparrow$ *Ans,*

Using Fig. 13.5*c* and the formulas of Table 13.3,

$M_3 = \dfrac{19{,}290(12)}{8} = 28{,}940 \text{ lb-ft}$ ↻ *Ans*

$M_4 = M_3 = 28{,}940 \text{ lb-ft}$ ↺ *Ans*

From statics,

$R_3 = R_4 = 9645 \text{ lb}\uparrow$ *Ans*

PROBLEMS

13.38. Two steel beams are in contact at their midspans as shown in Fig. P13.38; the bottom of beam 1-2 rests on the top of beam 3-4. Beam 1-2 is fixed at both ends and has $I = 400$ in.^{4}; beam 3-4 is pin connected at each end and has $I = 300$ in.^{4} Determine the reactions of each beam due to a vertical 50-kip load applied 5 ft from reaction 1.

***13.39.** Two steel beams are in contact at their midspans as shown in Fig. P13.39; the bottom of beam 1-2 rests on the top of beam 3-4. Beam 1-2 is pin connected at each end and has $I = 300$ in.^{4}; beam 3-4 is fixed at both ends and has $I = 200$ in.^{4} Determine the reactions of each beam due to a vertical 40-kip load applied 6 ft from reaction 2.

13.40. Two steel beams are in contact at their midspans as shown in Fig. P13.40; the bottom of beam 3-4 rests on the top of beam 1-2. Both beams are restrained at each end. For beam 1-2, $I = 400$ in.^{4}; for beam 3-4, $I = 800$ in.^{4} Determine the reactions of each beam due to a vertical 75-kip load applied 5 ft from reaction 3.

***13.41.** The free end of cantilever beam 3-4 rests on top of simply supported beam 1-2 as shown in Fig. P13.41. For beam 1-2, $I = 400$ in.^{4}; for beam 3-4, $I = 300$ in.^{4} Both beams are of steel. Determine the reactions of each beam due to a vertical 30-kip load applied 4 ft from reaction 3.

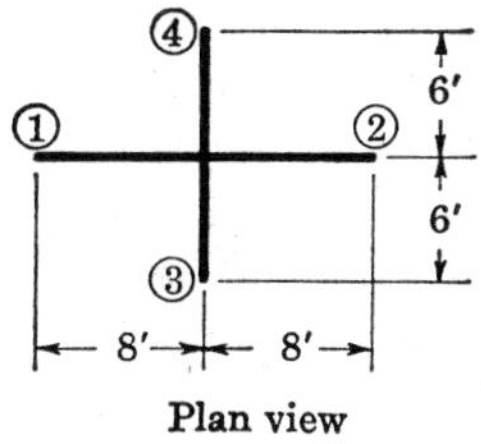

FIG. P13.38, P13.39, P13.40

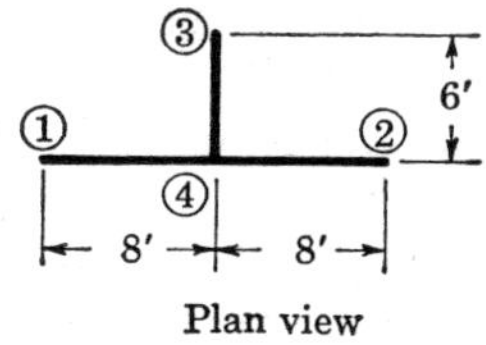

FIG. P13.41, P13.42, P13.43

13.42. The free end of cantilever beam 3-4 rests on the top of restrained beam 1-2 as shown in Fig. P13.42. Both beams are of steel and have moments of inertia of 400 in.[4] Determine the reactions of each beam due to a vertical uniform load of 5 kips/ft applied over the entire length of beam 3-4.

13.43. The midspan of simply supported beam 1-2 rests on the free end of cantilever beam 3-4 as shown in Fig. P13.43. Both beams are of steel. For beam 1-2, $I = 400$ in.^{4}; for beam 3-4, $I = 800$ in.^{4} Determine the reactions of each beam due to a vertical 30-kip load applied 6 ft from reaction 2.

***13.44.** Two spans of a steel beam are connected by a hinge as shown in Fig. P13.44. For span 1-2, $I = 800$ in.^{4}; for span 2-3, $I = 400$ in.^{4} Determine the reactions due to a 25-kip vertical load at 8 ft from reaction 1.

13.45. Two spans of a steel beam are connected by a hinge as shown in Fig. P13.45. For span 1-2, $I = 300$ in.^{4}; for span 2-3, $I = 200$ in.^{4} Determine the reactions due to a 15-kip vertical load at 6 ft from reaction 3.

13.46. Two spans of a steel beam are connected by a hinge as shown in Fig. P13.46. For span 1-2, $I = 400$ in.^{4}; for span 2-3, $I = 300$ in.^{4} Determine the reactions due to a uniform load of 2 kips/ft applied to the entire length of span 1-2.

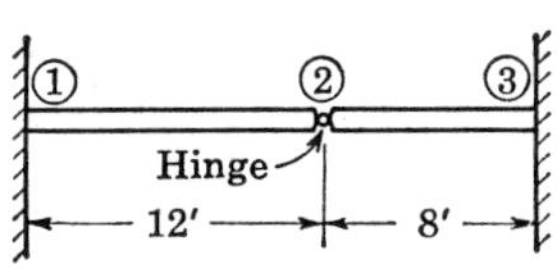

FIG. P13.44, P13.45, P13.46

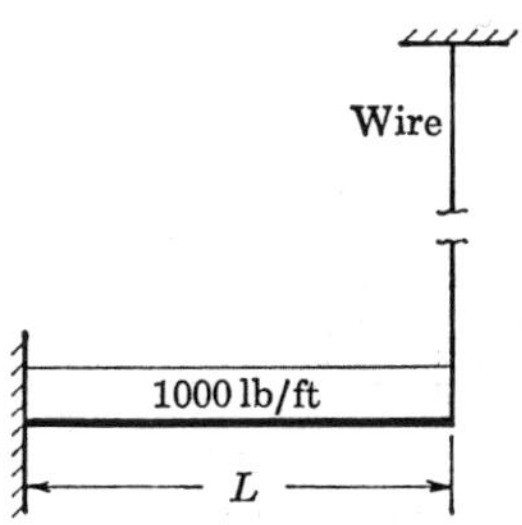

FIG. P13.47, P13.48

★**13.47.** A 6-ft steel beam is fixed at one end and supported by a $\frac{3}{8}$-in.-diam steel wire at the other as shown in Fig. P13.47. The wire is 10 ft long; I for the beam is 25 in.^{4} Determine the beam reactions and the stress in the wire due to the uniform load shown.

13.48. A 10-ft steel beam is fixed at one end and supported by a $\frac{1}{2}$-in.-diam aluminum ($E = 10{,}000{,}000$ psi) wire at the other as shown in Fig. P13.48. The wire is 5 ft long; I for the beam is 100 in.^{4} Determine the beam reactions and the stress in the wire due to the uniform load shown.

13.49. A 30-ft aluminum ($E = 10{,}000{,}000$ psi) beam having $I = 15$ in.^{4} is supported at three points as shown in Fig. P13.49. The modulus of the spring used as the middle support is 1000 lb/in. Determine the beam reactions due to equal loads of $P = 1000$ lb applied at 10 ft from each end.

13.50. A 24-ft steel beam having $I = 10$ in.^{4} is supported at three points as shown in Fig. P13.50. The modulus of the spring used as the middle support is 1000 lb/in. Determine the beam reactions due to equal loads of $P = 800$ lb applied at 7 ft from each end.

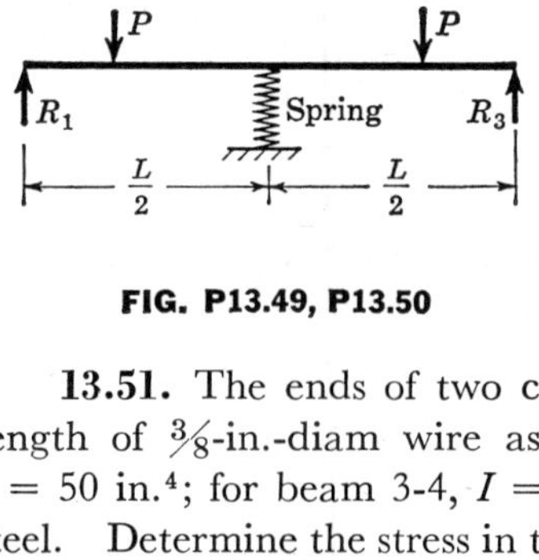

FIG. P13.49, P13.50

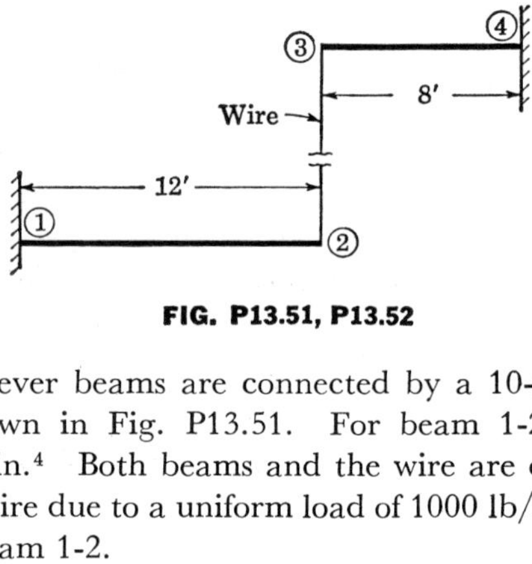

FIG. P13.51, P13.52

13.51. The ends of two cantilever beams are connected by a 10-ft length of $\frac{3}{8}$-in.-diam wire as shown in Fig. P13.51. For beam 1-2, $I = 50$ in.^{4}; for beam 3-4, $I = 30$ in.^{4} Both beams and the wire are of steel. Determine the stress in the wire due to a uniform load of 1000 lb/ft applied over the entire length of beam 1-2.

13.52. The ends of two cantilever beams are connected by a 15-ft length of $\frac{1}{4}$-in.-diam wire as shown in Fig. P13.52. For beam 1-2, $I = 30$ in.^{4}; for beam 3-4, $I = 50$ in.^{4} Both beams and the wire are of steel. Determine the stress in the wire due to a concentrated load of 5 kips applied at the midspan of beam 1-2.

14

CONTINUOUS BEAMS

14.1 INTRODUCTION

A continuous beam is one which is supported at more than two points. Beams of this type are generally used because the restraint at the interior supports results in smaller critical moments and deflections. In the case of reinforced-concrete construction, however, continuous beams are generally used in order to avoid the difficulty of making moment-free connections. The advantage of beam continuity in reducing critical moments is illustrated in Fig. 14.1. It can be noted that, for this particular loading condition, the critical moment for the continuous beam is 25 per cent less than that of an equivalent series of simply supported beams.

Since, by definition, continuous beams have more than two reaction unknowns, they are statically indeterminate. However, if the internal moments at the interior supports are determined, the various reactions can be computed by applying the equations of equilibrium to free bodies of the separate spans. Although the methods of the three previous chapters can be used to determine these internal moments, they generally become laborious and impractical when applied to beams having more than two or three spans. For this reason, the support

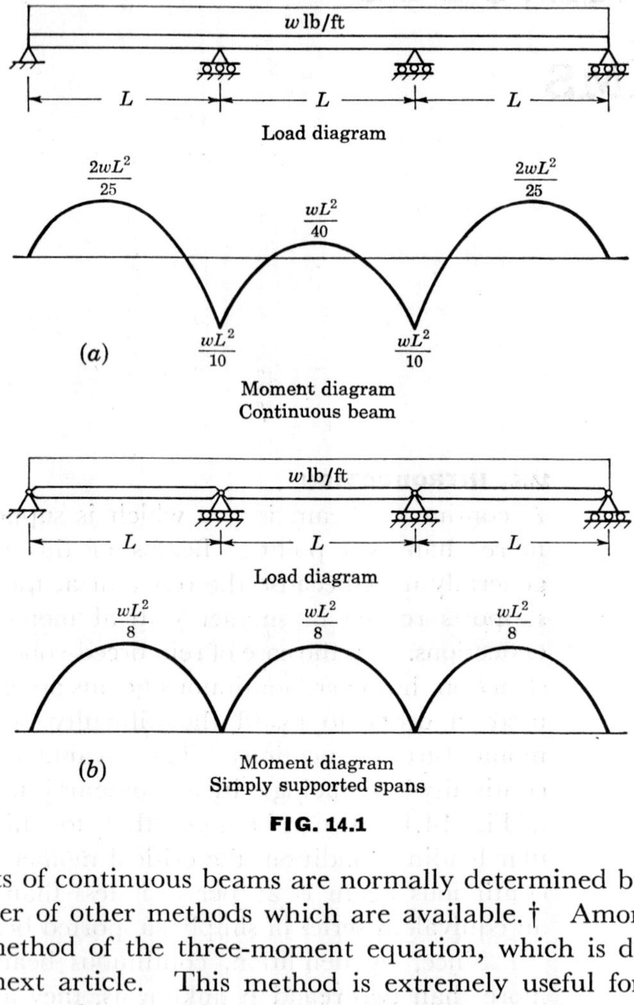

FIG. 14.1

moments of continuous beams are normally determined by one of a number of other methods which are available.† Among these is the method of the three-moment equation, which is described in the next article. This method is extremely useful for beams having two, three, or four spans. However, for beams having

† Most of these methods are described in any of the texts on indeterminate structures. See, for example, J. S. Kinney, "Indeterminate Structural Analysis," Addison-Wesley Publishing Company, Inc., Reading, Mass., 1957.

more than about five spans, it generally becomes quite cumbersome because of the large number of simultaneous equations that must be solved.

14.2 THE THREE-MOMENT EQUATION

The three-moment equation defines the relationship among the bending moments at any three consecutive supports of a continuous beam. It will be derived with reference to spans L_1 and L_2 of the constant-cross-section continuous beam shown in Fig. 14.2*a*. Owing to the general loading indicated, the elastic curve for the two spans is assumed to be as shown in Fig. 14.2*b*. Assuming that the elastic limit is not exceeded, the elastic curve will be continuous over the two spans; therefore the slope at B for span L_2 must be equal to the slope at B for span L_1. Thus in equation form, at reaction B, $\theta_2 = \theta_1$. For each span, the slope at B can be considered as resulting from the superposition of three loading conditions applied to a simply supported beam. Thus in Fig. 14.2*c*, α denotes the slope at support B due to the applied loading and β and γ denote the slopes at support B due to the internal moments at the supports. Since $\theta_1 = \alpha_1 + \beta_1 + \gamma_1$ and $\theta_2 = \alpha_2 + \beta_2 + \gamma_2$, and since $\theta_1 = \theta_2$,

$$\alpha_1 + \beta_1 + \gamma_1 = \alpha_2 + \beta_2 + \gamma_2 \qquad (a)$$

The values of β and γ can be expressed in terms of the end moments from previously determined equations. Thus, with the use of the formulas of Table 13.2 for values and Fig. 14.2*c* for signs, Eq. (*a*) becomes

$$\alpha_1 + \frac{M_B L_1}{3EI} + \frac{M_A L_1}{6EI} = -\alpha_2 - \frac{M_B L_2}{3EI} - \frac{M_C L_2}{6EI} \qquad (b)$$

Multiplying through Eq. (*b*) by $6EI$ and collecting terms,

$$M_A L_1 + 2M_B(L_1 + L_2) + M_C L_2 = -6EI\alpha_1 - 6EI\alpha_2 \qquad (14.1)$$

It should be noted that in the derivation, positive bending moments were assumed at all supports. Therefore, when Eq. (14.1) is used, plus values of M indicate positive bending moments and minus values negative bending moments. Also, because of the derivation, the signs of α_1 and α_2 are correct for downward-applied loads.

In using the three-moment equation, it is convenient to compute values of $6EI\alpha$ from formulas which have previously been derived. Such formulas, for a variety of load conditions, are nor-

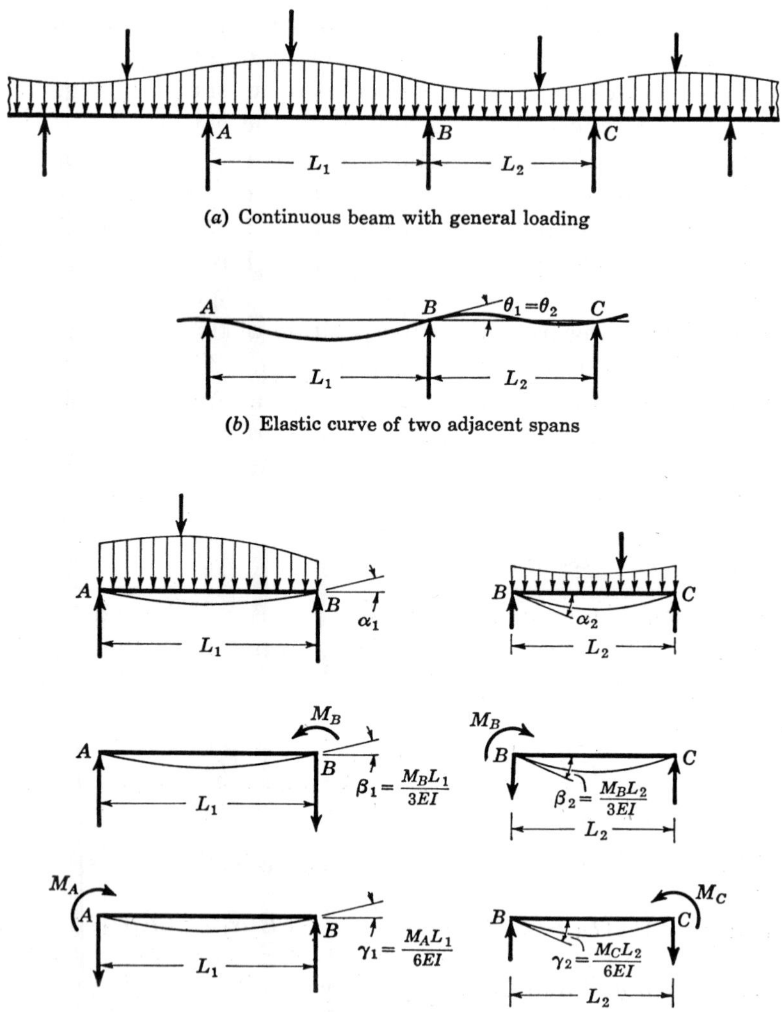

(*a*) Continuous beam with general loading

(*b*) Elastic curve of two adjacent spans

(c) Slope at *B* by parts

FIG. 14.2

mally available from tables in handbooks and other sources;† the $6EI\alpha$ formulas for the more commonly encountered load conditions are given in Table 14.1. Of course, if such tables are not available, tables which give end slopes for different load conditions of simply supported beams can be used almost as easily.

TABLE 14.1 Formulas for Use with Three-moment Equation

<table>
<tr><th>Loading</th><th>$6EI\alpha_1$</th><th>$6EI\alpha_2$</th></tr>
<tr><td>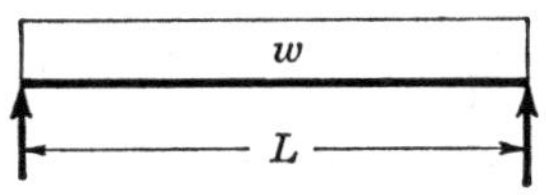
</td><td>$\frac{wL^3}{4}$</td><td>$\frac{wL^3}{4}$</td></tr>
<tr><td>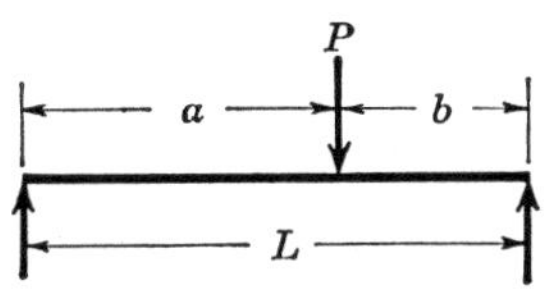
</td><td>$\frac{Pa}{L}(L^2-a^2)$</td><td>$\frac{Pb}{L}(L^2-b^2)$</td></tr>
<tr><td>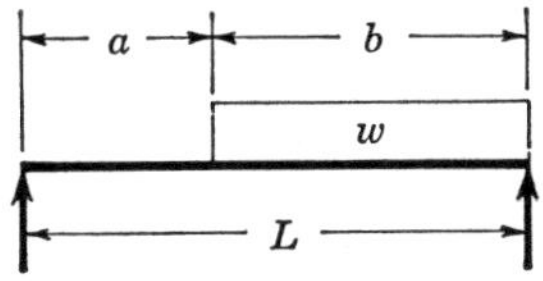
</td><td>$\frac{w}{4L}(L^2-a^2)^2$</td><td>$\frac{wb^2}{4L}(2L^2-b^2)$</td></tr>
<tr><td>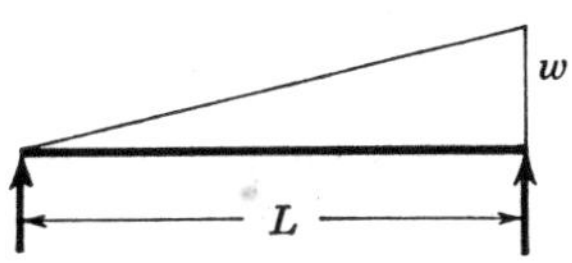
</td><td>$\frac{8}{60}wL^3$</td><td>$\frac{7}{60}wL^3$</td></tr>
</table>

As will be seen, if the ends of a two-span continuous beam are not restrained, the moment at the exterior supports will be known and the moment at the middle support can be determined directly from the three-moment equation. However, for beams having more than two spans, it is necessary to write the equation more than once. Thus, in the general case, a beam of n spans will have

† Most such tables list values of $6Ax/L$ which are equivalent to values of $6EI\alpha$. $6Ax/L$ terms occur in the three-moment equation when it is derived by the area-moment method.

unknown moments at $(n - 1)$ supports, and the three-moment equation can be written for $(n - 1)$ pairs of adjacent spans. The general procedure is to start at the left and write the equation for the first two spans, then for the second and third spans, and so on, until the right span has been included. The equations thus obtained are then solved simultaneously for the required moments.

Example 14.1. Determine the bending moments at the supports of the continuous beam shown in Fig. 14.3.

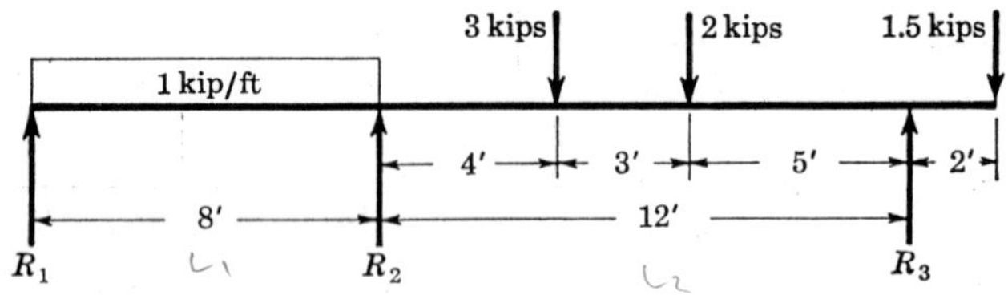

FIG. 14.3

The bending moments at the end supports are determined directly from the load diagram. Thus

$$M_1 = 0 \qquad Ans$$

and

$$M_3 = -1.5(2) = -3 \text{ kip-ft} \qquad Ans$$

The bending moment at the middle support is found from the three-moment equation. Using Table 14.1 for values of $6EI\alpha$ for each load and the known values of M_1 and M_3,

$$M_AL_1 + 2M_B(L_1 + L_2) + M_CL_2 = -6EI\alpha_1 - \Sigma 6EI\alpha_2$$

$$0(8) + 2M_2(8 + 12) + (-3)(12) = -\frac{1(8)^3}{4} - \frac{3(8)}{12}(12^2 - 8^2) - \frac{2(5)}{12}(12^2 - 5^2)$$

$$40M_2 - 36 = -128 - 160 - 99.2$$

$$M_2 = -\frac{351.2}{40} = -8.78 \text{ kip-ft} \qquad Ans$$

Example 14.2. Determine the bending moment at the supports of the beam shown in Fig. 14.4.

Directly from the load diagram, $M_1 = -800(3) = -2400$ lb-ft and $M_4 = 0$ (*Ans*).

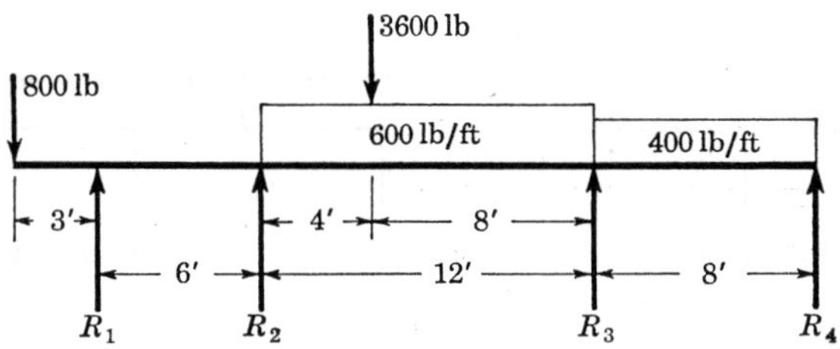

FIG. 14.4

Applying the three-moment equation to the first and second spans,

$$M_A L_1 + 2M_B(L_1 + L_2) + M_C L_2 = -\Sigma 6EI\alpha$$

$$-2400(6) + 2M_2(6 + 12) + M_3(12) = -\frac{600(12)^3}{4} - \frac{3600(8)}{12}(144 - 64)$$

$$-14{,}400 + 36M_2 + 12M_3 = -259{,}200 - 192{,}000$$

$$36M_2 + 12M_3 = -436{,}800 \qquad (a)$$

Applying the three-moment equation to the second and third spans and noting that M_A, M_B, and M_C represent M_2, M_3, and M_4, respectively,

$$M_2(12) + 2M_3(12 + 8) + 0(8) = -\frac{600(12)^3}{4} - \frac{3600(4)}{12}(144 - 16) - \frac{400(8)^3}{4}$$

$$12M_2 + 40M_3 = -259{,}200 - 153{,}600 - 51{,}200$$

$$12M_2 + 40M_3 = -464{,}000 \qquad (b)$$

Solving Eqs. (*a*) and (*b*) simultaneously,

$M_2 = -9185$ lb-ft and $M_3 = -8840$ lb-ft *Ans*

PROBLEMS

14.1–14.10. For the continuous beam loaded as shown, determine the moment at each support. (*14.1, *14.4, *14.7, *14.10)

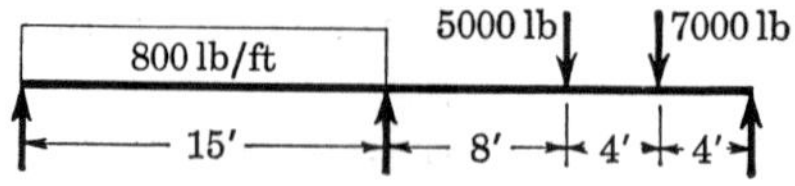

FIG. P14.1

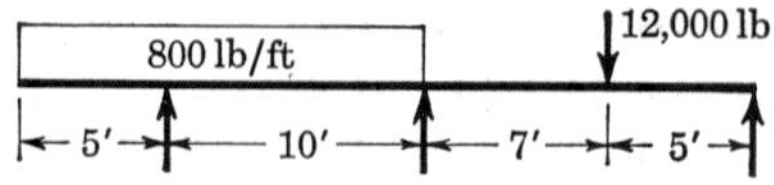

FIG. P14.2

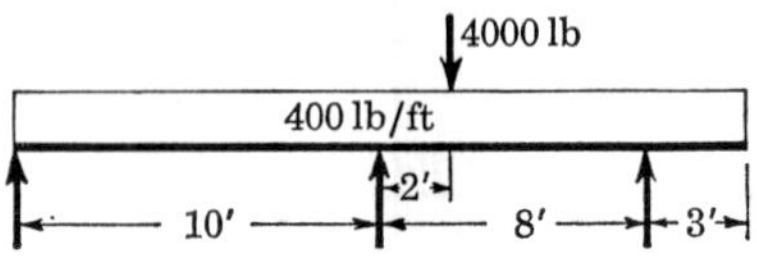

FIG. P14.3

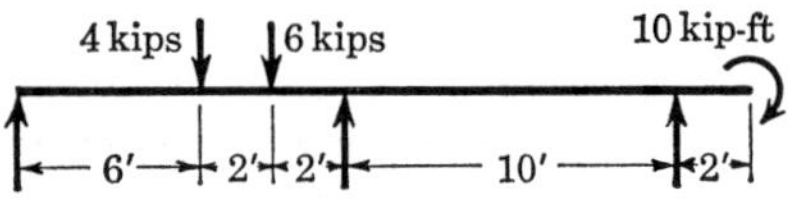

FIG. P14.4

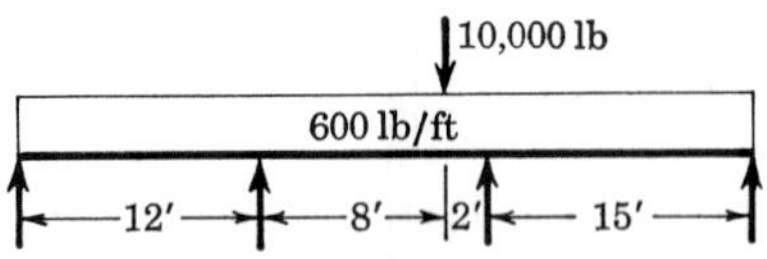

FIG. P14.5

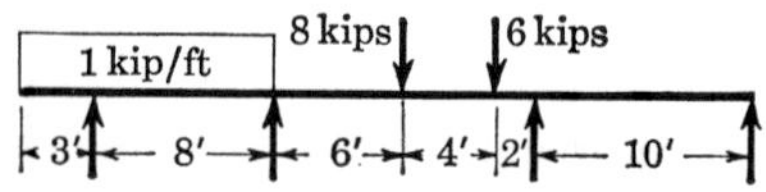

FIG. P14.6

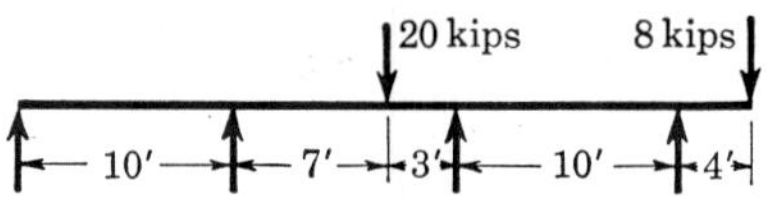

FIG. P14.7

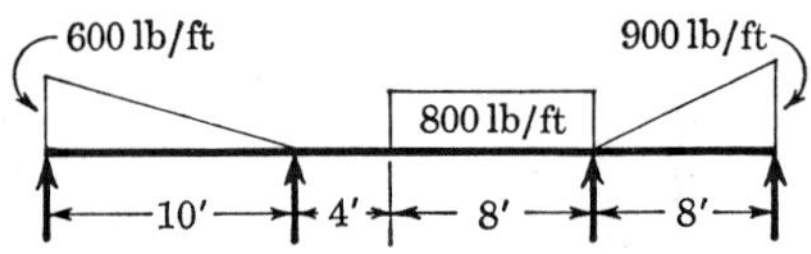

FIG. P14.8

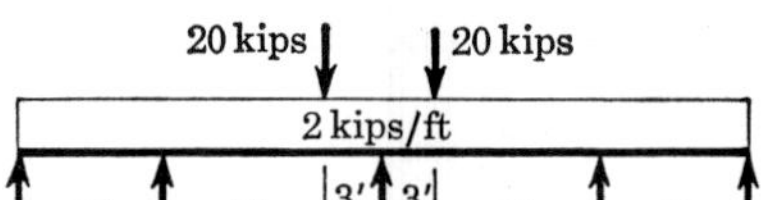

FIG. P14.9

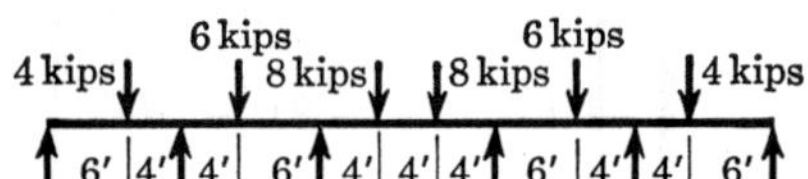

FIG. P14.10

14.3 REACTIONS

Once the moments at the supports are known, the reactions of a continuous beam can be easily determined from the free-body diagrams of the individual spans. One method of doing this will be illustrated by the example that follows.

Example 14.3. Determine the reactions of the continuous beam shown in Fig. 14.5*a*. From Example 14.2, it is known that $M_2 = -9185$ lb-ft and $M_3 = -8840$ lb-ft.

The free-body diagrams of the various spans are drawn as shown in Fig. 14.5*b*. The directions of the internal moments M_2 and M_3 are determined from the bending-moment sign convention (that is, since they are both negative, they must act in such a way as to produce tension in the top fibers of the various spans). As can be seen, R_1 and R_4 can be determined directly from the free bodies of the end spans. Once

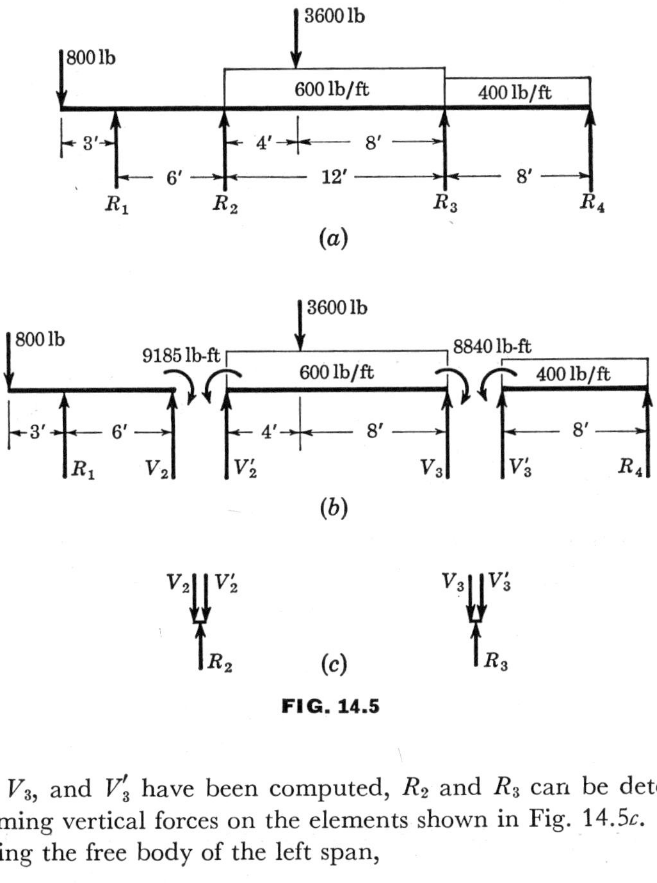

FIG. 14.5

V_2, V'_2, V_3, and V'_3 have been computed, R_2 and R_3 can be determined by summing vertical forces on the elements shown in Fig. 14.5*c*.

Using the free body of the left span,

$$\Sigma M_2 = 0 \qquad +6R_1 - 800(9) + 9185 = 0$$

$$R_1 = -330 = 330 \text{ lb}\downarrow \qquad \textit{Ans}$$

$$\Sigma M_1 = 0 \qquad +6V_2 + 800(3) - 9185 = 0$$

$$V_2 = 1130 \text{ lb}\uparrow$$

$$\Sigma F_y = 0 \qquad -800 - 330 + 1130 = 0 \qquad \textit{Check}$$

Using the free body of the middle span,

$$\Sigma M_3 = 0 \qquad +12V_2' - 600(12)(6) - 3600(8) - 9185 + 8840 = 0$$
$$V_2' = 6030 \text{ lb}\uparrow$$
$$\Sigma M_2 = 0 \qquad +12V_3 - 600(12)(6) - 3600(4) + 9185 - 8840 = 0$$
$$V_3 = 4770 \text{ lb}\uparrow$$
$$\Sigma F_y = 0 \qquad -600(12) - 3600 + 6030 + 4770 = 0 \qquad \textit{Check}$$

Using the free body of the right span,

$$\Sigma M_4 = 0 \qquad +8V_3' - 400(8)(4) - 8840 = 0$$
$$V_3' = 2705 \text{ lb}\uparrow$$
$$\Sigma M_3 = 0 \qquad +8R_4 - 400(8)(4) + 8840 = 0$$
$$R_4 = 495 \text{ lb}\uparrow \qquad \textit{Ans}$$
$$\Sigma F_y = 0 \qquad -400(8) + 2705 + 495 = 0 \qquad \textit{Check}$$

Since $\Sigma F_y = 0$ on the elements shown in Fig. 14.5*c*,

$$R_2 = V_2 + V_2' = 1130 + 6030 = 7160 \text{ lb}\uparrow \qquad \textit{Ans}$$
$$R_3 = V_3 + V_3' = 4770 + 2705 = 7475 \text{ lb}\uparrow \qquad \textit{Ans}$$

As a further arithmetic check, the sum of the vertical forces on the entire beam must be equal to zero,

$$-800 - 600(12) - 3600 - 400(8) - 330 + 7160 + 7475 + 495 = 0$$
$$-15{,}130 + 15{,}130 = 0 \qquad \textit{Check}$$

Once the principles used in Example 14.3 are understood, a slightly modified procedure for computing reactions can be employed to advantage. This procedure entails fewer and simpler computations and is less subject to arithmetic errors. By this method, the end shears for the individual spans are computed "by parts": one part due to end moments and one additional part due to each applied load. Example 14.4 illustrates this modified method of computing reactions.

Example 14.4. Determine the reactions of the beam shown in Fig. 14.6*a*. From Example 14.2, it is known that $M_2 = -9185$ lb-ft and $M_3 = -8840$ lb-ft.

The end shears by parts for each span and for the overhang are as shown in Fig. 14.6*b*. The top line shows the values of the end shears due to end moments; subsequent lines show end shears due to the individual applied loads.

The individual values of end shear can usually be listed directly without formal written computations or auxiliary diagrams. As an example, the computations for the middle span will be explained. As shown in Fig. 14.6*c*, the end shears due to the end moments must constitute a couple equal to 9185 − 8840 or 345 lb-ft; therefore, these end shears are $^{345}/_{12}$ = 29, say 30 lb (up at the left and down at the right).

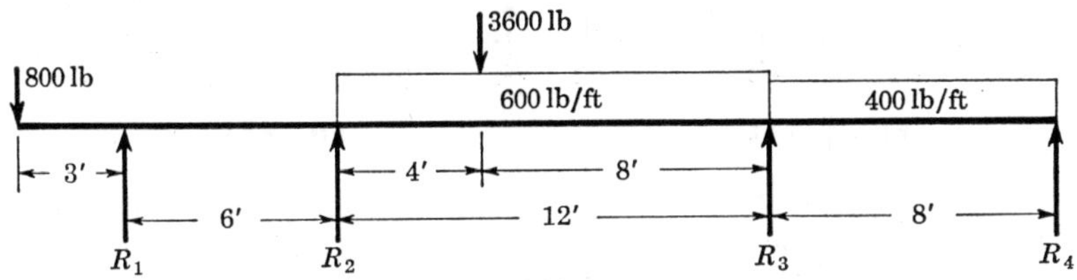

(*a*) Load diagram

	1130↓	1130↑	30↑	30↓	1105↑	1105↓
800↑			3600↑	3600↑	1600↑	1600↑
			2400↑	1200↑		
800↑	1130↓	1130↑	6030↑	4770↑	2705↑	495↑

(*b*) End shears by parts

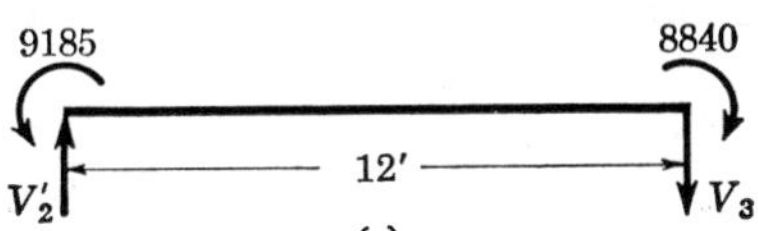

(*c*)

FIG. 14.6

It may be noted that a diagram such as Fig. 14.6*c* need not be drawn, since the directions of end moments and end shears can be easily visualized. From the load diagram, Fig. 14.6*a*, the end shears due to the uniform load are each 600(6) or 3600 lb up; the end shears due to the 3600 lb concentrated load are $(^8/_{12})(3600)$ or 2400 lb up on the left and $(^4/_{12})(3600)$ or 1200 lb up on the right.

After the end shears on each side of the reactions have been summed, the reactions are determined as

$R_1 = 800 + (-1130) = 330 \text{ lb}\downarrow$ *Ans*
$R_2 = 1130 + 6030 = 7160 \text{ lb}\uparrow$ *Ans*
$R_3 = 4770 + 2705 = 7475 \text{ lb}\uparrow$ *Ans*
$R_4 = 495 \text{ lb}\uparrow$ *Ans*

PROBLEMS

***14.11.** Determine the reactions of the continuous beam shown in Fig. P14.11. It is known that $M_2 = +300$ lb-ft and $M_3 = -1360$ lb-ft.

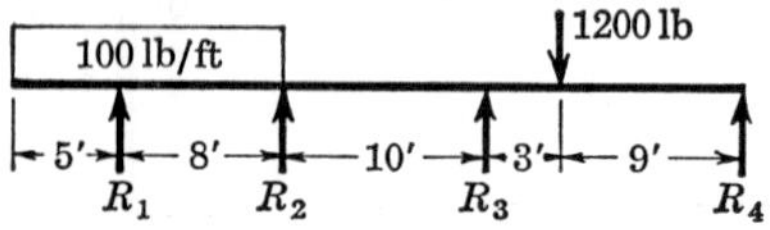

FIG. P14.11

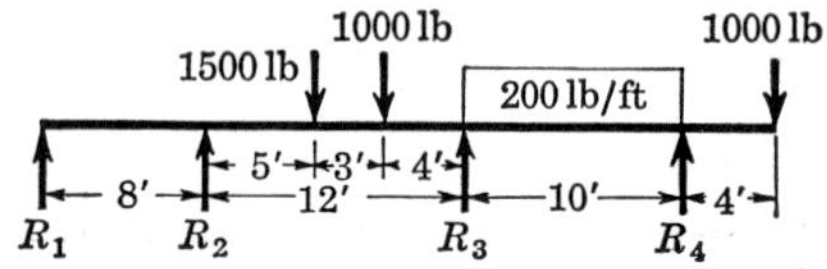

FIG. P14.12

14.12. Determine the reactions of the continuous beam shown in Fig. P14.12. It is known that $M_2 = -2400$ lb-ft and $M_3 = -2480$ lb-ft.

14.13. Determine the reactions of the continuous beam shown in Fig. P14.13. It is known that $M_2 = -9.20$ kip-ft and $M_3 = -1.56$ kip-ft.

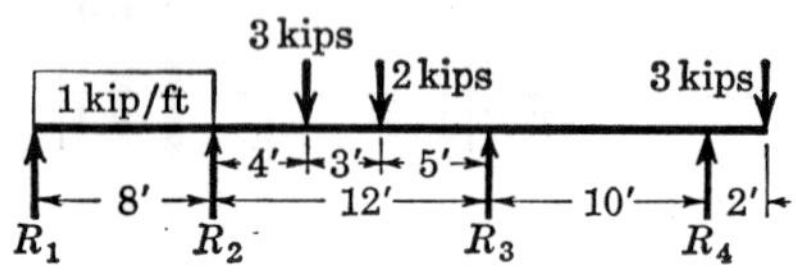

FIG. P14.13

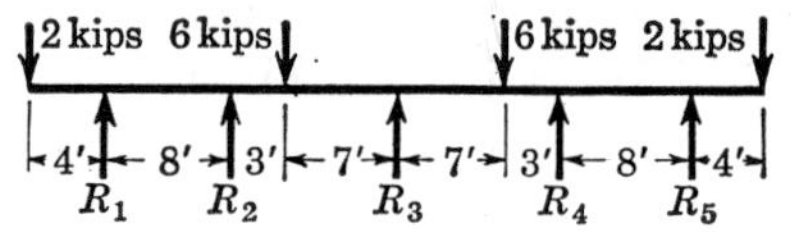

FIG. P14.14

***14.14.** For the beam shown in Fig. P14.14, show that $M_2 = M_4 = -2.20$ kip-ft and $M_3 = -7.09$ kip-ft and determine the reactions.

14.15. For the beam shown in Fig. P14.15, show that $M_2 = M_5 = -16.0$ kip-ft and $M_3 = M_4 = -20.2$ kip-ft and determine the reactions.

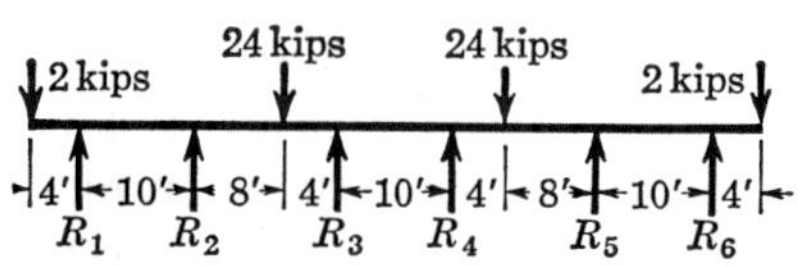

FIG. P14.15

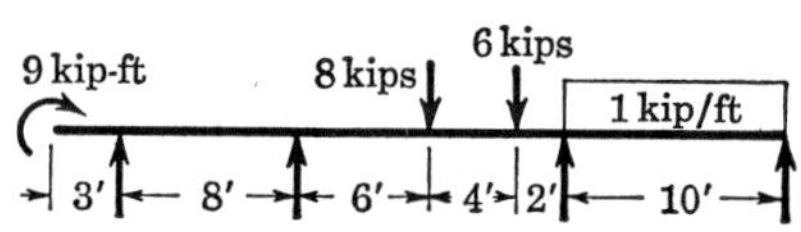

FIG. P14.17

14.16. A continuous beam having five equal spans, each of length L, supports a uniform load of w lb/ft over its entire length. Show that the moments at the interior supports are $-\frac{4}{38}wL^2$ and $-\frac{3}{38}wL^2$, and determine the reactions.

14.17–14.19. For the continuous beam shown in Figs. P14.17 to P14.19, (*a*) determine the reactions, (*b*) draw the complete shear and moment diagrams. (*14.17)

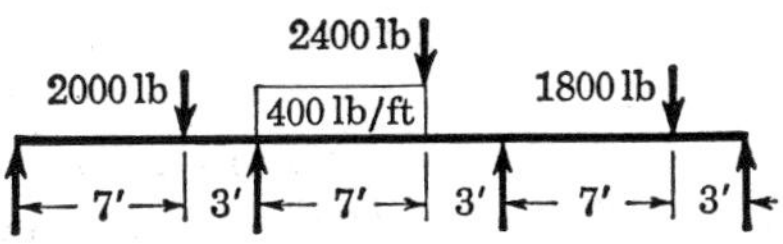

FIG. P14.18

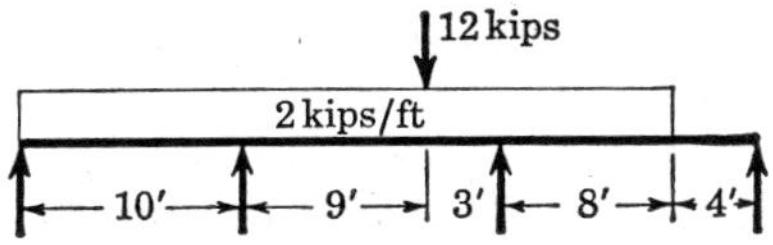

FIG. P14.19

14.4 RESTRAINED ENDS

If the ends of a continuous beam are restrained, an additional unknown moment will exist at each point of restraint; therefore, an additional equation will be required for each restrained end. If the end is completely restrained (i.e., fixed), the additional equation can be obtained from the fact that the slope of the elastic curve at the point of restraint is zero. Thus, for the portion of a continuous beam shown in Fig. 14.7*a*, the slope θ at B must equal zero. If this slope is considered as resulting from the superposition of the three load conditions used for the derivation of the three-moment equation (see Art. 14.2 and Fig. 14.2*c*), the slope of the elastic curve at the point of restraint can be written as

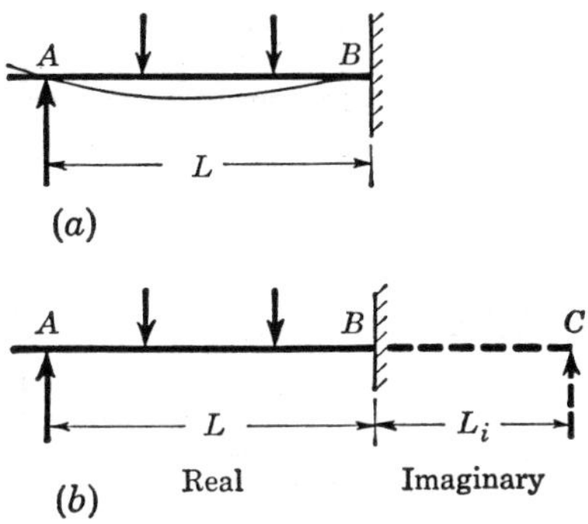

FIG. 14.7

$$\alpha + \beta + \gamma = \theta = 0$$

Therefore

$$\alpha + \frac{M_B L}{3EI} + \frac{M_A L}{6EI} = 0$$

or

$$M_A L + 2M_B L = -6EI\alpha \tag{14.2}$$

As a matter of convenience, Eq. (14.2) can be considered as the equivalent of the three-moment equation applied to one real and one imaginary span having a common support at the point of restraint. This fact can be verified by adding an imaginary span to Fig. 14.7*a* as shown in Fig. 14.7*b*, applying the three-moment equation, and setting the loading and length of the imaginary span equal to zero.

$$M_A L_1 + 2M_B(L_1 + L_2) + M_C L_2 = -6EI\alpha_1 - 6EI\alpha_2$$
$$M_A L + 2M_B(L + 0) + 0 = -6EI\alpha$$

which is identical with Eq. (14.2). Thus if imaginary spans are added at the fixed ends of continuous beams, the unknown moments can be determined using the three-moment equation exclusively.

Example 14.5. Determine the reactions of the beam shown in Fig. 14.8*a*.

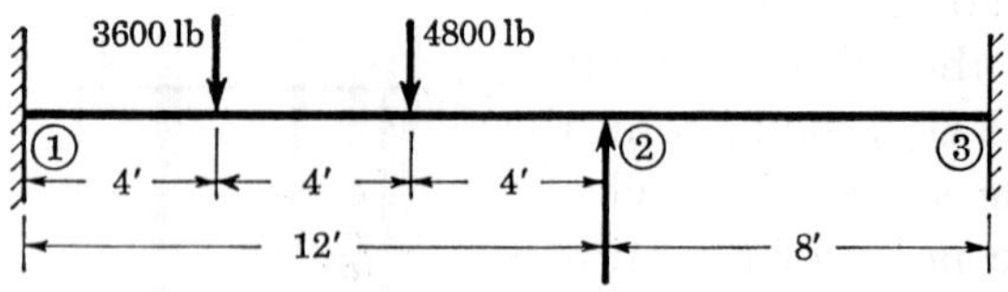

(*a*) Load diagram

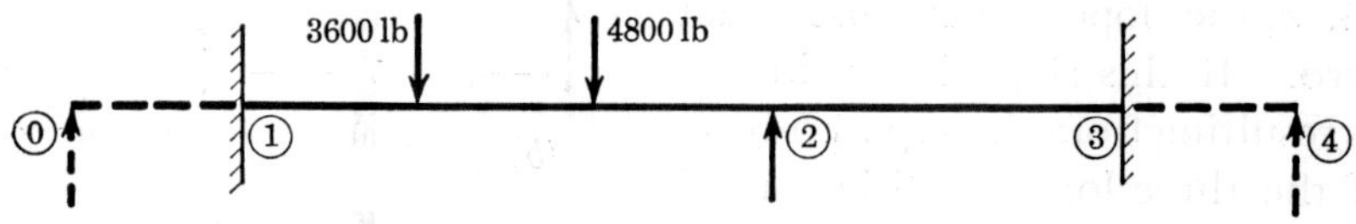

(*b*) Load diagram with imaginary spans

498↑	498↓	1320↑	1320↓
2400↑	1200↑		
1600↑	3200↑		
4498↑	3902↑	1320↑	1320↓

(*c*) End shears by parts

FIG. 14.8

The beam is considered as having imaginary spans adjacent to each fixed end, as shown in Fig. 14.8*b*. Applying the three-moment equation successively to spans 0-1 and 1-2, 1-2 and 2-3, and finally to spans 2-3 and 3-4,

$$0 + 2M_1(0 + 12) + M_2(12) = -\frac{3600(8)}{12}(144 - 64) - \frac{4800(4)}{12}(144 - 16)$$

$$24M_1 + 12M_2 = -396{,}800 \qquad (a)$$

$$M_1(12) + 2M_2(12 + 8) + M_3(8)$$

$$= -\frac{3600(4)}{12}(144 - 16) - \frac{4800(8)}{12}(144 - 64)$$

$$12M_1 + 40M_2 + 8M_3 = -409{,}600 \qquad (b)$$

$$M_2(8) + 2M_3(8 + 0) + 0 = 0$$

$$M_2 + 2M_3 = 0 \qquad (c)$$

Solving Eqs. (*a*), (*b*), and (*c*) simultaneously, the bending moments at the supports are $M_1 = -13{,}010$, $M_2 = -7040$, and $M_3 = +3520$. Therefore, the reaction moments are

$M_1 = 13{,}010$ lb-ft ↻ and $M_3 = 3520$ lb-ft ↺ *Ans*

The vertical reactions are computed from the end shears as shown in Fig. 14.8*c*.

$R_1 = 4498$ lb↑ *Ans*
$R_2 = 3902 + 1320 = 5222$ lb↑ *Ans*
$R_3 = 1320$ lb↓ *Ans*

PROBLEMS

14.20–14.25. For the beam shown in Figs. P14.20 to P14.25, (*a*) determine the moment at each support, (*b*) draw the complete shear and moment diagrams. (*14.20, *14.23)

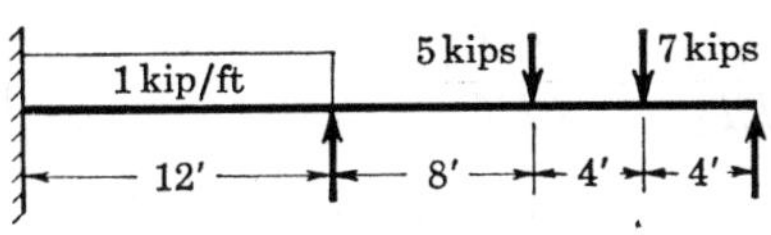

FIG. P14.20

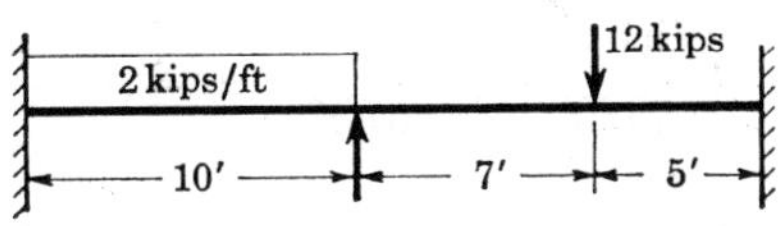

FIG. P14.21

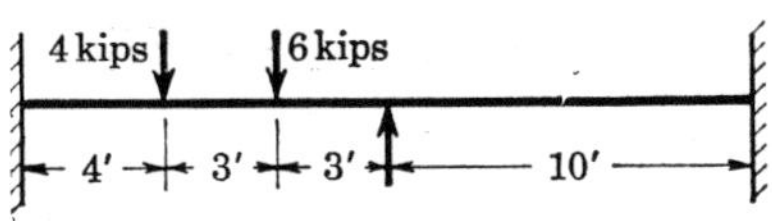

FIG. P14.22

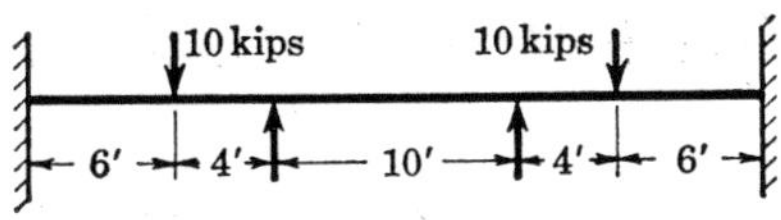

FIG. P14.23

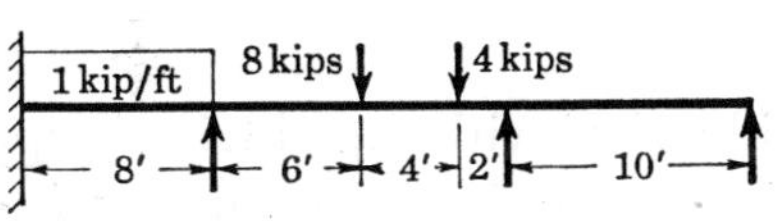

FIG. P14.24

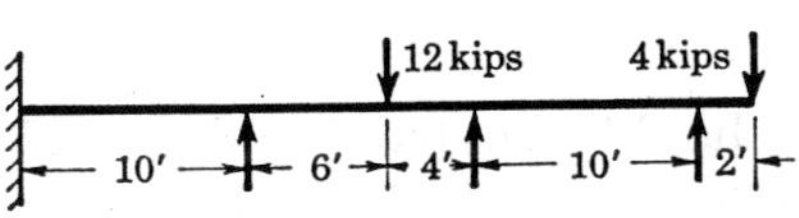

FIG. P14.25

14.26. A beam of length L is fixed at the left end and supported on rollers at the other. Using the three-moment equation, determine the reactions due to a uniform load of w lb/ft applied over the entire span.

★14.27. A 12-ft beam is fixed at both ends. It is loaded with 24 kips at 3 ft from the left end and 40 kips at 5 ft from the right end. Using the three-moment equation, determine the reactions.

14.28. A 12-ft beam is fixed at the right end and supported on rollers at the other. It is loaded with 2 kips at 3 ft from the left end and 3 kips at 4 ft from the right end. Using the three-moment equation, determine the reactions.

14.5 SPANS OF DIFFERENT CROSS SECTION

In Art. 14.2, the three-moment equation was derived for beams of constant cross section. For beams having spans with different cross sections, a somewhat different and more general version of the equation can be derived. Thus, referring to the previous derivation starting with Eq. (*a*) and letting I_1 and I_2 designate the moments of inertia of spans AB and BC, respectively,

$$\alpha_1 + \beta_1 + \gamma_1 = \alpha_2 + \beta_2 + \gamma_2$$

Paralleling the previous derivation,

$$\alpha_1 + \frac{M_B L_1}{3EI_1} + \frac{M_A L_1}{6EI_1} = -\alpha_2 - \frac{M_B L_2}{3EI_2} - \frac{M_C L_2}{6EI_2}$$

Multiplying through by $6EI_1$, the equation becomes

$$M_A L_1 + 2M_B L_1 + 2M_B L_2 \frac{I_1}{I_2} + M_C L_2 \frac{I_1}{I_2} = -6EI_1\alpha_1 - 6EI_1\alpha_2$$

or

$$M_A L_1 + 2M_B \left(L_1 + L_2 \frac{I_1}{I_2}\right) + M_C L_2 \frac{I_1}{I_2} = -6EI_1\alpha_1 - 6EI_2\alpha_2 \frac{I_1}{I_2} \qquad (14.3)$$

It should be noted that Eq. (14.3) is identical with Eq. (14.1) except for the ratio I_1/I_2 at three places; therefore, it can be used with tabular values of $6EI\alpha$, such as are listed in Table 14.1.

PROBLEMS

★14.29. Solve Prob. 14.6 if the moments of inertia for the left, middle, and right spans were 60, 120, and 40 in.^{4}, respectively.

14.30. Solve Prob. 14.9 if the moments of inertia were 300 in.^{4} for the two 12-ft spans and 100 in.^{4} for the two 8-ft spans.

14.31. Solve Prob. 14.22 if the moments of inertia were 60 in.^{4} for the left span and 40 in.^{4} for the right span.

15

COLUMNS

15.1 INTRODUCTION

Compression members which are relatively long in proportion to their cross-sectional dimensions fail because of instability rather than by fracture or overall yielding; such members are known as columns. This basic difference between columns and short compression members is illustrated in Fig. 15.1, which shows two compression members made of the same material and having identical cross sections. As indicated, the short member will fail by yielding at a load P_1, while the column will fail by becoming unstable and buckling at a substantially smaller load P_2. It has been seen that the load-carrying capacity of an axially loaded short compression member depends essentially on its cross-sectional area and on the strength of the material of which it is made. The load-carrying capacity of an axially loaded column, however, is dependent on several other (or additional) factors, namely, the geometry of the cross section, the length of the member, and the stiffness of the material. The maximum load which a column can support before becoming unstable is known as its *critical load*. Thus, for most practical purposes, the critical load is the largest load which a column can support.†

If a column becomes unstable while in a state of elastic stress, it is known as a *slender column* (or *long column*) and the condition at failure is described as *elastic instability*. Conversely, if a column becomes unstable after having been stressed beyond the proportional limit, it is known as an *intermediate column* and the condition at failure is described as *inelastic instability*. Compression members which fail by fracture or general yielding without becoming unstable are known as *posts* (or simply as *short compression members*).

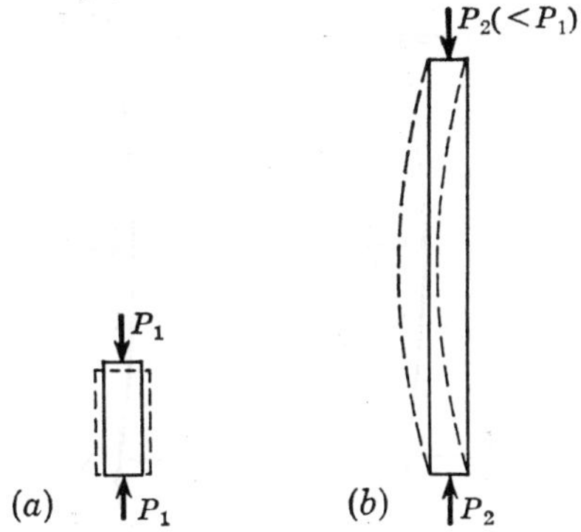

FIG. 15.1 Failure of short compression member and column of identical cross section and material.

15.2 SLENDER COLUMNS; EULER'S EQUATION

The understanding of column instability and the meaning of critical load can be furthered by the consideration of a fictitious member known as an ideal slender column. As the name implies, such a member is perfectly homogeneous, perfectly straight, and free to rotate in any direction at the ends. Referring to the round ideal slender column shown in Fig. 15.2*a* supporting an axial load less than the critical, a small transverse force F has been applied, causing a small deflection. Since the column represents a stable load condition, if the transverse force were withdrawn, the column would straighten out. However, if the column were supporting its critical

† Of academic interest is the fact that a column may be able to support in the unstable condition a load much larger than its critical. This is possible only if the column is slender enough so that its critical load causes stresses well below the yield point (strength) of the material. See S. P. Timoshenko, "Theory of Elastic Stability," pp. 69–75, McGraw-Hill Book Company, Inc., New York, 1936.

load P_{cr} and the same force F was applied (Fig. 15.2*b*) and then withdrawn, the column would retain its bent shape as indicated

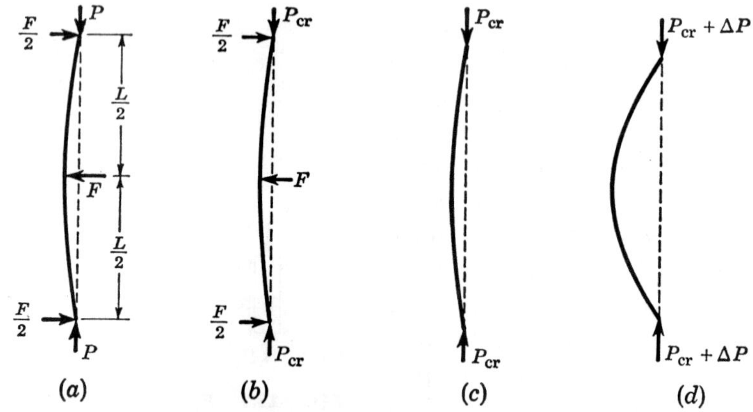

FIG. 15.2

in Fig. 15.2*c*. From this condition, a transverse force in the opposite direction could be used (depending on the time of application) either to straighten the column out or to give it a slight deflection in the opposite direction. The column would also straighten out from the condition in Fig. 15.2*c* if the axial load were decreased to any value less than the critical. Conversely, if the load were increased even slightly, the bending of the column would increase a disproportionate amount. The column would then be in an unstable condition in that any additional axial load or any slight lateral load would cause it to bend excessively as shown in Fig. 15.2*d*. This excessive bending deformation, resulting from column instability, is known as buckling.

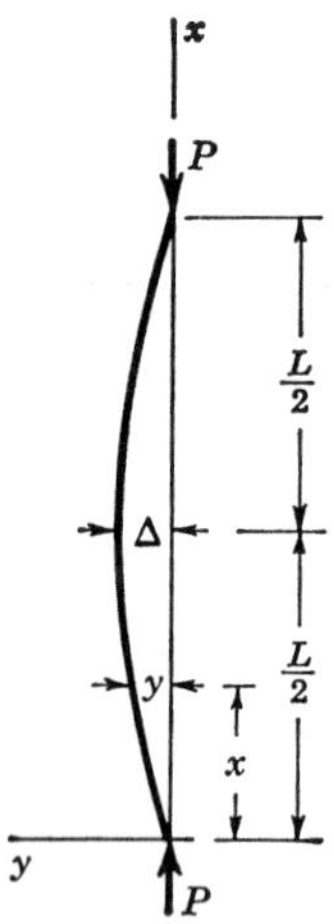

FIG. 15.3

The critical load for slender columns is defined by Euler's equation,† the derivation of which is based on the elastic curve of an ideal slender column. The column is given a small midlength deflection Δ, so that its elastic curve under the critical load P is as shown in Fig. 15.3.

† Named after the Swiss mathematician Leonard Euler (1707–1783) who first derived it in a paper published in 1757.

The x and y axes are taken as shown as a matter of convenience in using the equation of the elastic curve. The equation of the elastic curve is

$$EI \frac{d(dy/dx)}{dx} = -Py$$

Multiplying both sides of the equation by $2dy$ in order to obtain a convenient form for integration,

$$EI\,2\,\frac{dy}{dx}\,d\frac{dy}{dx} = -2Py\,dy$$

Integrating,

$$EI\left(\frac{dy}{dx}\right)^2 = -Py^2 + C_1$$

Evaluating C_1: at $y = \Delta$, $dy/dx = 0$. Therefore

$$C_1 = +P\Delta^2 \qquad \text{and}$$

$$EI\left(\frac{dy}{dx}\right)^2 = -Py^2 + P\Delta^2 \qquad \text{or}$$

$$\left(\frac{dy}{dx}\right)^2 = \frac{P}{EI}(\Delta^2 - y^2)$$

Taking the square root of both sides and separating the variables,

$$\frac{dy}{dx} = \sqrt{\frac{P}{EI}}\sqrt{\Delta^2 - y^2} \qquad \text{and}$$

$$\frac{dy}{\sqrt{\Delta^2 - y^2}} = \sqrt{\frac{P}{EI}}\,dx$$

Integrating a second time,

$$\arcsin\frac{y}{\Delta} = \sqrt{\frac{P}{EI}}\,x + C_2$$

Evaluating C_2: at $x = 0$, $y = 0$. Therefore, $C_2 = 0$. Thus, under the critical load, the column is bent into the shape of the sine curve defined by the equation

$$\arcsin\frac{y}{\Delta} = \sqrt{\frac{P}{EI}}\,x$$

Since $y = \Delta$ at $x = L/2$,

$$\frac{\pi}{2} = \sqrt{\frac{P}{EI}}\frac{L}{2}$$

Therefore the critical load of a slender column is defined by the equation

$$P_{cr} = \frac{\pi^2 EI}{L^2} \tag{15.1a}$$

Thus the critical load as defined by Euler's equation is a function of the stiffness of the material E, the stiffness of the section I, and the length of the member L. It is not dependent on the strength of the material. Since Euler's equation is based on elastic properties, it should be emphasized that it is valid only if the proportional limit of the material is not exceeded. Also, since a column will buckle about the weakest axis of its section, I in the equation represents the least moment of inertia of the cross-sectional area.

A more useful form of Euler's equation is obtained by expressing I as Ak^2 (where k is known as the *radius of gyration*) and rearranging terms. Thus

$$P_{cr} = \frac{\pi^2 E(Ak^2)}{L^2} \qquad \text{or}$$

$$\left(\frac{P}{A}\right)_{cr} = \frac{\pi^2 E}{(L/k)^2} \tag{15.1b}$$

The term $(P/A)_{cr}$ represents the *average stress* in the column under the critical load. Although it is not a maximum stress, it is referred to as the *critical stress* and is sometimes designated as s_{cr}. The term L/k is known as the slenderness ratio; it serves as a measure of compression member flexibility. For example, compression members are normally classified as slender columns, intermediate columns, or posts on a slenderness-ratio basis.

For any particular material, Euler's equation can be represented by a plot of critical stress $(P/A)_{cr}$ versus the slenderness ratio. Such plots are shown in Fig. 15.4 for structural steel and for a structural aluminum alloy. Since the relationship among the variables is dependent on Young's modulus, the curves are not

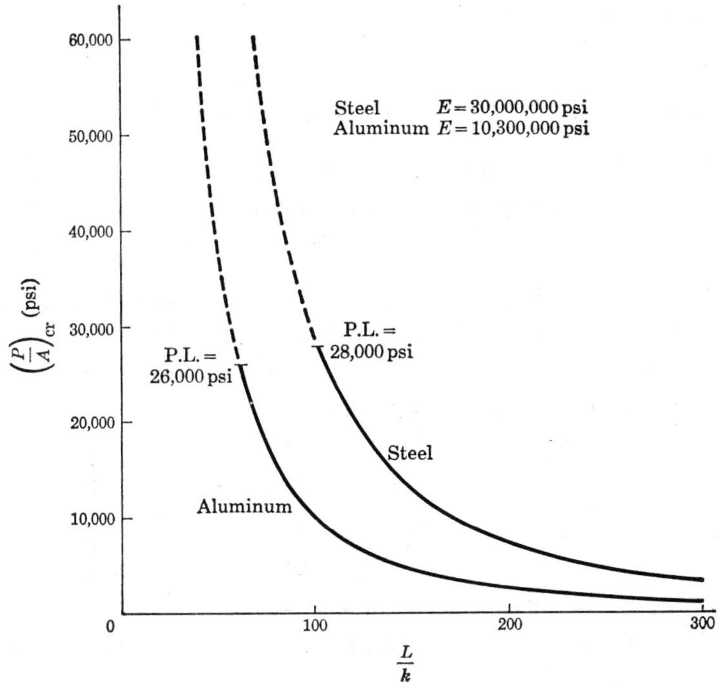

FIG. 15.4 Plot of Euler's equation.

valid for values of $(P/A)_{cr}$ that exceed the proportional limit. Thus, the invalid portion of each curve is represented by a dotted line.

PROBLEMS

15.1. A certain alloy steel has a proportional limit of 60,000 psi and a Young's modulus of 30,000,000 psi. For this material, plot a curve of critical stress versus slenderness ratio for all values for which Euler's equation applies up to an L/k of 300.

15.2. A certain magnesium alloy has a proportional limit of 30,000 psi and a Young's modulus of 6,500,000 psi. For this material, plot a curve of critical stress versus slenderness ratio for all values for which Euler's equation applies up to an L/k of 300.

15.3. A certain beryllium copper has a proportional limit of 120,000 psi and a Young's modulus of 18,000,000 psi. For this material, plot a curve of critical stress versus slenderness ratio for all values for which Euler's equation applies up to an L/k of 300.

★**15.4.** A 1-in.-diam steel bar (proportional limit = 30,000 psi; E = 30,000,000 psi) is to act as a pin-connected column. Determine its critical load if it is (*a*) 2.5 ft long, (*b*) 5 ft long.

15.5. A 1- by 3-in. aluminum bar (proportional limit = 35,000 psi; E = 10,000,000 psi) is to act as a pin-connected column. Determine its critical load if it is (*a*) 2 ft long, (*b*) 4 ft long.

15.6. An 8WF31 section is to act as a pin-connected column. Based on a factor of safety of 2.5, determine its allowable load if it is (*a*) 20 ft long, (*b*) 30 ft long.

15.7. A 20-ft pin-connected timber column (proportional limit = 4000 psi; E = 1,600,000 psi) has the cross section shown in Fig. P15.7. Determine its allowable load based on a factor of safety of 3.

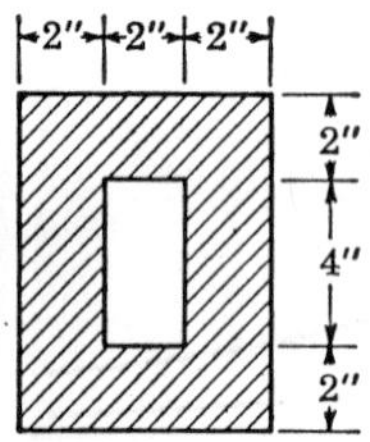

FIG. P15.7

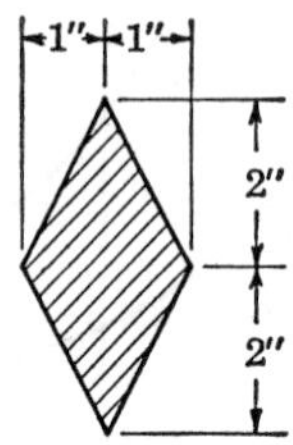

FIG. P15.8

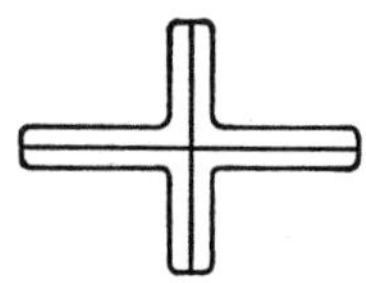

FIG. P15.9

15.8. A 5-ft pin-connected aluminum column (proportional limit = 35,000 psi; E = 10,000,000 psi) has the cross section shown in Fig. P15.8. Determine its allowable load based on a factor of safety of 2.

★**15.9.** A 15-ft pin-connected column is made up of four 4- by 3- by ½-in. angles arranged as shown in Fig. P15.9. (*a*) Determine its critical load; (*b*) determine the critical load of one angle acting separately.

15.3 EFFECTIVE LENGTH

Since Euler's equation was derived for a column having ends which were prevented from translating but which were free to rotate, L in the equation represents the length of column between points of zero moment. Therefore, for columns having different end conditions, the length used in Euler's equation should also be equal to the distance between points of zero moment. This distance is known as the *effective length*. The effective lengths, designated as L_e, for columns having several different combinations of end supports are shown in Fig. 15.5. As shown, a column of length L which is fixed at both ends will have inflection points at $L/4$ from each support; its effective length therefore will be the middle one-half of its actual length. As is also shown, a column which is pin connected at one end and fixed at the other will have an effective length very nearly equal to 0.7 its actual length. If a column is fixed at one end and completely free at the other, its elastic curve will be one-half that of a pin-connected column of twice its length; therefore, the effective length of such a column will be twice its actual length. Thus, it should be apparent that the manner in

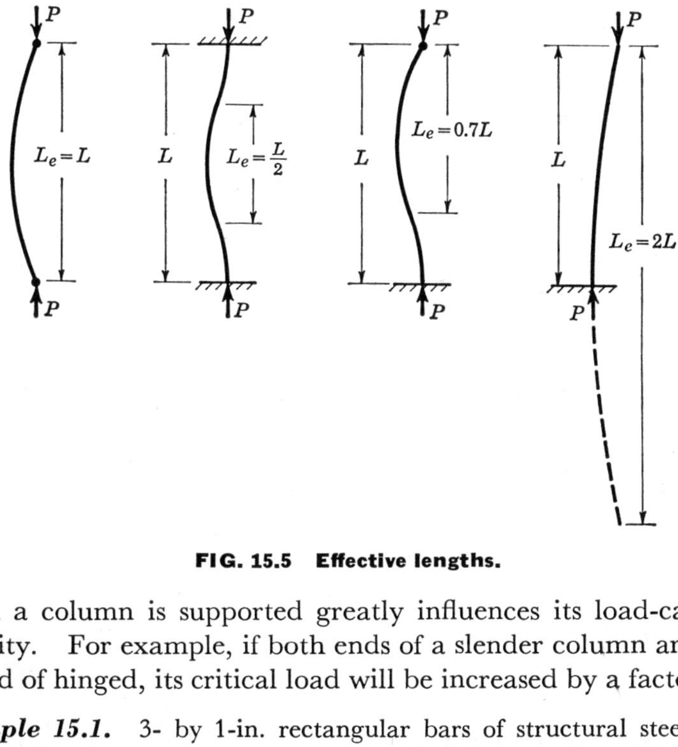

FIG. 15.5 Effective lengths.

which a column is supported greatly influences its load-carrying capacity. For example, if both ends of a slender column are fixed instead of hinged, its critical load will be increased by a factor of 4.

Example 15.1. 3- by 1-in. rectangular bars of structural steel (E = 30,000,000 psi, proportional limit = 28,000 psi) are to be used as compression members having the following lengths and end supports:

(*a*) 5 ft, pinned at each end
(*b*) 4 ft, fixed at each end
(*c*) 3 ft, fixed at one end and completely free at the other

Determine the critical load for the members to which Euler's equation applies.

Computing the least radius of gyration of the section,

$$k = \sqrt{\frac{I}{A}} = \sqrt{\frac{bd^3/12}{bd}} = \frac{d}{\sqrt{12}} = \frac{1}{\sqrt{12}} = 0.289 \text{ in.}$$

(*a*) For the 5-ft column pinned at both ends,

$$\frac{L}{k} = \frac{5(12)}{0.289} = 208$$

$$\left(\frac{P}{A}\right)_{cr} = \frac{\pi^2 E}{(L/k)^2} = \frac{9.87(30{,}000{,}000)}{(208)^2} = 6840 \text{ psi}$$

Since $(P/A)_{cr}$ is less than the proportional limit, Euler's equation applies and

$$P_{cr} = \left(\frac{P}{A}\right)_{cr} A = 6840(3) = 20{,}520 \text{ lb} \qquad \textit{Ans}$$

(*b*) For the 4-ft column fixed at both ends,

$$\frac{L}{k} = \frac{0.5(4)(12)}{0.289} = 83$$

$$\left(\frac{P}{A}\right)_{cr} = \frac{9.87(30{,}000{,}000)}{(83)^2} = 43{,}000 \text{ psi}$$

Since the computed value of $(P/A)_{cr}$ is greater than the proportional limit, it is invalid and Euler's equation does not apply (*Ans*).

(*c*) For the 3-ft column fixed at one end and free at the other,

$$\frac{L}{k} = \frac{2(3)(12)}{0.289} = 249$$

$$\left(\frac{P}{A}\right)_{cr} = \frac{9.87(30{,}000{,}000)}{(249)^2} = 4780 \text{ psi}$$

Therefore, since $(P/A)_{cr}$ is less than the proportional limit,

$$P_{cr} = 4780(3) = 14{,}340 \text{ lb} \qquad \textit{Ans}$$

Example 15.2. An aluminum ($E = 10{,}000{,}000$ psi; proportional limit $= 40{,}000$ psi) column 2 by 3 in. in cross section and 10 ft long is fixed at both ends. In order to increase its load-carrying capacity, it is braced at midlength in the weak direction as shown in Fig. 15.6*a*. Determine the critical load.

As shown in Fig. 15.6*b*, the effective lengths with respect to the principal axes are different. Therefore, the slenderness ratio must be computed for each case and the larger value used to compute the critical load.

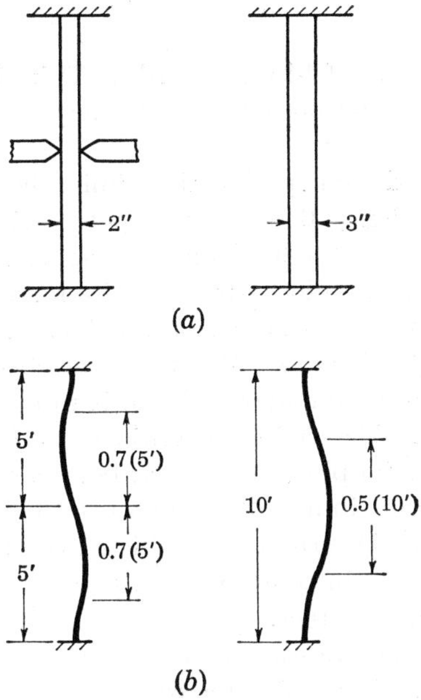

FIG. 15.6

With respect to the minor principal axis, $k = 2/\sqrt{12} = 0.577$ in. The corresponding effective length is $0.7(5)(12) = 42$ in., and

$$\frac{L}{k} = \frac{42}{0.577} = 72.8$$

With respect to the major principal axis, $k = 3/\sqrt{12} = 0.866$ in. The corresponding effective length is $0.5(10)(12) = 60$ in., and

$$\frac{L}{k} = \frac{60}{0.866} = 69.3$$

Using the larger value of L/k,

$$\left(\frac{P}{A}\right)_{cr} = \frac{\pi^2 E}{(L/k)^2} = \frac{9.87(10{,}000{,}000)}{(72.8)^2} = 18{,}620 \text{ psi}$$

$$P = 18{,}620(6) = 111{,}700 \text{ lb} \qquad \textit{Ans}$$

PROBLEMS

NOTE: For all problems of this group, use the following values for proportional limit and Young's modulus: structural steel: $s_{pl} = 30{,}000$ psi, $E = 30{,}000{,}000$ psi; aluminum alloy: $s_{pl} = 35{,}000$ psi, $E = 10{,}000{,}000$ psi; magnesium alloy: $s_{pl} = 25{,}000$ psi, $E = 6{,}500{,}000$ psi.

★15.10. A 30-ft length of 4-in. standard steel pipe is to act as a column fixed at both ends. Based on a factor of safety of 2.25, determine the allowable load.

15.11. A certain design specification limits the slenderness ratio of any member to 350 or less. Based on this specification, select the lightest standard steel pipe to act as a 40-ft outdoor flagpole.

15.12. An aluminum-alloy bar 2 by ½ in. in cross section was tested as a column with partially restrained ends. The bar buckled elastically at a load of 22,000 lb. What was the effective length?

★15.13. A 12-ft aluminum-alloy column is to have a 3- by 5-in. rectangular cross section. Determine the critical load if the column is to be (*a*) fixed at both ends, (*b*) pin connected at both ends.

15.14. A 5-ft magnesium-alloy column is to have a 1.5- by 2-in rectangular cross section. Determine the critical load if the column is to be (*a*) fixed at one end and pin connected at the other, (*b*) fixed at one end and completely free at the other.

15.15. A 10-ft structural-steel column is to have a 2- by 3-in. rectangular cross section. Determine the critical load if the column is to be (*a*) fixed at both ends, (*b*) fixed at one end and pin connected at the other.

★15.16. A 10WF49 section is used as a 40-ft pin-end column. In order to increase its load-carrying capacity, it is braced in the weak direction (only) at 24 ft from one end. Based on a factor of safety of 2.25, determine the allowable load.

15.17. A 12WF40 section is used as a 50-ft column fixed at the bottom and pin connected at the top. In order to increase its load-carrying capacity, it is braced in the weak direction (only) at 30 ft from the bottom. Based on a factor of safety of 2.25, determine the allowable load.

15.18. A 5-in. standard steel pipe is to be used as a 30-ft column fixed at the bottom and completely free at the top. It is braced (in all directions) at a point 10 ft from the top. Based on a factor of safety of 2.25, determine the allowable load.

15.4 INTERMEDIATE COLUMNS; TANGENT-MODULUS EQUATION

As previously stated, an intermediate column is one which buckles from a condition of inelastic stress. The critical load of an intermediate column is defined by a modification of Euler's equation known as the *Euler-Engesser*† or *tangent-modulus equation.* Since in the plastic range the stress-strain relationship is nonlinear, the modification is accomplished by replacing Young's modulus in Euler's equation with a variable modulus E_t. The equation thus becomes

$$\left(\frac{P}{A}\right)_{cr} = \frac{\pi^2 E_t}{(L/k)^2} \tag{15.2}$$

where E_t is defined as $ds/d\epsilon$, the slope of the stress-strain curve for the particular critical stress in the plastic range. Since E_t is a variable, Eq. (15.2) cannot be solved directly. However, if the stress-strain curve of the material is known, values of E_t for corresponding values of stress can be determined from the slopes of the curve. The equation can then be represented graphically and used in this form. The variable modulus E_t is known as the *tangent modulus* because it is the slope of the tangent to the stress-strain curve for any particular value of stress.

For any particular material, the curve representing the plot of the tangent-modulus equation will merge with the curve representing Euler's equation. The continuous curve will thus define the critical load for the complete range of slenderness ratios. Such a continuous curve for structural steel is shown in Fig. 15.7; also indicated is the classification of compression members of this material on an L/k basis. Thus for structural steel, slender columns are members having slenderness ratios of about 100 or more while intermediate columns have slenderness ratios in the range of about 30 to 100. Compression members of any material with slenderness ratios of 30 or less are normally assumed to fail without buckling and are classified as posts.

The tangent-modulus equation is seldom used for practical applications because of the difficulty in obtaining sufficiently precise values of E_t. Most intermediate columns are designed and checked on the basis of empirical formulas.

† First proposed in 1889 by the German engineer Friedrich Engesser (1848–1931). The equation was rejected by his peers on incorrect grounds until 1946, when the American engineer F. R. Shanley provided an adequate and satisfactory explanation of the theory related to the equation. Thus it is seen that the authorities in a particular field can be wrong and can be so for a considerable length of time.

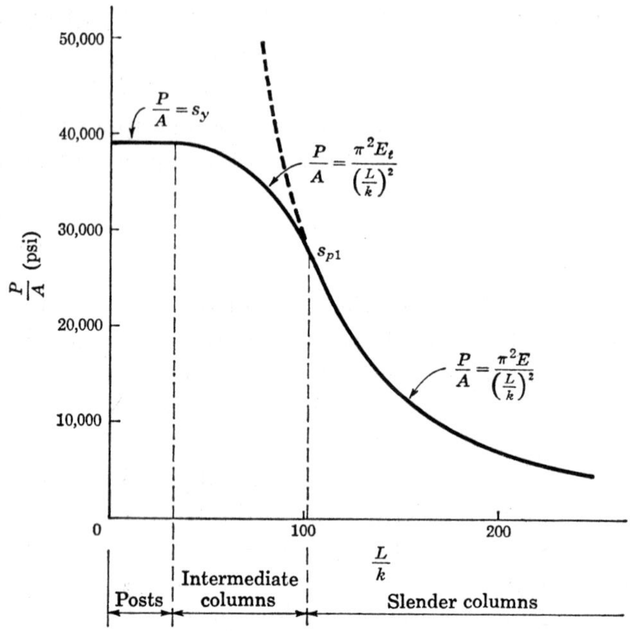

FIG. 15.7

15.5 EMPIRICAL FORMULAS

Many design specifications require the use of column formulas which are essentially empirical. These formulas are normally based on the testing to failure of a large number of specimen columns. The results of the tests are represented by a plot of corresponding values of critical stress $(P/A)_{cr}$ versus slenderness ratio L/k. The formulas are simply the equations of the smooth curves which best fit the test results. Since the $(P/A)_{cr}$ versus L/k relationship for any material is normally best represented by a *series* of curves, any empirical formula will be valid only for a particular range of slenderness ratios. Thus empirical equations have both upper and lower limits of L/k for which they are applicable. It is essential that these limits be carefully observed. Most empirical formulas fall into the three categories discussed in separate paragraphs below.

Straight-line Formulas. These formulas are based on the assumption that, for a particular range of slenderness ratios, the relationship between $(P/A)_{cr}$ and L/k is linear. They can be expressed as equations of the form

$$\left(\frac{P}{A}\right)_{cr} = C_1 - C_2\frac{L}{k} \tag{15.3}$$

where C_1 and C_2 are constants obtained from the plot of test data.

Parabolic Formulas. These formulas are based on the assumption that, for a particular range of slenderness ratios, a parabolic relationship exists between $(P/A)_{cr}$ and L/k. They can be expressed as equations of the form

$$\left(\frac{P}{A}\right)_{cr} = C_1 - C_2\left(\frac{L}{k}\right)^2 \tag{15.4}$$

C_1 and C_2 are constants selected to make the formula for a particular material conform to the plot of test data.

Gordon-Rankine Formulas. This group of formulas can be expressed as equations of the form

$$\left(\frac{P}{A}\right)_{cr} = \frac{C_1}{1 + C_2(L/k)^2} \tag{15.5}$$

C_1 and C_2 are constants selected to make the formula for a particular material conform to the plot of test data.†

In order to cover the desired range of slenderness ratios, a typical design specification will normally include two or more column formulas. Straight-line and parabolic formulas are normally used for intermediate columns, while the use of the Gordon-Rankine formula is normally restricted to the slender-column range. A specification may also require the use of Euler's equation for the high range of (L/k)'s and one or more empirical formulas for the lower range. Many column formulas in specifications include a factor of safety and thus define the average allowable stress P/A rather than the critical stress $(P/A)_{cr}$.

† The Gordon-Rankine formula was originally derived as a theoretical equation for the critical stress of slender columns. However, since the theory was based on a fallacious assumption, it is now generally classified and used as an empirical formula.

15.6 STRUCTURAL-STEEL COLUMNS

For structural-steel building construction, the most commonly used column formulas are those of the American Institute of Steel Construction (AISC).† The specifications of this organization include a series of three column formulas applicable to secondary members (such as are used for bracing) as well as main members. The formulas include a factor of safety and thus define the allowable average stress P/A. These equations together with their limits are as follows:

For main and secondary members having (L/k)'s from 0 to 120,

$$\frac{P}{A} = 17{,}000 - 0.485\left(\frac{L}{k}\right)^2 \tag{15.6}$$

For secondary members having (L/k)'s from 121 to 200,

$$\frac{P}{A} = \frac{18{,}000}{1 + [(L/k)^2/18{,}000]} \tag{15.7}$$

For main members not subject to shock loading or vibration and having (L/k)'s from 121 to 200,

$$\frac{P}{A} = \frac{18{,}000}{1 + [(L/k)^2/18{,}000]}\left(1.6 - \frac{L/k}{200}\right) \tag{15.8}$$

The specifications prohibit the use of any compression member having a slenderness ratio greater than 200. They also specify that main members which are subject to vibration or shock loading shall not have a slenderness ratio greater than 120. It is further specified that the length L used for the above equations shall be the overall unbraced column length.

Example 15.3. A 12WF53 is to be used as a 36-ft main column not subject to vibration or shock loading. Using AISC specifications determine the allowable load if the column is to be (*a*) supported only at the ends, (*b*) supported at the ends and fully braced at midlength.

From Table A.4, a 12WF53 has an area of 15.59 in.^{2} and a least radius of gyration of 2.48 in.

† For bridges and related structures, the column formulas which are normally used are those of the American Association of State Highway Officials (AASHO) and the American Railway Engineering Association (AREA).

(*a*) If the column is to be supported only at the ends,

$$\frac{L}{k} = \frac{36(12)}{2.48} = 174$$

Therefore, using Eq. (15.8),

$$\frac{P}{A} = \frac{18{,}000}{1 + [(174)^2/18{,}000]}\left(1.6 - \frac{174}{200}\right) = 4900 \text{ psi}$$

$$P = 4900(15.59) = 76{,}400 \text{ lb} \qquad \textit{Ans}$$

(*b*) If the column is to be braced at midlength,

$$\frac{L}{k} = \frac{18(12)}{2.48} = 87$$

Therefore, using Eq. (15.6),

$$\frac{P}{A} = 17{,}000 - 0.485(87)^2 = 13{,}330 \text{ psi}$$

$$P = 13{,}330(15.59) = 208{,}000 \text{ lb} \qquad \textit{Ans}$$

PROBLEMS

NOTE: Use AISC specifications for all problems of this group; assume no vibration or shock loading.

★**15.19.** Determine the allowable load for a 4- by 3- by ⅜-in. angle used as a 9-ft secondary compression member.

15.20. An 8WF31 is to be used as a 30-ft secondary compression member. Determine the allowable load if it is (*a*) supported at the ends only, (*b*) supported at the ends and braced at midlength.

15.21. Determine the allowable load for a 10I25.4 section used as a 15-ft secondary compression member.

★**15.22.** A 12WF190 is to be used as a 30-ft main column. Determine the allowable load.

15.23. A 12WF120 is to be used as a 40-ft main column. Determine the allowable load.

15.24. A 10WF49 is to be used as a 38-ft main column. Determine the allowable load if it is (*a*) supported at the ends only, (*b*) supported at the ends and fully braced at midlength.

★**15.25.** A 14WF74 is to be used as a 39-ft main column. Determine the allowable load if it is (*a*) supported at the ends only, (*b*) supported at the ends and fully braced at 13 ft from each end.

15.26. A 14WF61 is to be used as a 28-ft main column. Determine the allowable load if it is (*a*) supported at the ends only, (*b*) supported at the ends and braced in the weak direction (only) at midlength.

15.27. A 12WF85 is to be used as a 36-ft main column. Determine the allowable load if it is (*a*) supported at the ends only, (*b*) supported at the ends and braced in the weak direction (only) at midlength.

15.7 ALUMINUM COLUMNS

Aluminum columns are most often designed on the basis of formulas recommended by the manufacturers of the material. Since aluminum is so often used in aircraft structures, where the factors of safety are not standardized, the formulas normally define the critical rather than the allowable stress. The factor of safety to be assigned is thus left to the judgment of the engineer in charge. For nonaircraft structures a factor of safety of 2.5 is frequently used with aluminum columns.

One of the aluminum alloys most commonly used for structures is designated as 2014-T6. For this alloy the following column formula is recommended.†

For members having (L/k)'s from 0 to 62

$$\left(\frac{P}{A}\right)_{cr} = 78{,}000 - 840\,\frac{L}{k} \tag{15.9}$$

In this equation L represents the effective length of the column.

For columns having (L/k)'s greater than 62, Euler's equation (with an E of 10,300,000 psi) is recommended.

PROBLEMS

***15.28.** A 2014-T6 aluminum compression member is to have a 1- by 2-in. rectangular cross section and a length of 2 ft. Based on a factor of safety of 1.5, determine the allowable load if it is (*a*) fixed at both ends, (*b*) fixed at one end and completely free at the other.

15.29. A 2014-T6 aluminum compression member is to be made of an 8-ft length of pipe having inside and outside diameters of 2.5 and 3.0 in.

† "Aluminum Structural Design," Reynolds Metals Company, Louisville, Ky., 1951.

Determine the critical load if the column is (*a*) pin connected at the ends, (*b*) pin connected at the ends and fully braced at 5 ft from one end.

***15.30.** A 2014-T6 aluminum pipe having inside and outside diameters of 3.25 and 4.00 in. is to be used as a pin-end compression member. Based on a factor of safety of 2.5, determine the maximum allowable length of the member if it is to support a load of (*a*) 72 kips, (*b*) 36 kips.

15.31. A 2014-T6 aluminum pin-end compression member is to be made of a 4- by 3- by ½-in. angle (having the same dimensions as the same size structural-steel angle). Based on a factor of safety of 2.5, determine the allowable load if the member is (*a*) 3 ft long, (*b*) 5 ft long.

15.8 TIMBER COLUMNS

The most commonly used formulas for timber columns are those developed by the United States Forest Products Laboratory.† Since most timber columns are rectangular in cross section, the formulas are expressed in terms of L/b, the ratio of the unsupported length to the least transverse dimension (instead of in terms of L/k). The formulas are applicable to any type of wood for which the constants designated as c, K, and E have been determined. In the equations, c designates the allowable compressive stress for posts, K represents the minimum L/b ratio for which Euler's equation applies, and E represents Young's modulus. The formulas include a factor of safety (of 3) and therefore define the allowable average compressive stress. Values of c, K, and E for several of the most commonly used structural timbers (in the dry condition) are listed in Table 15.1.

Members having (L/b)'s of 11 or less are considered as posts governed by

$$\frac{P}{A} = c \tag{15.10}$$

Members having (L/b)'s between 11 and K are considered as intermediate columns governed by

$$\frac{P}{A} = c\left[1 - \frac{1}{3}\left(\frac{L/b}{K}\right)^4\right] \tag{15.11}$$

† "Wood Handbook," Agriculture Handbook 72, prepared by the Forest Products Laboratory, Superintendent of Documents, Government Printing Office, Washington, D.C., 1955.

TABLE 15.1 Properties for Use with Column Formulas of the Forest Products Laboratory
(For normal loading conditions of seasoned lumber in the dry condition)

Species (typical structural grade)	c	E	K†
Douglas fir, coast type	1450	1,600,000	21.3
Southern pine	1450	1,600,000	21.3
Oak, red or white	1350	1,500,000	21.3
White pine	1000	1,000,000	20.2
Hemlock, eastern	950	1,100,000	21.8

† Computed from $K = (\pi/2)\sqrt{E/6c}$, on the assumption that the proportional limit is two-thirds of c.

Members having (L/b)'s from K to 50 are considered as slender columns governed by

$$\frac{P}{A} = \frac{\pi^2 E}{36(L/b)^2} \tag{15.12}$$

This formula is simply Euler's equation in terms of L/b and with a factor of safety of 3; the upper limit of $L/b = 50$ is set for practical considerations.

Although the above formulas were designed for use with rectangular sections, they can also be applied to round columns. This is done by determining the dimension b of a rectangular column which has a least radius of gyration equivalent to that of the round column. Thus, since the least radii of gyration for a rectangle and circle are $b/\sqrt{12}$ and $D/4$, respectively,

$$\frac{b}{\sqrt{12}} = \frac{D}{4} \qquad \text{or}$$

$$b = 0.866D$$

Therefore, the timber formulas of this article can be used for round columns provided that b is used as 0.866 times the diameter.

Example 15.4. Determine the allowable loads of standard 8 by 10 southern pine compression members which are 6, 12, and 20 ft long.

From Table A.1, $b = 7.5$ in. and $A = 71.3$ in.^{2}; from Table 15.1, $c = 1450$ psi, $E = 1{,}600{,}000$ psi, and $K = 21.3$.

For the 6-ft length

$$\frac{L}{b} = \frac{6(12)}{7.5} = 9.6$$

Therefore

$$\frac{P}{A} = c = 1450 \text{ psi} \qquad \text{and}$$

$$P = 1450(71.3) = 103{,}400 \text{ lb} \qquad Ans$$

For the 12-ft length

$$\frac{L}{b} = \frac{12(12)}{7.5} = 19.2$$

Therefore

$$\frac{P}{A} = c\left[1 - \frac{1}{3}\left(\frac{L/b}{K}\right)^4\right] = 1450\left[1 - \frac{1}{3}\left(\frac{19.2}{21.3}\right)^4\right]$$

$$\frac{P}{A} = 1450(0.78) = 1130 \text{ psi} \qquad \text{and}$$

$$P = 1130(71.3) = 80{,}600 \text{ lb} \qquad Ans$$

For the 20-ft length

$$\frac{L}{b} = \frac{20(12)}{7.5} = 32$$

Therefore

$$\frac{P}{A} = \frac{\pi^2 E}{36(L/b)^2} = \frac{9.87(1{,}600{,}000)}{36(32)^2} = 428 \text{ psi} \qquad \text{and}$$

$$P = 428(71.3) = 30{,}500 \text{ lb} \qquad Ans$$

PROBLEMS

***15.32.** A standard 4 by 6 section of southern pine is to be used as a 12-ft pin-end column. Determine the allowable load if the member is (*a*) unbraced, (*b*) fully braced at midlength.

15.33. A standard 6 by 10 section of Douglas fir is to be used as an unbraced pin-end column. Determine the maximum allowable length of the column and the corresponding allowable load.

15.34. Standard 8 by 12 sections of red oak are to be used as columns 10 and 18 ft long. Determine the allowable load for each member.

★**15.35.** A standard 6 by 14 section of white pine is used as a 22-ft pin-end column. Determine the allowable load if the column is braced in the weak direction (only) at midlength.

15.36. A standard 2 by 8 section of eastern hemlock is used as a 10-ft pin-end column. Determine the allowable load if the column is braced in the weak direction (only) at the quarter points of the length.

15.37. Determine the allowable load of a 12-in.-diam Douglas fir column which is (*a*) 10 ft long, (*b*) 15 ft long.

15.9 DESIGN OF AXIALLY LOADED COLUMNS

Column design usually consists of selecting a suitable cross section for a member of specified loading and effective length. This process is complicated by the fact that the allowable average stress is dependent on the slenderness ratio while the slenderness ratio is dependent on the cross section to be selected. For this reason, column design is normally most efficiently performed by following a logical trial-and-check procedure. One such procedure is outlined below:

1. An allowable P/A, within the limits of the column equations to be used, is assumed and used to *estimate* the required cross-sectional area.

2. On the basis of cross-sectional area and least radius of gyration, a section is selected for consideration (to determine if it would be suitable). This section is referred to as the *first trial section.*

3. The L/k for the first trial section is computed and used in the appropriate formula to compute the allowable stress and the area which would be required with this allowable stress.

4. The area required is compared with the actual area of the first trial section. Of course, if the actual area is equal to or greater than the area required, the section would be safe to use. However, if the actual area is substantially larger than the area required, it is likely that a more economical section could be used.

5. If the first trial section is either unsafe or uneconomical, its area required is used as a guide in selecting a second trial section.

6. The trial-and-check procedure is continued until a section which is both safe and economical has been determined. The

selection of any trial section is guided by the area required of the previous trial section.

Example 15.5. Select the lightest WF section to act as a 25-ft main column supporting a 300-kip axial load. Use AISC specifications, and assume that the column will be subjected to some vibration.

The design must be based on the equation $P/A = 17{,}000 - 0.485\,(L/k)^2$ within limits of L/k from 0 to 120. Assuming a P/A of 12,000, the assumed area required (for the first trial section) will be

$$\frac{300{,}000}{12{,}000} = 25 \text{ in.}^2$$

Trying a 14WF87 having an area of 25.56 in.² and a minimum k of 3.70 in. (this section has the largest minimum k of any available section having an area of approximately 25 in.²),

$$\frac{L}{k} = \frac{25(12)}{3.70} = 81$$

$$\frac{P}{A} = 17{,}000 - 0.485(81)^2 = 13{,}820 \text{ psi}$$

$$A_r = \frac{300{,}000}{13{,}820} = 21.71 \text{ in.}^2 \text{ (safe)}$$

Although this section is safe, its actual area is substantially greater than that required; therefore it is likely that a lighter section can be used.

Trying a 14WF78 having an area of 22.94 in.² and a minimum k of 3.00 in.,

$$\frac{L}{k} = \frac{25(12)}{3.00} = 100$$

$$\frac{P}{A} = 17{,}000 - 0.485(100)^2 = 12{,}150 \text{ psi}$$

$$A_r = \frac{300{,}000}{12{,}150} = 24.69 \text{ in.}^2 \text{ (unsafe)}$$

Trying a 14WF84 having an area of 24.71 in.² and a minimum k of 3.02 in.,

$$\frac{L}{k} = \frac{25(12)}{3.02} = 99$$

$$\frac{P}{A} = 17{,}000 - 0.485(99)^2 = 12{,}250 \text{ psi}$$

$$A_r = \frac{300{,}000}{12{,}250} = 24.49 \text{ in.}^2 \text{ (safe)}$$

A further check of the table reveals that all other available sections which could be considered (on an area basis) have smaller minimum k's. Therefore use 14WF84 (*Ans*).

PROBLEMS

***15.38.** A 20-ft southern pine compression member is to support an axial load of 22 kips. Determine the lightest standard section which may be used.

15.39. A 15-ft red oak compression member is to support an axial load of 110 kips. Determine the lightest standard section which may be used.

15.40. A 2014-T6 aluminum rectangular compression member is to have an effective length of 36 in. and support an axial 125-kip load. The cross-sectional width is to be twice the thickness. Thicknesses of the material are available in 0.1-in. increments from 1 in. up. Based on a factor of safety of 2.5, determine the lightest section which can be used.

***15.41.** A 2014-T6 aluminum round compression member is to have an effective length of 8 ft and support an axial 8500-lb load. Based on a factor of safety of 2.5, determine the smallest diameter (to the nearest 0.1 in.) which can be used.

15.42. Using AISC specifications, select the lightest equal-leg angle to act as a 14-ft secondary member under a compressive load of 24 kips.

15.43–15.46. For the length L and axial load P listed, select the lightest WF section to act as a main column. Use AISC specifications, and assume that the column will not be subject to vibration or shock loading.

15.43. $P = 425$ kips; $L = 20$ ft.
*15.44. $P = 300$ kips; $L = 25$ ft.
15.45. $P = 185$ kips; $L = 22$ ft.
15.46. $P = 100$ kips; $L = 30$ ft.

15.10 ECCENTRIC LOADING; SECANT FORMULA

Most columns are intended to be or are assumed to be axially loaded. Practical considerations, however, frequently necessitate the use of columns which must support loading of definite and significant eccentricity. The critical load for such columns is defined by a theoretical equation known as the secant formula. The derivation of this formula, which closely parallels that of

Euler's equation, will be made with reference to the pin-end column shown in Fig. 15.8. Letting P designate the critical load causing a midlength deflection Δ, the equation of the elastic curve is

$$EI \frac{d(dy/dx)}{dx} = -Py \qquad \text{or}$$

$$EI\, 2 \frac{dy}{dx}\, d\frac{dy}{dx} = -2Py\, dy$$

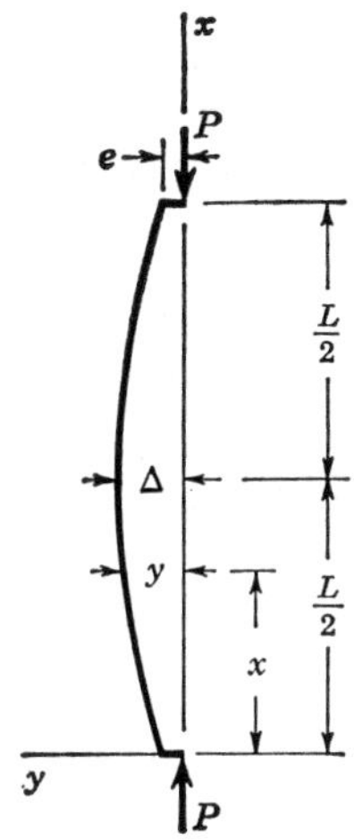

FIG. 15.8

Integrating,

$$EI \left(\frac{dy}{dx}\right)^2 = -Py^2 + C_1$$

Since $dy/dx = 0$ at $y = \Delta$, $C_1 = +P\Delta^2$ and

$$EI \left(\frac{dy}{dx}\right)^2 = -Py^2 + P\Delta^2$$

Therefore

$$\frac{dy}{dx} = \sqrt{\frac{P}{EI}}\sqrt{\Delta^2 - y^2} \qquad \text{or}$$

$$\frac{dy}{\sqrt{\Delta^2 - y^2}} = \sqrt{\frac{P}{EI}}\, dx$$

Integrating again,

$$\arcsin \frac{y}{\Delta} = \sqrt{\frac{P}{EI}}\, x + C_2$$

Since $y = e$ at $x = 0$, $C_2 = \arcsin e/\Delta$; also since $y = \Delta$ at $x = L/2$, the equation becomes

$$\frac{\pi}{2} = \sqrt{\frac{P}{EI}}\frac{L}{2} + \arcsin \frac{e}{\Delta}$$

Solving for Δ,

$$\frac{e}{\Delta} = \sin\left(\frac{\pi}{2} - \frac{L}{2}\sqrt{\frac{P}{EI}}\right) = \cos \frac{L}{2}\sqrt{\frac{P}{EI}}$$

Thus

$$\Delta = e \sec \frac{L}{2} \sqrt{\frac{P}{EI}} = e \sec \sqrt{\frac{PL^2}{4EI}}$$

As defined by Eq. (9.1), the maximum compressive stress s occurs at the right of the section at midlength. Noting that the moment at midlength is $P\Delta$ and expressing I as Ak^2,

$$s = \frac{P}{A} + \frac{P\Delta c}{Ak^2}$$

Thus

$$s = \frac{P}{A} + \frac{Pc}{Ak^2}\left(e \sec \sqrt{\frac{PL^2}{4EAk^2}}\right) \qquad \text{or}$$

$$s = \frac{P}{A}\left(1 + \frac{ec}{k^2} \sec \frac{L}{k} \sqrt{\frac{P/A}{4E}}\right)$$

Failure is assumed to occur when the maximum compressive stress reaches the yield point s_y; therefore the critical stress P/A for eccentrically loaded columns is defined by the equation

$$\frac{P}{A} = \frac{s_y}{1 + (ec/k^2)[\sec (L/k) \sqrt{(P/A)/4E}]} \tag{15.13a}$$

While the secant formula is theoretically correct for the assumptions made and is the basis for several design specifications, it is quite cumbersome for practical use. For design purposes, it is normally reduced to a series of curves (or tables) applicable to the particular material to be used; each curve of a series normally represents the plot of P/A versus L/k for a particular value of ec/k^2. When such curves (or tables) are not available, the secant formula can be more easily used for design purposes in the following form:

$$s_y = \frac{fP}{A} + \frac{fPe}{I/c} \sec L \sqrt{\frac{fP}{4EI}} \tag{15.13b}$$

where f represents the factor of safety with respect to yielding.

Example 15.6. An 18-ft column is to support a 200-kip load applied with a 5-in. eccentricity with respect to the major principal axis of the section. Based on the secant formula, a factor of safety of 2, and a yield point of 36,000 psi, determine if a 14WF87 section could be safely used.

The section will be safe if the right-hand part of Eq. (15.13*b*) does not exceed the yield point. From Table A.4, $A = 25.56$ in.^{2}, $I = 966.9$ in.^{4}, $I/c = 138.1$ in.^{3}

$$s = \frac{fP}{A} + \frac{fPe}{I/c} \sec L \sqrt{\frac{fP}{4EI}}$$

Letting $fP = 2(200{,}000) = 400{,}000$ lb,

$$s = \frac{400{,}000}{25.56} + \frac{400{,}000(5)}{138.1} \sec (18)(12) \sqrt{\frac{400{,}000}{4(30{,}000{,}000)(967)}}$$

$$s = 15{,}650 + 14{,}480 \sec 0.401$$
$$s = 15{,}650 + 15{,}730$$
$$s = 31{,}380 < 36{,}000 \text{ (safe)} \qquad \textit{Ans}$$

PROBLEMS

★15.47. Equation (15.13*a*) is to be represented as a series of curves applicable to a steel having a yield point of 36,000 psi and $E = 30{,}000{,}000$ psi. For the $ec/k^2 = 0.20$ curve, compute the value of L/k which corresponds to a critical stress of 30,000 psi.

15.48. Equation (15.13*a*) is to be represented as a series of curves applicable to a steel having a yield point of 36,000 psi and $E = 30{,}000{,}000$ psi. For the $ec/k^2 = 0.6$ curve, compute the value of L/k which corresponds to a critical stress of 15,000 psi.

★15.49. A 24-ft column is to support a 75-kip load applied with a 12-in. eccentricity with respect to the major principal axis of the section. Based on the secant formula, a factor of safety of 2, and a yield point of 36,000 psi, determine if a 14WF61 section could be safely used.

15.50. A 20-ft column is to support a 150-kip load applied with a 6-in. eccentricity with respect to the major principal axis of the section. Based on the secant formula, a factor of safety of 2, and a yield point of 36,000 psi, determine if a 12WF72 section could be safely used.

15.11 ECCENTRIC LOADING; EMPIRICAL ANALYSIS

Eccentrically loaded members are frequently designed and analyzed by methods which are largely empirical in nature. These methods are based on the assumption that the load-carrying capacity of any eccentrically loaded column is dependent on its ability to withstand axial loading and bending simultaneously. By one such method, a member is assumed to be safe if the sum of the ratios of computed stress to allowable stress for the two types of loading does not exceed unity. As an example, the AISC specifications state that members subject to both axial and bending stresses shall be so proportioned that

$$\frac{f_a}{F_a} + \frac{f_b}{F_b} \leqq 1 \tag{15.14}$$

in which F_a = allowable axial stress
F_b = allowable bending stress
f_a = computed axial stress
f_b = computed bending stress

It should be noted that this specification can also be applied to members which are subjected simultaneously to axial and transverse loads; such members are sometimes referred to as beam-columns.

Example 15.7. An 18-ft column is to support a 200-kip load applied with a 5-in. eccentricity with respect to the major principal axis of the section. Using AISC specifications, determine if a 14WF87 section could be safely used.

Determining the allowable bending stress (see Art. 6.5),

$$\frac{Ld}{bt} = \frac{[18(12)]14}{14.5(0.688)} = 303(<600)$$

Therefore, F_b = 20,000 psi.

Determining the allowable axial stress,

$$\frac{L}{k} = \frac{18(12)}{3.70} = 58$$

Therefore, from Eq. (15.6),

$$F_a = 17{,}000 - 0.485(58)^2 = 15{,}370 \text{ psi}$$

Determining the computed stress for the 14WF87 (A = 25.56 in.^{2} and I/c = 138.1 in.^{3}),

$$f_a = \frac{P}{A} = \frac{200{,}000}{25.56} = 7825 \text{ psi}$$

$$f_b = \frac{M}{I/c} = \frac{200{,}000(5)}{138.1} = 7240 \text{ psi}$$

Using Eq. (15.14),

$$\frac{f_a}{F_a} + \frac{f_b}{F_b} = \frac{7825}{15{,}370} + \frac{7240}{20{,}000}$$

$$\frac{f_a}{F_a} + \frac{f_b}{F_b} = 0.509 + 0.362 = 0.871 \text{ (safe)} \qquad \textit{Ans}$$

PROBLEMS

★15.51. A 24-ft column is to support a 75-kip load applied with a 12-in. eccentricity with respect to the major principal axis of the section. Using AISC specifications, determine if a 14WF61 section could be safely used.

15.52. A 20-ft column is to support a 150-kip load applied with a 6-in. eccentricity with respect to the major principal axis of the section. Using AISC specifications, determine if a 12WF72 section could be safely used.

★15.53. A 12-ft southern pine column is to support a 25-kip load applied with a 4-in. eccentricity with respect to the major principal axis of the section. On the basis of Eq. (15.14) determine if a standard 6 by 12 section could be safely used. Use $F_b = 1750$ psi.

15.54. A 15-ft white oak column is to support a 100-kip load applied with a 3-in. eccentricity with respect to the major principal axis of the section. On the basis of Eq. (15.14) determine if a standard 10 by 16 section could be safely used. Use $F_b = 1700$ psi.

16

DYNAMIC LOADING; ELASTIC ENERGY

16.1 INTRODUCTION

As has been seen, the behavior of statically loaded members is normally explained and predicted on the basis of the conditions of equilibrium and the proportionality of stress and strain. However, when a member is dynamically loaded, an additional concept, that of work energy, is required. Thus when load is applied to a member (either statically or dynamically), work is performed by the force, and energy is absorbed by the member. When a material is deformed elastically, the energy that it absorbs is recoverable and is known as *elastic energy*.

Although in this chapter the work-energy method will be applied primarily to dynamically loaded members, the basic concept has many useful applications for statically loaded structures. For example, the concept is basic to several methods of analysis applicable to trusses. These methods are frequently used to determine truss deflections. In the case of statically indeterminate trusses, they may also be used to determine truss reactions and internal forces.

The fundamental equation for elastic energy, due to normal stress, will be derived with respect to the elemental mass shown in Fig. 16.1. It is assumed that the element has been loaded at a uniform rate to a force of $dF = s\,dA$ and a total deformation of $d\delta = \epsilon\,dL$. Since the energy absorbed by the element must be equal to the work done on it,

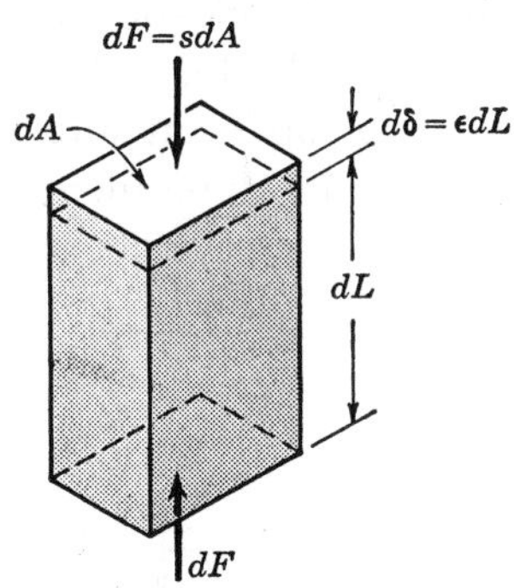

FIG. 16.1

dU = *average force* times *displacement*

$$dU = \frac{0 + dF}{2}\,d\delta = \frac{s\,dA}{2}\,(\epsilon\,dL)$$

Since it is assumed that the material has not exceeded its elastic limit,

$$dU = \frac{s\,dA}{2}\frac{s}{E}dL = \frac{s^2}{2E}\,dA\,dL$$

Thus elastic energy due to normal stress is defined by the equation

$$dU = \frac{s^2}{2E}\,dV \qquad (16.1)$$

The method of determining the maximum stress in a dynamically loaded member is essentially one of equating the energy absorbed to the energy applied. Since work is a product of force and displacement, the work done (and thus the energy applied) by any force is a function of the deformation of the member on which it acts. Thus, as shown in Fig. 16.2, if a weight W dropped from a height h caused a member to deform an amount δ (under the weight), the energy applied to the member would be

$$U = W(h + \delta)$$

Since the velocity of a freely falling body is $v = \sqrt{2gh}$, the energy applied could also be expressed as

$$U = W\left(\frac{v^2}{2g} + \delta\right)$$

where v designates the velocity of the weight W at the instant of impact.

It should be understood that the maximum elastic deformations from dynamic loading occur only instantaneously; any dynamically loaded member will vibrate with decreasing amplitude until it assumes a condition of static loading. It follows, therefore, that, if the elastic limit is not exceeded, the maximum stress in a dynamically loaded body will occur only instantaneously.

As has been stated, the general procedure for determining the stresses in dynamically loaded members is to equate the energy

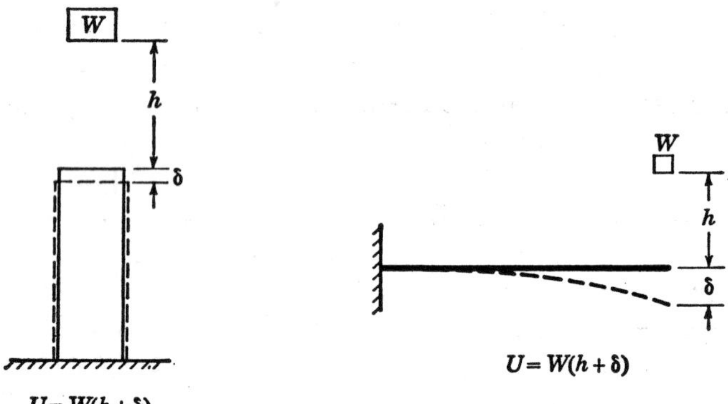

FIG. 16.2 Energy applied by falling weights.

applied to the energy absorbed (i.e., the strain energy). It should be emphasized that this procedure is based on the simplifying assumption that any losses of energy (such as to heat, to the supports, and to the mass applying the load) can be neglected. It will also be assumed that the energy absorbed will not be localized and that the stress distribution in a member from dynamic loading will be the same as from static loading. These assumptions are valid for practical purposes if the velocity at impact is not excessive.

16.2 AXIAL LOADING

For any axially loaded member, the normal stress on every cross section and, therefore, for every differential element is the same. Therefore, the energy absorbed can be defined by integrating both sides of Eq. (16.1) with s as a constant. Thus the elastic energy of an axially loaded member is

$$U = \frac{s^2}{2E} V \tag{16.2}$$

From Eq. (16.2), it can be noted that a member which is properly designed to resist static loading may be ill suited to resist dynamic loading. While static strength is dependent on a limiting stress and cross-sectional area, the dynamic strength of a member is dependent on Young's modulus and its *volume* as well as on a limiting stress. Thus, for example, an aluminum tension member could absorb approximately three times as much energy as a steel member of the same dimensions and the same elastic limit. Also, since the energy-absorbing capacity of a member is a direct function of its volume, the dynamic load-carrying capacity of either member could be increased by merely making it longer.

In designing for dynamic loading, it is good practice to make the net cross-sectional area of a member the same (or very nearly so) throughout its length. This should be done because any surplus of area above the least cross-sectional area actually *decreases* the energy-absorbing capacity of a member. Although the surplus material furnishes more volume, it serves to decrease the stress at the larger sections. Thus, since elastic energy is a function of the square of the stress, less energy can be absorbed. The detrimental effect of surplus material on the ability of a member to resist dynamic loading is illustrated by Example 16.1.

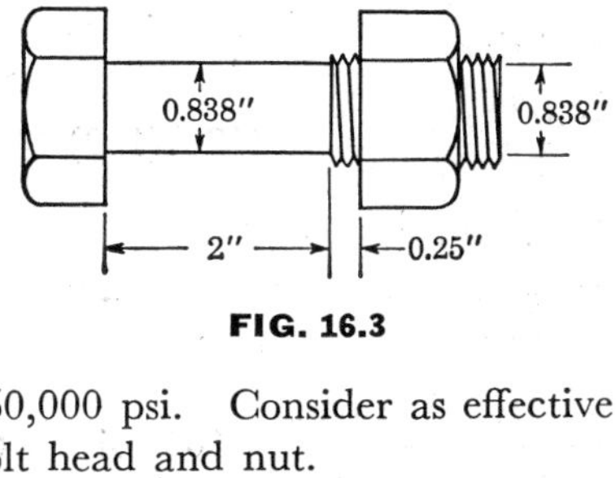

FIG. 16.3

Example 16.1. Compare the amount of energy which the two high-strength bolts shown in Fig. 16.3 can absorb without exceeding an elastic stress of 50,000 psi. Consider as effective only the 2.25-in. length between the bolt head and nut.

The 1-in. bolt will be considered as being composed of two segments: one segment 2 in. in length and 1 in. in diameter and the other 0.25 in.

in length and 0.838 in. in diameter (the thread material being neglected). From Table A.3, the areas of the two segments are 0.785 and 0.551 in.^{2}, respectively. The maximum stress, of 50,000 psi, will occur in the threaded segment. Since the stress is inversely proportional to the area, the corresponding stress for the unthreaded segment will be

$$50{,}000\,\frac{0.551}{0.785} = 35{,}100 \text{ psi}$$

Therefore, the energy which can be absorbed by this bolt without exceeding a stress of 50,000 psi is

$$U = \sum \frac{s^2}{2E} V = \frac{(35{,}100)^2}{2(30{,}000{,}000)}(2)(0.785) + \frac{(50{,}000)^2}{2(30{,}000{,}000)}(0.25)(0.551)$$

$$U = 32.2 + 5.7 = 37.9 \text{ in.-lb} \qquad \textit{Ans}$$

The 0.838-in. bolt, having a uniform net diameter, will be stressed to 50,000 psi throughout its effective length. Therefore, for this bolt

$$U = \frac{s^2}{2E} V = \frac{(50{,}000)^2}{2(30{,}000{,}000)}(2.25)(0.551)$$

$$U = 51.7 \text{ in.-lb} \qquad \textit{Ans}$$

Thus the smaller of the two bolts can absorb 36 per cent more energy.

Example 16.2. A 10-ft-long standard 12 by 12 timber post (E = 1,600,000 psi) is to support a weight of 14 kips. Determine the maximum stress and deformation in the post if the weight is (*a*) gradually applied, (*b*) suddenly applied, and (*c*) dropped from a height of 3 in.

(*a*) If the load is gradually applied, the work done on the post would be $(W/2)\delta = 14{,}000\delta/2$. Equating the energy applied to the energy absorbed,

$$\frac{14{,}000\delta}{2} = \frac{s^2 V}{2E}$$

and since $\delta = sL/E$,

$$\frac{14{,}000}{2}\,\frac{s120}{E} = \frac{s^2}{2E}(120)(132)$$

$$s = 106 \text{ psi} \qquad \textit{Ans}$$

Of course, a gradually applied load is a *static* load, so that the answer could have been obtained as

$$s = \frac{W}{A} = \frac{14{,}000}{132} = 106 \text{ psi} \qquad \textit{Check}$$

$$\delta = \frac{sL}{E} = \frac{106(120)}{1{,}600{,}000} = 0.00795 \text{ in.} \qquad \textit{Ans}$$

(*b*) A suddenly applied load means that the weight is brought into contact with the member and suddenly released; thus the work done on the post would be $W\delta$ or $14{,}000\delta$. Therefore

$$14{,}000\,\frac{s120}{E} = \frac{s^2}{2E}(120)(132)$$

$$s = 212 \text{ psi} \qquad Ans$$

$$\delta = \frac{sL}{E} = \frac{212(120)}{1{,}600{,}000} = 0.0159 \text{ in.} \qquad Ans$$

Thus, it can be seen that applying a load suddenly rather than gradually (i.e., statically) has the effect of doubling both the resulting stresses and deformations.

(*c*) If the load is dropped from a height of 3 in., the work done on the post would be $14{,}000(3 + \delta)$. Therefore

$$14{,}000\left(3 + \frac{s120}{1{,}600{,}000}\right) = \frac{s^2}{2(1{,}600{,}000)}(120)(132)$$

From which

$$s^2 - 212s - 8{,}480{,}000 = 0$$

$$s = 3020 \text{ psi} \qquad Ans$$

$$\delta = \frac{sL}{E} = \frac{3020(120)}{1{,}600{,}000} = 0.226 \text{ in.} \qquad Ans$$

PROBLEMS

NOTE: For the problems of this group, use the following values for Young's modulus: steel, 30,000,000 psi; aluminum, 10,000,000 psi; brass, 15,000,000 psi.

★**16.1.** A steel compression member has a cross-sectional area of 2 in.^{2} and a length of 6 in. Determine the energy load required to stress the member to (*a*) 20,000 psi, (*b*) 30,000 psi.

16.2. Two axially loaded steel members are described as follows: member *A* is 6 in. long and 2 by 1.5 in. in cross-section; member *B* is 18 in. long and 1.5 in. in diameter. Based on an allowable stress of 20,000 psi, compare the ability of the two members to resist (*a*) static loading, (*b*) dynamic loading.

16.3. An aluminum member and a steel member are identical in size, 2 in. in diameter and 12 in. long, and have the same elastic limit of 25,000 psi. Determine the amount of energy which each member can absorb elastically.

★**16.4.** ⅝-in.-diam aluminum bolts are to be used to connect the flanges of two sections of pipe as shown in Fig. P16.4. If each bolt must absorb 6 in.-lb of energy without exceeding a stress of 24,000 psi, determine the minimum distance b which the threads must extend beyond the inside face of the nut. Assume that the threads must extend at least ⅛ in. beyond the nut, and consider as effective (in absorbing energy) only the length $c = 1.25$ in. between the bolt head and the nut.

16.5. ¾-in.-diam steel bolts are to be used to connect the flanges of two sections of water main as shown in Fig. P16.5. If each bolt must absorb 4 in.-lb of energy without exceeding a stress of 24,000 psi, determine the minimum distance b which the threads must extend beyond the inside face of the nut. Assume that the threads must extend at least ⅛ in. beyond the nut, and consider as effective (in absorbing energy) only the length $c = 1.50$ in. between the bolt head and the nut.

16.6. ⅞-in.-diam brass bolts are to be used to connect the flanges of two sections of pipe as shown in Fig. P16.6. If each bolt must absorb 10 in.-lb of energy without exceeding a stress of 22,000 psi, determine the minimum distance b which the threads must extend beyond the inside face of the nut. Assume that the threads must extend at least ⅛ in. beyond the nut, and consider as effective (in absorbing energy) only the length $c = 2.00$ in. between the bolt head and the nut.

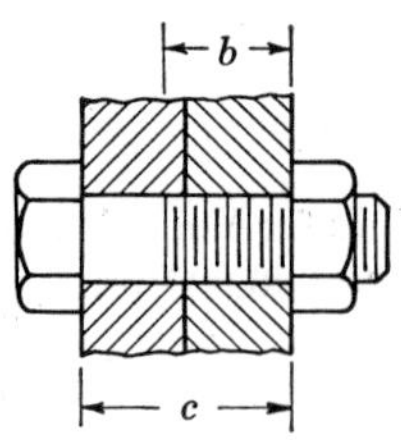

FIG. P16.4, P16.5, P16.6

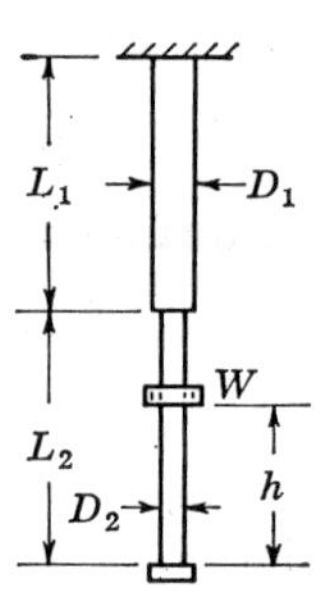

FIG. P16.7, P16.8, P16.9

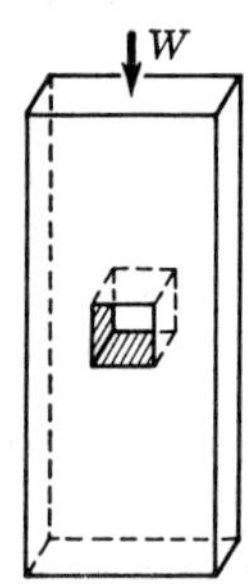

FIG. P16.10, P16.11, P16.12

★**16.7.** The aluminum member shown in Fig. P16.7 is symmetrical about the vertical axis and has diameters $D_1 = 1$ in. and $D_2 = $ ⅝ in. and lengths $L_1 = 4.5$ ft and $L_2 = 3$ ft. Determine the maximum weight W which can be dropped from a height $h = 15$ in. without stressing the member above 20,000 psi.

16.8. The steel member shown in Fig. P16.8 is symmetrical about the vertical axis and has diameters $D_1 = $ ⅞ in. and $D_2 = $ ½ in. and lengths $L_1 = 4$ ft and $L_2 = 2.5$ ft. Determine the maximum height h

from which the weight $W = 8$ lb can be dropped without stressing the member above 20,000 psi.

16.9. The aluminum member shown in Fig. P16.9 is symmetrical about the vertical axis and has diameters $D_1 = \frac{3}{4}$ in. and $D_2 = \frac{3}{8}$ in. and lengths $L_1 = 3.5$ ft and $L_2 = 2$ ft. Determine the maximum height h from which the weight $W = 15$ lb can be dropped without stressing the member above 20,000 psi.

***16.10.** A 1.5-ft-long standard 2 by 6 Douglas fir ($E = 1{,}600{,}000$ psi) post is to support an axial load W. If the maximum stress is not to exceed 1450 psi, determine the allowable load if it is (*a*) statically applied and (*b*) applied with a velocity of 3 fps. (*c*) What would be the allowable value of W for the load conditions of (*a*) and (*b*) if a centrally located 2-in.-square hole (actual size) were cut out of the member as shown in Fig. P16.10?

16.11. A 2.5-ft-long standard 3 by 8 red oak ($E = 1{,}500{,}000$ psi) post is to support an axial load W. If the maximum stress is not to exceed 1350 psi, determine the allowable load if it is (*a*) statically applied and (*b*) applied with a velocity of 4 fps. (*c*) What would be the allowable value of W for the load conditions of (*a*) and (*b*) if a centrally located 3-in.-square hole (actual size) were cut out of the member as shown in Fig. P16.11?

16.12. A 4-ft-long standard 4 by 10 white pine ($E = 1{,}000{,}000$ psi) post is to support an axial load W. If the maximum stress is not to exceed 1000 psi, determine the allowable load if it is (*a*) statically applied and (*b*) applied with a velocity of 5 fps. (*c*) What would be the allowable value of W, for the load conditions of (*a*) and (*b*), if a centrally located 4-in.-square hole (actual size) were cut out of the member as shown in Fig. P16.12?

16.3 EQUIVALENT STATIC LOAD; LOAD FACTOR

Frequently, the stresses and deformations due to dynamic loading can be most easily computed by replacing (for analysis purposes) the dynamic load with an *equivalent static load*. As the name implies, an equivalent static load is a static load which would apply the same energy and therefore produce the same stresses and deformations as would result (instantaneously) from a given dynamic load. The ratio of the equivalent static load to the applied load is known as the *load factor* and will be designated as n. Thus, any equivalent static load can be expressed as the product of the (dynamically) applied load and the appropriate load factor; for example, if a 100-lb load were applied with a load factor of 3, the equivalent static load would be 300 lb. For some conditions, load factors can be computed; for others, they must be determined empirically. The load factor for freely falling bodies will be derived with respect

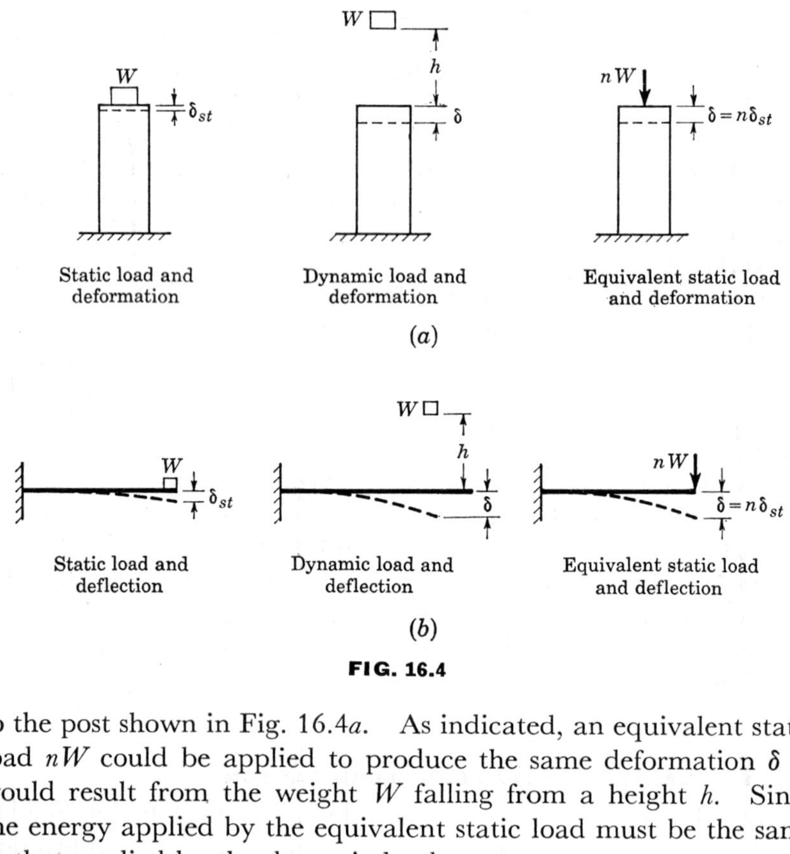

FIG. 16.4

to the post shown in Fig. 16.4*a*. As indicated, an equivalent static load nW could be applied to produce the same deformation δ as would result from the weight W falling from a height h. Since the energy applied by the equivalent static load must be the same as that applied by the dynamic load,

$$\frac{0 + nW}{2}(n\delta_{st}) = W(h + \delta)$$

From which

$$\frac{n^2\delta_{st}}{2} = h + n\delta_{st} \qquad \text{or}$$

$$n^2\delta_{st} - 2n\delta_{st} - 2h = 0$$

Solving for the load factor,

$$n = 1 + \sqrt{1 + \frac{2h}{\delta_{st}}} \tag{16.3a}$$

As can be seen from Fig. 16.4*b*, the identical derivation could have been made with respect to a member in bending. In fact, the load factor defined by Eq. (16.3*a*) can be used for any member having a linear load-deformation relationship; for example, it can be used for torsional loading of cylindrical members. As long as the elastic limit is not exceeded, the load factor can be used to determine dynamic stresses and deformations directly; thus dynamic stresses and deformations can be expressed as ns_{st} and $n\delta_{st}$. Since the velocity of a freely falling body is equal to $\sqrt{2gh}$, a load factor can also be written in terms of the velocity v at the instant of impact; thus

$$n = 1 + \sqrt{1 + \frac{v^2}{g\delta_{st}}} \tag{16.3b}$$

Many specifications are written in terms of an *impact factor*, which defines the difference between the effect of a load dynamically applied and statically applied. Thus an impact factor is equivalent to the load factor minus one. Most such factors are empirically determined on the basis of tests and experience. For freely falling bodies the impact factor would be the equivalent of the radical term of the load factor.†

Example 16.3. A 10-ft-long standard 12 by 12 timber post (E = 1,600,000 psi) is to support a weight of 14 kips. Determine the maximum (instantaneous) stress and deformation in the post if the weight is (*a*) suddenly applied and (*b*) dropped from a height of 3 in.

(*a*) The load factor for suddenly applied loads is

$$n = 1 + \sqrt{1 + 0} = 2$$

therefore

$$s = \frac{nW}{A} = \frac{2(14{,}000)}{132} = 212 \text{ psi} \qquad \textit{Ans}$$

$$\delta = \frac{sL}{E} = \frac{212(120)}{1{,}600{,}000} = 0.0159 \text{ in.} \qquad \textit{Ans}$$

† Unfortunately, there is some confusion regarding the meaning of impact factor. Many engineers and authors use it synonymously with load factor.

(*b*) Computing the load factor if the weight were dropped from a height of 3 in.,

$$\delta_{st} = \frac{WL}{AE} = \frac{14{,}000(120)}{132(1{,}600{,}000)} = 0.00795 \text{ in.}$$

$$n = 1 + \sqrt{1 + \frac{2h}{\delta_{st}}} = 1 + \sqrt{1 + \frac{2(3)}{0.00795}} = 28.5$$

Therefore

$$s = \frac{nW}{A} = \frac{28.5(14{,}000)}{132} = 3020 \text{ psi} \qquad \textit{Ans}$$

$$\delta = n\delta_{st} = 28.5(0.00795) = 0.226 \text{ in.} \qquad \textit{Ans}$$

Example 16.4. An 8WF17 section is to be used as a 12-ft simply supported beam loaded with a 500-lb weight at midspan. Determine the maximum bending stress and deflection if the weight is dropped from a height of 4 in.

For an 8WF17, $I = 56.4$ in.^{4} and $I/c = 14.1$ in.^{3} Using Table 13.2, the deflection under the load if it were statically applied would be

$$\delta_{st} = \frac{WL^3}{48EI} = \frac{500(12)^3(1728)}{48(30{,}000{,}000)(56.4)} = 0.01838 \text{ in.}$$

Therefore

$$n = 1 + \sqrt{1 + \frac{(2)(4)}{0.01838}} = 21.9$$

$$M = \frac{nWL}{4} = \frac{21.9(500)(12)}{4} = 32{,}850 \text{ lb-ft}$$

$$s = \frac{M}{I/c} = \frac{32{,}850(12)}{14.1} = 28{,}000 \text{ psi} \qquad \textit{Ans}$$

$$\delta = n\delta_{st} = 21.9(0.01838) = 0.402 \text{ in.} \qquad \textit{Ans}$$

PROBLEMS

NOTE: For the problems of this group, use the following values for Young's modulus: steel, 30,000,000 psi; timber, 1,600,000 psi.

***16.13.** A cable hoist is used to lower a 1000-lb weight at a uniform velocity of 3 fps. The cable has a cross-sectional area of 0.19 in.^{2} and an effective E of 15,000,000 psi. Determine the stress in the cable if the hoist drum is suddenly stopped (locked) when the weight is 40 ft below the center of the drum.

16.14. A cable hoist is used to lower a 2-kip weight at a uniform velocity of 5 fps. The cable has a cross-sectional area of 0.27 in.^{2} and an effective E of 15,000,000 psi. Determine the stress in the cable if the drum is suddenly stopped (locked) when the weight is 60 ft below the center of the drum.

***16.15.** The assembly shown in Fig. P16.15 consists of a $\frac{3}{8}$-in.-diam steel rod having a length $L = 2.5$ ft and a spring having a modulus of 120 lb/in. The weight $W = 6$ lb is dropped from a height $h = 15$ in. and strikes the stop at the end of the rod. (*a*) Determine the maximum instantaneous values of stress in the rod and deflection of the spring. (*b*) What would be the maximum instantaneous stress in the rod if the spring were not used?

16.16. The assembly shown in Fig. P16.16 consists of a $\frac{1}{2}$-in.-diam steel rod having a length $L = 3.5$ ft, and a spring having a modulus of 320 lb/in. The weight $W = 8$ lb is dropped from a height $h = 1.5$ ft and strikes the stop at the end of the rod. (*a*) Determine the maximum instantaneous values of stress in the rod and deflection of the spring. (*b*) What would be the maximum instantaneous stress in the rod if the spring were not used?

16.17. The assembly shown in Fig. P16.17 consists of a $\frac{5}{8}$-in.-diam steel rod having a length $L = 4$ ft and a spring having a modulus of 200 lb/in. The weight $W = 10$ lb is dropped from a height $h = 2.5$ ft and strikes the stop at the end of the rod. (*a*) Determine the maximum instantaneous values of stress in the rod and deflection of the spring. (*b*) What would be the maximum instantaneous stress in the rod if the spring were not used?

***16.18.** A 10-ft-long standard 4 by 8 timber section is used as a beam restrained at both ends (with the nominal 8-in. dimension as the depth). What maximum instantaneous stress and deflection would be caused by dropping a 20-lb weight 12 in. to the midspan of the beam?

16.19. A 7I15.3 section is used as a 6-ft cantilever beam. If a 40-lb weight is dropped 12 in. to the midspan of the beam, what would be the maximum instantaneous stress and deflection?

***16.20.** A 5I10.0 section is used as a 4-ft cantilever beam. What maximum instantaneous stress and deflection would be caused by a 30-lb weight striking the beam at 1 ft from the free end with a velocity of 9 fps?

16.21. A 12-ft-long standard 4 by 10 timber section is to be used as a beam (with the nominal 10-in. dimension as the depth) supported on springs at each end. If the modulus of each spring is 200 lb/in., determine the maximum beam stress and deflection due to a 30-lb weight dropped 30 in. to the midspan of the beam.

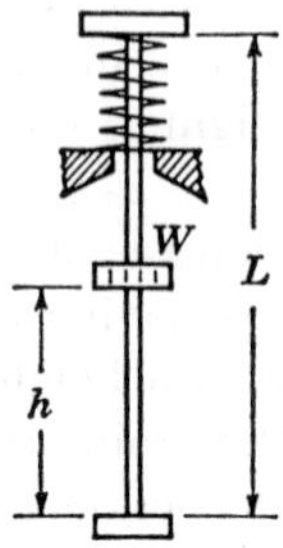

FIG. P16.15, P16.16, P16.17

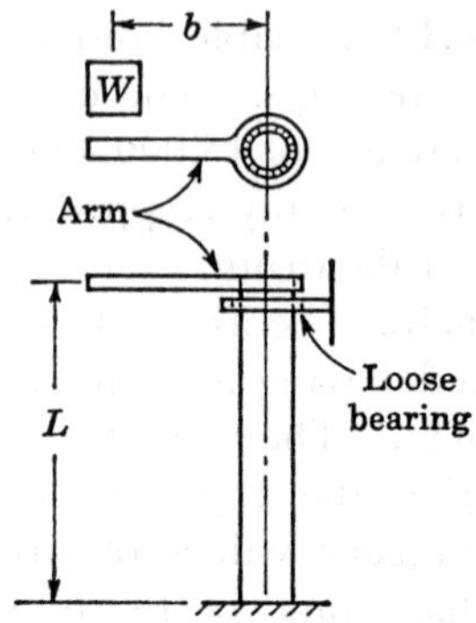

FIG. P16.22, P16.23

★16.22. A $\frac{3}{4}$-in. standard steel pipe is fixed at one end and fitted with a rigid transverse arm as shown in Fig. P16.22; a loose bearing is provided (at the end with the rigid arm) for lateral support. If $L = 6$ ft and $b = 10$ in., determine the maximum instantaneous shearing stress caused by the weight $W = 5$ lb striking the arm with a velocity of 6 fps. Use $E_s = 12{,}000{,}000$ psi.

16.23. A 1-in. standard steel pipe is fixed at one end and fitted with a rigid transverse arm as shown in Fig. P16.23; a loose bearing is provided (at the end with the rigid arm) for lateral support. If $L = 4$ ft and $b = 12$ in., determine the maximum instantaneous shearing stress caused by a weight $W = 15$ lb striking the arm with a velocity of 9 fps. Use $E_s = 12{,}000{,}000$ psi.

16.4 ELASTIC ENERGY FROM BENDING

Several of the many methods used for solving statically indeterminate structures and determining beam deformations are based on

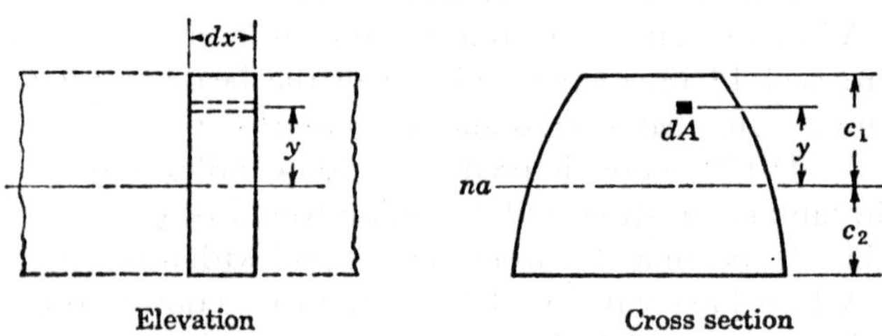

FIG. 16.5

the elastic energy absorbed by a member in bending. An equation defining this energy will be derived using the differential segment of beam shown in Fig. 16.5. Considering the element indicated, having a volume of $dA\,dx$, the elastic energy due to bending [from Eq. (16.1)] would be $dU = (s^2/2E)\,dA\,dx$. Based on the same assumptions as were made for the flexure formula and expressing bending stress as $s = My/I$, the elastic energy of the entire beam would be

$$U = \iint \frac{s^2}{2E}\,dA\,dx = \iint \left(\frac{My}{I}\right)^2 \frac{dA\,dx}{2E} \qquad \text{or}$$

$$U = \int \frac{M^2\,dx}{2EI^2} \int y^2\,dA = \int \frac{M^2\,dx}{2EI}$$

Thus the elastic energy due to bending can be expressed as

$$U = \frac{1}{2EI} \int M^2\,dx \tag{16.4}$$

Example 16.5. An aluminum beam (E = 10,000,000 psi; I = 200 in.^{4}) is loaded as shown in Fig. 16.6. Determine the elastic energy due to bending and the deflection under the load.

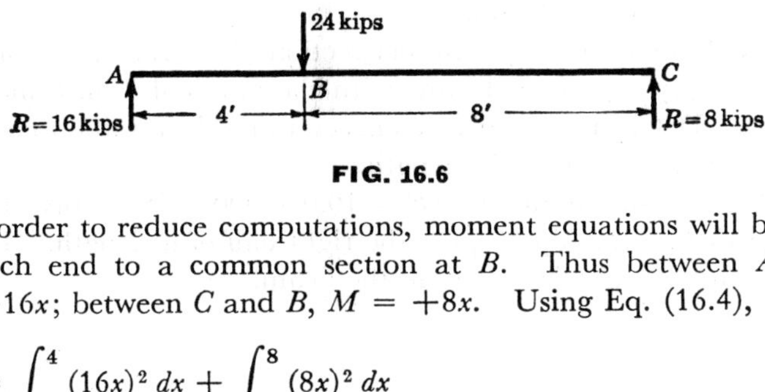

FIG. 16.6

In order to reduce computations, moment equations will be written from each end to a common section at B. Thus between A and B, $M = +16x$; between C and B, $M = +8x$. Using Eq. (16.4),

$$2EIU = \int_0^4 (16x)^2\,dx + \int_0^8 (8x)^2\,dx$$

$$2EIU = \left[\frac{256x^3}{3}\right]_0^4 + \left[\frac{64x^3}{3}\right]_0^8$$

$$2EIU = 5460 + 10{,}920 = 16{,}380 \text{ ft}^3\text{-kip}^2$$

$$U = \frac{16{,}380(1728)}{2(10{,}000)(200)} = 7.08 \text{ in.-kip} \qquad \textit{Ans}$$

Equating the energy applied to the energy absorbed,

$$\frac{0 + 24}{2}\,\delta_B = 7.08$$

$$\delta_B = 0.59 \text{ in.} \qquad \textit{Ans}$$

PROBLEMS

★16.24. In terms of limiting stress s, Young's modulus E, and volume V, evaluate the elastic energy which can be absorbed by a rectangular member used as a cantilever beam loaded with a couple at the free end. How does this compare with the energy which could be absorbed by the member if it were axially loaded?

16.25. In terms of limiting stress s, Young's modulus E, and volume V, evaluate the elastic energy which can be absorbed by a rectangular member used as a cantilever beam loaded with a concentrated load at the free end. How does this compare with the energy which could be absorbed by the member if it were axially loaded?

16.26. In terms of limiting stress s, Young's modulus E, and volume V, evaluate the elastic energy which can be absorbed by a rectangular member used as a simply supported beam with a midspan concentrated load. How does this compare with the energy which could be absorbed by the member if it were axially loaded?

★16.27. An 8WF17 section (E = 30,000,000 psi) is used as a 9-ft cantilever beam supporting 500 lb/ft over its entire length. Determine the elastic energy in the beam.

16.28. A standard 4 by 8 timber section (E = 1,600,000 psi) is used as a 12-ft simply supported beam (with the nominal 8-in. dimension as the depth). Determine the elastic energy in the beam due to a uniform load of 200 lb/ft over its entire length.

16.29. An 8-ft aluminum (E = 10,000,000 psi) simply supported beam is loaded with 2 kips/ft over the right half of its length. If $I = 10$ in.^{4}, determine the elastic energy in the beam.

16.5 ELASTIC ENERGY FROM TORSION

Equation (16.1) was derived to define the elastic energy resulting from normal stress. The elastic energy resulting from shearing stress can be similarly derived to be

$$dU = \frac{s_s^{\,2}}{2E_s}\, dV \tag{16.5}$$

This equation can be used as the basis for defining the elastic energy in a cylindrical member subjected to torsion. For the elemental mass indicated in Fig. 16.7, the elastic energy due to torsion would be

$$dU = \frac{s_s^2}{2E_s}\,dA\,dx = \left(\frac{T\rho}{J}\right)^2 \frac{dA\,dx}{2E_s}$$

The elastic energy in the entire member would be

$$U = \iint \frac{T^2\rho^2}{J^2}\frac{dA\,dx}{2E_s} = \int \frac{T^2\,dx}{2E_sJ^2}\int \rho^2\,dA$$

Thus the elastic energy in any cylindrical member subjected to torsion can be expressed as

$$U = \frac{1}{2E_sJ}\int T^2\,dx \tag{16.6}$$

If T is constant over the entire length L of the member, the above

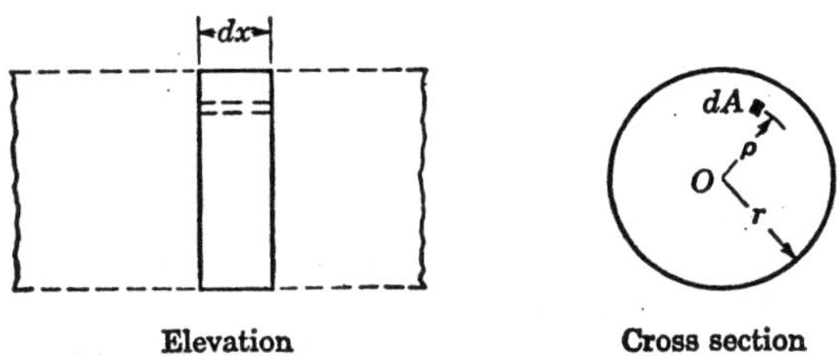

FIG. 16.7

equation becomes $U = T^2L/2E_sJ$ or, in terms of the angle of twist, $U = T\theta/2$.

PROBLEMS

★**16.30.** In terms of the limiting stress s_s, modulus of rigidity E_s, and volume V, evaluate the elastic energy which can be absorbed by a cylindrical member subjected to a constant torque.

16.31. A hollow cylindrical member having inner and outer radii r and R, respectively, is subjected to a constant torque. Evaluate the elastic energy which the member can absorb in terms of the limiting stress s_s, modulus of rigidity E_s, volume V, and the radii r and R.

16.32. An 8-ft length of 4-in. standard steel pipe (E_s = 12,000,000 psi) is fixed at both ends and subjected to a torque at midlength. If the energy applied by the torque is 2500 in.-lb, determine the maximum shearing stress.

16.33. A 5-ft length of 2-in.-diam aluminum (E_s = 4,000,000 psi) rod is fixed at both ends and subjected to a torque at 2 ft from one end. If the energy applied by the torque is 1250 in.-lb, determine the maximum shearing stress.

17

PLASTIC ANALYSIS

17.1 INTRODUCTION

In the application of elastic theory to design, the load-carrying capacity of any structural member is assumed to be limited by the occurrence of yielding at any point. However, even with elastic design, certain localized yielding, such as that due to stress concentration, is frequently neglected. The extent to which such yielding can be disregarded is, of course, dependent on the ductility of the material being used. Inasmuch as ductile materials do not usually fail until an appreciable amount of yielding has occurred, the load-carrying capacity of such members can be more realistically determined by extending the stress analysis into the plastic range.

Fundamental to the understanding of plastic analysis is the fact that, as a member (or system of members) is stressed beyond the elastic limit, a redistribution of internal forces and moments may occur. Thus the load-carrying capacity of any system is not necessarily limited by general yielding of any one member. Likewise, the load-carrying capacity of any single member is not necessarily limited by general yielding at any single section.

Plastic analysis of ductile materials is facilitated by the assumption of an *idealized* stress-strain curve such as is shown in Fig. 17.1. As indicated, by such a curve, it is assumed that the proportional limit and the yield point (or yield strength) are equal and that the stress-strain behavior of the material in the yield range can be represented by a horizontal line. It should be noted that, although the material could sustain additional stress beyond the yield range, such additional stress would be preceded by excessive strain. For this reason, even by plastic analysis, it is common practice to limit stresses to the yield point of the material being used. Structural steel, which has a well-defined yield point, is an outstanding example of a material which can be assumed to behave in this manner. However, the assumption can also be made for certain nonferrous alloys which do not have a definite yield point but which are sufficiently ductile. Materials for which an idealized stress-strain curve, such as in Fig. 17.1, can be assumed are frequently termed *elastic-plastic* materials.

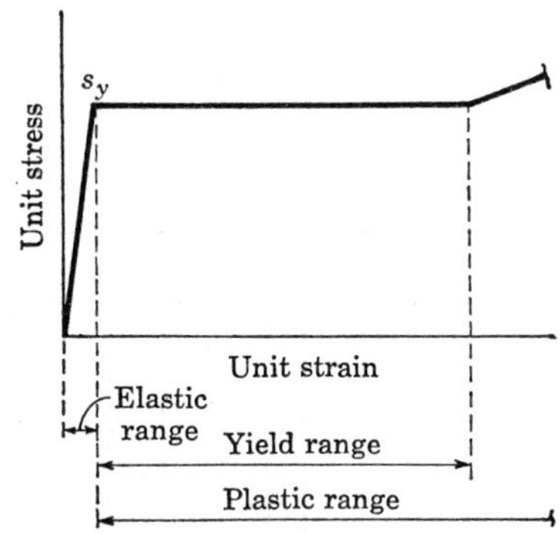

FIG. 17.1 Idealized stress-strain curve.

Since plastic analysis is generally a more realistic approach to the determination of structural strength, its use will frequently result in substantial savings of materials. It will be seen that these savings are particularly significant when the method is applied to statically indeterminate structures. It can also be noted that the method of plastic analysis generally involves fewer and simpler computations than would be required by elastic analysis.

17.2 AXIAL LOADING

By elastic analysis the maximum load for any statically indeterminate system of axially loaded members is limited by incipient yielding of any one member. Actually, however, such a system will not fail until substantial yielding has occurred in all the members or until the system has been made unstable by yielding (of one or more members). For this reason the actual load-carrying capacity of any statically indeterminate system can be more closely predicted by plastic rather than elastic analysis.

Example 17.1. A uniform rigid block is to be supported on three bars as shown in Fig. 17.2. Bars A and C are of aluminum (E = 10,000,000 psi; s_y = 30,000 psi); each is 12.000 in. long and 1.5 in. square. Bar B is of steel (E = 30,000,000 psi; s_y = 33,000 psi), 7.990 in. long and 1.0 in. square. Based on a factor of safety of 1.75, determine the allowable weight of the block.

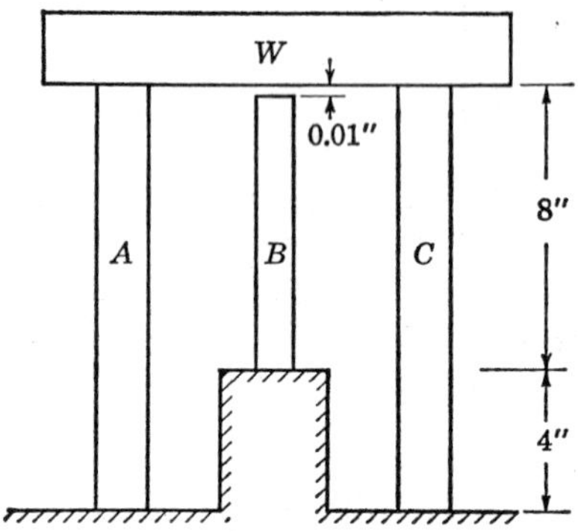

FIG. 17.2

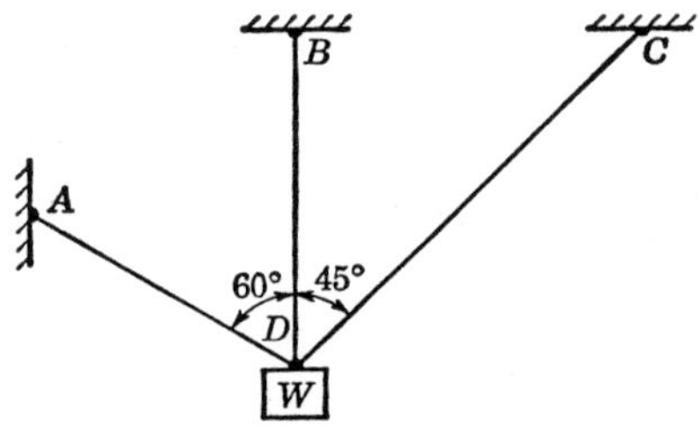

FIG. 17.3

Determining the weight W which would cause yielding in all three bars,

$$W = P_A + P_B + P_C = 2P_A + P_B$$
$$W = 2(30{,}000)(2.25) + 33{,}000(1)$$
$$W = 135{,}000 + 33{,}000 = 168{,}000 \text{ lb}$$

Therefore, based on the prescribed factor of safety, the allowable weight is

$$W = \frac{168{,}000}{1.75} = 96{,}000 \text{ lb} \qquad \textit{Ans}$$

Example 17.2. A weight W is to be supported by three pin-connected steel bars as shown in Fig. 17.3. Each bar is 0.25 in.^{2} in cross section. Based on a yield point of 33,000 psi and a factor of safety of 2, determine the allowable value of W.

It is noted that, in order for equilibrium to be maintained, the horizontal components of the forces in DA and DC must be equal. Thus from a free-body diagram of pin D

$$\Sigma H = 0 \qquad 0.707P_{DC} = 0.866P_{DA}$$

and

$$P_{DC} = 1.225P_{DA}$$

Therefore, for member DA to yield, member DC would have to be stressed well beyond the yield point. Thus the limiting internal forces for the three members are

$$P_{DB} = P_{DC} = 0.25(33{,}000) = 8250$$

and

$$P_{DA} = \frac{8250}{1.225} = 6730$$

From the free-body diagram of pin D, the weight which would cause failure is

$$\Sigma V = 0 \qquad W = 6730(0.5) + 8250 + 8250(0.707)$$
$$W = 17{,}450 \text{ lb}$$

The allowable weight, based on the prescribed factor of safety, is

$$W = \frac{17{,}450}{2} = 8725 \text{ lb} \qquad \textit{Ans}$$

PROBLEMS

NOTE: Assume all materials in the following problems to be elastic-plastic.

17.1. An aluminum rod 15.006 in. long is to be placed inside a steel pipe 15.000 in. long and loaded with a weight W as shown in Fig. P17.1. Cross-sectional areas are aluminum = 4 in.^{2}, steel = 1.5 in.^{2} Based on s_y = 27,000 psi for aluminum, s_y = 33,000 psi for steel, and a factor of safety of 1.8, determine the allowable load.

17.2. A brass rod 10.005 in. long is to be placed inside an aluminum pipe 10.000 in. long and loaded with a weight W as shown in Fig. P17.2. Cross-sectional areas are brass = 2 in.^{2}, aluminum = 1.2 in.^{2} Based on s_y = 40,000 psi for brass, s_y = 30,000 psi for aluminum, and a factor of safety of 2, determine the allowable load.

★**17.3.** A rigid horizontal beam is supported by a hinge and two wires as shown in Fig. P17.3. The left wire is of aluminum (s_y = 27,000 psi) and has a cross-sectional area of 1 in.^{2}; the right wire is of steel (s_y = 33,000 psi) and has a cross-sectional area of 0.6 in.^{2} If a = 4 ft, b = c = 2 ft, determine the allowable load Q based on a factor of safety of 1.8.

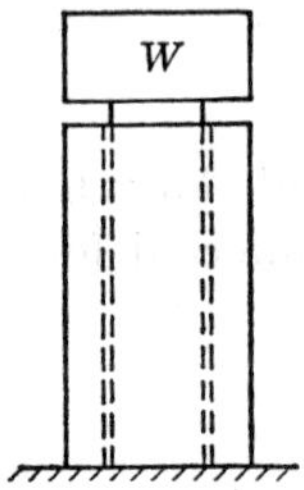

FIG. P17.1, P17.2

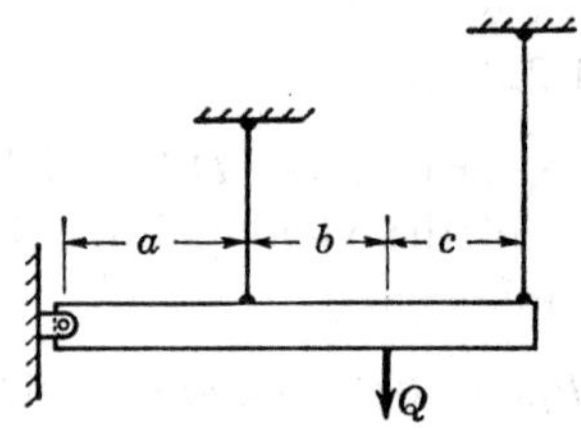

FIG. P17.3, P17.4

17.4. A rigid horizontal beam is supported by a hinge and two wires as shown. The left wire is of steel (s_y = 33,000 psi) and has a cross-sectional area of 0.8 in.^{2}; the right wire is of aluminum (s_y = 30,000 psi) and has a cross-sectional area of 0.6 in.^{2} If a = 2 ft, b = 1 ft, c = 3 ft, determine the allowable load Q based on a factor of safety of 1.9.

★**17.5.** A weight W is to be supported by three ¼-in.-diam steel wires as shown in Fig. P17.5. If $\theta = 30°$, determine the allowable weight based on s_y = 33,000 psi and a factor of safety of 1.65.

17.6. A weight W is to be supported by three wires as shown in Fig. P17.6. The middle wire is of aluminum (s_y = 27,000 psi) and is ⅜ in. in diameter; the two outer wires are of steel (s_y = 33,000 psi) and are $\frac{3}{16}$ in. in diameter. If $\theta = 20°$, determine the allowable weight based on a factor of safety of 1.8.

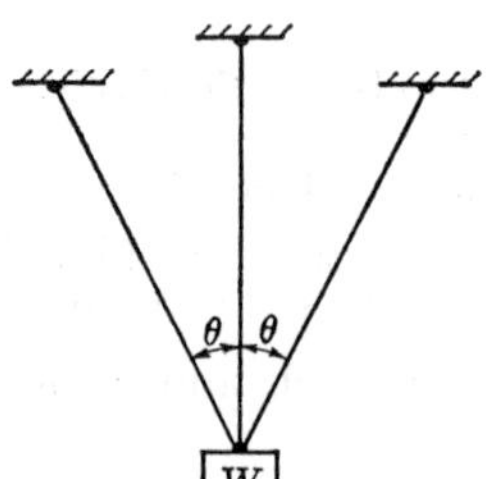

FIG. P17.5, P17.6

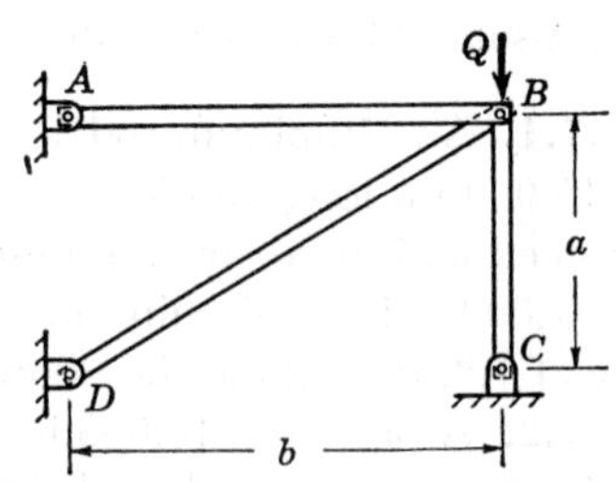

FIG. P17.7, P17.8

★**17.7.** For the framework shown in Fig. P17.7 let a = 3 ft and b = 4 ft. Cross-sectional areas of members AB, BC, and BD are 4, 3, and 4 in.^{2}, respectively. Determine the allowable load Q based on $s_y = 33{,}000$ psi and a factor of safety of 1.65.

17.8. For the framework shown in Fig. P17.8 let $a = 5$ ft and $b = 12$ ft. Cross-sectional areas for members AB, BC, and BD are 6, 4, and 8 in.^{2}, respectively. Determine the allowable load Q based on $s_y = 33{,}000$ psi and a factor of safety of 1.65.

17.3 PLASTIC ACTION OF BEAMS

Any ductile flexural member can be loaded until the section of maximum bending moment has been yielded throughout; such a section is said to constitute a *plastic hinge*. The significance of a plastic hinge and the manner in which it is formed in an elastic-plastic material will be explained with reference to Fig. 17.4. The explanation will be based on the assumptions made in the derivation of the flexure formula and on the further assumption that the yield point of the material is the same in tension and compression. As the bending moment at any section is gradually increased, a linear stress distribution will occur until the maximum bending stress reaches the yield point s_y; this stress distribution is shown in Fig. 17.4*a*. With further increase of moment, the material which has reached the yield point will maintain a constant stress, although continuing to deform. Therefore, in order to furnish additional resisting moment, progressively more material must be stressed to the yield point. This progressive yielding can continue, as indi-

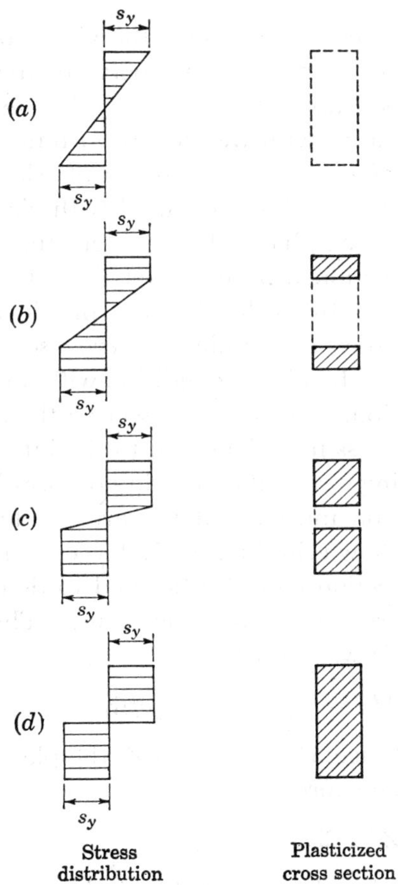

FIG. 17.4 Development of plastic hinge.

cated by Fig. 17.4*b* and *c*, until complete plastification of the cross section is approached. The limit of the process of progressive yielding is shown in Fig. 17.4*d*. Any section with such a stress distribution is termed a plastic hinge, since, for practical purposes, it could resist no additional moment. Any further increase of loading would merely cause the beam to rotate at the section without any change of stress (and, therefore, without any change of internal moment).

The moment in a plastic hinge M_p can be expressed as

$$M_p = s_y Z \tag{17.1}$$

where s_y represents the yield point (or yield strength) of the material and Z is a property of the cross section known as the *plastic modulus*. As a matter of interest, the above equation can be compared with the flexure formula, which can be written in the form $M = s(I/c)$. Thus the plastic modulus Z is analogous to the section modulus I/c used with elastic analysis of bending. It should be emphasized, however, that the plastic modulus of any cross section will be greater than its section modulus.

In evaluating the plastic modulus of a section it should be noted that, unless the cross section is symmetrical, the neutral axis for the plastic condition will not be the same as for the elastic condition. Since at any section the internal force due to tensile bending stress must be equal to the internal force due to compressive bending stress, the area in compression at a completely plasticized section must equal the area in tension. Thus for the cross section shown in Fig. 17.5, for example, a_c must be equal to a_t. This section can also be used to define the plastic modulus in terms of cross-sectional dimensions. The plastic moment of the section can be written as

$$M_p = (s_y a_c)\bar{y}_c + (s_y a_t)\bar{y}_t$$

Since M_p/s_y defines Z, the plastic modulus of any cross section can be expressed as

$$Z = \Sigma a\bar{y} \tag{17.2}$$

or as the sum of the *absolute values* of the first moments of both tension and compression areas with respect to the neutral axis (of the plasticized section). For example, the plastic modulus of any rectangular section of width b and depth h would be

$$Z = \frac{bh}{2}\frac{h}{4} + \frac{bh}{2}\frac{h}{4} = \frac{bh^2}{4}$$

This can be compared with the corresponding section modulus I/c of $bh^2/6$. The ratio of the plastic modulus Z of any section to its section modulus I/c is known as its *shape factor*. Thus the shape factor for a rectangular section is 1.5. The *average* shape factor for the wide-flange sections† most commonly used as beams is 1.12.

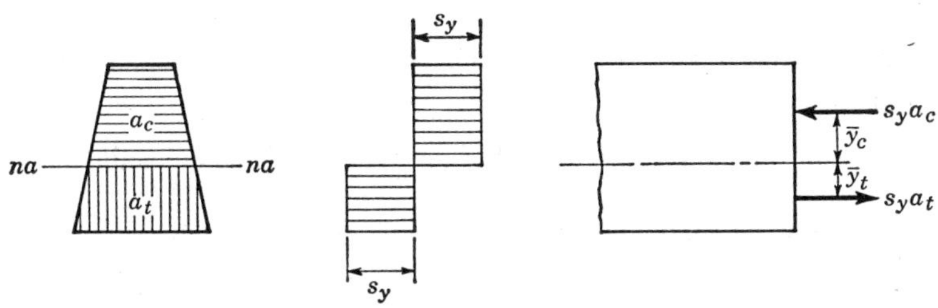

FIG. 17.5 Plastic hinge moment.

By plastic analysis, failure of any statically determinate beam is assumed when a plastic hinge is developed at any section. Although some additional loading could be sustained, it would be accompanied by excessive beam deflection.

Example 17.3. The simply supported beam shown in Fig. 17.6 is made of structural steel having $s_y = 33{,}000$ psi. Based on a factor of safety of 1.65, determine the allowable uniform load.

By inspection, the neutral axis of the plasticized section will be at the bottom of the flange. Therefore

$$Z = \Sigma a\bar{y} = 6(1)(\tfrac{1}{2}) + 6(1)(3) = 21 \text{ in.}^3$$

If the loading were gradually increased, a plastic hinge would first develop at midspan, where the moment would be a maximum and equal to $8w$.

† Values of plastic modulus for various standard steel beam sections are tabulated in the appendix of Lynn S. Beedle, "Plastic Design of Steel Frames," John Wiley & Sons, Inc., New York, 1958.

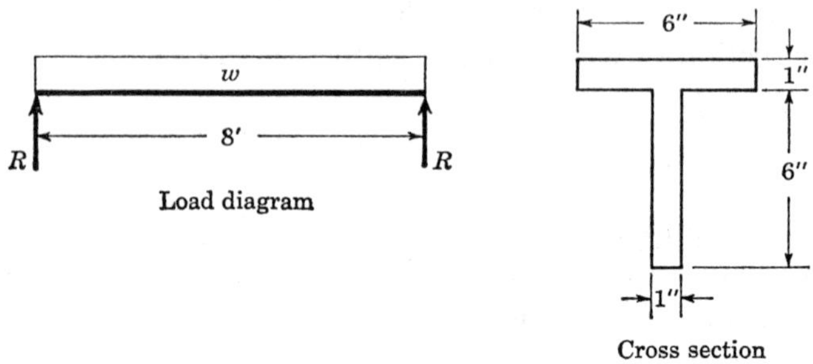

FIG. 17.6

Therefore the uniform load which would cause failure is determined as

$$M_p = s_y Z \qquad (8w)(12) = 33{,}000(21)$$
$$w = 7220 \text{ lb/ft}$$

and, based on the prescribed factor of safety, the allowable uniform load is

$$w = \frac{7220}{1.65} = 4380 \text{ lb/ft} \qquad \textit{Ans}$$

PROBLEMS

NOTE: Assume all materials in the following problems to be elastic-plastic.

★17.9. Determine the plastic modulus and shape factor for a cylindrical beam of radius R.

17.10. Determine the plastic modulus of a hollow cylindrical beam having inner and outer radii of r and R.

17.11. Determine the plastic modulus and shape factor of a hollow cylindrical beam having inside and outside diameters of 5 and 6 in.

★17.12. Neglecting fillets, determine the plastic modulus and shape factor of a 16WF40 section.

17.13. Neglecting fillets, determine the plastic modulus and shape factor of a 27WF177 section.

★17.14. A 16-ft simply supported beam is loaded with 16 kips at 4 ft from the left end and 12 kips at 6 ft from the right end. Based on a yield point of 33,000 psi and a factor of safety of 1.65, determine the plastic modulus required.

17.15. A 16-ft simply supported beam is loaded with 20 kips at midspan and with 10 kips at 4 ft from each end. Based on a yield point of 33,000 psi and a factor of safety of 1.65, determine the plastic modulus required.

17.16. An aluminum (s_y = 28,000 psi) beam having the load diagram shown in Fig. P17.16 is to have a rectangular cross section with the depth equal to three times the width. Based on a factor of safety of 1.5, determine the minimum cross-sectional dimensions required.

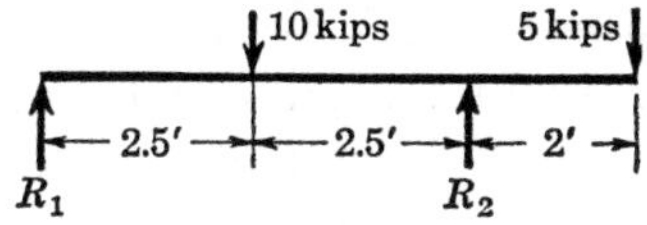

FIG. P17.16

17.17. An aluminum (s_y = 28,000 psi) beam is to have the load diagram and cross section shown in Fig. P17.17. Based on a factor of safety of 1.6, determine the allowable value of *P*.

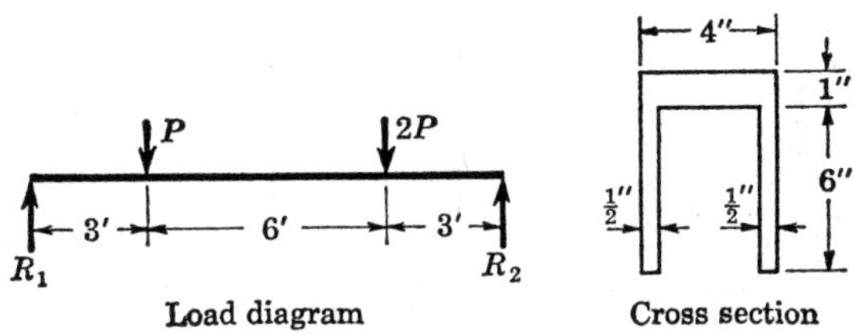

FIG. P17.17

★**17.18.** A steel (s_y = 33,000 psi) beam is to have the cross section and load diagram shown in Fig. P17.18. Based on a factor of safety of 1.65, determine the allowable uniform load *w*.

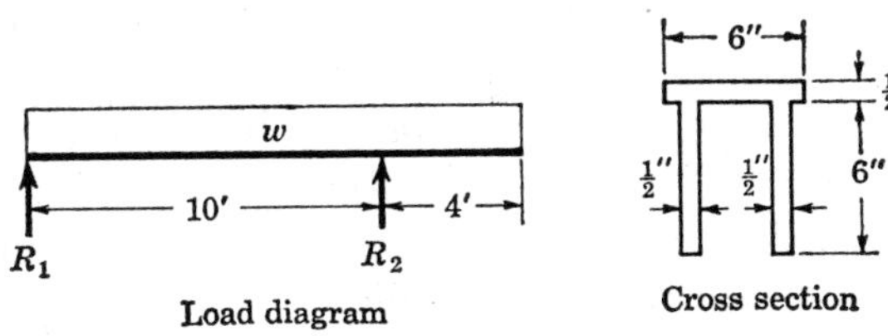

FIG. P17.18

17.4 RESTRAINED BEAMS

As has been stated, any statically determinate beam will fail whenever it develops a single plastic hinge. Statically indeterminate beams, however, will not fail until plastic hinges have been developed at more than one section. In fact, failure will occur only when the number of plastic hinges exceeds the number of redundant reaction components (reaction components which are not needed for stability) by one.† Thus the fixed beam shown in Fig. 17.7, which has two redundant reaction components, would not fail until three plastic hinges were developed. Similarly the propped beam shown in Fig. 17.8, which has only one redundant reaction component, would fail after the development of only two plastic hinges.

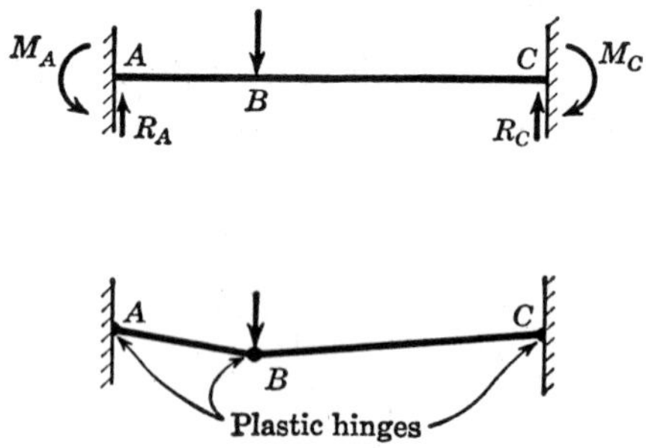

FIG. 17.7 Plastic failure of restrained beam.

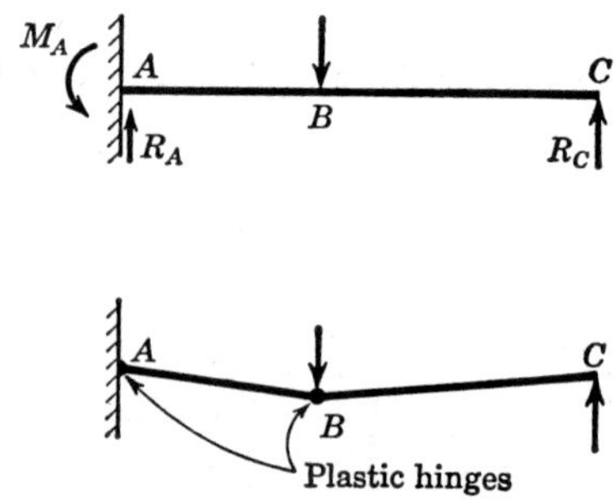

FIG. 17.8 Plastic failure of propped beam.

It should be noted that, as the loading of a statically indeterminate beam is increased, a "redistribution" of moment will occur upon the formation of each plastic hinge (except the last). Thus, after a plastic hinge has been formed, the moment at that section remains constant and the section acts as a real hinge with respect to further loading. This process is continued, with the formation of each hinge, until the final hinge forms; although some additional loading could be sustained, it would be accompanied by excessive deflection. It will be seen that the recognition and utilization of this redistribution-of-moment principle permit much higher allowable loads than would be permitted by elastic analysis.

Example 17.4. A 12WF31 section is to be used as a beam having the load diagram shown in Fig. 17.9*a*. Based on a yield point of 33,000 psi and a factor of safety of 1.65, determine the allowable load by (*a*) elastic analysis and (*b*) plastic analysis. Assume a shape factor of 1.12 for the section.

† Of course this rule does not necessarily apply to members subjected to axial loading in addition to bending.

(*a*) Using values from Table 13.3, the elastic-moment diagram is sketched as shown in Fig. 17.9*b*. Since, by elastic analysis, failure is assumed when the yield stress is reached at the section of maximum moment,

$$s = \frac{M}{I/c} \qquad 33{,}000 = \frac{27w(12)}{39.4}$$
$$w = 4010$$

And based on the prescribed factor of safety, the allowable uniform load is

$$w = \frac{4010}{1.65} = 2430 \text{ lb/ft} \qquad \textit{Ans}$$

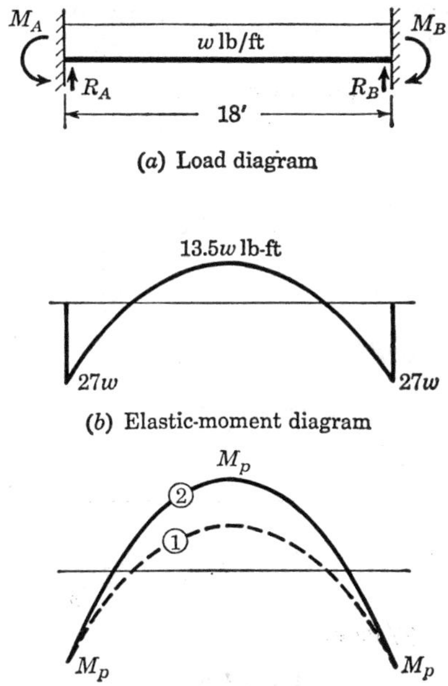

(*a*) Load diagram

(*b*) Elastic-moment diagram

(*c*) Plastic-moment diagram

FIG. 17.9

(*b*) If the loading were increased beyond 4010 lb/ft, plastic hinges would form at the supports; upon the formation of these hinges the moment diagram would be as shown by curve 1 of Fig. 17.9*c*. Failure would not occur, however, until the loading was further increased and an additional plastic hinge formed at midspan; the moment diagram at failure is shown as curve 2 of Fig. 17.9*c*. To express M_p in terms of the load, the moment at midspan $+M_p$ is written in terms of the moment at either support:

$$+M_p = -M_p + 9w(9) - 9w(4.5) \qquad \text{or}$$
$$M_p = 20.25w$$

Using $Z = 1.12(I/c) = 1.12(39.4) = 44.1$,

$$M_p = s_y Z \qquad (20.25w)(12) = 33{,}000(44.1)$$
$$w = 5990$$

and based on the prescribed factor of safety, the allowable uniform load is

$$w = \frac{5990}{1.65} = 3630 \text{ lb/ft} \qquad \textit{Ans}$$

Thus, with the same factor of safety, the allowable load is 49 per cent greater by plastic analysis than by elastic analysis. Of this 49 per cent, 12 per cent is due to plastification of the section and the remainder due to redistribution of bending moment.

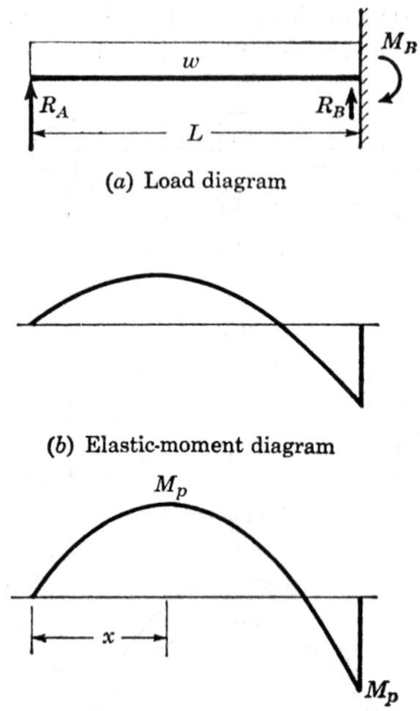

FIG. 17.10

Example 17.5. For a propped beam supporting a uniform load over its entire length, as shown in Fig. 17.10*a*, determine the critical load in terms of the plastic moment.

The elastic-moment diagram is sketched as shown in Fig. 17.10*b*. Loading can be increased until two plastic hinges are formed as indicated in Fig. 17.10*c*.

From a free body of the entire span,

$$M_p + R_A L - \frac{wL^2}{2} = 0 \qquad (a)$$

From a free body of the segment to the left of the section of maximum positive bending moment,

$$-M_p + R_A x - \frac{wx^2}{2} = 0 \qquad (b)$$

Also, since the shear is zero at the section of maximum positive bending moment,

$$+R_A - wx = 0 \qquad (c)$$

Solving the above three equations simultaneously gives

$$x = 0.414L \qquad R_A = 0.414wL \qquad \text{and} \qquad M_p = 0.0858wL^2$$

Therefore

$$w = 11.66\frac{M_p}{L^2} \qquad Ans$$

PROBLEMS

NOTE: For the following problems, use s_y = 33,000 psi and a factor of safety of 1.65.

★**17.19.** A 20-ft beam is to be restrained at both ends; equal loads of 36 kips are to be applied 5 ft from each end. Determine the plastic modulus required.

17.20. A 12-ft beam is to be restrained at both ends and is to support a 64-kip load at 3 ft from the left end. Determine the plastic modulus required.

17.21. A 12-ft beam having a plastic modulus of 100 in.^{3} is fixed at the left end and supported on rollers at the right end. Determine the allowable midspan load.

★**17.22.** A 20-ft beam having a plastic modulus of 150 in.^{3} is fixed at the right end and supported on rollers at the left end. Equal concentrated loads are to be applied 5 ft from each end. Determine the allowable loads.

17.23. A 20-ft beam having a plastic modulus of 120 in.^{3} is restrained at both ends. Equal concentrated loads are to be applied 5 ft from each end. Determine the allowable load.

17.24. A 24-ft beam having a plastic modulus of 200 in.^{3} is restrained at both ends. It is to support a uniform load of w lb/ft over its entire length and a concentrated midspan load of $10w$. Determine the allowable value of w.

17.5 CONTINUOUS BEAMS

Plastic analysis is especially advantageous for application to continuous beams in that the computations involved are much simpler than would be required by elastic analysis. This simplicity arises from the fact that, because of the plastic hinges which form at the supports, the collapse loading for each span can be computed independently.

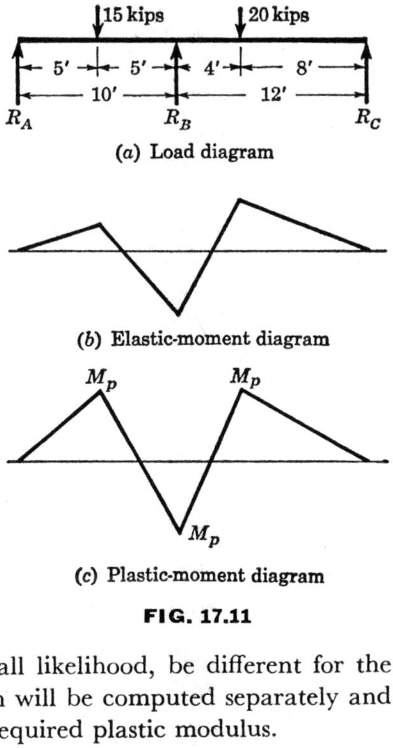

FIG. 17.11

Example 17.6. A continuous steel beam is to have a constant cross section and is to be loaded as shown in Fig. 17.11*a*. Based on a yield point of 33,000 psi and a factor of safety of 1.65, determine the plastic modulus required.

The elastic-moment diagram is sketched (for shape only) as shown in Fig. 17.11*b*. Assuming that the two spans would collapse simultaneously, the plastic-moment diagram would be as shown in Fig. 17.11*c*. Actually the plastic moment will, in all likelihood, be different for the two spans; therefore, M_p for each span will be computed separately and the larger value used to compute the required plastic modulus.

To evaluate M_p for span AB, free bodies of the entire span and of the left half of the span are used. Thus

$+10R_A - 15(5) + M_p = 0$ and
$+5R_A - M_p = 0$

Solving these two equations simultaneously,

$M_p = 25$ kip/ft

To evaluate M_p for span BC, free bodies of the entire span and of the right 8 ft of the span are used. Thus

$+12R_C - 20(4) + M_p = 0$ and
$+8R_C - M_p = 0$

Solving these two equations simultaneously,

$M_p = 32$ kip-ft

The required plastic modulus is governed by the plastic moment of span BC; hence, with the prescribed factor of safety,

$$Z = \frac{M_p}{s_y/1.65} = 1.65\,\frac{32(12)}{33} = 19.2 \text{ in.}^3 \qquad Ans$$

Example 17.7. A continuous beam fixed at both ends is to be loaded as shown in Fig. 17.12*a*. The plastic moduli for spans AB and BC are 36 and 48 in.^{3}, respectively. Based on a yield point of 33,000 psi and a factor of safety of 1.65, determine the allowable load.

The elastic-moment diagram is sketched as shown in Fig. 17.12*b*; the plastic-moment diagram is sketched as shown in Fig. 17.12*c*. It should be noted that, because the plastic moduli for the two spans are different, the plastic moments M_{p1} and M_{p2} are not equal; also, since $M_p = s_yZ$,

$M_{p1} = {}^{36}\!/_{48}M_{p2} = 0.75M_{p2}$

Noting that span AB in the plastic condition is symmetrically loaded and that therefore $R_A = 6w$, a free body of the left half of the span can be used to define the plastic moment in terms of loading. Thus

$+6w(6) - M_{p1} - 6w(3) - M_{p1} = 0$ or
$M_{p1} = 9w$

Determining the loading which would fail the left span,

$M_p = s_yZ \qquad 9w(12) = 33{,}000(36)$
$w = 11{,}000$ lb/ft

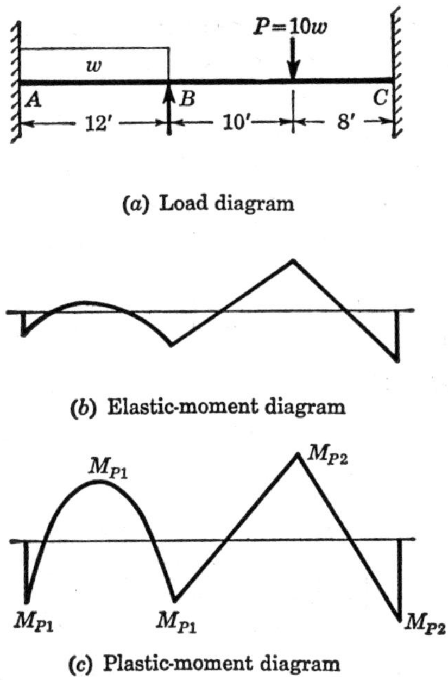

FIG. 17.12

To evaluate M_{p2} in terms of the loading of span BC, free-body diagrams of the entire span and of the right 8 ft of the span are used. Thus

$$+18R_C - M_{p2} - 10w(10) + M_{p1} = 0 \qquad \text{and}$$
$$+8R_C - M_{p2} - M_{p2} = 0$$

Setting $M_{p1} = 0.75M_{p2}$ and solving the above equations simultaneously,

$$M_{p2} = 23.5w$$

Determining the loading which would fail the right span,

$$M_p = s_y Z \qquad 23.5w(12) = 33{,}000(48)$$
$$w = 5620 \text{ lb/ft}$$

The allowable loading, therefore, is limited by the strength of span BC and with the prescribed factor of safety

$$w = \frac{5620}{1.65} = 3400 \text{ lb/ft} \qquad \textit{Ans}$$

$$P = 3400(10) = 34{,}000 \text{ lb} \qquad \textit{Ans}$$

PROBLEMS

NOTE: For the following problems use $s_y = 33{,}000$ psi and a factor of safety of 1.65.

17.25–17.27. A continuous beam of constant cross section is to have the load diagram shown in Figs. P17.25 to P17.27. Determine the plastic modulus required. (*17.25)

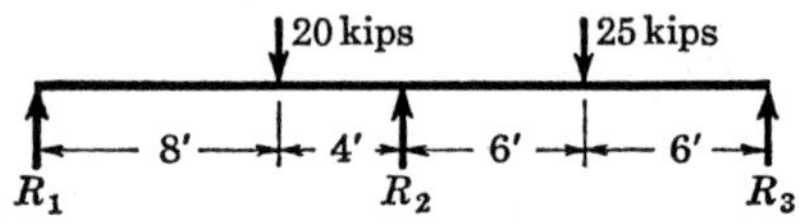

FIG. P17.25

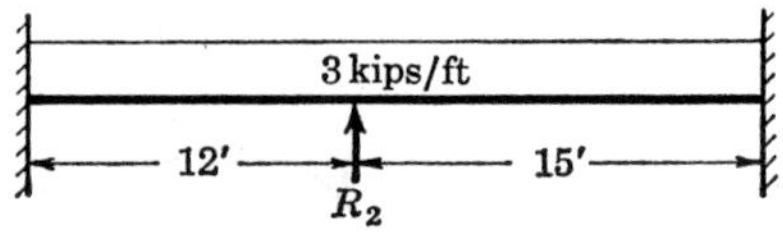

FIG. P17.26

17.28. A continuous beam of constant cross section is to have the load diagram shown in Fig. P17.28. If the plastic modulus is 60 in.^{3}, determine the allowable load.

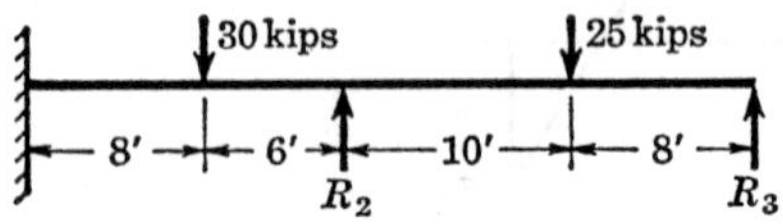

FIG. P17.27

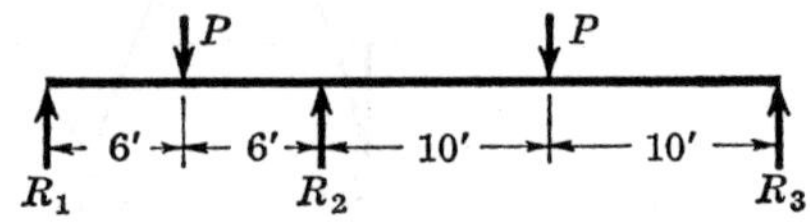

FIG. P17.28, P17.29

17.29. A continuous beam is to have the load diagram shown in Fig. P17.29. The plastic modulus of the left span is 20 in.^{3}; the plastic modulus of the right span is 40 in.^{3} Determine the allowable load.

★17.30. A continuous beam of constant cross section is to have the load diagram shown in Fig. P17.30. If the plastic modulus is 60 in.^{3}, determine the allowable load.

17.31. A continuous beam is to have the load diagram shown in Fig. P17.31. The plastic modulus of the left span is 50 in.^{3}; the plastic modulus of the right span is 40 in.^{3} Determine the allowable load.

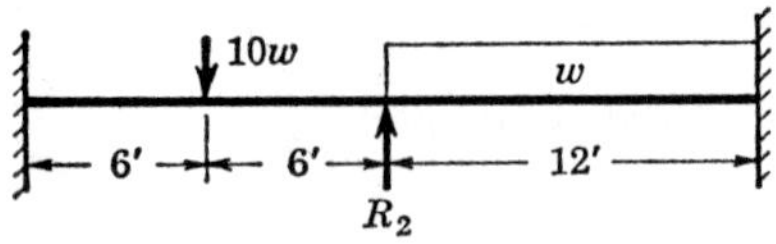

FIG. P17.30, P17.31

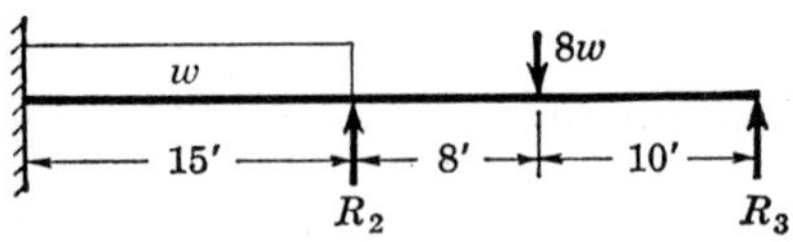

FIG. P17.32, P17.33

17.32. A continuous beam of constant cross section is to have the load diagram shown in Fig. P17.32. If the plastic modulus is 100 in.^{3}, determine the allowable load.

★17.33. A continuous beam is to have the load diagram shown in Fig. P17.33. The plastic modulus of the left span is 60 in.^{3}; the plastic modulus of the right span is 80 in.^{3} Determine the allowable load.

APPENDIX

The tables herein are all selected from the *Steel Construction Manual of the American Institute of Steel Construction* and are reproduced by permission. Some of the tables are abridged from the original source.

TABLES

TABLE A.1

TIMBER
AMERICAN STANDARD SIZES
PROPERTIES FOR DESIGNING
NATIONAL LUMBER MANUFACTURERS ASSOCIATION

Nominal Size	American Standard Dressed Size	Area of Section	Weight per Foot	Moment of Inertia	Section Modulus
In.	In.	In.^{2}	Lb.	In.^{4}	In.^{3}
2 x 4	1⅝x3⅝	5.89	1.64	6.45	3.56
6	5⅝	9.14	2.54	24.1	8.57
8	7½	12.2	3.39	57.1	15.3
10	9½	15.4	4.29	116	24.4
12	11½	18.7	5.19	206	35.8
14	13½	21.9	6.09	333	49.4
16	15½	25.2	6.99	504	65.1
18	17½	28.4	7.90	726	82.9
3 x 4	2⅝x3⅝	9.52	2.64	10.4	5.75
6	5⅝	14.8	4.10	38.9	13.8
8	7½	19.7	5.47	92.3	24.6
10	9½	24.9	6.93	188	39.5
12	11½	30.2	8.39	333	57.9
14	13½	35.4	9.84	538	79.7
16	15½	40.7	11.3	815	105
18	17½	45.9	12.8	1172	134
4 x 4	3⅝x3⅝	13.1	3.65	14.4	7.94
6	5⅝	20.4	5.66	53.8	19.1
8	7½	27.2	7.55	127	34.0
10	9½	34.4	9.57	259	54.5
12	11½	41.7	11.6	459	79.9
14	13½	48.9	13.6	743	110
16	15½	56.2	15.6	1125	145
18	17½	63.4	17.6	1619	185
6 x 6	5½x5½	30.3	8.40	76.3	27.7
8	7½	41.3	11.4	193	51.6
10	9½	52.3	14.5	393	82.7
12	11½	63.3	17.5	697	121
14	13½	74.3	20.6	1128	167
16	15½	85.3	23.6	1707	220
18	17½	96.3	26.7	2456	281
20	19½	107.3	29.8	3398	349
8 x 8	7½x7½	56.3	15.6	264	70.3
10	9½	71.3	19.8	536	113
12	11½	86.3	23.9	951	165
14	13½	101.3	28.0	1538	228
16	15½	116.3	32.0	2327	300
18	17½	131.3	36.4	3350	383
20	19½	146.3	40.6	4634	475
22	21½	161.3	44.8	6211	578

Nominal Size	American Standard Dressed Size	Area of Section	Weight per Foot	Moment of Inertia	Section Modulus
In.	In.	In.^{2}	Lb.	In.^{4}	In.^{3}
10x10	9½x9½	90.3	25.0	679	143
12	11½	109	30.3	1204	209
14	13½	128	35.6	1948	289
16	15½	147	40.9	2948	380
18	17½	166	46.1	4243	485
20	19½	185	51.4	5870	602
22	21½	204	56.7	7868	732
24	23½	223	62.0	10274	874
12x12	11½x11½	132	36.7	1458	253
14	13½	155	43.1	2358	349
16	15½	178	49.5	3569	460
18	17½	201	55.9	5136	587
20	19½	224	62.3	7106	729
22	21½	247	68.7	9524	886
24	23½	270	75.0	12437	1058
14x14	13½x13½	182	50.6	2768	410
16	15½	209	58.1	4189	541
18	17½	236	65.6	6029	689
20	19½	263	73.1	8342	856
22	21½	290	80.6	11181	1040
24	23½	317	88.1	14600	1243
16x16	15½x15½	240	66.7	4810	621
18	17½	271	75.3	6923	791
20	19½	302	83.9	9578	982
22	21½	333	92.5	12837	1194
24	23½	364	101	16763	1427
18x18	17½x17½	306	85.0	7816	893
20	19½	341	94.8	10813	1109
22	21½	376	105	14493	1348
24	23½	411	114	18926	1611
26	25½	446	124	24181	1897
20x20	19½x19½	380	106	12049	1236
22	21½	419	116	16150	1502
24	23½	458	127	21089	1795
26	25½	497	138	26945	2113
28	27½	536	149	33795	2458
24x24	23½x23½	552	153	25415	2163
26	25½	599	166	32472	2547
28	27½	646	180	40727	2962
30	29½	693	193	50275	3408

All properties and weights given are for dressed size only.
The weights given above are based on assumed average weight of 40 pounds per cubic foot.

TABLE A.2

STEEL PIPE

DIMENSIONS							COUPLINGS			PROPERTIES		
Nom. Dia. In.	Outside Dia. In.	Inside Dia. In.	Thickness In.	Weight per Foot Lb. Plain Ends	Weight per Foot Lb. Thread & Cplg.	Threads per Inch	Outside Dia. In.	Length In.	Weight Lb.	I In.^{4}	A In.^{2}	k In.
STANDARD												
1/8	.405	.269	.068	.24	.25	27	.562	7/8	.03	.001	.072	.12
1/4	.540	.364	.088	.42	.43	18	.685	1	.04	.003	.125	.16
3/8	.675	.493	.091	.57	.57	18	.848	1 1/8	.07	.007	.167	.21
1/2	.840	.622	.109	.85	.85	14	1.024	1 3/8	.12	.017	.250	.26
3/4	1.050	.824	.113	1.13	1.13	14	1.281	1 5/8	.21	.037	.333	.33
1	1.315	1.049	.133	1.68	1.68	11 1/2	1.576	1 7/8	.35	.087	.494	.42
1 1/4	1.660	1.380	.140	2.27	2.28	11 1/2	1.950	2 1/8	.55	.195	.669	.54
1 1/2	1.900	1.610	.145	2.72	2.73	11 1/2	2.218	2 3/8	.76	.310	.799	.62
2	2.375	2.067	.154	3.65	3.68	11 1/2	2.760	2 5/8	1.23	.666	1.075	.79
2 1/2	2.875	2.469	.203	5.79	5.82	8	3.276	2 7/8	1.76	1.530	1.704	.95
3	3.500	3.068	.216	7.58	7.62	8	3.948	3 1/8	2.55	3.017	2.228	1.16
3 1/2	4.000	3.548	.226	9.11	9.20	8	4.591	3 5/8	4.33	4.788	2.680	1.34
4	4.500	4.026	.237	10.79	10.89	8	5.091	3 5/8	5.41	7.233	3.174	1.51
5	5.563	5.047	.258	14.62	14.81	8	6.296	4 1/8	9.16	15.16	4.300	1.88
6	6.625	6.065	.280	18.97	19.19	8	7.358	4 1/8	10.82	28.14	5.581	2.25
8	8.625	8.071	.277	24.70	25.00	8	9.420	4 5/8	15.84	63.35	7.265	2.95
8	8.625	7.981	.322	28.55	28.81	8	9.420	4 5/8	15.84	72.49	8.399	2.94
10	10.750	10.192	.279	31.20	32.00	8	11.721	6 1/8	33.92	125.9	9.178	3.70
10	10.750	10.136	.307	34.24	35.00	8	11.721	6 1/8	33.92	137.4	10.07	3.69
10	10.750	10.020	.365	40.48	41.13	8	11.721	6 1/8	33.92	160.7	11.91	3.67
12	12.750	12.090	.330	43.77	45.00	8	13.958	6 1/8	48.27	248.5	12.88	4.39
12	12.750	12.000	.375	49.56	50.71	8	13.958	6 1/8	48.27	279.3	14.58	4.38
EXTRA STRONG												
1/8	.405	.215	.095	.31	.32	27	.582	1 1/8	.05	.001	.093	.11
1/4	.540	.302	.119	.54	.54	18	.724	1 3/8	.07	.004	.157	.15
3/8	.675	.423	.126	.74	.75	18	.898	1 5/8	.13	.009	.217	.20
1/2	.840	.546	.147	1.09	1.10	14	1.085	1 7/8	.22	.020	.320	.25
3/4	1.050	.742	.154	1.47	1.49	14	1.316	2 1/8	.33	.045	.433	.32
1	1.315	.957	.179	2.17	2.20	11 1/2	1.575	2 3/8	.47	.106	.639	.41
1 1/4	1.660	1.278	.191	3.00	3.05	11 1/2	2.054	2 7/8	1.04	.242	.881	.52
1 1/2	1.900	1.500	.200	3.63	3.69	11 1/2	2.294	2 7/8	1.17	.391	1.068	.61
2	2.375	1.939	.218	5.02	5.13	11 1/2	2.870	3 5/8	2.17	.868	1.477	.77
2 1/2	2.875	2.323	.276	7.66	7.83	8	3.389	4 1/8	3.43	1.924	2.254	.92
3	3.500	2.900	.300	10.25	10.46	8	4.014	4 1/8	4.13	3.894	3.016	1.14
3 1/2	4.000	3.364	.318	12.51	12.82	8	4.628	4 5/8	6.29	6.280	3.678	1.31
4	4.500	3.826	.337	14.98	15.39	8	5.233	4 5/8	8.16	9.610	4.407	1.48
5	5.563	4.813	.375	20.78	21.42	8	6.420	5 1/8	12.87	20.67	6.112	1.84
6	6.625	5.761	.432	28.57	29.33	8	7.482	5 1/8	15.18	40.49	8.405	2.20
8	8.625	7.625	.500	43.39	44.72	8	9.596	6 1/8	26.63	105.7	12.76	2.88
10	10.750	9.750	.500	54.74	56.94	8	11.958	6 5/8	44.16	211.9	16.10	3.63
12	12.750	11.750	.500	65.42	68.02	8	13.958	6 5/8	51.99	361.5	19.24	4.34
DOUBLE—EXTRA STRONG												
1/2	.840	.252	.294	1.71	1.73	14	1.085	1 7/8	.22	.024	.504	.22
3/4	1.050	.434	.308	2.44	2.46	14	1.316	2 1/8	.33	.058	.718	.28
1	1.315	.599	.358	3.66	3.68	11 1/2	1.575	2 3/8	.47	.140	1.076	.36
1 1/4	1.660	.896	.382	5.21	5.27	11 1/2	2.054	2 7/8	1.04	.341	1.534	.47
1 1/2	1.900	1.100	.400	6.41	6.47	11 1/2	2.294	2 7/8	1.17	.568	1.885	.55
2	2.375	1.503	.436	9.03	9.14	11 1/2	2.870	3 5/8	2.17	1.311	2.656	.70
2 1/2	2.875	1.771	.552	13.70	13.87	8	3.389	4 1/8	3.43	2.871	4.028	.84
3	3.500	2.300	.600	18.58	18.79	8	4.014	4 1/8	4.13	5.992	5.466	1.05
3 1/2	4.000	2.728	.636	22.85	23.16	8	4.628	4 5/8	6.29	9.848	6.721	1.21
4	4.500	3.152	.674	27.54	27.95	8	5.233	4 5/8	8.16	15.28	8.101	1.37
5	5.563	4.063	.750	38.55	39.20	8	6.420	5 1/8	12.87	33.64	11.34	1.72
6	6.625	4.897	.864	53.16	53.92	8	7.482	5 1/8	15.18	66.33	15.64	2.06
8	8.625	6.875	.875	72.42	73.76	8	9.596	6 1/8	26.63	162.0	21.30	2.76

LARGE O. D. PIPE

Pipe 14″ and larger is sold by actual O. S. diameter and thickness.
Sizes 14″, 15″, and 16″ are available regularly in thicknesses varying by 1/16″ from 1/4″ to 1″, inclusive.

TABLE A.3

SCREW THREADS

American National Form
American Standard, B 1.1—1935.

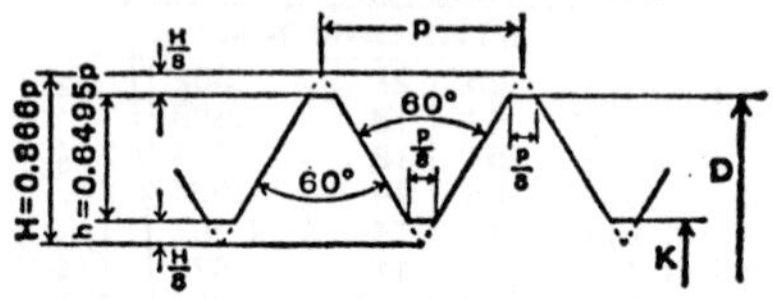

DIAMETER Total D In.	DIAMETER Net K In.	AREA Total Dia., D Sq. In.	AREA Net Dia., K Sq. In.	Number of Threads per Inch
¼	.185	.049	.027	20
⅜	.294	.110	.068	16
½	.400	.196	.126	13
⅝	.507	.307	.202	11
¾	.620	.442	.302	10
⅞	.731	.601	.419	9
1	.838	.785	.551	8
1⅛	.939	.994	.693	7
1¼	1.064	1.227	.890	7
1⅜	1.158	1.485	1.054	6
1½	1.283	1.767	1.294	6
1¾	1.490	2.405	1.744	5
2	1.711	3.142	2.300	4½
2¼	1.961	3.976	3.021	4½
2½	2.175	4.909	3.716	4
2¾	2.425	5.940	4.619	4

DIAMETER Total D In.	DIAMETER Net K In.	AREA Total Dia., D Sq. In.	AREA Net Dia., K Sq. In.	Number of Threads per Inch
3	2.675	7.069	5.621	4
3¼	2.925	8.296	6.720	4
3½	3.175	9.621	7.918	4
3¾	3.425	11.045	9.214	4
4	3.675	12.566	10.608	4
4¼	3.798	14.186	11.330	2⅞
4½	4.028	15.904	12.741	2¾
4¾	4.255	17.721	14.221	2⅝
5	4.480	19.635	15.766	2½
5¼	4.730	21.648	17.574	2½
5½	4.953	23.758	19.268	2⅜
5¾	5.203	25.967	21.262	2⅜
6	5.423	28.274	23.095	2¼

Sizes over 4" are old U. S. Standard; there is no American Standard.
Dimensions are maximum; specify "Free Fit—Class 2." For Bolts from 2½" to 6" diameter it is always necessary to bill the number of threads per Inch.

TABLE A.4

WF SHAPES

PROPERTIES FOR DESIGNING

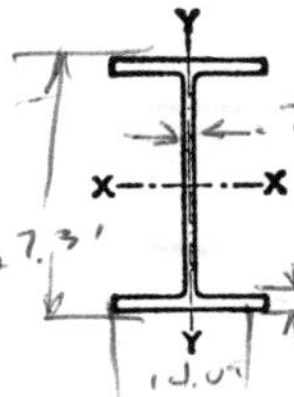

Nominal Size	Weight per Foot	Area	Depth	Flange Width	Flange Thickness	Web Thickness	Axis X-X I	Axis X-X I/c	Axis X-X k	Axis Y-Y I	Axis Y-Y I/c	Axis Y-Y k
In.	Lb.	In.^{2}	In.	In.	In.	In.	In.^{4}	In.^{3}	In.	In.^{4}	In.^{3}	In.
27 x 14	177	52.10	27.31	14.090	1.190	.725	6728.6	492.8	11.36	518.9	73.7	3.16
	160	47.04	27.08	14.023	1.075	.658	6018.6	444.5	11.31	458.0	65.3	3.12
	145	42.68	26.88	13.965	.975	.600	5414.3	402.9	11.26	406.9	58.3	3.09
27 x 10	114	33.53	27.28	10.070	.932	.570	4080.5	299.2	11.03	149.6	29.7	2.11
	102	30.01	27.07	10.018	.827	.518	3604.1	266.3	10.96	129.5	25.9	2.08
	94	27.65	26.91	9.990	.747	.490	3266.7	242.8	10.87	115.1	23.0	2.04
24 x 14	160	47.04	24.72	14.091	1.135	.656	5110.3	413.5	10.42	492.6	69.9	3.23
	145	42.62	24.49	14.043	1.020	.608	4561.0	372.5	10.34	434.3	61.8	3.19
	130	38.21	24.25	14.000	.900	.565	4009.5	330.7	10.24	375.2	53.6	3.13
24 x 12	120	35.29	24.31	12.088	.930	.556	3635.3	299.1	10.15	254.0	42.0	2.68
	110	32.36	24.16	12.042	.855	.510	3315.0	274.4	10.12	229.1	38.0	2.66
	100	29.43	24.00	12.000	.775	.468	2987.3	248.9	10.08	203.5	33.9	2.63
24 x 9	94	27.63	24.29	9.061	.872	.516	2683.0	220.9	9.85	102.2	22.6	1.92
	84	24.71	24.09	9.015	.772	.470	2364.3	196.3	9.78	88.3	19.6	1.89
	76	22.37	23.91	8.985	.682	.440	2096.4	175.4	9.68	76.5	17.0	1.85
21 x 13	142	41.76	21.46	13.132	1.095	.659	3403.1	317.2	9.03	385.9	58.8	3.04
	127	37.34	21.24	13.061	.985	.588	3017.2	284.1	8.99	338.6	51.8	3.01
	112	32.93	21.00	13.000	.865	.527	2620.6	249.6	8.92	289.7	44.6	2.96
21 x 9	96	28.21	21.14	9.038	.935	.575	2088.9	197.6	8.60	109.3	24.2	1.97
	82	24.10	20.86	8.962	.795	.499	1752.4	168.0	8.53	89.6	20.0	1.93
21 x 8¼	73	21.46	21.24	8.295	.740	.455	1600.3	150.7	8.64	66.2	16.0	1.76
	68	20.02	21.13	8.270	.685	.430	1478.3	139.9	8.59	60.4	14.6	1.74
	62	18.23	20.99	8.240	.615	.400	1326.8	126.4	8.53	53.1	12.9	1.71

TABLE A.4 (Continued)

WF SHAPES

PROPERTIES FOR DESIGNING

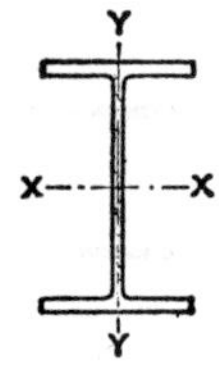

Nominal Size	Weight per Foot	Area	Depth	Flange Width	Flange Thickness	Web Thickness	Axis X-X I	Axis X-X I/c	Axis X-X k	Axis Y-Y I	Axis Y-Y I/c	Axis Y-Y k
In.	Lb.	In.²	In.	In.	In.	In.	In.⁴	In.³	In.	In.⁴	In.³	In.
18 x 11¾	114	33.51	18.48	11.833	.991	.595	2033.8	220.1	7.79	255.6	43.2	2.76
	105	30.86	18.32	11.792	.911	.554	1852.5	202.2	7.75	231.0	39.2	2.73
	96	28.22	18.16	11.750	.831	.512	1674.7	184.4	7.70	206.8	35.2	2.71
18 x 8¾	85	24.97	18.32	8.838	.911	.526	1429.9	156.1	7.57	99.4	22.5	2.00
	77	22.63	18.16	8.787	.831	.475	1286.8	141.7	7.54	88.6	20.2	1.98
	70	20.56	18.00	8.750	.751	.438	1153.9	128.2	7.49	78.5	17.9	1.95
	64	18.80	17.87	8.715	.686	.403	1045.8	117.0	7.46	70.3	16.1	1.93
18 x 7½	60	17.64	18.25	7.558	.695	.416	984.0	107.8	7.47	47.1	12.5	1.63
	55	16.19	18.12	7.532	.630	.390	889.9	98.2	7.41	42.0	11.1	1.61
	50	14.71	18.00	7.500	.570	.358	800.6	89.0	7.38	37.2	9.9	1.59
16 x 11½	96	28.22	16.32	11.533	.875	.535	1355.1	166.1	6.93	207.2	35.9	2.71
	88	25.87	16.16	11.502	.795	.504	1222.6	151.3	6.87	185.2	32.2	2.67
16 x 8½	78	22.92	16.32	8.586	.875	.529	1042.6	127.8	6.74	87.5	20.4	1.95
	71	20.86	16.16	8.543	.795	.486	936.9	115.9	6.70	77.9	18.2	1.93
	64	18.80	16.00	8.500	.715	.443	833.8	104.2	6.66	68.4	16.1	1.91
	58	17.04	15.86	8.464	.645	.407	746.4	94.1	6.62	60.5	14.3	1.88
16 x 7	50	14.70	16.25	7.073	.628	.380	655.4	80.7	6.68	34.8	9.8	1.54
	45	13.24	16.12	7.039	.563	.346	583.3	72.4	6.64	30.5	8.7	1.52
	40	11.77	16.00	7.000	.503	.307	515.5	64.4	6.62	26.5	7.6	1.50
	36	10.59	15.85	6.992	.428	.299	446.3	56.3	6.49	22.1	6.3	1.45
14 x 14½	95	27.94	14.12	14.545	.748	.465	1063.5	150.6	6.17	383.7	52.8	3.71
	87	25.56	14.00	14.500	.688	.420	966.9	138.1	6.15	349.7	48.2	3.70
14 x 12	84	24.71	14.18	12.023	.778	.451	928.4	130.9	6.13	225.5	37.5	3.02
	78	22.94	14.06	12.000	.718	.428	851.2	121.1	6.09	206.9	34.5	3.00
14 x 10	74	21.76	14.19	10.072	.783	.450	796.8	112.3	6.05	133.5	26.5	2.48
	68	20.00	14.06	10.040	.718	.418	724.1	103.0	6.02	121.2	24.1	2.46
	61	17.94	13.91	10.000	.643	.378	641.5	92.2	5.98	107.3	21.5	2.45

TABLE A.4 (Continued)

WF SHAPES

PROPERTIES FOR DESIGNING

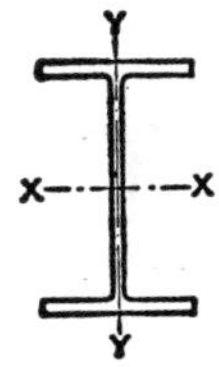

Nominal Size	Weight per Foot	Area	Depth	Flange Width	Flange Thickness	Web Thickness	Axis X-X I	Axis X-X I/c	Axis X-X k	Axis Y-Y I	Axis Y-Y I/c	Axis Y-Y k
In.	Lb.	$In.^2$	In.	In.	In.	In.	$In.^4$	$In.^3$	In.	$In.^4$	$In.^3$	In.
14 x 8	53	15.59	13.94	8.062	.658	.370	542.1	77.8	5.90	57.5	14.3	1.92
	48	14.11	13.81	8.031	.593	.339	484.9	70.2	5.86	51.3	12.8	1.91
	43	12.65	13.68	8.000	.528	.308	429.0	62.7	5.82	45.1	11.3	1.89
14 x 6¾	38	11.17	14.12	6.776	.513	.313	385.3	54.6	5.87	24.6	7.3	1.49
	34	10.00	14.00	6.750	.453	.287	339.2	48.5	5.83	21.3	6.3	1.46
	30	8.81	13.86	6.733	.383	.270	289.6	41.8	5.73	17.5	5.2	1.41
12 x 12	190	55.86	14.38	12.670	1.736	1.060	1892.5	263.2	5.82	589.7	93.1	3.25
	161	47.38	13.88	12.515	1.486	.905	1541.8	222.2	5.70	486.2	77.7	3.20
	133	39.11	13.38	12.365	1.236	.755	1221.2	182.5	5.59	389.9	63.1	3.16
	120	35.31	13.12	12.320	1.106	.710	1071.7	163.4	5.51	345.1	56.0	3.13
	106	31.19	12.88	12.230	.986	.620	930.7	144.5	5.46	300.9	49.2	3.11
	99	29.09	12.75	12.190	.921	.580	858.5	134.7	5.43	278.2	45.7	3.09
	92	27.06	12.62	12.155	.856	.545	788.9	125.0	5.40	256.4	42.2	3.08
	85	24.98	12.50	12.105	.796	.495	723.3	115.7	5.38	235.5	38.9	3.07
	79	23.22	12.38	12.080	.736	.470	663.0	107.1	5.34	216.4	35.8	3.05
	72	21.16	12.25	12.040	.671	.430	597.4	97.5	5.31	195.3	32.4	3.04
	65	19.11	12.12	12.000	.606	.390	533.4	88.0	5.28	174.6	29.1	3.02
12 x 10	58	17.06	12.19	10.014	.641	.359	476.1	78.1	5.28	107.4	21.4	2.51
	53	15.59	12.06	10.000	.576	.345	426.2	70.7	5.23	96.1	19.2	2.48
12 x 8	50	14.71	12.19	8.077	.641	.371	394.5	64.7	5.18	56.4	14.0	1.96
	45	13.24	12.06	8.042	.576	.336	350.8	58.2	5.15	50.0	12.4	1.94
	40	11.77	11.94	8.000	.516	.294	310.1	51.9	5.13	44.1	11.0	1.94
12 x 6½	36	10.59	12.24	6.565	.540	.305	280.8	45.9	5.15	23.7	7.2	1.50
	31	9.12	12.09	6.525	.465	.265	238.4	39.4	5.11	19.8	6.1	1.47
	27	7.97	11.95	6.500	.400	.240	204.1	34.1	5.06	16.6	5.1	1.44

TABLE A.4 (Continued)

WF SHAPES

PROPERTIES FOR DESIGNING

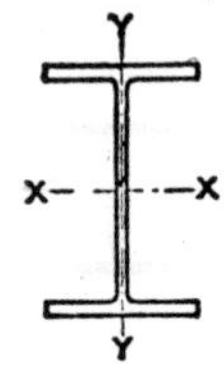

Nominal Size	Weight per Foot	Area	Depth	Flange		Web Thickness	Axis X-X			Axis Y-Y		
				Width	Thickness		I	I/c	k	I	I/c	k
In.	Lb.	$In.^2$	In.	In.	In.	In.	$In.^4$	$In.^3$	In.	$In.^4$	$In.^3$	In.
10 x 10	112	32.92	11.38	10.415	1.248	.755	718.7	126.3	4.67	235.4	45.2	2.67
	100	29.43	11.12	10.345	1.118	.685	625.0	112.4	4.61	206.6	39.9	2.65
	89	26.19	10.88	10.275	.998	.615	542.4	99.7	4.55	180.6	35.2	2.63
	77	22.67	10.62	10.195	.868	.535	457.2	86.1	4.49	153.4	30.1	2.60
	72	21.18	10.50	10.170	.808	.510	420.7	80.1	4.46	141.8	27.9	2.59
	66	19.41	10.38	10.117	.748	.457	382.5	73.7	4.44	129.2	25.5	2.58
	60	17.66	10.25	10.075	.683	.415	343.7	67.1	4.41	116.5	23.1	2.57
	54	15.88	10.12	10.028	.618	.368	305.7	60.4	4.39	103.9	20.7	2.56
	49	14.40	10.00	10.000	.558	.340	272.9	54.6	4.35	93.0	18.6	2.54
10 x 8	45	13.24	10.12	8.022	.618	.350	248.6	49.1	4.33	53.2	13.3	2.00
	39	11.48	9.94	7.990	.528	.318	209.7	42.2	4.27	44.9	11.2	1.98
	33	9.71	9.75	7.964	.433	.292	170.9	35.0	4.20	36.5	9.2	1.94
10 x 5¾	29	8.53	10.22	5.799	.500	.289	157.3	30.8	4.29	15.2	5.2	1.34
	25	7.35	10.08	5.762	.430	.252	133.2	26.4	4.26	12.7	4.4	1.31
	21	6.19	9.90	5.750	.340	.240	106.3	21.5	4.14	9.7	3.4	1.25
8 x 8	67	19.70	9.00	8.287	.933	.575	271.8	60.4	3.71	88.6	21.4	2.12
	58	17.06	8.75	8.222	.808	.510	227.3	52.0	3.65	74.9	18.2	2.10
	48	14.11	8.50	8.117	.683	.405	183.7	43.2	3.61	60.9	15.0	2.08
	40	11.76	8.25	8.077	.558	.365	146.3	35.5	3.53	49.0	12.1	2.04
	35	10.30	8.12	8.027	.493	.315	126.5	31.1	3.50	42.5	10.6	2.03
	31	9.12	8.00	8.000	.433	.288	109.7	27.4	3.47	37.0	9.2	2.01
8 x 6½	28	8.23	8.06	6.540	.463	.285	97.8	24.3	3.45	21.6	6.6	1.62
	24	7.06	7.93	6.500	.398	.245	82.5	20.8	3.42	18.2	5.6	1.61
8 x 5¼	20	5.88	8.14	5.268	.378	.248	69.2	17.0	3.43	8.5	3.2	1.20
	17	5.00	8.00	5.250	.308	.230	56.4	14.1	3.36	6.7	2.6	1.16

TABLE A.5

AMERICAN STANDARD BEAMS

PROPERTIES FOR DESIGNING

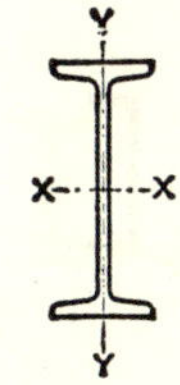

Nominal Size	Weight per Foot	Area	Depth	Flange Width	Flange Thickness	Web Thickness	Axis X-X I	Axis X-X I/c	Axis X-X k	Axis Y-Y I	Axis Y-Y I/c	Axis Y-Y k
In.	Lb.	In.²	In.	In.	In.	In.	In.⁴	In.³	In.	In.⁴	In.³	In.
24 x 7⅞	120.0	35.13	24.00	8.048	1.102	.798	3010.8	250.9	9.26	84.9	21.1	1.56
	105.9	30.98	24.00	7.875	1.102	.625	2811.5	234.3	9.53	78.9	20.0	1.60
24 x 7	100.0	29.25	24.00	7.247	.871	.747	2371.8	197.6	9.05	48.4	13.4	1.29
	90.0	26.30	24.00	7.124	.871	.624	2230.1	185.8	9.21	45.5	12.8	1.32
	79.9	23.33	24.00	7.000	.871	.500	2087.2	173.9	9.46	42.9	12.2	1.36
20 x 7	95.0	27.74	20.00	7.200	.916	.800	1599.7	160.0	7.59	50.5	14.0	1.35
	85.0	24.80	20.00	7.053	.916	.653	1501.7	150.2	7.78	47.0	13.3	1.38
20 x 6¼	75.0	21.90	20.00	6.391	.789	.641	1263.5	126.3	7.60	30.1	9.4	1.17
	65.4	19.08	20.00	6.250	.789	.500	1169.5	116.9	7.83	27.9	8.9	1.21
18 x 6	70.0	20.46	18.00	6.251	.691	.711	917.5	101.9	6.70	24.5	7.8	1.09
	54.7	15.94	18.00	6.000	.691	.460	795.5	88.4	7.07	21.2	7.1	1.15
15 x 5½	50.0	14.59	15.00	5.640	.622	.550	481.1	64.2	5.74	16.0	5.7	1.05
	42.9	12.49	15.00	5.500	.622	.410	441.8	58.9	5.95	14.6	5.3	1.08
12 x 5¼	50.0	14.57	12.00	5.477	.659	.687	301.6	50.3	4.55	16.0	5.8	1.05
	40.8	11.84	12.00	5.250	.659	.460	268.9	44.8	4.77	13.8	5.3	1.08
12 x 5	35.0	10.20	12.00	5.078	.544	.428	227.0	37.8	4.72	10.0	3.9	.99
	31.8	9.26	12.00	5.000	.544	.350	215.8	36.0	4.83	9.5	3.8	1.01
10 x 4⅝	35.0	10.22	10.00	4.944	.491	.594	145.8	29.2	3.78	8.5	3.4	.91
	25.4	7.38	10.00	4.660	.491	.310	122.1	24.4	4.07	6.9	3.0	.97
8 x 4	23.0	6.71	8.00	4.171	.425	.441	64.2	16.0	3.09	4.4	2.1	.81
	18.4	5.34	8.00	4.000	.425	.270	56.9	14.2	3.26	3.8	1.9	.84
7 x 3⅝	20.0	5.83	7.00	3.860	.392	.450	41.9	12.0	2.68	3.1	1.6	.74
	15.3	4.43	7.00	3.660	.392	.250	36.2	10.4	2.86	2.7	1.5	.78
6 x 3⅜	17.25	5.02	6.00	3.565	.359	.465	26.0	8.7	2.28	2.3	1.3	.68
	12.5	3.61	6.00	3.330	.359	.230	21.8	7.3	2.46	1.8	1.1	.72
5 x 3	14.75	4.29	5.00	3.284	.326	.494	15.0	6.0	1.87	1.7	1.0	.63
	10.0	2.87	5.00	3.000	.326	.210	12.1	4.8	2.05	1.2	.82	.65
4 x 2⅝	9.5	2.76	4.00	2.796	.293	.326	6.7	3.3	1.56	.91	.65	.58
	7.7	2.21	4.00	2.660	.293	.190	6.0	3.0	1.64	.77	.58	.59
3 x 2⅜	7.5	2.17	3.00	2.509	.260	.349	2.9	1.9	1.15	.59	.47	.52
	5.7	1.64	3.00	2.330	.260	.170	2.5	1.7	1.23	.46	.40	.53

TABLE A.6

AMERICAN STANDARD CHANNELS

PROPERTIES FOR DESIGNING

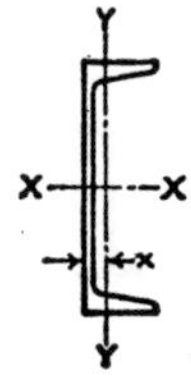

Nominal Size	Weight per Foot	Area	Depth	Flange Width	Flange Average Thickness	Web Thickness	Axis X-X I	Axis X-X I/c	Axis X-X k	Axis Y-Y I	Axis Y-Y I/c	Axis Y-Y k	Axis Y-Y x
In.	Lb.	In.^{2}	In.	In.	In.	In.	In.^{4}	In.^{3}	In.	In.^{4}	In.^{3}	In.	In.
15 x 3⅜	50.0	14.64	15.00	3.716	.650	.716	401.4	53.6	5.24	11.2	3.8	.87	.80
	40.0	11.70	15.00	3.520	.650	.520	346.3	46.2	5.44	9.3	3.4	.89	.78
	33.9	9.90	15.00	3.400	.650	.400	312.6	41.7	5.62	8.2	3.2	.91	.79
12 x 3	30.0	8.79	12.00	3.170	.501	.510	161.2	26.9	4.28	5.2	2.1	.77	.68
	25.0	7.32	12.00	3.047	.501	.387	143.5	23.9	4.43	4.5	1.9	.79	.68
	20.7	6.03	12.00	2.940	.501	.280	128.1	21.4	4.61	3.9	1.7	.81	.70
10 x 2⅝	30.0	8.80	10.00	3.033	.436	.673	103.0	20.6	3.42	4.0	1.7	.67	.65
	25.0	7.33	10.00	2.886	.436	.526	90.7	18.1	3.52	3.4	1.5	.68	.62
	20.0	5.86	10.00	2.739	.436	.379	78.5	15.7	3.66	2.8	1.3	.70	.61
	15.3	4.47	10.00	2.600	.436	.240	66.9	13.4	3.87	2.3	1.2	.72	.64
9 x 2½	20.0	5.86	9.00	2.648	.413	.448	60.6	13.5	3.22	2.4	1.2	.65	.59
	15.0	4.39	9.00	2.485	.413	.285	50.7	11.3	3.40	1.9	1.0	.67	.59
	13.4	3.89	9.00	2.430	.413	.230	47.3	10.5	3.49	1.8	.97	.67	.61
8 x 2¼	18.75	5.49	8.00	2.527	.390	.487	43.7	10.9	2.82	2.0	1.0	.60	.57
	13.75	4.02	8.00	2.343	.390	.303	35.8	9.0	2.99	1.5	.86	.62	.56
	11.5	3.36	8.00	2.260	.390	.220	32.3	8.1	3.10	1.3	.79	.63	.58
7 x 2⅛	14.75	4.32	7.00	2.299	.366	.419	27.1	7.7	2.51	1.4	.79	.57	.53
	12.25	3.58	7.00	2.194	.366	.314	24.1	6.9	2.59	1.2	.71	.58	.53
	9.8	2.85	7.00	2.090	.366	.210	21.1	6.0	2.72	.98	.63	.59	.55
6 x 2	13.0	3.81	6.00	2.157	.343	.437	17.3	5.8	2.13	1.1	.65	.53	.52
	10.5	3.07	6.00	2.034	.343	.314	15.1	5.0	2.22	.87	.57	.53	.50
	8.2	2.39	6.00	1.920	.343	.200	13.0	4.3	2.34	.70	.50	.54	.52
5 x 1¾	9.0	2.63	5.00	1.885	.320	.325	8.8	3.5	1.83	.64	.45	.49	.48
	6.7	1.95	5.00	1.750	.320	.190	7.4	3.0	1.95	.48	.38	.50	.49
4 x 1⅝	7.25	2.12	4.00	1.720	.296	.320	4.5	2.3	1.47	.44	.35	.46	.46
	5.4	1.56	4.00	1.580	.296	.180	3.8	1.9	1.56	.32	.29	.45	.46
3 x 1½	6.0	1.75	3.00	1.596	.273	.356	2.1	1.4	1.08	.31	.27	.42	.46
	5.0	1.46	3.00	1.498	.273	.258	1.8	1.2	1.12	.25	.24	.41	.44
	4.1	1.19	3.00	1.410	.273	.170	1.6	1.1	1.17	.20	.21	.41	.44

TABLE A.7

ANGLES
EQUAL LEGS
PROPERTIES FOR DESIGNING

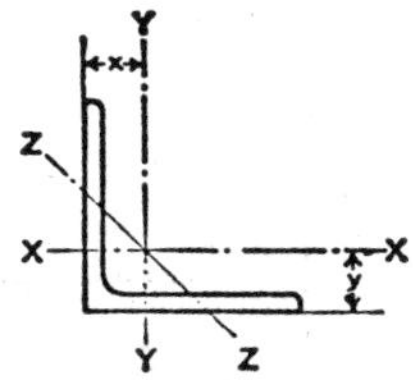

Size	Thickness	Weight per Foot	Area	AXIS X-X AND AXIS Y-Y				AXIS Z-Z
				I	I/c	k	x or y	k
In.	In.	Lb.	$In.^2$	$In.^4$	$In.^3$	In.	In.	In.
6 x 6	1	37.4	11.00	35.5	8.6	1.80	1.86	1.17
	7/8	33.1	9.73	31.9	7.6	1.81	1.82	1.17
	3/4	28.7	8.44	28.2	6.7	1.83	1.78	1.17
	5/8	24.2	7.11	24.2	5.7	1.84	1.73	1.18
	9/16	21.9	6.43	22.1	5.1	1.85	1.71	1.18
	1/2	19.6	5.75	19.9	4.6	1.86	1.68	1.18
	7/16	17.2	5.06	17.7	4.1	1.87	1.66	1.19
	3/8	14.9	4.36	15.4	3.5	1.88	1.64	1.19
	5/16	12.5	3.66	13.0	3.0	1.89	1.61	1.19
5 x 5	7/8	27.2	7.98	17.8	5.2	1.49	1.57	.97
	3/4	23.6	6.94	15.7	4.5	1.51	1.52	.97
	5/8	20.0	5.86	13.6	3.9	1.52	1.48	.98
	1/2	16.2	4.75	11.3	3.2	1.54	1.43	.98
	7/16	14.3	4.18	10.0	2.8	1.55	1.41	.98
	3/8	12.3	3.61	8.7	2.4	1.56	1.39	.99
	5/16	10.3	3.03	7.4	2.0	1.57	1.37	.99
4 x 4	3/4	18.5	5.44	7.7	2.8	1.19	1.27	.78
	5/8	15.7	4.61	6.7	2.4	1.20	1.23	.78
	1/2	12.8	3.75	5.6	2.0	1.22	1.18	.78
	7/16	11.3	3.31	5.0	1.8	1.23	1.16	.78
	3/8	9.8	2.86	4.4	1.5	1.23	1.14	.79
	5/16	8.2	2.40	3.7	1.3	1.24	1.12	.79
	1/4	6.6	1.94	3.0	1.1	1.25	1.09	.80
3½ x 3½	1/2	11.1	3.25	3.6	1.5	1.06	1.06	.68
	7/16	9.8	2.87	3.3	1.3	1.07	1.04	.68
	3/8	8.5	2.48	2.9	1.2	1.07	1.01	.69
	5/16	7.2	2.09	2.5	.98	1.08	.99	.69
	1/4	5.8	1.69	2.0	.79	1.09	.97	.69
3 x 3	1/2	9.4	2.75	2.2	1.1	.90	.93	.58
	7/16	8.3	2.43	2.0	.95	.91	.91	.58
	3/8	7.2	2.11	1.8	.83	.91	.89	.58
	5/16	6.1	1.78	1.5	.71	.92	.87	.59
	1/4	4.9	1.44	1.2	.58	.93	.84	.59
	3/16	3.71	1.09	.96	.44	.94	.82	.59

TABLE A.8

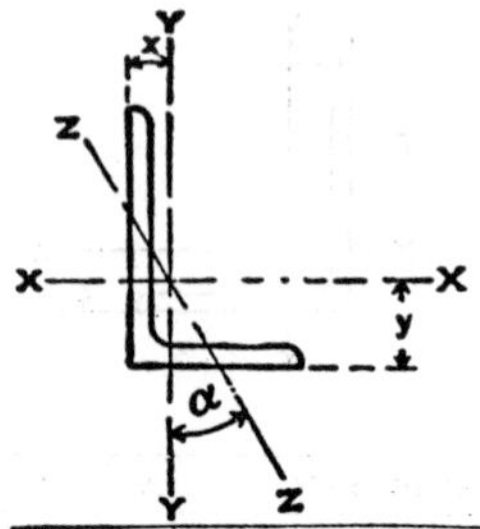

ANGLES
UNEQUAL LEGS

PROPERTIES FOR DESIGNING

Size	Thick-ness	Weight per Foot	Area	AXIS X-X				AXIS Y-Y				AXIS Z-Z	
				I	$\frac{I}{c}$	k	y	I	$\frac{I}{c}$	k	x	k	Tan α
In.	In.	Lb.	$In.^2$	$In.^4$	$In.^3$	In.	In.	$In.^4$	$In.^3$	In.	In.	In.	
6 x 4	⅞	27.2	7.98	27.7	7.2	1.86	2.12	9.8	3.4	1.11	1.12	.86	.421
	¾	23.6	6.94	24.5	6.3	1.88	2.08	8.7	3.0	1.12	1.08	.86	.428
	⅝	20.0	5.86	21.1	5.3	1.90	2.03	7.5	2.5	1.13	1.03	.86	.435
	9/16	18.1	5.31	19.3	4.8	1.90	2.01	6.9	2.3	1.14	1.01	.87	.438
	½	16.2	4.75	17.4	4.3	1.91	1.99	6.3	2.1	1.15	.99	.87	.440
	7/16	14.3	4.18	15.5	3.8	1.92	1.96	5.6	1.9	1.16	.96	.87	.443
	⅜	12.3	3.61	13.5	3.3	1.93	1.94	4.9	1.6	1.17	.94	.88	.446
	5/16	10.3	3.03	11.4	2.8	1.94	1.92	4.2	1.4	1.17	.92	.88	.449
6 x 3½	½	15.3	4.50	16.6	4.2	1.92	2.08	4.3	1.6	.97	.83	.76	.344
	⅜	11.7	3.42	12.9	3.2	1.94	2.04	3.3	1.2	.99	.79	.77	.350
	5/16	9.8	2.87	10.9	2.7	1.95	2.01	2.9	1.0	1.00	.76	.77	.352
	¼	7.9	2.31	8.9	2.2	1.96	1.99	2.3	0.85	1.01	.74	.78	.355
5 x 3½	¾	19.8	5.81	13.9	4.3	1.55	1.75	5.6	2.2	.98	1.00	.75	.464
	⅝	16.8	4.92	12.0	3.7	1.56	1.70	4.8	1.9	.99	.95	.75	.472
	½	13.6	4.00	10.0	3.0	1.58	1.66	4.1	1.6	1.01	.91	.75	.479
	7/16	12.0	3.53	8.9	2.6	1.59	1.63	3.6	1.4	1.01	.88	.76	.482
	⅜	10.4	3.05	7.8	2.3	1.60	1.61	3.2	1.2	1.02	.86	.76	.486
	5/16	8.7	2.56	6.6	1.9	1.61	1.59	2.7	1.0	1.03	.84	.76	.489
	¼	7.0	2.06	5.4	1.6	1.61	1.56	2.2	.83	1.04	.81	.76	.492
5 x 3	½	12.8	3.75	9.5	2.9	1.59	1.75	2.6	1.1	.83	.75	.65	.357
	7/16	11.3	3.31	8.4	2.6	1.60	1.73	2.3	1.0	.84	.73	.65	.361
	⅜	9.8	2.86	7.4	2.2	1.61	1.70	2.0	.89	.84	.70	.65	.364
	5/16	8.2	2.40	6.3	1.9	1.61	1.68	1.8	.75	.85	.68	.66	.368
	¼	6.6	1.94	5.1	1.5	1.62	1.66	1.4	.61	.86	.66	.66	.371

TABLE A.8 (Continued)

ANGLES
UNEQUAL LEGS
PROPERTIES FOR DESIGNING

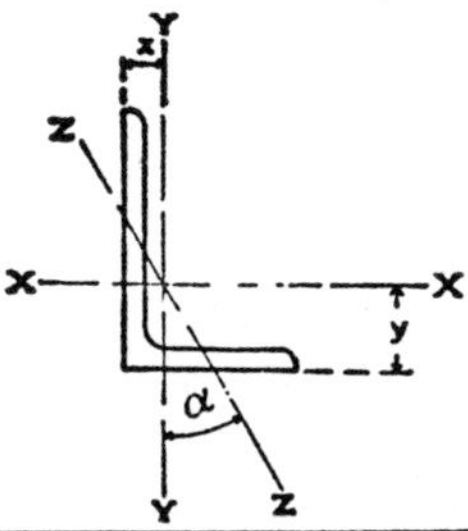

Size	Thick-ness	Weight per Foot	Area	AXIS X-X				AXIS Y-Y				AXIS Z-Z	
				I	I/c	k	y	I	I/c	k	x	k	Tan α
In.	In.	Lb.	In.²	In.⁴	In.³	In.	In.	In.⁴	In.³	In.	In.	In.	
4 x 3½	⅝	14.7	4.30	6.4	2.4	1.22	1.29	4.5	1.8	1.03	1.04	.72	.745
	½	11.9	3.50	5.3	1.9	1.23	1.25	3.8	1.5	1.04	1.00	.72	.750
	7/16	10.6	3.09	4.8	1.7	1.24	1.23	3.4	1.4	1.05	.98	.72	.753
	⅜	9.1	2.67	4.2	1.5	1.25	1.21	3.0	1.2	1.06	.96	.73	.755
	5/16	7.7	2.25	3.6	1.3	1.26	1.18	2.6	1.0	1.07	.93	.73	.757
	¼	6.2	1.81	2.9	1.0	1.27	1.16	2.1	.81	1.07	.91	.73	.759
4 x 3	⅝	13.6	3.98	6.0	2.3	1.23	1.37	2.9	1.4	.85	.87	.64	.534
	½	11.1	3.25	5.1	1.9	1.25	1.33	2.4	1.1	.86	.83	.64	.543
	7/16	9.8	2.87	4.5	1.7	1.25	1.30	2.2	1.0	.87	.80	.64	.547
	⅜	8.5	2.48	4.0	1.5	1.26	1.28	1.9	.87	.88	.78	.64	.551
	5/16	7.2	2.09	3.4	1.2	1.27	1.26	1.7	.73	.89	.76	.65	.554
	¼	5.8	1.69	2.8	1.0	1.28	1.24	1.4	.60	.90	.74	.65	.558
3½ x 3	½	10.2	3.00	3.5	1.5	1.07	1.13	2.3	1.1	.88	.88	.62	.714
	7/16	9.1	2.65	3.1	1.3	1.08	1.10	2.1	.98	.89	.85	.62	.718
	⅜	7.9	2.30	2.7	1.1	1.09	1.08	1.9	.85	.90	.83	.62	.721
	5/16	6.6	1.93	2.3	.95	1.10	1.06	1.6	.72	.90	.81	.63	.724
	¼	5.4	1.56	1.9	.78	1.11	1.04	1.3	.59	.91	.79	.63	.727
3½x2½	½	9.4	2.75	3.2	1.4	1.09	1.20	1.4	.76	.70	.70	.53	.486
	7/16	8.3	2.43	2.9	1.3	1.09	1.18	1.2	.68	.71	.68	.54	.491
	⅜	7.2	2.11	2.6	1.1	1.10	1.16	1.1	.59	.72	.66	.54	.496
	5/16	6.1	1.78	2.2	.93	1.11	1.14	.94	.50	.73	.64	.54	.501
	¼	4.9	1.44	1.8	.75	1.12	1.11	.78	.41	.74	.61	.54	.506
3 x 2½	½	8.5	2.50	2.1	1.0	.91	1.00	1.3	.74	.72	.75	.52	.667
	7/16	7.6	2.21	1.9	.93	.92	.98	1.2	.66	.73	.73	.52	.672
	⅜	6.6	1.92	1.7	.81	.93	.96	1.0	.58	.74	.71	.52	.676
	5/16	5.6	1.62	1.4	.69	.94	.93	.90	.49	.74	,68	.53	.680
	¼	4.5	1.31	1.2	.56	.95	.91	.74	.40	.75	.66	.53	.684

ANSWERS TO SELECTED PROBLEMS

1.1. $DE = 17.8$ ksi tension; $EJ = 15.1$ ksi compression; $DJ = 16.7$ ksi compression
1.3. $BC = 16.0$ ksi compression; $CH = 18.75$ ksi tension; $HI = 16.0$ ksi tension
1.5. (*a*) 16,000 psi; (*b*) 16,000 psi; (*c*) 12,000 psi; (*d*) 12,000 psi
1.8. (*a*) 3050 psi; (*b*) 159 psi
1.11. (*a*) 0.0008 in./in.; (*b*) 0.0008 ft/ft; (*c*) 800 μ in./in.
1.13. (*a*) 0.0006 in./in.; (*b*) 0.0006 radian
1.15. (*a*) 20,000 psi; (*b*) 7600 psi
1.18. (*a*) 240,000 lb; (*b*) 80,000 lb; (*c*) 52,000 lb
1.21. 9.8 in. **1.22.** -0.0176 in.; $+0.000588$ in.
1.24. 0.30; 16,000,000 psi **1.26.** -0.177 in.
1.28. -0.000165 in.; -0.00432 in.; $+0.00998$ in.
1.30. 4.47 in. **1.33.** 12,460 psi; 24,920 lb
1.35. (*a*) 196°F; (*b*) 18,750 psi; (*c*) 25,000 psi
1.38. 0.0050 in. **1.40.** 0.0000416 in.
2.1. (*b*) 59,000 psi; (*e*) 31,300,000 psi
2.4. (*d*) 4,400,000 psi; (*e*) 4.4 in.-lb/in.^{3}
2.14. (*a*) 19.7 in.-lb/in.^{3}, 22,600 in.-lb/in.^{3}; (*b*) 45.7 in.
2.18. 0.29 **3.1.** 0.125 in.
3.4. (*a*) 17,280 lb; (*b*) 1.42 in., 1.31 in.
3.7. 0.092 in.^{2}, 0.31 in. **3.10.** 48,600 lb **3.13.** 14,400 lb
3.16. Aluminum 8000 psi, steel 12,000 psi
3.19. (*a*) 6.55 in.; (*b*) aluminum 9090 psi, steel 16,360 psi
3.22. Timber 882 psi, steel 10,600 psi **3.25.** Steel 800 psi, brass 7100 psi

3.28. Steel 19,100 psi, aluminum 3930 psi
3.31. 12,270 psi tension, 14,400 psi compression
3.34. Aluminum 8070 psi, steel 12,100 psi
3.37. 11,450 psi tension (in each material)
3.40. Steel 5320 psi tension, brass 4700 psi compression
3.43. −474°F
3.46. +205°F
3.49. (*a*) 33,000 psi; (*b*) fracture
3.52. (*a*) 33,000 psi; (*b*) 40,900 psi
3.55. 0.103 in.
3.58. (*a*) 0.30 in.; (*b*) 4.04 in.^{2}
3.60. 450 psi, 59
3.62. 67 psi
4.2. $V = -150x$; $V = -2100$
4.4. $V = +3600$; $V = +6800 - 800x$
4.7. $V_{max} = 1240$; $V_{min} = -1560$
4.10. $V_{max} = 12$; $V_{min} = -21$
4.17. $M = +800x$; $M = +800x - 150(x - 4)^2$
4.20. $M = -x^2$; $M = -8(x - 2) - 5(x - 4)$
4.31. $V = -8$ kips; $M = -6$ kip-ft
4.34. $V = +1600$ lb; $M = +34{,}800$ lb-ft
4.37. $V_C = +1000$ lb; $V_D = -1400$ lb
4.39. $V_{max} = +2000$; $M_{max} = +5600$
4.42. $M_{max} = +68{,}000$ at 6.78 ft
4.44. $V = 0$ and $M = 0$ at 6 ft
4.47. $M_{min} = -23{,}800$
4.49. $M = 0$ at 2.76 and 7.24 ft
4.54. $M_{max} = +1670$; $M_{min} = -9600$
4.57. $M = +30$ and -10 at 5 ft
4.60. $M_{max} = +9088$ at 6.6 ft
4.62. $M_{max} = +7275$ at 4.5 ft
5.1. (*a*) 1600 psi tension and compression; (*b*) 1070 psi compression
5.4. 13,530 lb-ft
5.7. (*a*) 9510 psi tension and compression; (*b*) 1690 psi compression
5.10. 943 lb/ft
5.13. (*a*) 5990 lb/ft; (*b*) 4660 lb/ft
5.16. 305,000 lb-ft
5.19. (*a*) 16,000 psi tension, 8000 psi compression; (*b*) 4800 psi tension
5.21. 2990 psi tension, 2250 psi compression
5.25. 19,000 lb, 25,200 lb
5.27. 1400 lb/ft, 4900 lb/ft
5.29. 10.47 ft, 10.95 ft
5.30. Max = 234 psi
5.33. 2.04 by 4.08 in.
5.36. $P = 10.61$ kips
5.40. 17.92 ft, 14.3 ft
5.42. 17.7 kips, 22.3 kips, 106.1 kips
5.46. 1195 lb/ft, 2295 lb/ft
5.50. 24,850 psi, 39,100 psi, 4900 psi
5.55. 2680 lb/ft
5.58. 6030 lb
5.60. Aluminum 8950 psi, timber 1274 psi, steel 17,000 psi
6.1. 24WF84
6.4. 24WF94
6.7. 8 by 16
6.10. 6 by 10
6.13. 1 in., 22.8 ft
6.15. Web t = ⅜ in.; flange t = 2 in.
6.19. (*a*) 8540 psi; (*b*) 17,080 psi
6.21. 8000 psi, 12,200 psi
7.1. 0.144 radian
7.3. 4.5 ft
7.5. 1.38 in.
7.7. 13,900 psi
7.10. 5.88 in.
7.13. 7850 lb-in., 3.42 in.
7.17. (*a*) 27,500 lb-ft; (*b*) 19,200 lb-ft
7.22. (*a*) 1.02 in., 1.28 in.; (*b*) 3 ft

7.26. 13,600 psi
7.29. Steel 11,900 psi, aluminum 2380 psi
7.32. Aluminum 12,950 psi, brass 7300 psi
7.35. 53,500 psi, 6.67 in. **7.38.** (*a*) 5/8 in.; (*b*) 5.3
7.41. 6.6 in.
7.44. (*a*) 2360 lb-ft; (*b*) 2860 lb-ft **7.47.** 1855 lb/in., 9280 psi
7.51. 114,500 lb-ft **7.53.** $e = 1.97$ in. **7.57.** $e = 0.82$ in.
8.1. 451 psi **8.4.** 0.44 **8.7.** 1.79 in.
8.10. 0.49 **8.15.** 46,900 lb, 6 **8.18.** 119,200 lb, 12
8.21. 168,000 lb-in. **8.24.** 9.17 in. **8.26.** 5.95 in.
8.28. 6.36 kips **8.30.** 13.52 kips **8.32.** 6.72 kips
8.34. 6.35 in. **8.37.** 9.6 in. **8.40.** 19.4 in., 20.3 in.
8.43. 3/16 in. **8.46.** 1/4 in. **8.49.** 3/8 in.
8.51. 1/4 in. **8.53.** 3/16 in.
9.1. 5330 psi tension, 6670 psi compression
9.4. 1753 lb
9.7. 706 psi tension, 686 psi compression
9.10. 17,200 psi, 17,200 psi, 1910 psi **9.13.** 8410 lb
9.16. 18.03 ksi tension, 18.96 ksi compression
9.19. 798 psi tension, 930 psi compression
9.23. 1606 psi tension, 1678 psi compression
9.26. 1584 psi tension, 1750 psi compression
9.29. 0.479 ksf, 2.145 ksf, 2.521 ksf, 0.855 ksf
9.32. 14,400 psi tension, 14,550 psi compression
9.35. 4.85 in.
9.38. 5210 psi compression, 1040 psi compression, 1040 psi compression, 3120 psi tension
9.41. 154.5 kips **9.44.** 600,000 lb and 1.89 in.
10.1. 8550 psi tension, 5160 psi
10.4. 21,700 psi compression, 6730 psi
10.7. 920 psi tension, 8770 psi
10.14. 6500 psi, 19,000 psi tension, 6000 psi tension
10.17. 10,200 psi, 3800 psi compression, 24,200 psi compression
10.20. 10,970 psi, 18,470 psi tension, 3470 psi compression
10.23. 5 psi tension, 333 psi compression
10.26. 2760 psi tension, 2760 psi compression
10.29. 7200 psi tension, 1240 psi compression
10.32. 10,350 psi tension, 1150 psi compression
10.35. 3 psi tension, 1337 psi compression
11.1. −0.187 in.
11.3. (*a*) −0.257 in.; (*b*) 0.00686 radian
11.5. (*a*) −2.49 in.; (*b*) −1.78 in. **11.7.** (*a*) −1.85 in.; (*b*) 0.0238 radian
11.9. −1.27 in. **11.13.** −0.768 in.
11.15. (*a*) −1.08 in.; (*b*) −0.338 in. **11.19.** −0.145 in., −0.39 in.
11.23. −0.546 in., −1.58 in. **11.26.** −0.87 in.

11.29. −0.30 in. **11.30.** +1.96 in. **11.34.** −0.187 in.
11.35. −0.267 in., −0.265 in.
11.38. −0.422 in. **11.41.** −0.222 in. **11.43.** −0.324 in.
11.46. +0.108 in., −0.249 in. **11.48.** +0.766 in., −2.89 in.
11.54. −0.594 in.
11.55. (*a*) 36 kips, 135 kip-ft; (*b*) −0.144 in.
11.57. (*a*) 9800 lb, 37,600 lb-ft
11.59. (*a*) 6000 lb, 14,400 lb-ft, 2100 lb, 7200 lb-ft; (*b*) −0.18 in.
11.62. (*a*) 5850 lb, 9900 lb-ft, 1350 lb, 4500 lb-ft
11.64. (*a*) 3070 lb-ft, 2227 lb, 173 lb **11.70.** 5760 lb-ft, −1440 lb, +7200 lb
11.74. 4.3 kip-ft, −5.18 kips, 11.5 kip-ft, +5.18 kips
12.1. (*a*) −0.257 in.; (*b*) 0.00686 radian
12.3. (*a*) −2.49 in.; (*b*) −1.78 in. **12.5.** (*a*) −0.138 in.; (*b*) −0.235 in.
12.7. −0.714 in. **12.11.** −0.768 in.
12.13. −0.544 in., −1.58 in. **12.15.** −0.30 in.
12.17. +0.00417 radian, −0.39 in. **12.20.** −0.87 in.
12.22. −0.693 in. **12.23.** −0.0297 radian, +1.96 in.
12.25. −0.187 in. **12.26.** −0.267 in. **12.29.** −0.422 in.
12.32. −0.222 in. **12.34.** −0.321 in. **12.36.** −0.594 in.
12.37. +0.108 in., −0.249 in. **12.40.** (*a*) −0.76 in.; (*b*) +0.53 in.
12.43. −0.49 in.
12.48. (*a*) 36 kips, 135 kip-ft; (*b*) −0.144 in.
12.50. (*a*) 24 kips, 55.5 kip-ft
12.52. (*a*) 6000 lb, 14,400 lb-ft, 2100 lb, 7200 lb-ft; (*b*) −0.18 in.
12.54. 12.33 kips, 35.7 kip-ft, 17.67 kips, 40.7 kip-ft
12.56. 2227 lb, 3070 lb-ft, 173 lb
12.61. 16.5 kips, 31.5 kip-ft, 19.5 kips
12.65. −5.18 kips, 4.3 kip-ft, +5.18 kips, 11.5 kip-ft
12.68. −0.57 in. **12.72.** −1.63 in.
13.1. 2.49 in., 0.0276 radian **13.3.** 0.00417 radian, 0.39 in.
13.5. 0.69 in. **13.8.** 0.32 in. **13.9.** 0.138 in.
13.14. 0.76 in. **13.16.** 55.5 kip-ft, 24 kips
13.19. M_1 = 60.8 kip-ft; M_2 = 63.5 kip-ft; y = 0.136 in.
13.22. M_1 = 82.5 kip-ft; M_2 = 79.1 kip-ft; y = 0.086 in.
13.25. M_1 = 72 kip-ft; R_1 = 30 kips; R_2 = 18 kips
13.29. 112 in.^{3}
13.32. M_1 = 6 kip-ft; R_1 = 7.5 kips; R_2 = 31.5 kips
13.35. $R_1 = R_3$ = 18 kips, R_2 = 60 kips
13.37. M_1 = 13.5 kip-ft; R_1 = 5.91 kips; R_2 = 7.25 kips; M_3 = 28.5 kip-ft; R_3 = 6.84 kips
13.39. R_1 = −0.8 kip; R_2 = 9.2 kips; $M_3 = M_4$ = 47.4 kip-ft; $R_3 = R_4$ = 15.8 kips
13.41. $R_1 = R_2$ = 4.12 kips; M_3 = 70.6 kip-ft; R_3 = 21.76 kips
13.44. M_1 = 102.3 kip-ft; R_1 = 16.86 kips; M_3 = 65.1 kip-ft; R_3 = 8.14 kips
13.47. M_1 = 6920 lb-ft; R_1 = 4153 lb, s = 16,710 psi

14.1. $M_2 = -25,400$ lb-ft **14.4.** $M_2 = -5.66$ kip-ft
14.7. $M_2 = -11.93$ kip-ft; $M_3 = -6.87$ kip-ft
14.10. $M_2 = -5.52$ kip-ft; $M_3 = -16.33$ kip-ft
14.11. $R_1 = 1094$ lb; $R_2 = 40$ lb; $R_3 = 1179$ lb; $R_4 = 187$ lb
14.14. $R_1 = 2.73$ kips; $R_2 = 2.98$ kips; $R_3 = 4.58$ kips
14.17. $R_1 = -2.48$ kips; $R_2 = 6.92$ kips; $R_3 = 16.31$ kips; $R_4 = 3.25$ kips
14.20. $M_1 = -6.84$ kip-ft; $M_2 = -22.32$ kip-ft
14.23. $M_1 = -14.4$ kip-ft; $M_2 = -4.8$ kip-ft
14.27. $M_1 = -89.1$ kip-ft; $R_1 = 35.3$ kips; $M_2 = -81.6$ kip-ft; $R_2 = 28.7$ kips
14.29. $M_2 = -12.2$ kip-ft; $M_3 = -6.0$ kip-ft
15.4. (*a*) 16,140 lb; (*b*) 4030 lb **15.9.** (*a*) 171,000 lb; (*b*) 12,160 lb
15.10. 29,500 lb **15.13.** (*a*) 215,000 lb; (*b*) 53,700 lb
15.16. 148,500 lb **15.19.** 17,300 lb **15.22.** 616,000 lb
15.25. (*a*) 86,000 lb; (*b*) 328,000 lb **15.28.** (*a*) 57,500 lb; (*b*) 4920 lb
15.30. (*a*) 4.6 ft; (*b*) 7.5 ft **15.32.** (*a*) 5670 lb; (*b*) 22,100 lb
15.35. 35,400 lb **15.38.** 8 by 8 **15.41.** 2.5 in.
15.44. 14WF84 **15.47.** $L/k = 0$
15.49. Yes (29,570 $<$ 36,000 psi) **15.51.** Safe (0.915 $<$ 1)
15.53. Not safe (1.09 $>$ 1) **16.1.** (*a*) 80 in.-lb; (*b*) 180 in.-lb
16.4. 0.61 in. **16.7.** 23.4 lb
16.10. (*a*) 13,250 lb; (*b*) 64.5 lb; (*c*) 8540 lb, 28.4 lb
16.13. 29,300 psi
16.15. (*a*) 1385 psi, 1.275 in.; (*b*) 40,400 psi
16.18. 0.147 in., 1461 psi **16.20.** 34,600 psi, 0.297 in.
16.22. 6800 psi **16.24.** $s^2V/6E$, $\frac{1}{3}$ **16.27.** 377 in.-lb
16.30. $s_s{}^2V/4E_s$ **17.3.** 24,670 lb **17.5.** 2680 lb
17.7. 108,000 lb **17.9.** $4r^3/3$, 1.70 **17.12.** 71.7 in.^{3}, 1.11
17.14. 41.4 in.^{3} **17.18.** 3120 lb/ft **17.19.** 54 in.^{3}
17.22. 62,500 lb **17.25.** 30 in.^{3} **17.30.** 6670 lb/ft
17.33. 5.3 kips/ft

INDEX